The Art and Science of Microelectronic Circuit Design

Anatoly Belous • Vitali Saladukha

The Art and Science of Microelectronic Circuit Design

Springer

Anatoly Belous
Integral
Minsk, Belarus

Vitali Saladukha
Integral
Minsk, Belarus

ISBN 978-3-030-89856-4 ISBN 978-3-030-89854-0 (eBook)
https://doi.org/10.1007/978-3-030-89854-0

This Springer imprint is published by the registered company Springer Nature Switzerland AG
The registered company address is: Gewerbestrasse 11, 6330 Cham, Switzerland

Preface

This book is intended primarily for students, teachers, undergraduates, engineers and technicians specializing in the field of microelectronics and its numerous applications. In addition, the contents of the book can be useful to scientists and specialists in the field of design, organization of production, and operation of radio-electronic devices and systems for household, industrial and special (space and military) purposes. In fact, the materials of this book are a practical guide (handbook) for the design of modern silicon digital microcircuits and systems-on-a-chip (SoC).

At the time of this book's publication, there are quite a lot of texts devoted to this area of focus, for example, *Handbook of Digital CMOS Technology and System* by Abbas Karim; *Handbook of Digital Technology for High Speed Design* by Tom Granberg; and *Design Techniques for High-Frequency CMOS Integrated Circuits: From 10 GHz to 100 GHz* by Zhiming Deng.

Here it is worth recalling a number of major "classic" publications on this subject matter written many years ago, which can still be seen in the workplaces of today's "electronics" engineers.

The publication closest to the topic discussed here and widely known to students, *The Art of Circuit Design* monograph, a classic textbook on digital and analogue circuitry, was written by American practical scientists Paul Horowitz from Harvard University and Winfield Hill from Rowland Institute for Science, Cambridge, Massachusetts. Its first English edition (*Cambridge University Press*) was published in 1980 and has subsequently gone through dozens of editions and is still in demand by students today.

This excitement around the book written by American specialists and its ongoing popularity among a wide range of readers can be explained, on the first hand, by broad coverage of its subject area, that is, the basics of designing radio electronic circuits, and by extensive background information on the element base ("bricks" that make up radio electronic devices) at the time of writing the book.

And, on the second hand, in contrast to classical textbooks with an abundance of mathematical calculations and physical formulas, the authors outlined here all the basic (at that time) aspects of designing radio electronic devices, in simple language and with a large number of clear practical examples, at a level that is understandable even to "poorly trained" readers.

For its popularity, extraordinary for such publications, among students and electronics engineers, the book received a well-deserved informal title in the 1990s, *"the bible of electronics"*.

It is obvious that over the past 40(!) years since writing of this book, the element base of microcircuits, radio electronic devices and systems, subject to the well-known Moore's law, has changed fundamentally. Those "bricks", brilliantly described in this "bible of electronics", have long ago been incorporated into larger "building blocks" (IP-blocks or "Intellectual properties") that modern IC and SoC are "assembled" of; new elements that simply could not be implemented technologically in the past have also appeared; and basic elements that operate on entirely new physical principles and mechanisms have emerged.

A major advantage of the book, proposed by the authors, is specifically a detailed description of the principles of operation and rules for application of these present-day basic elements in microelectronic devices. For example, elements implemented on modern bipolar field technology (BiCMOS) just did not exist at the time of publication of the last English edition of the "bible"; the same can be said about micropower CMOS basic functional components.

Until now, *The Semiconductor Circuitry: Handbook by Tietze U., Schenk Ch.* is also used in the training courses of many universities. In Germany and Russia, this book has survived more than 20 editions. Although this book deals only with the structures of the simplest semiconductor elements, which are practically not used in microelectronic devices today (except for elements of power electronics, studied in detail in that book), nevertheless, regular versions of this reference manual are still periodically issued by publishers and are in demand by specialists and students, since more "up-to-date" publications have not been available on sale until now.

Today in the US, UK and European book markets, there are a number of other books devoted to circuit design of modern microelectronic devices. Most of them, however, consider only individual components of a complex design problem and describe particular technologies (methods for reducing power dissipation, increasing performance, simulation and protection against spurious effects) in relation to specific technology basis, such as CMOS, bipolar, BICMOS and SOI.

The motivation of the authors to write these chapters on "circuit design" was the desire to help a wide range of students, teachers and engineers, specializing in the design and operation of various microelectronic devices, to understand physical mechanisms of processes occurring inside these "bricks" used to build modern ICs and systems-on-a-chip.

This book is based on the seminar materials of the lecture courses read by the authors for many years in Russian and Belarusian universities and academic institutions for students, graduate students, undergraduates and teachers of the following specialties: electronics and microelectronics; design and technology of electronic means; instrument engineering; computer science and computer engineering;

electronic engineering, radio engineering and communication; and automation and control.

In addition, we used the results of our own research, previously published in monographs (including those published abroad), patents and articles, as well as the results of our practical activities in the field of design and application of microelectronic devices.

Figuratively speaking, the modern students and engineers, specializing in the sphere of microelectronics and its applications, can derive from these books, articles and social nets the detailed and comprehensive description of some solitary "trees", but they will have to apply the strenuous efforts to gain the whole picture of the "forest".

Therefore, the authors have aspired for the quite ambitious challenge – to create the integral picture of such "microelectronic forest", consisting of individual "trees" (chapters). Each of the chapters is consistently dealing with major milestones of modern microcircuit design, ranging from description of physical concepts of transistor operation, basic libraries and design flows, methods of logic designing of low power consumption microcircuits, assembly technologies of microcircuits and systems on chip to quality management methods based upon special test structures.

Meanwhile, the authors have taken guidance from the following ***principles*** of construction of the chapter material, which was rather easy to formulate, but quite hard to encompass in the process of writing the book:

1. For the sake of being a popular edition amongst the vast circle of readers (engineers and students), the book should meet simultaneously the integral functions of both the classic manual and a brief reference book, being with the finesse of the captivating reading.
2. The book should comprise only those methods, schematic and technological solutions whose effectiveness was proven earlier with the practice of their application.
3. In the textual part of the book, it is necessary to use the maximum possible scope of the graphic material, reflecting the effectiveness of the various working scenarios.

The following introduction provides a brief summary of each of the chapters of this book.

Circuit Design The largest scope of material (four chapters) in our book is devoted to the fundamentals of circuit engineering of modern microelectronic devices (microcircuits and systems-on-a-chip). This is due to the fact that our book eliminates a number of "gaps", obvious to the specialists in this field, in a large volume of existing scientific and technical literature on issues relating to the analysis of operating peculiarities, design methods and fundamentals of practical application of digital microcircuits in modern microelectronic devices.

Here, a large set of effective ***circuit design solutions*** of basic elements is presented to meet the requirements for modern complex-functional, high-performance and reliable microelectronic devices.

This topic deserves a more detailed explanation.

As you know, the process of developing any digital microcircuit consists of two main interrelated stages: *logic design*, which determines the logical organization (architecture), command system, interface, structure of control and data processing devices, including timing diagram of operation, and *circuit design*, which includes a set of objectives for selecting a technological basis, converting logic circuits into electrical circuits at a transistor level, choosing circuitry solutions for basic elements and synchronization methods, and designing power supply circuits and devices for protection against external and internal interference and static electricity charges.

While methodology and ways of solving problems of the *logic design* stage are widely enough reviewed in numerous foreign and domestic publications, this is unfortunately not the case with the *circuit design* stage.

So, in numerous educational scientific and technical literature, existing at present, methods of designing various functional blocks of combinational (decoders, multiplexers, demultiplexers, adders, multipliers, etc.) and sequential types (automata with memory as flip-flops, registers, counters, etc.) are described in detail, and various methods and means of their computer-aided design are also discussed.

In this case, these nodes and blocks are represented at the level of "small squares" described in the language of Boolean algebra ("AND", "NOT", "AND-NOT", "OR-NOT", etc.) or in the form of graphical symbols (D-trigger, R-S-trigger, DV-trigger, etc.).

Of course, this procedure is a mandatory and integral initial stage of the end-to-end design process of any microelectronic device. However, both the designer and the end user of a microelectronic device must understand what is "inside" these blocks and nodes. The designer needs this in order to ensure the required values of electrical and dynamic parameters of the device under design by selecting the appropriate elements (transistors) and their connections.

The user or operator of this microelectronic device needs to know the "stuffing" of these blocks in order to understand specific features of a particular microelectronic device functioning in its various operating modes.

After all, even the structure of such a simple "brick" as an internal memory device of a microcircuit, D-trigger, can be implemented by dozens of different circuitry solutions for interconnecting its component transistors. And a modern student should clearly understand how this synthesized block (a set of "squares") "turns into" the layout of the corresponding area of an IC's semiconductor chip, where placement of transistors on chip surface by organizing appropriate links and interconnections of these transistors and their connections with other blocks allows you to implement a preset algorithm for the functioning of the block (node).

But this D-trigger can be implemented using various technologies like CMOS, BiCMOS and bipolar (STTL, ECL, I^2L), each with its own “subtle aspects”, and this is how these four chapters on “circuit design” appeared.

The four chapters of this book on “circuit design” address such a complex problem: numerous examples are given for the basic blocks of modern microelectronic devices at the level of transistors and their interconnections. It is shown, for example, that the simplest D-trigger, depending on its circuit design, will provide different numerical values required by the designer, such as response time, load capacity, noise immunity and power consumption.

An additional peculiarity of these four chapters on “circuit design” is a detailed description of various types of matching devices (elements) – input and output ones, which provide electrical and time correlation during microcircuit operation in a finished radio electronic device, as well as methods and circuitry solutions for reducing energy consumption of modern microcircuits, as an ever-relevant problem.

After all, it is the circuitry solutions of microcircuit basic elements that finally define numerical values of electric, static and dynamic characteristics, power consumption, response time, noise immunity, and even the area of an IC chip.

Design Libraries for Submicron Microcircuits

Another separate “training” chapter focuses on the contents and basic rules for the development and application of design libraries for submicron microcircuits, known in the design communities as “design kits” or PDK (Process Design Kits).

Recently, due to the emergence of separate (independent of microcircuit designers) semiconductor fabrication plants engaged in custom serial production of microcircuits (Integrated Circuits Foundry, ICF) and in connection with the objective need to share standard design tools and “purchased” IP blocks (intellectual properties) from other companies, it is the design libraries (PDKs) that became the main link between designers and manufacturers of microcircuits.

This chapter details the typical design flow and PDK structure, minimum contents of a basic library and minimum list of standard elements (cells), level translators, current source models, input and output buffers, and typical data files of design library. Specific instructional (educational) PDK by Synopsys is also described here.

Design Flows Used for Submicron Microcircuits This chapter deals with the study of design flows of modern submicron microcircuits. Here, the specifics of selecting a particular design flow in dependence on initial requirements of a microcircuit customer are considered. The contents of the basic stages of design flow are described, that is, system, functional and physical design as well as design verification. Specific features of designing microelectronic products at a higher level of complexity, that is, systems-on-a-chip (SoC), are examined in detail: development trends, detailed design flows, design methodologies, significant peculiarities of SoC system design stage and basic components of CAD tools for the system level. To "reinforce" the studied material, clear practical examples of SoC simulation process, described in the Cadence Incisive Simulation Environment, will be given here with the author's comments.

Fundamentals of Low-Power CMOS-Microcircuits Logic Design This chapter deals with the peculiarities of low-power CMOS microcircuit *logic* design. The developmental challenges of modern submicron technology force IC designers to seek ever-new techniques and methods of design, aimed at reducing power consumption caused by leakage currents, whose importance and contribution increases significantly with the decrease in design rules. This chapter reviews new methods of logic design of CMOS microcircuits with reduced power consumption, based on the mathematical apparatus of probabilistic evaluation of various optimization options by the predicted so-called "switching activity" of main blocks (nodes) of the designed microcircuit. The basic stages of such a logic design are presented here, starting from selection of required logic element basis, logic synthesis in this chosen basis, optimization procedure for two-level logic circuits, optimization of multi-level logic circuits built on both two-input and multi-input logic gates, and finishing with procedures for "technology mapping" and evaluation of power consumption of the designed (synthesized) microcircuit both at logic and circuit design levels.

At the end of the chapter, the applicable hardware-software complex for the logic design of micro-power CMOS IC is described in detail.

Quality Management Systems for IC Manufacturing It is known that a developer-designed microcircuit is then usually sent to a semiconductor fabrication plant to manufacture pilot samples (experimental prototypes). Global practice of the last decade shows that only about 40% of designed microcircuits meet the requirements of original specification (terms of reference) after manufacture; therefore, sometimes full compliance of products with the technical design assignment and, beyond that, the "targeted" yield can be achieved only after a number of successive iterations (corrections).

In order to improve efficiency and reduce the time for analysis and identification of the causes of failure to achieve the specified characteristics or yield percentage of the designed microcircuit, the process flow often uses special test structures placed on the chip (wafer) of the designed IC. Understanding physical mechanisms of IC failures is necessary not only for proper selection of microcircuit design solutions, taking into account permissible and critical levels of current density in active semiconductor structures and interconnections, and electric field strength in dielectric layers, but it is also required for developing appropriate measures to identify and reject potentially unreliable microcircuits at various stages of serial production.

This chapter will discuss the key principles of formation of such semiconductor test structures "built in a chip (or a wafer)", the definition of their list and functions, the foundations of step-by-step reliability prediction of the designed IC based on the results of statistical processing of numerical values of the measured parameters of such test structures. Mathematical models linking obtained statistical distribution of technical parameter values of the microcircuits with specific reliability indexes are presented and explained herein; specific practical examples of process optimization with the help of special test wafers are also given.

Minsk, Belarus

Anatoly Belous
Vitali Saladukha

Acknowledgements

The authors would like to express their gratitude to the colleagues who actively participated in the preparation and discussion of the contents of this book, whose critical comments, advice and additions contributed to improving both the structure of this book and the material presented in it: V. Borisenko, V. Bondarenko, V. Stempitsky, A. Silin and L. Lynkov.

The authors extend their special thanks to the colleagues who provided the original materials of their own theoretical and practical research on the subject matter of this book, whose inclusion in the book greatly enhanced the practical focus of this work: P. Bibilo, L. Cheremisinova, A. Petlitsky and T. Petlitskaya.

The authors enclose their gratitude to the reviewer V. Labunov, academician at the National Academy of Sciences of Belarus and foreign elected member of the Academy of Sciences of the Russian Federation, whose specific comments and suggestions have significantly contributed to the formation of the final shape of this book, offered to the reader.

The authors are also grateful to O. Antipenko for her high-quality execution of a large amount of work on technical design of the book's manuscript. They also offer thanks to E. Kartashova and K. Gaivoronsky for their contribution in preparation of the graphic images, to the employees of the publishing house TEKHNOSPHERE, to O. Kuleshova and S. Orlov for the providing graphic materials, and to V. Chikilev, M. Biryukova, L. Ignatovich, Yu. Sizov, N. Mazurina, I. Kurshatsova and V. Ilyenkov for assistance in translation of the textual materials of the book.

Contents

Chapter 1
Standard Characteristics of Digital Microcircuits

This chapter provides well-marshaled and interpreted essential technical characteristics of up-to-date digital microcircuits: summarized general structure of microcircuits, architecture of internal standard cells and matching components, and system of essential parameters (functional, electrical, and dynamic parameters).

There are elaborated key options of schematic implementation of basic logic elements of digital microcircuits, impact of destabilizing factors to their serviceability (tolerance to electrical and temperature overloads, impacts produced by external and internal electrical noises, including noises by supply lines and common lincs).

The chapter concludes with the insight into the main parasitic elements and parasitic effects in digital microcircuits (parasitic transistors, latch-up effects, Miller effect, etc.)

1.1 Structure of Digital Microcircuits

1.1.1 General Structure of Digital Microcircuits

It is a common fact that one of the main trends in microelectronics is constant persistent increase of functional complexity of digital microcircuit which is attained essentially through increase in the number of elements being integrated into a semiconductor chip and ongoing improvement of schematic solutions implemented in standard cells. Main trends leading the way to increase of functional complexity of digital microcircuits are reduction of cell linear dimensions (scaling) at simultaneous reduction of the applied supply voltage value which results in reduction of the total value of their loading capacitances. However, marginal growth of cell number on a chip is limited by maximum tolerable chip area (S_{KP}) and maximum tolerable power dissipated by a package of a digital microcircuit ($P_{package}$). Reduction of cell power consumption and miniaturization of their sizes tend to cut down their output currents and also noticeably affect their load-carrying capacitance and interference immunity

A. Belous, V. Saladukha, *The Art and Science of Microelectronic Circuit Design*,
https://doi.org/10.1007/978-3-030-89854-0_1

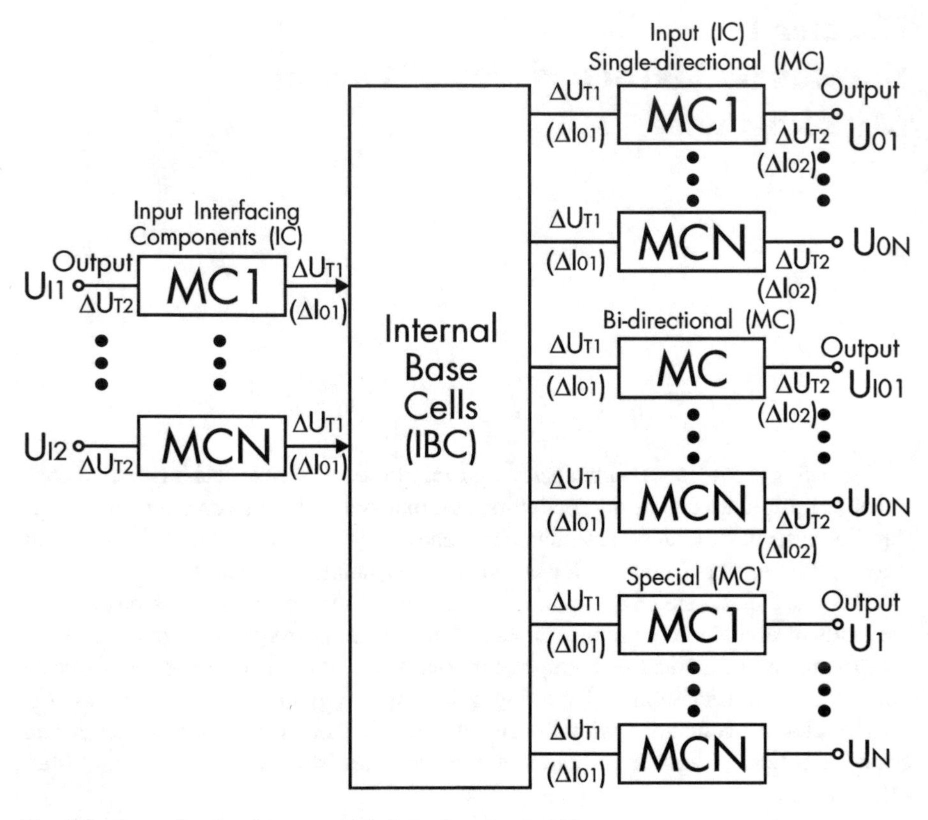

Fig. 1.1 Generalized architecture of digital microcircuits [1]

of microcircuits. Whereas internal interferences inside of microcircuits are comparatively small and loading capacities of microcircuit internal cells are not high, using low-power cells with low output currents as the internal cells does not normally create any problems, making it possible to decrease the value of logic voltage drop and to obtain in the process high-speed performance, low power consumption, and larger integration scale of digital microcircuits.

However, when such cells are used to receive the signals external for microcircuits, the value of aggregated interference immunity of microcircuits is being reduced, and rather low load-carrying capacitance of the cells not infrequently prevents from obtaining the specified high-speed performance of microcircuits when numerous external loading is being controlled. As a consequence, in digital microcircuits, the architecture depicted in [1] is widely used where elements having two values of logic voltage (currents) swing are used ΔU_T (ΔI_O). The architecture of digital microcircuits is shown in Fig. 1.1 and contains two standard types of cells: internal base cells (IBC), with input (output) logic swing ΔU_{TI} (ΔI_{OI}) of low value $\Delta U_{TI} \approx 0.2 \div 1.5$ V ($\Delta I_{0I} \approx 0.1 \div 1$ mA), and external matching components (MC), with input logic swing ΔU_{T2} of higher value (switching threshold $U_T = 1.5 \div 15$ V) and output matching components with logic input swing equal to logic swing of IBC ΔU_{TI} and output matching components with input logic swing equal to logic swing of IBC ΔU_{TI} and output logic swing ΔU_{T2} (ΔI_{O2}) of higher value $\Delta U_{T2} \approx 2 \div 15$ V ($\Delta I_{O2} \approx 1 \div 50$ mA).

Thus, in such a way, internal base cells with small voltage (current) swing ΔU_{TI} (ΔI_{OI}) and output **matching components (MCs)** with higher output currents ΔI_{O2} support high-speed performance of microcircuits, and high switching threshold ΔU_{T2} of input MCs contributes to better interference immunity of microcircuits. Whereas the values of internal ΔU_{TI} (ΔI_{OI}) and external ΔU_{T2} (ΔI_{O2}) logic swing levels considerably differ, the input MCs are performing as converters of logic levels from external ΔU_{T2} into internal ΔU_{TI} (ΔI_{OI}), and output matching components (MCs) generate the external levels with swing ΔU_{T2} (ΔI_{O2}) from the internal ones with swing ΔU_{TI} (ΔI_{OI}).

Along with the matching components of the said types, digital microcircuits may apply bidirectional matching components which both receive external levels with their further conversion into the internal ones and also generate external levels from the internal ones. Moreover in microcircuits, there may be applied some special matching components for the purpose of connecting the external components (quartz resonators, capacities, etc.) not implemented in the microcircuits, providing reference voltages, receipt of analog signals, etc.

1.1.2 Architecture of Internal Cells of Digital Microcircuits

Internal base cells are divided into two main groups of cells: logic cells (LC) and memory cells (MC).

Base Logic Cells Base logic cells (LC) of digital microcircuits are used for logic conversion of the input data provided in binary or in any other code and generating potential (or current) levels of output signals with electrical characteristics which correspond to the data being coded at LC output and support the LCs being joined together [1].

The most elementary architecture of base logic cell incorporates two components (Fig. 1.2): switching (toggling) component, (SC) converting input data U_I, and loading component (LC), which affords generating the relevant level of output signals U_O. Loading component (LC) may be either input signal U_I controlled (bar line in Fig. 1.2) or non-controllable one. Depending on the method of data transmission, logic cells may be divided into two big groups: asynchronous or static and synchronous or dynamic ones. The architecture of logic cells falling into the first group is shown in Fig. 1.2, and here the time of output signal U_O generation is determined by intrinsic delay time t_P of logic cell. Architecture of logic cells falling into the second group is shown in Fig. 1.3 [1], and here time t_P of output signal generation U_O is quantized at specific moments of time set by the signal frequency delivered to clock C input. Option with logic cells falling into the second group may

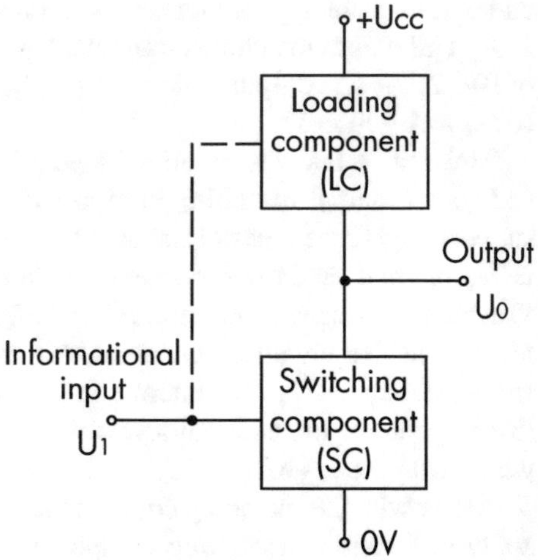

Fig. 1.2 General structure of internal base static logic cell [1]

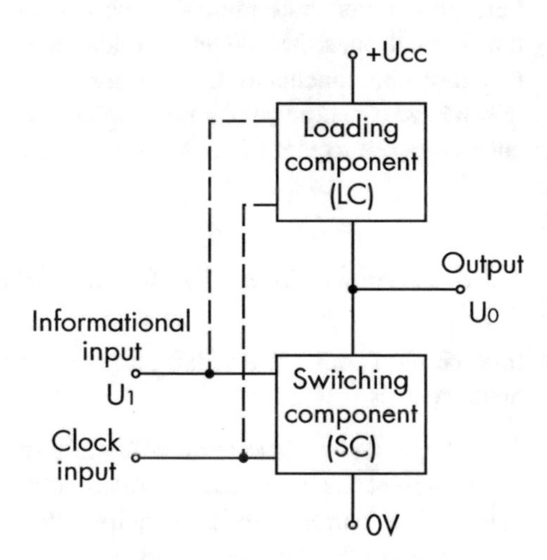

Fig. 1.3 Architecture of internal base dynamic logic cell

be implemented with a few clock inputs which is called polysynchronous. In such a logic cell, the time of input signal generation depends on the string of clock signals. Architecture of logic cell shown in Fig. 1.2 has one data input; hence it performs the simplest logic conversions of input data signal U_I. Therefore, to increase the number of logic functions, the logic cell circuit is supplemented with an additional logic component (LC).

Logic component (LC) may be introduced either at LC input (Fig. 1.4a) (LC of type CMOS, TTLS) or at LC output (Fig. 1.4b) (LC of I^2L type), and it allows generating logic function of *N* input data signals or N output functions of single input signal. The most function-ridden is the logic cell containing the logic component both at input (LC1) and output (LC2) LC (Fig. 1.4c) which enables generation of N output logic functions of M input data signals. For such LCs, however, generation of

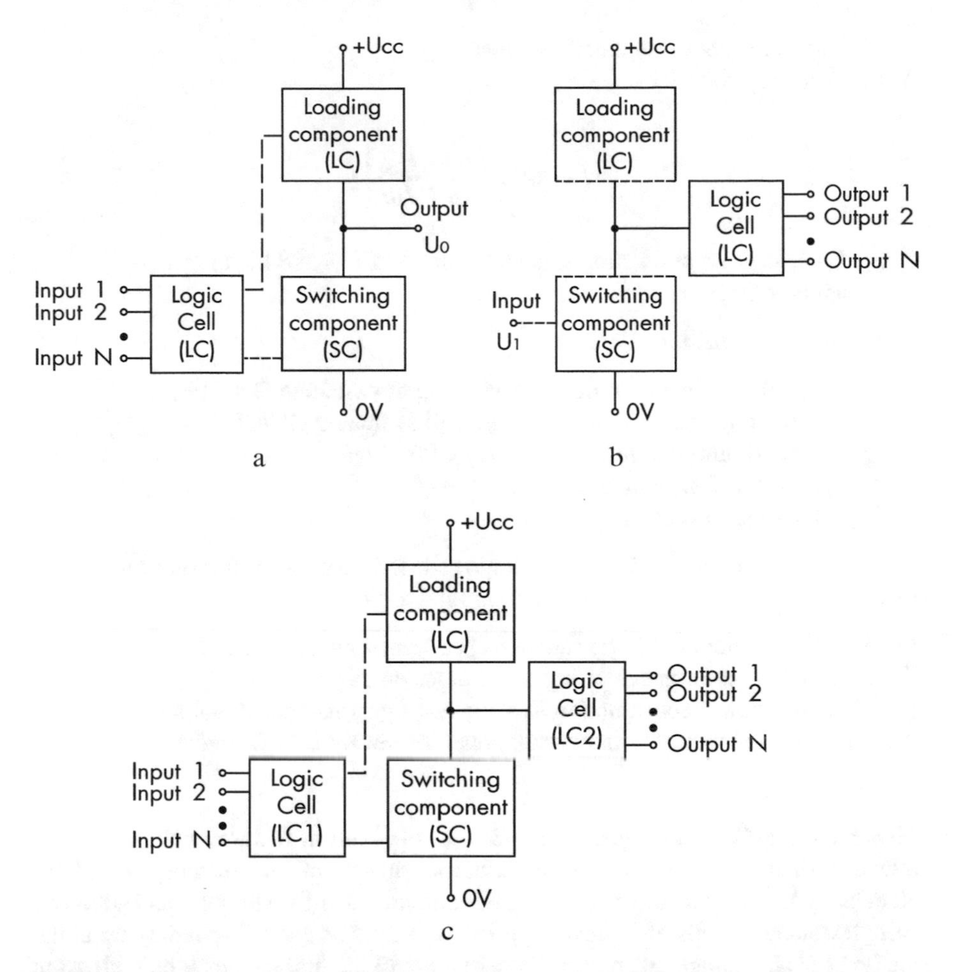

Fig. 1.4 Architecture of internal base logic cells: (**a**) multi-input; (**b**) multi-output; (**c**) multi-input\multi-output base logic cell

appropriate levels of signal and logic swing ΔU_T. appears to be quite a tricky problem.

System of parameters of internal LC incorporates:

1. *Static parameters*.

U_{OH}, U_{OL}–output voltage of signal high and low levels.
$\Delta U_T = U_{OH}$ -U_{OL}–logic voltage swings.
I_{OH}, I_{OL} ~ output current of high and low level.
U_T–threshold switching voltage.
ΔU_T^+, ΔU_T^-,–reserve of interference immunity to positive and negative interferences.

I_{IH}, I_{IL}–input current of high and low levels.
N–load-carrying capacity

$$N = \min\left\{\frac{I_{OL}}{I_{IL}}; \frac{I_{OH}}{I_{IH}}\right\};$$

I^{c}_{CCL}, I^{c}_{CCH}–consumption current in static states of low and high levels.
U_{CC}–supply voltage.

2. *Dynamic parameters.*

   t_{pLH}, t_{pHL}–delay time of logic cell switching at switching ON/OFF,
   t_{HL} (τ_r), t_{LH} (τ_f)–front duration of logic cell switching ON/OFF,
   τ_{IH}, τ_{IL}–maximum duration of input signal front/cut,
   F–maximum LC switching frequency.
   I_{CCF}–dynamic LC consumption current.

Since in digital microcircuits, base logic cells (LC) are responsible for major logic loading; the requirements as below are applied to them:

(a) Maximum number of logic functions performed by a single LC.
(b) High-speed performance of logic and other functions.
(c) Minimum power consumption in static and dynamic operational modes.
(d) Minimum number of circuitry components relevant for LC implementation.

Memory Base Cells In digital microcircuits the built-in memory cells (MC) are intended for saving and storage of data. In the capacity of such memory cells, both elementary memory cells with control circuits intended for storing data bulks and complex memory cells of "clock' triggers" type may be used. Depending upon the method of data storage, memory cells pertaining to the first type may be both static and dynamic ones. In MC of static type, the data saved may be stored as long as desired. The body of static memory cell is a bi-stable cell formed by cross-joining of two inverting logic cells. Block diagram of such memory cell by the example of CMOS logic cell is shown in Fig. 1.5a. Dynamic memory cells (MC) (Fig. 1.5b) incorporate one LC and an additional memory cell (MC), arranged as per charge accumulation principle with writing circuits. Having in mind that over the time on the storage component loss of charge may occur, i.e., some data may be lost, such memory cells need recurrent restoration (regeneration) of the state. Methods of MC data formation when LC of various types are in use are elaborated in [2–4].

An extensive group of memory cells used in contemporary digital microcircuits is formed by trigger circuits with two stable states which are set at providing the respective signal combination to the control inputs and held within the time set by the designer after the effect created by these signals expires. In digital microcircuits, flip-flops RS, D, JK, and T have gotten the most widespread use. Depending upon

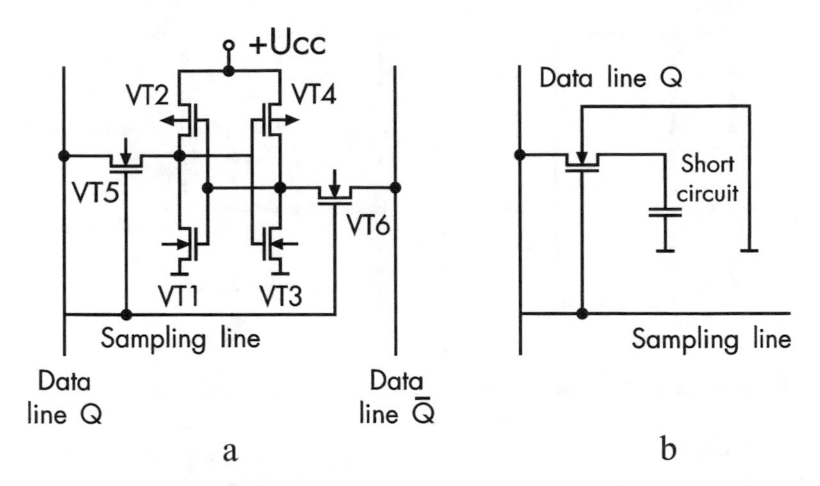

Fig. 1.5 Block diagram of the elementary memory cells (MC) of static (**a**) and dynamic (**b**) types

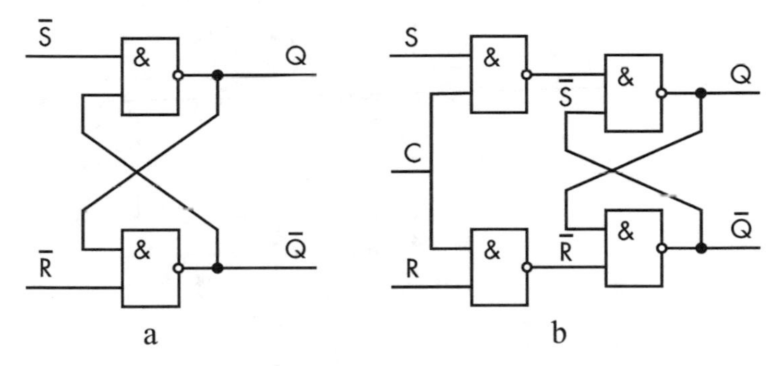

Fig. 1.6 Block diagram of unclocked RS flip-flop (**a**) and clocked flip-flop (**b**) with application of logic cell of type "NAND" [1]

the types of memory components used in flip-flops, they are divided into static, dynamic, or combined static-dynamic. If the state of flip- flop alters over entering the clock signal, then such flip-flops are referred to as "clocked flip-flops." In unclocked (asynchronous) flip-flops, toggling occurs when the relevant combination of input signals enters the control inputs. In digital microcircuits, clocked flip-flops synchronized by either signal level or signal front have gained the most extensive use.

Conventional unclocked RS flip-flop is made as the simplest bi-stable cell controlled by R&S input (Fig. 1.6a). The problem for RS triggers is that occurrence of even short pulses at inputs R, S may lead to erroneous setting of the flip-flop. Therefore, when working with RS flip-flops, one as a rule uses an additional clock signal *C*, limiting the time when inputs R, S are in active state (Fig. 1.6b).

The flip-flops clocked by signal front as illustrated by an example of Dt-type flip-flop tend to alter their state when the respective clock signal front arrives at clock input: positive or negative. At static levels of clock signal, the state of flip-flop is

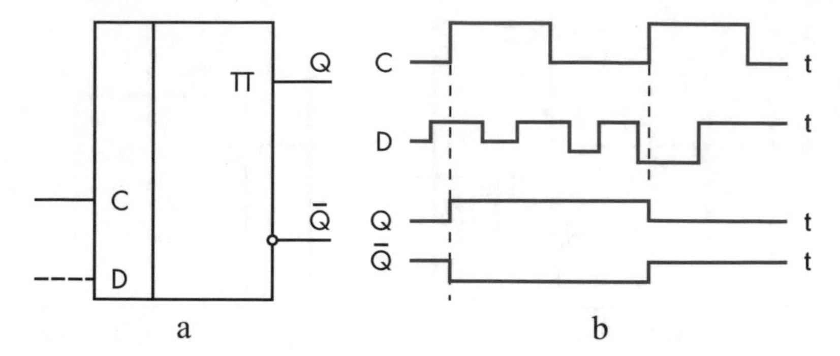

Fig. 1.7 Designation (**a**) and operational time diagrams (**b**) of signal front clocked Dt-type flip-flop

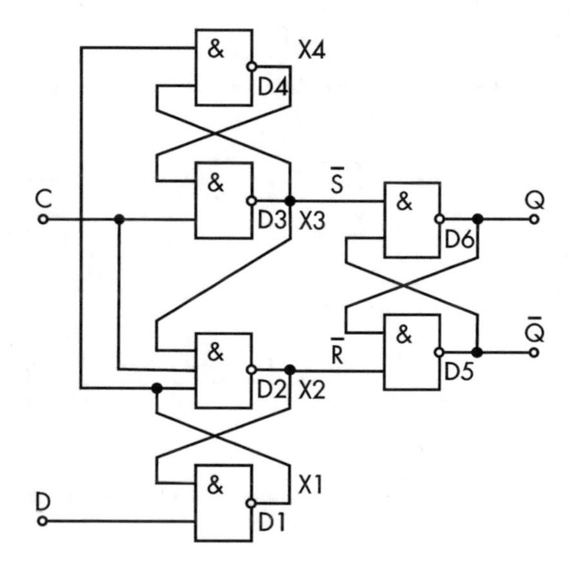

Fig. 1.8 Example of functional diagram for signal front clocked Dt flip-flop

retained irrespective of the levels of input signals. Time diagram showing operation of such flip-flop via example of Dt-type flip-flop is shown in Fig. 1.7b. An example of functional diagram of front clocked Dt-flip-flop on "NAND" logic cell is shown in Fig. 1.8 [1]. The state of flip-flops clocked by clock signal level may vary during the time when EN clock pulse is in effect when data signals enter D input. Within the pause, when EN clock signal level varies, the state of triggers does not depend on input signal levels. Designation of D-type flip-flop is available in Fig. 1.9a, and its operational time diagram is shown in Fig. 1.9b. If in the flip-flop pertaining to the said group only one output is in use, sometimes it is called "latch." An example of D-type flip-flop functional circuit clocked by signal level on the base of logic cell with function "NAND" is shown in Fig. 1.9c.

Other types of flip-flops may be generated on the base of clocked RS-type flip-flop via joining into various configurations. Thus, JK-type flip-flop is formed through sequential joining ("master-slave" type) of two RS-type flip-flops

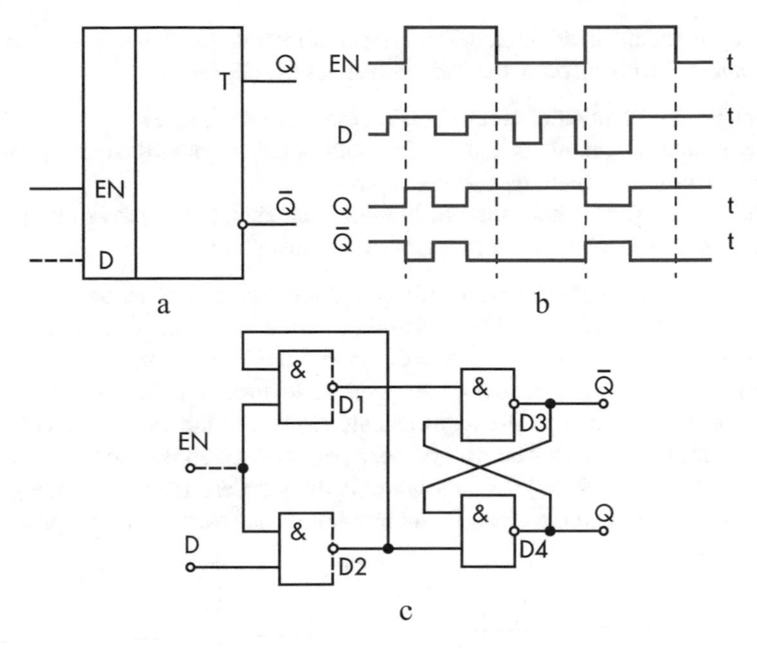

Fig. 1.9 Designation (**a**) time diagrams (**b**) and an example of functional diagram (**c**) of signal level clocked D flip-flop

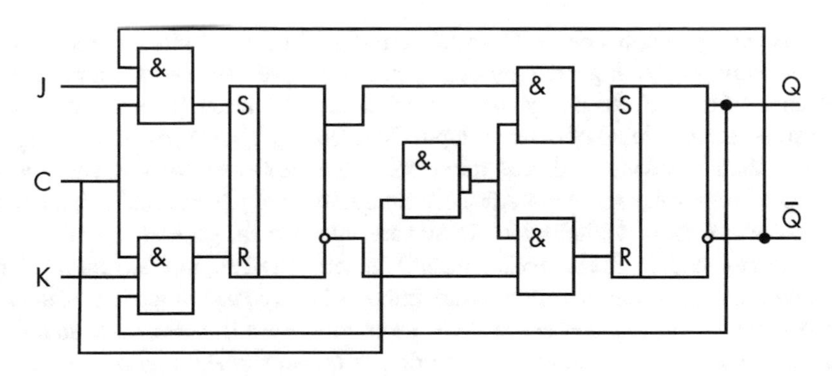

Fig. 1.10 JK flip-flop block diagram

(Fig. 1.10) by means of establishing feedbacks from outputs Q,Q of output RS-type flip-flop ("slave") to inputs R,S of input RS flip-flop ("master").

T flip-flop is created from JK flip-flop with recourse to introducing one T input linked with both outputs J, K.

In greater depth, the methods of flip-flops forming by means of logic cells of diverse types are covered in literature [1, 2, 4].

The system of memory cell (MC) static parameters is similar to the system of static parameters of internal logic cells (LC) of digital microcircuits. System of MC dynamic parameters in addition to the system of internal LC dynamic parameters of digital microcircuits incorporates the parameters as follows:

t_L, t_H–minimum duration of clock signals of low and high levels,
t_{SU}–preset time, i.e., minimum time of providing data signal till clock signal when stable writing into memory cell takes place,
t_H–holding time p minimum time of holding data signal after providing the clock signal when steady writing in memory cell takes place.

For Memory Cells (MC) of dynamic type there is an additional parameter t_{REC} – that is maximum period of MC refreshing signal when no data is lost in the MC. Whereas in majority of cases MC circuits are built on the basis of internal base logic cells, the key requirements applied to them for further use in digital microcircuits are similar to the requirements applied to the base logic cells (LC). However, unlike base LCs where high-speed performance appears to be more critical parameter, for MC, where passive storage is the standard mode (excluding signal digital processing microcircuits), the more relevant parameter is static consumption current.

1.1.3 Architecture of Digital Microcircuit Matching Elements

Input matching components (MC) in modern digital microcircuits are intended for the following purposes: generation of internal logic levels for digital microcircuits, protection of internal chains against external electrical impacts (interferences, static discharges, etc.), gaining of external input electrical signals, storing of input signals levels at their receiving in digital microcircuits, enhancement of sensitivity level, high-speed performance, and interference immunity of digital microcircuits. Output interface elements of digital microcircuits are intended for generating external for microcircuits output voltage levels (current), generation of appropriate output signal front durations, gaining of microcircuits internal LC signals, protection of internal microcircuits chains against external electrical influences (interferences, static discharge, overloads, etc.), and temporal storing of output signal levels at their outputting from microcircuits.

Input Matching Components The major elements of input marching component are shown in Fig. 1.11.

The fundamental part of input matching component is comprised of signal level conversion circuit which transforms external input levels into internal ones, which are distinguished by microcircuit internal logic cells (LC). Therefore, input matching components may be split into two groups:

(a) Input matching components (MC) with input levels which coincide with the levels of microcircuit internal LCs (TTL input levels, TTL levels of internal LC; CMOS input levels, CMOS levels of internal LC; etc.)

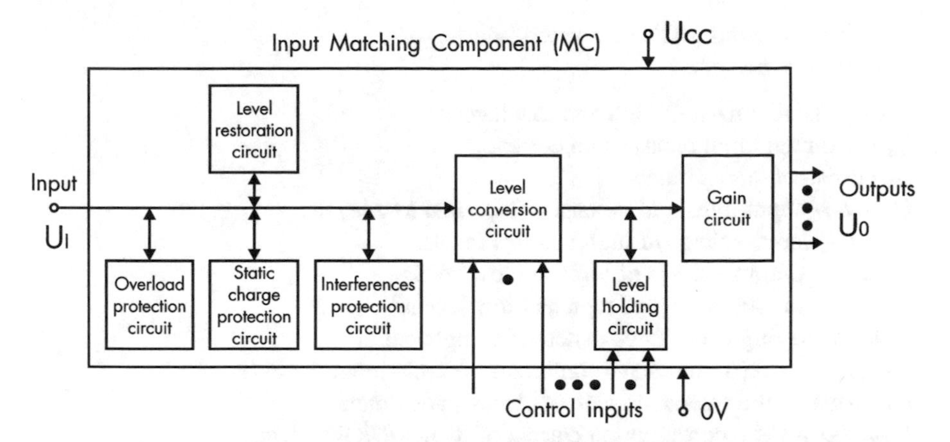

Fig. 1.11 Standard structure of input matching component (MC) of digital microcircuit [4]

(b) Input MCs with conversion of signal levels (for instance, input TTL levels, CMOS levels of internal LC; input ECL levels, TTL levels of internal LC; etc.)

Whereas there exist three most widely applied and fundamentally different systems of logic signal levels, TTL. CMOS, ECL, and I2L, hereinafter, the conversion circuits of input MC levels of the listed level types are addressed.

At MC's output, there may be introduced signal gain circuit which amplifies the input levels and generates high output currents to control greater number of LC internal inputs. Outputs of matching components may be single ones (may generate either direct or inverse signal to input one) and multiple (they may generate a few direct and inverse signals to input one). If it is necessary to store the levels of input signals, it is possible to complete input matching component with a memory cell responsible for temporal storage of input levels and their provision to matching component output by control signal. In the capacity of storage circuits, there may be used memory cells clocked by front of control signal level. For protection of the internal components from the external influences on the MC circuit there can be introduced the protection circuit from the static electricity and the protection circuit from the extreme values of currents / voltages at the MC inputs. If the MC input is connected with the signal source, that generates the noises, the special protection circuits are used, reducing sensitivity to these noises. Here the best case efficiency may be attained at implementing MC's circuit with transfer characteristic of hysteresis type.

At application of digital microcircuits, it may happen so that no fixed input levels are provided to microcircuit inputs, i.e., to microcircuit inputs an unknown potential is supplied (that is particularly dangerous for CMOS microcircuits). In such a case at MC's input there may be introduced a level recovery circuit, eliminating the "floating potential" effect at MC' input.

System of input MC's parameters consists of:
Static parameters:

I_{IH}, I_{IL}–input current of high and low levels.
l_{IA}–maximum input break down current,
I_D–current of clamp diode.
U_{TH}, U_{TL}–input threshold voltage of high and low levels.
U_{IH}, U_{IL}–input voltage of high and low levels.
U_{0H}, U_{OL}–output voltage of high and low levels.
I_{0H}, I_{OL} ~ output current of high and low levels.
N–load-carrying capacity of matching component.
$U_{I\ max}$, $U_{I\ min}$–maximum and minimum tolerable input voltages.
U_{ESD}–maximum tolerable value of electrostatic potential.
I_{CCL}, I_{CCH}–static consumption current of IC in ON/OFF state.
U_{CC}, U_{SS}–supply voltage.
0 V, U_{dd}–zero potential voltage (ground).

Dynamic parameters

t_{pHL}, t_{pLH}–delay time of MC's triggering at switching ON/OFF,
t_{HL} (τ_r), t_{LH} (τ_f)–MC's switching ON/OFF front duration,
τ_{IH}, τ_{IL}–maximum duration of input signal front/cut,
F–maximum frequency of input signal.
I_{CCF}–MC's dynamic consumption current.

Considering that in digital microcircuits input matching components do not have logic content and are intended mainly for matching purposes, they should be in compliance with the following major requirements applied to them:

(a) Maximum reserves of interference immunity against stresses produced by various noises.
(b) Minimum number of circuit elements contributing to reduction of the area taken by the matching component on a chip of digital microcircuits and to increase in the number of matching components located along the chip perimeter.
(c) Minimum conversion and matching time, i.e., delay times t_{pLH}, t_{pHL} and durations of fronts t_{HL}, t_{LH} should be of minimum value, which helps minimize the influence produced by high-speed performance of input matching components at aggregated high-speed performance of microcircuits.
(d) Minimum consumption current in static I_{CCL}, I_{CCH} and dynamic modes I_{CCF}.

The diagram of input matching component available in Fig. 1.11 contains a maximum number of functional blocks used in MC. However, depending upon modes of input MC's application in digital microcircuits, the block diagram of matching component may be simplified at the expense of excluding some separate functional blocks and circuits depending upon application conditions of microcircuits.

Input Matching Major components of output MC's architecture are shown in Fig. 1.12.

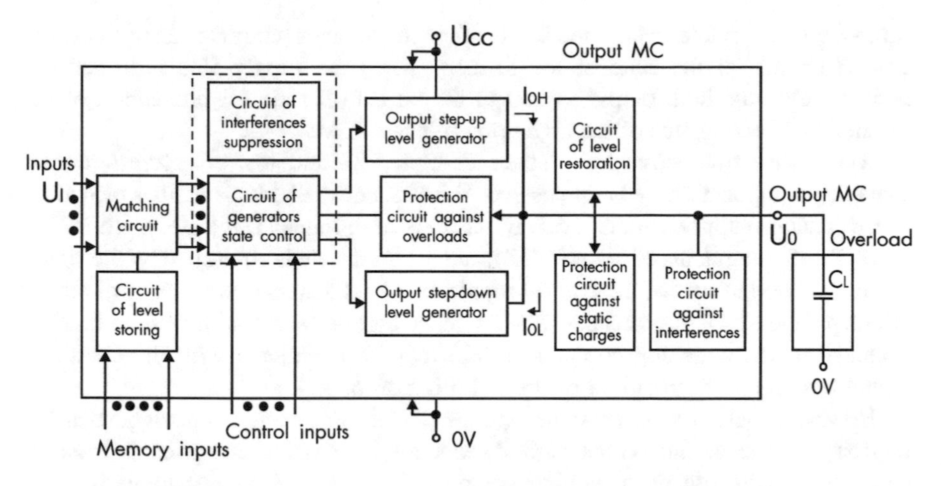

Fig. 1.12 Standard architecture of output IC of digital microcircuits [1]

Output MC primarily consists of two signal-level generators connected in parallel manner: step-up generator and step-down generator. Step-down output generator amplifies internal output signal of logic cell and generates low-level output signal U_{OL}, charging load capacitor C_L. Step-up output generator also amplifies the internal output signal of LC and generates output signal of high-level U_{OH}, charging load capacity C_L. Interfacing components of such type are called active output (AO). Since at switching this generator ON the second generator needs be switched off, the order (sequence) of their switching on is controlled by generator state control circuit. At the input of the latter, matching circuit is switched on, and it receives the internal signals of microcircuits and converts them into levels relevant to control signal output generators. Noise balancing circuit (interference inverter) for the interferences occurring either in supply chains U_{CC} or 0 V or at output pin U_O at MC toggling may be connected to output generator state control circuit. This circuit is able to quiescent "let-through" currents in supply line U_{CC} of matching components (MC) to control output current rise rate dI_O/dt, to accord automatically output resistance of MC with load, etc.

Apart from internal interference absorption circuit at IC output, there may be connected an additional protection circuit against the interferences occurring at output signals. Protection circuit against static discharges supports limiting of external electrostatic voltages which enter MC's output and form the leakage circuits for static charges. At application of microcircuits, it may happen so that to MC's output found at some definite level there enters opposite signal level, i.e., there occurs output overload by current resulting to deteriorating of output IC components. For protection of MC against such stresses, output IC may be completed with overloading protection circuit which analyses output signal level (voltage, current) and form the signal according to which control circuit put output signal generators into switched-off state or limit the levels of output voltages (currents). In this case the control circuit for the state of the formers, when the supply Ucc is turned on, switches over the both output former to the turned off state for prevention of the possibility of emergence of the high current levels in MC.

Along with two active logic states of high U_{OH} and low U_{OL} levels, output matching component may be in passive "third" state when MC is neither providing nor receiving output currents, i.e., output IC is high impedance state. Such MC is called "with three states of output" (TS), and the third state is controlled by the signal being provided to one of the control inputs of output generator state control circuit which put the MC into switched-off state. Since here at the output indefinite floating potential is set, matching component (IC) may be completed with the circuit of signal-level recovery circuit at the output (high or low).

However, there may exist some options of matching components where in the capacity of one of the output signal generators, a passive component (resistor) external or built into the matching component is used. If the resistor is used as step-up generator, the MC is called "matching component with open collector" (OC) in case of bipolar circuits and "with open drain" or passive high level"–in the case of MOS circuits. If resistor is used as step-down generator, matching component is called "with open emitter" (OE) in case of bipolar circuits and with "open drain" or " with passive low level"–in the case of MOS circuits. The merit of such matching components is that their outputs may be directly linked together at arranging the buses receiving the systems from many sources. On buses of digital microcircuits, such integration ensures logic function of "wiring AND-OR" without application of additional circuit components.

The system of output ICs' parameters incorporates:

Static parameters:

U_{OH}, U_{OL}–output voltage of high and low levels.
I_{OH}, I_{OL}–output currents of high and low levels.
I_{CCL}, I_{CCH}–MC's static consumption current in two logic states.
U_{IH}, U_{IL}–input voltage of high and low levels.

For output MCs of "three states output" type, the additional parameters are as follows:

I_{OZH}, I_{OZL} ~ output current in the third state at high and low voltage levels at output.
I_{CCZ}–static consumption current of MC in the third state.

Output MCs with "passive high level" in place of parameter U_{OH} are specified by parameter $I^{H}{}_{LO}$–output leakage current of closed matching component at high level of output voltage.

Output matching components with passive low level in place of parameter U_{OL} are specified by parameter I_{LO}–output leakage current of closed matching component at low level of output voltage.

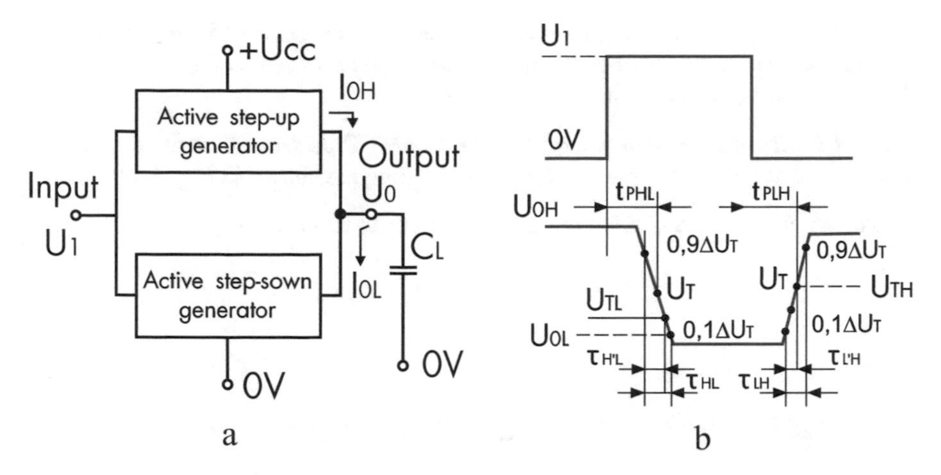

Fig. 1.13 Architecture of output MC of type "active output" (**a**) and time diagrams of signals (**b**)

Dynamic parameters:

t_{pHL}, t_{pLH}–delay time of MC toggling at switching ON/OFF. Output matching component of "three state" type in additional is specified by dynamic parameters at transfer into the third state and backward,

t_{pHZ}, t_{pLZ}–delay time of IC toggling into the third state from the states of high and low levels,

t_{pZH}, t_{pZL}–delay time of IC toggling from the third state into the states of high and low levels (Fig. 1.13).

Apart from the listed above, output matching components are specified also by the parameters having different dimensionalities: F, maximum switching frequency of output signal; I_{CCF}, dynamic consumption current of matching component(IC).

Similar to input ones, output MCs do not have logic content and are intended for performing functions of coupling and level generating; therefore, the main requirements applicable to the matching components in digital microcircuits are as follows:

(a) Minimum number of circuit components permitting to reduce the area taken by matching component on a chip of digital microcircuits and also to increase the number of matching components located along the die perimeter of digital microcircuits.
(b) Minimum time of output level generation, i.e., minimum value of times t_{pHL}, t_{pLH} and front duration τ_{HL}, τ_{LH} needed to reduce the influence of output matching elements on the total high-speed performance of digital microcircuits.
(c) Minimum consumption currents I_{CC} in all modes needed to reduce the total consumption current of digital microcircuits.
(d) High load-carrying capacity, i.e., maximum currents I_{OH}, I_{OL} for direct control of digital microcircuits with higher load-carrying capacity.

Block diagram of output matching component shown in Fig. 1.12 consists of maximum number of functional blocks and circuits applied in an interfacing component (IC). Depending however upon application conditions and load of the output MC, its circuit may be simplified at the expense of removal of some separate functional blocks and circuits.

Output Matching Component of "Active Input" Type (AI) Circuit of matching component and time diagram of signals at toggling are shown in Fig. 1.13b.

At switching on and attaining output level in the IC [1, 2]:

$$U_O(t) \approx U_{TL}; \quad t = \tau'_{HL},$$

we get:

$$\tau'_{HL} \approx \tau_L \ln\left[{}^{(U_{OH}-U_{OL})}/_{(U_{TL}-U_{OL})}\right] \approx R^{AB}_{OL} C'_L \ln\left[{}^{(U_{OH}-U_{OL})}/_{(U_{TL}-U_{OL})}\right],$$

where:

U_{TL}–threshold voltage of loading circuit.
${R^{AB}}_{OL}$–output resistance of open step-down generator.
$C'_L = C_O + C_L$–total load capacitance.
Co–output capacity of interfacing component.
C_L–external load capacitance.

Output resistance R_{OL} is determined by output current I_{OL} of output step-down generator. Assuming that in the active step-down generator output transistor is completely open and within the voltage range from U_{OH} to U_{TL} capacitance C'_L will be discharged by direct current I_{OL}; as a consequence, ${R^{AB}}_{OL}$ has the form of a variable value. For evaluation purposes, it is possible to use mean value of output resistance.

$$R^{AB}_{OL} \approx \frac{1}{2} \quad \frac{U_{OH} + U_{TL}}{I_{OL}},$$

where I_{OH} is the maximum output current of step-down generator.
At deactivating and attaining output level [1, 2]:

$$U_O(t) \approx U_{TH}; \quad t = \tau'_{LH},$$

$$\tau'_{LH} \approx R^{AB}_{OH} C'_L \ln\left[{}^{(U_{OH}-U_{OL})}/_{(U_{OH}-U_{TH})}\right],$$

where:

U_{TH} is the threshold voltage of load circuit.

$$\tau'_H \approx R^{AB}_{OH} C'_L;$$

${R^{AB}}_{OH}$ ~ output resistance of open step-up generator.

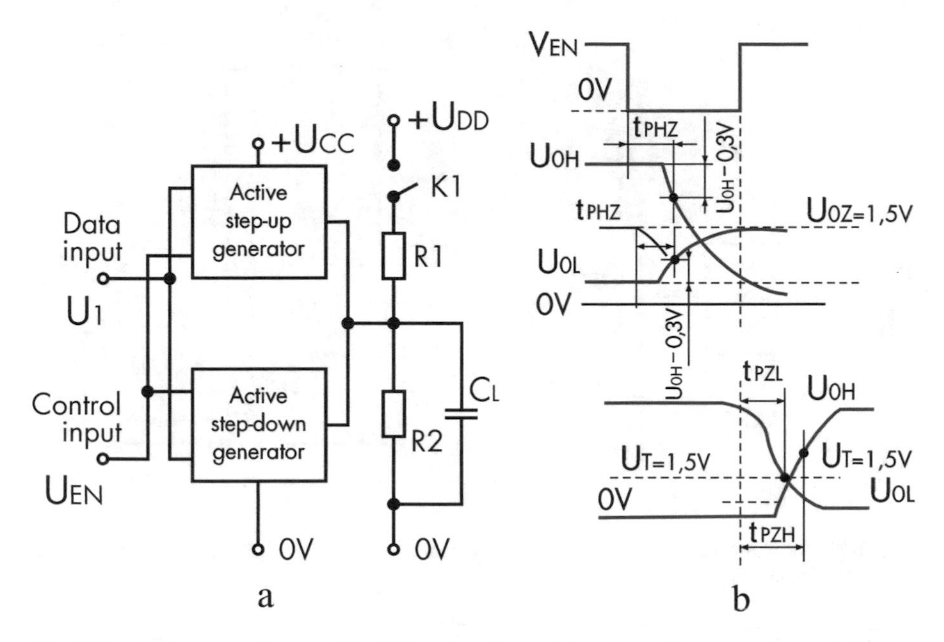

Fig. 1.14 Architecture of output IC of "three state" type (**a**) and time diagrams of signals (**b**)

Output resistance R_{OH} is determined by output current I_{OH} of open step-down generator of matching component (MC). Assuming that at active step-up generator output transistor is completely open, then within voltage range from U_{OL} to U_{TH}, load capacitance will be charged by direct current I_{OH}; therefore, $R^{AB}{}_{OH}$ is also a variable value. For estimation calculations, it is possible to use approximate formulae:

$$R^{AB}_{OH} \approx \frac{1}{2} \quad \frac{U_{OH} + U_{TH}}{I_{OH}},$$

where I_{OH} maximum output current of step-up generator.

Output Interfacing Component with "Three States: (TS) Circuit of matching component and time diagram of the signals are shown in Fig. 1.14a, b.

Whereas circuit of matching component of "three states" type in active states is similar to circuit of AB type, then in these states, the duration of activating/inactivating τ'_{HL}, τ'_{LH} is similar to the same parameters of matching component of "active input" type. Apart from the said parameters output matching component is specified by duration of transition into the "third state" and backward: τ_{HZ}, τ_{LZ}– duration of transition into the "third state" from the state of high and low levels of the signals; τ_{ZH}, τ_{ZL}–duration of transition from the third state into the state of low and high levels.

These parameters are actually exhaustively defined by load of the matching component. Having in mind the diversity of level latches at outputs of "three states"

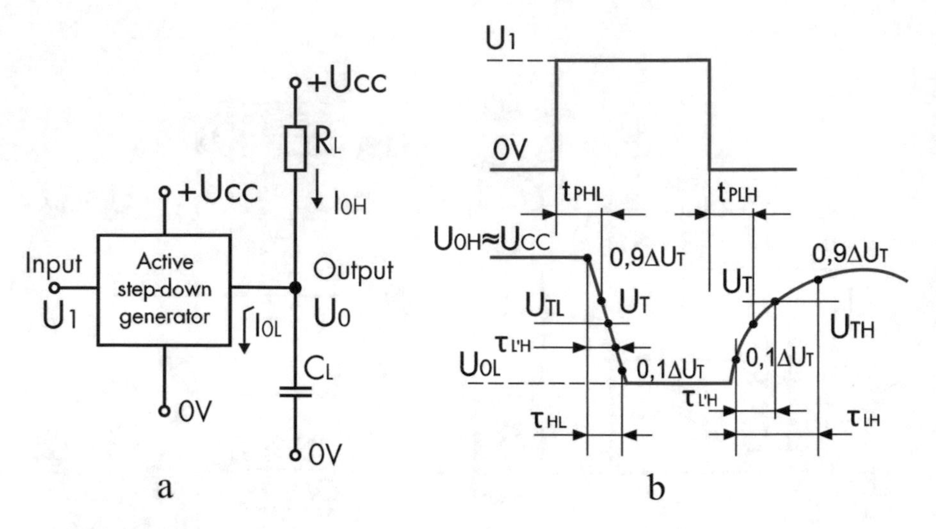

Fig. 1.15 Architecture of output matching component with passive high level (**a**) and time diagrams of signals (**b**)

type, no specific formulae to calculate the durations 'τ_{HZ}, τ_{LZ}, τ_{ZH}, τ_{ZL} are provided. If in the capacity of level latch at MC's output a resistor connected to supply pin U_{CC}, or resistor connected to common pin 0 V IC, is used, these parameters may be defined by the formula of front duration for output MC of type "passive low" and "passive high level."

Output Matching Component with "Passive High Level (*OK*)" Diagram of matching component and time diagrams of signals are shown in Fig. 1.15a, b. For this type of matching component at activation, it is essential to take into consideration that an additional current is flowing through resistor R_L into step-down generator.

$$I_R \approx (U_{CC} - U_O)/R_L,$$

which reduces discharge current I_{OL} of load capacitance C'_L and increases output resistance $R^{ПB}{}_{OL}$ of step-down generator. Then duration of activation:

$$\tau'_{LH} \approx R^{ПB}_{OL} C'_L \ln \left[{}^{(U_{CC}-U_{OL})}\!/\!{}_{(U_{TL}-U_{OL})} \right];$$

where $R^{ПB}_{OL} \approx \frac{1}{2} \left(\frac{U_{CC}}{I_{OL}} + \frac{U_{TL}}{I_{OL}-(U_{CC}-U_{TL})/R_L} \right)$ is the output resistance of open step-down generator.

I_{OL} – maximum output current of step-down generator.

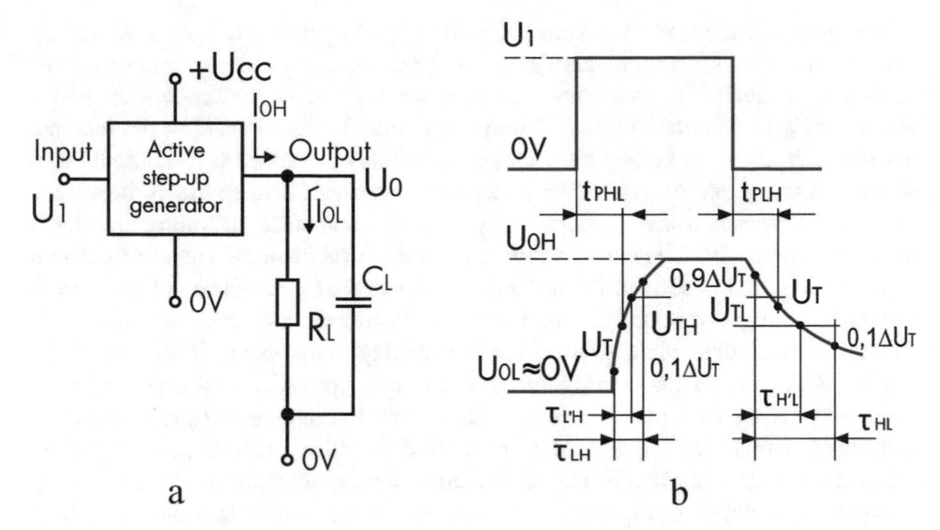

Fig. 1.16 Architecture of output matching component with passive low level (**a**) and time diagrams of signals (**b**)

$$\tau'_{LH} \approx R_L C'_L \ln\left[{}^{(U_{CC}-U_{OL})}/_{(U_{CC}-U_{TH})}\right],$$

where $R_{OL}^{ПВ} - R_L$ is the duration of inactivating.

Output Matching Component with "Passive Low Level" Diagram of matching component and time diagrams of signal are shown in Fig. 1.16a, b. As at transition into low-level state output resistance of matching component is stable and equal to $R^{ПВ}{}_{OL} = R_L$, then duration is:

$$\tau'_{LH} \approx R_L C'_L \ln\left[{}^{U_{OH}}/_{U_{TL}}\right].$$

At transition of matching component into high-level state, it is essential to take into consideration that the step-down generator supplies some portion of current to load resistor R_L: $I_R \approx U_O/R_L$, which reduces output current I_{OH} and charging load capacitance C'_L. Then duration is:

$$\tau'_{LH} \approx R_{OL}^{ПВ} C'_L \ln\left[{}^{U_{OH}}/_{(U_{OH}-U_{TH})}\right],$$

where $R_{OLПВ} \approx 0,5\ U_{TH}/(I_{OH} - U_{TH}/R_L)$ is the output resistance of step-up generator.

I_{OH} – maximum output current of step- up generator.

Matching Components of Other Types In digital microcircuits the maximum number of external package pins is limited, therefore on chip in order to eliminate this default rather often bi-directional matching components are used which are functioning both for receipt and generating the signals of digital microcircuits. However, in such ICs, there may be used output matching components which allow supply of external voltages to output and may be disconnected from loading. To such circuits, there belong matching components of "three states of output" type or those having one output active generator: step-down or step-up. Bidirectional matching components are specified by systems of parameters, activating both input and output parameters. The total value of parameters (at identical values of external voltages/currents) is defined by algebraic summation of parametric values of individual matching components. Apart from bidirectional matching components in digital microcircuits, there exist specific matching components having no logic content (load) and do not form signal levels. To such types of matching components, there may be referred pins for connecting external components (quartz resistors, capacities, inductances, etc.) not implemented in microcircuits, pins to supply reference voltages (currents), etc. Considering that application of such matching components is noted for diversity, and their types depend upon specific application conditions, this work does not cover these types of ICs.

1.2 System of Main Parameters and Chracteristics of Digital Microcircuits

1.2.1 Functional Parameters of Digital Microcircuits

At designing of any radio-electronic system as well as at designing and analyzing of digital microcircuits, profound understanding of hierarchy incorporating key parameters and characteristics of digital microcircuits is an essential prerequisite for successful problem-solving. There are three groups of parameters of digital microcircuits: functional, electrical, and dynamic. Let's look into the main parameters in each of these groups, being guided by commonly adopted international designations [1].

The main functional parameter of digital microcircuits is operational efficiency K measured by a number of operations completed for a unit of time (e.g., MIPS–million instructions per second). Depending upon the type of operations, one may differentiate a few values of operational efficiency:

(a) In the format of operations register-register K_{RR}.
(b) In the format of operations register-memory K_{RM}.
(c) In multiplication operation format K_{MPY}.
(d) In mixed operation format K_{MIX}.

Out of the other functional parameters, the following are used: (a) digit of address word n_A, -number of digits in address word; (b) digit of data word n_D, -number of digit in data word; (c) digit of command word n_{NS}, -number of digits in command word (micro-command); and (d) Q_{NS}, -total number of commands (micro-commands) implemented by digital microcircuits.

1.2.2 Electical Parameters of Digital Microcircuits

Electrical parameters of digital microcircuits are specified by voltages, current, and power and define main technical characteristics of microprocessor systems.

It is well-known that if at designing of such systems only functional capabilities of digital microcircuits are taken into account, such systems as a rule are not going to be operable.

At scrutinizing the characteristics of digital microcircuits, there are widely used notions of fan-in and fan-out coefficients. Fan-in coefficient is understood as the number of individual inputs by means of which the circuits may be connected to the outputs of the circuits of the same type. Similarly, fan-out coefficient is understood as the number of inputs in the circuits of the same type, which may be connected to a single microcircuit output. These definitions do not consider any specific values of input and output currents. For digital microcircuits, one of the critical electrical parameters is load-carrying capacity specifying current load of microcircuit. The factor of load-carrying capacity determines the number of single loads (inputs of analog digital microcircuits) which may be connected to an output at the specified operational conditions in the set operational modes. Dynamic loads (input and output capacities of microcircuit – loads) may limit the quantity of individual static loads which may be connected to one output of microcircuits. Let's review briefly the key electrical parameters of digital microcircuits by splitting them into three groups depending on the units of their parameters. Let's confine to just their enumeration and short definition [1].

Parameters Measured in Volts Threshold voltage of high-level U_{TH} is the minimum value of high-level voltage at microcircuits input when one steady state of digital microcircuits is stored.

Threshold voltage of low-level U_{TL} is the maximum value of low-level voltage at microcircuit input when one steady state of digital microcircuit is stored.

Input voltage of high-level U_{IH} is the value of high -level voltage at microcircuit input assuring the specified reserve of interference immunity.

Input voltage of low-level U_{IL} is the value of low-level voltage at microcircuit input ensuring the required reserve of interference immunity.

Input voltage of high-level U_{OH}.

Output voltage of low-level U_{OL}.

Interference immunity at input high-level voltage ΔU_T–absolute value of difference between input voltage of high level and input threshold voltage of high level.

Interference immunity at input low level of voltage ΔU_T–absolute value of difference between input voltage of low level of microcircuit and input threshold voltage of low level.

Voltage of supply source U_{CC} is the value of supply source voltage supporting microcircuit operation in the specified mode.

Voltage of protection diodes U_D–value of forward voltage drop at input protection diointerferences.

Numerical values of the parameters are set for each specific type of digital microcircuits.

Parameters Measured as Current Static consumption current I_{CCO} is the value of current consumed by microcircuit from supply source in static mode.

Dynamic consumption current I_{CCF} is the consumption current of microcircuits in toggling mode at the set operational frequency.

Consumption current of microcircuit at low (I_{CCL}) or high (I_{CCH}) level of output voltage.

Input current of low level I_{IL} is the value of input current at low-level voltage U_{IL} at microcircuit input.

Input current of high-level I_{IH} is the value of input current at the voltage high-level U_{IH} at microcircuit input.

Leakage current of low level at input I_{LIL} is the current in the input circuit at input voltage of low level and at the set modes at other pins of microcircuits.

Leakage current of high-level at input I_{LIH} is current in the input circuit at input voltages of high level and set modes at other pins.

Output current of high-level I_{OH} is the value of output current at high-level voltage U_{OH} at microcircuit output.

Output current of low-level I_{OL} is the value of output current at low-level voltage U_{OH} at microcircuit output.

Output current in OFF state I_{OZ} is output current of microcircuits with three states of output at OFF output.

Leakage current of low level at input I_{LOL} is leakage current of microcircuit in output circuit, at closed output state, and at output voltage of low level and set modes at other pins.

Leakage current of high level at output I_{LOH} is leakage current of microcircuits in output circuit at closed output state, at output voltage of high level, and with set modes at other pins.

"Short circuit" current I_{OS} is the value of output microcircuits current at output "short circuited" to common pin.

Direction of current flow (sink current or source current in the circuit) and numerical values of the currents are identified schematic solutions of input and output cascades.

Parameters Measured in Power Units Static consumption current P_{CCO} is the value of power consumed by digital microcircuit from source (sources) in static mode.

Dynamic consumption current P_{CCF} is the value of power consumed in digital microcircuits in the set dynamic mode (at the set clock frequency).

Dynamic consumption current P_{CCF} is of great importance in the first turn at calculation of power characteristics of microprocessor systems on the base of CMOS microcircuits (BiCMOS) for which there exists unique dependence of power upon frequency of microcircuits toggling.

Apart from the power consumed actually by digital microcircuits from supply sources, in the majority of cases at designing of microprocessor systems, one should take into consideration the power dissipated at input and output circuits of digital microcircuits to pins. These are circuits of load (output), clocking and control (output), etc. Such power augmentation (the so-called "power of losses") may turn out to be quite significant; therefore, it is defined by the sum of products of currents by voltage drop at input and output circuits multi-pin modern digital microcircuits (number of digital microcircuits pins may range from 124 to 172 pcs.)

1.2.3 Dynamic Parameters of Digital Microcircuits

A crucial characteristic of any digital microcircuit is the time diagram (TD) of its operation. Time diagram describes the deployed in-time sequence of providing data, clock and ancillary signals having control effect to digital microcircuits, and order of reading results of processing from microcircuits. Time diagram (TD) shows the type of these signals and indicates tolerable time correlations between them, tolerable durations, etc. Let's review the critical dynamic parameters most frequently used at analyzing the operation of digital microcircuits.

Clock signal repetition period T_C is time interval between start and end of each next clock pulses continuously following in sequence.

Pulse delay time t_P -is time interval between fronts of input and output pulse signals of digital microcircuits measured at the specified level of voltage or current.

Delay time of toggling from high-level to low-level t_{PHL} is time interval between fronts of input and output pulse signals measured at the level of threshold voltage U_{TL}.

Delay time of toggling from low-level into high-level t_{PLH} is time interval between front of input and output pulse signals measured at the level of threshold voltage U_{TH}.

Sometimes, for calculation, such parameter as average time of toggling delay is used:

$$t_a = \frac{(t_{PHL} + t_{PLH})}{2}$$

Turn-on time (front duration) t_{HL} is time interval during which the voltage at the specified output of microcircuits shifts from high level to low level measured at levels 0.1 and 0.9 or the specified values of voltage.

Turn-on time (cut duration) t_{LH} is time interval over which the voltage at the specified microcircuit output is shifted from low level to high level measured at levels 0.1 and 0.9 or at the specified voltage values.

Signal preset time t_{SU} is the time interval between start of two specified input signals at various inputs of microcircuits.

Time interval between end of two specified input signals at various microcircuit inputs is called signal holding (retention) time t_H.

Sampling time t_{CS} is the time interval between providing corresponding sampling signal to microcircuits input and obtaining of data signals at microcircuit outputs.

Data retention time t_{SG} is time interval over which a microcircuit being in the specified operational modes stores the data.

Recovery time t_{REC} is the time interval between the end of the specified signal at microcircuit pin and start of the specified signal of next cycle.

Duration of low-level signal τ_L is the time interval from the moment of signal transfer along the specified microcircuit pin from high-level state into low-level state till the time of its transfer from low-level state into high-level state measured at the specified voltage level.

Duration of high-level signal τ_H is the time interval from the time of signal going along the specified microcircuit pin from low-level state to high-level state till the time of its going from high-level state to low-level state measured at the specified voltage level.

Cycle time t_{CY} is the signal period duration at one of the control pins over which a digital microcircuit performs one of the functions.

Frequency of clock signal pulses t_C is the value reciprocal of period T_c of timing clock signal.

The majority of the enumerated dynamic parameters are interlinked through simple formula, for example, $T_C = \tau_L + \tau_H$, and extensively used at analyzing time diagrams of digital microcircuits operating.

Apart from the investigated essential electrical, dynamic, and power parameters at reviewing the operation of digital microcircuits, it is appropriate to use a range of other characteristics. Among them there are reliability and performance characteristics, degree of digital microcircuits immunity to the impact of penetrating radiation and electrostatic potential, and some others.

It should be particularly emphasized the relevance of impact produced by the operation modes and conditions at the parameters of digital microcircuits described normally in the format of parametric functions by the way of experiments for each type of digital microcircuits and entered then into the technical documentation (technical specifications) on the digital microcircuits as reference functions.

These are temperature functions of all parameters (electrical and dynamic) within the operational temperatures and other impacts produced by the ambient, dependence of these parameters of supply voltage (current) and also of the levels of input and output impacts (voltages and currents).

1.3 Schematic Implementation of Digital Microcircuits

1.3.1 Power Characteristics of Standard Logic Cells of Digital Microcircuits

Standard logic cells of digital microcircuits in terms of power have the form of the cluster of microelectronic components joined by the system of electronic circuits for transmission, accumulation, and dissipation of electrical power consumed from supply source. The cluster of components should definitely incorporate switches (transistors) controlled by input signals which jointly with other components (passive or active) provide digital data in the form of voltage levels or current at the cost of power stream (power flotation) between poles of supply sources. Along with swapping of the input data, configuration of standard LC circuits is changing, and respectively power streams inside it are subject to change, and then there occurs power accumulation in some components of standard LC and dissipation in other ones.

The dose and amount of power required for consumption, accumulation, and dissipation by standard logic cells are determined by:

- Operational fundamentals of essential electronic components and other microelectronic components of standard LC.
- Types of their connection in electronic circuits of standard LCs and methods of their linking to power supplies.
- Methods of microelectronic components formation and interconnections between them and power supplies within an integral.

Taken collectively, these factors define both the amount of power and rate of its consumption required to attain certain rate of digital data processing.

Classification of Standard Logic Cells with Reference to Power Consumption In compliance with the classification offered in [6], standard LC properties specifying their power characteristics (Fig. 1.17) are used as classification characteristics.

As the first classification characteristic, it is appropriate to use the type of power supplies (PS) digital circuits are consuming power from, which is essential for digital data processing. By this characteristic all types of basic LCs may be divided into three classes:

- Standard LCs powered by conventional power supplies of man- made origin: secondary sources of regulated voltage, batteries, electric accumulators, and others used for the prevailing number of digital circuits.
- Standard LCs which are powered by ionizing irradiation, light from man-made light sources, daylight, etc.

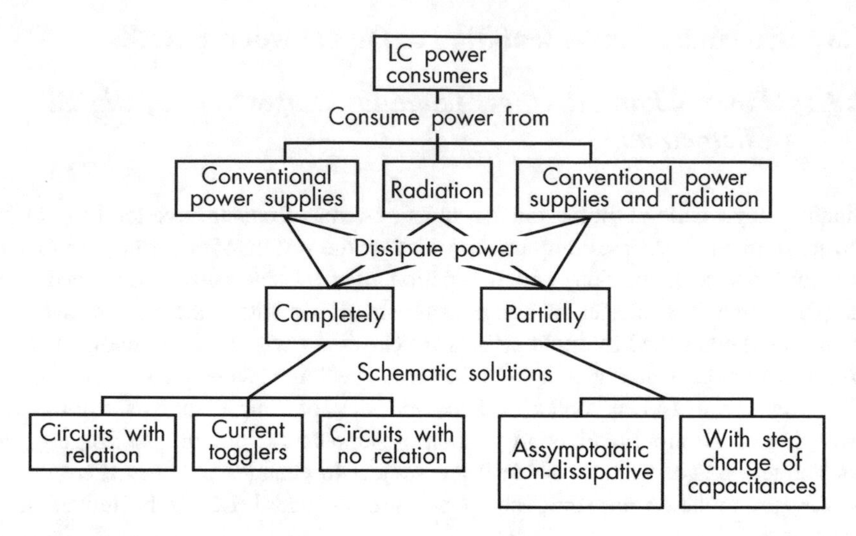

Fig. 1.17 Classification of digital logic cells

- LCs of the third class is fundamentally different from two previous ones in terms of availability of special tools essential for transfer of electric power supply mode from conventional power supplies to irradiation.

Conventional power supplies for energy transfer shall require two lines, at least: power supply line U_{CC} actually and common line OV. The lines are connected to each logic cell of digital device and occupy considerable part of the chip area. This circumstance in many ways determines the layout of the integrated circuit and inhibits interconnections inside of microcircuit. Increased length of lines and area of lines shall also have negative impact on reliability of the devices and density of their packing. Power supply of digital microcircuits with application of conventional interfacing components is a centralized one with its intrinsic imperfections and defects:

- Any single defect causing short circuit of power lines shall result in catastrophic failure of the device in general.
- Passing of the currents along power lines is responsible for transfer of some portion of the provided power supply into heat. Unavoidable power losses at delivery from the supply source to microcircuits components make conceptually impossible creation of completely adiabatic circuits, that is, the circuits not dissipating the power at all.
- As one more classification indicator, sometimes the used dissipated power percentage is applied when the information is processed. For a long span of time, there have existed logic cells dissipating the power completely only, where the power in the process of data processing was completely dissipated turning thus into heat evolved at the resistive components digital devices designs.

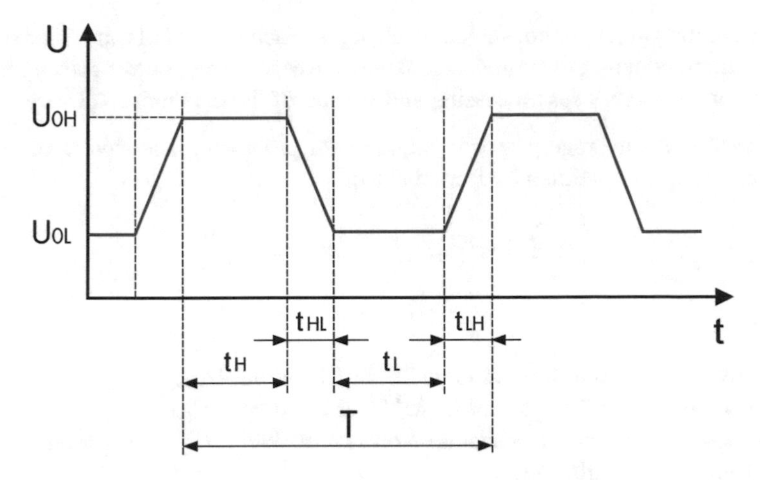

Fig. 1.18 Time diagram of voltage at standard logic cell output

Comparatively recently there have emerged the logic cells dissipating the power partially only when only portion of the power is used at processing of digital data, whereas the other is being returned into power supply and used repeatedly.

As the third classification characteristic, it is suggested that methods of LC schematic implementation be used which are different in the ways of binary data presentation and techniques of processing. In the fully dissipating circuits:

- In the circuits with relation to digital data processing level U_{OL} is generated due to drop at resistor with power dissipation (LC of TTLS type, etc.).
- In the circuits with no relation level U_{OH} is generated through accumulation of the energy at parasitic output capacitance with power dissipation on resistive elements at its transfer to capacitance (LC of CMOS type).
- Mechanism of LC operation at current switches resides in toggling the current of direct current generator between two circuits having a common major node connected to current generator. Depending upon the state, the constant current may pass into a common line along one of two possible circuits, generating thus levels U_{OL}/U_{OH} (LC of type logic matching component and I^2L).

As for partially dissipating logic cells, they are addressed in depth in [1, 6].

General Issues of Standard Logic Cell Energy Performance and Physics In the operation of any standard logic cell there may be distinguished four spaces of time which vary in terms of energy supply and performance. In Fig. 1.18, there is shown an epure (distribution diagram or profile) of voltage change at logic cell output depending upon the time where these time intervals are indicated. Over the time spaces t_L and t_H, logic cell is in non-active, so-called stationary (steady), states. Over the time spaces t_{HL} and t_{LH}, logic cell is found in the process of transition from one state into another one and vice versa. During exactly these time spaces, digital data is actually processed. In steady modes, the data states represented by the formula of voltages at the output are not subject to changes. Each one out of four states of logic gates is supplied with power in different ways. Respectively, power consumption or expenditure of energy for processing and storing of the data unit shall vary.

Generally, the average power W required for processing and storing of a single one data bit may be defined by formulae [6]:

$$W = \left(W^{H}_{cp} + W^{L}_{cm} + W^{HL}_{\partial} + W^{LH}_{\partial}\right),$$

where:

$W^{H}{}_{cp}$–power required to present logic “1” at LC output U_{OH}.
$W^{L}{}_{cm}$—power required to present logic “0” at LC output U_{OL}.
$W^{HL}{}_{\partial}$—power required to generate voltage of logic U_{OL}, or power ensuring “unlatching” of logic cell.
$W^{LH}{}_{\partial}$—power required to generate at LC output logic “1”U_{OH}, or power ensuring LC “latch-up”.

Power $W^{H}{}_{cp}$, supporting the generation of logic “1” U_{OH}, may be accumulated and stored in output capacitance C_O, making part of electronic circuits of LC with no relation (output capacitance); then this power will define U_{OH} as per the formula:

$$U_{OH} = \sqrt{\frac{2W^{H}_{cp}}{C_O}}$$

In logic cell with relation, maintaining the steady state of U_{OH} may be ensured at current passage, i.e., at transfer of the energy from the power supply through LC output.

Power $W^{L}{}_{cm}$, that supports generation of logic “0” U_{OL} may be accumulated and stored at input capacitance C_I, making part of electronic circuit (input capacitance).

In LC with relation (dependence), maintaining logic “0” U_{OL} shall be ensured at the cost of power transfer from power supply through open switches.

Transfer of electrical power as per the law by Joule-Lenz is being accompanied by its transformation into heat; therefore, even on steady states, the energy from power supplies is consumed. Amounts of energy $W^{H}{}_{cp}$ and $W^{L}{}_{cm}$ for LOC with relation depend on time spaces t_H and t_L when they reside in steady states.

At data processing in LC of the first type power $W^{HL}{}_{\partial}$ is accumulated with input capacitance C_I, and power $W^{LH}{}_{\partial}$ is accumulated at output capacitor C_O. Power is accumulated through its transfer from the power source along electrical circuits with parasitic resistors and therefore is accompanied by losses.

Steady power components $W^{H}{}_{cp}$ and $W^{L}{}_{cm}$ of energy in the majority of gate types are defined not only by appropriate voltage U_{OH} and U_{OL} but also by various parasitic effects in the electronic circuits. To such parasitic effects, first of all, there should be referred power dissipation at resistive elements of switches and connections and parasitic effects caused by current leakages through closed switches.

Relevant power characteristic of logic cell is the rate P(t) of power change in different times:

$$P(t) = \frac{dW}{dt}$$

With account of various momentary values of power over various time spaces of data generation, the common formula describing power may be as follows:

$$W = 0,5\left(\int_{t_L}^{t_H} P_{cm}^{L}(t)dt + \int_{t_H}^{t_{HL}} P_{\partial}^{H}(t)dt + \int_{t_H+t_{HL}}^{t_H+t_{HL}+t_L} P_{\partial}^{HL}(t)dt + \int_{t_H+t_{HL}+t_L}^{T} P_{\partial}^{LH}(t)dt\right)$$

where:

$P_{cm}^{L}(t)$–power in open state of LC or level presentation power U_{OL}
$P_{cm}^{H}(t)$–power in closed LC state or level presentation power U_{OH}
$P_{cm}^{HL}(t)$–level generation power U_{OL}
$P_{cm}^{LH}(t)$–level generation power U_{OH}.

Steady constituents of power P_{cm}^{L} and P_{cm}^{H} determine power consumption rates by power source at data providing and storage in the format of levels U_{OL} и U_{OH}, respectively. Dynamic constituents P_{∂}^{HL} and P_{∂}^{LH} determine power consumption rates at processing of new data by logic cell under impact produced by input signal.

Limits of integration in the expression for the energy of single switching are illustrated by time diagram of output voltage shown in Fig. 1.18. *W* approximately defines average value of single switching power of logic cell operating at input signal regularly changing.

Total amount of energy consumed by LC over the operation time *t*, at frequency *f* of regularly changing input signal, is defined by the following formula:

$$W_{\sum} = W \cdot f \cdot t$$

This energy is entirely dissipated in LC of conventional micro-schematics and induces the problem of heat removal. In Fig. 1.19 there is shown time dependence of microprocessor power density, illustrating the gravity of heat removal issue in digital devices [6]. Efficient heat removal is critical for thermal stabilization of digital devices because temperature rise is the main destabilizing factor at data processing with high rates. Therefore, digital circuits are characterized by the value of maximum rated P_{add} which specified minimum possible rate of heat removal from integrated circuit die.

Value $P_{sup.}$ is determined by IC package design, conditions, and design tools of heat sinker.

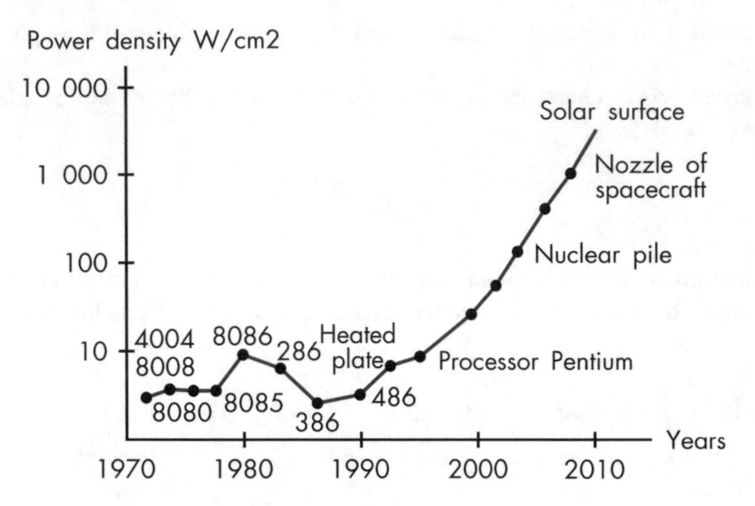

Fig. 1.19 Time dependence of microprocessor power density

The technique of data processing by means of electronic circuits involves generating diverse electrical circuits over various time intervals. Electrical circuits shall by no means have resistors incorporated. Switches alternatively connect LC output through these circuits either to power line U_{CC} or to common line OV. Along the said circuits, the currents are flowing and in compliance with Joule-Lenz's law, electric power, which is transformed into heat, shall be defined by the formula:

$$W = I^2R$$

Dissipation of power provided by power supply unavoidably occurs in all types of LC available in the classification diagram in Fig. 1.17.

Amount of energy consumed and rate of its consumption (power) from supply sources are crucial power characteristics of logic circuits.

The fundamental parameter universally accepted and specifying LC in terms of energy is the energy of single switching W_o, defined approximately as average power of data unit processing set specified approximately by the formula:

$$W_0 = P * \tau_{cp}$$

where:

P–power of LC.
τ_{cp}–average time of LC switching delay.

Power of single switching is the constant value for LC of certain defined schematics type. Power of single switching is the constant value for LC certain defined schematic type, design, physical structure and technology. The value w_0 does not depend on the operational modes of the logic element.

Rate of energy consumption from supply source (power P) may vary within broad ranges, but increase of power shall inevitably result in decrease of value τ_{cp}, and conversely, thus, this value of single switch power shall be constant, i.e., $W_0 = \text{const}$.

This parameter is a reliable and true-to-life proof of LC efficiency and perfections and therefore frequently used for process comparison purposes.

Generally in conventional integrated digital devices with full power dissipation, the entire energy consumed from supply sources is dissipated at resistive elements of electronic circuit transforming whereby into heat.

For them the formula below is true [5]:

$$Q = \sum_{i=1}^{N} W_{cp}^{1} \cdot f_i \cdot t,$$

where:

Q–heat amount emitted by die of integrated digital device for time t.
W_{av}^{1}–LC average switching power
f_i–LC switching frequency (toggle rate).

In the vast majority of up-to-date logic circuits, the power is consumed in full which gives rise to one of the most crucial problems, that is, availability of heat sink. Stabilization of temperature involves application of efficient design schematic tools. It makes the apparatus too sophisticated, increases its weight and overall dimensions, and worsens its operational reliability. For up-to-date electronic devices, thus, the issue of reducing gate togging power tends to become the first priority issue. Dire need to solve this problem gives impetus to search of new more energy-efficient solutions at the level of elementary logic gates. One of the possible ways to increase power efficiency is designing schematic solutions for gates where the energy provided by supply source is used in the process of operation with only partial transformation into heat.

Speed of Data Processing Speed of data processing at logic cell level is determined by its schematic configuration, layout, and predominantly high-speed performance of elementary transistors.

High-speed performance is characterized by times of toggling from open state to closed state and in reverse is defined by the mechanism of operation, physical architecture, and layout features.

In order to support high-speed performance of elementary switches, the transistors should be suitable for commutating the highest possible currents at lowest possible control voltages.

In terms of quantity, such as the transistor's function, switch amplifier shall be defined by conversion transconductance [5]:

$$K = \frac{\partial I_O}{\partial U_I}$$

Also some types of physical processes are deterministic for such critical properties of toggle as resistance between current conducting electrodes in open R_L and closed R_H states.

The distinctive features of bipolar transistor as a toggling tool resulting from the concept of its operation is high value of conversion conductance, low resistances in open state, and high resistances in closed state.

Conversion conductance shall be defined by the formula below:

$$K_{HE} = \frac{\partial Ic}{\partial Ube},$$

where:

I_C–collector current.
U_{BE}–base-emitted voltage.

The high value of the conversion steepness is determined by the exponential dependence of current on voltage. Small values of resistance between current conducting electrodes of emitter and collector in open state are associated with the so-called saturation mode. Under such mode resistance of ideal bipolar transistor is zero. Therefore, an open bipolar transistor is able to pass between electrodes of emitter and collector high currents with no noticeable voltage drop between them.

In monopole transistors, unlike bipolar ones, the mechanism of operation is based upon control of carriers of the same sign or electrons, or holes by input impact (field).

Current in field transistors, since they appear to be monopole transistors, is transferred by major carriers only, and parasitic effect of accumulation of minority carriers in them is not observed. With exclusion of is not observed, with exclusion of field transistors with control p-n junction. This is the first property of field transistors as toggles stemming from the physical concept of operation. High-speed performance of field transistors is determined by the channel resistance in direct proportion to its length and parasitic capacitances. Channel length stipulates also conversion conductance value which is defined here by the formula [5]:

$$K_{\Pi \mathrm{T}} = \frac{\partial I_C}{\partial U_3}$$

where:

I_d–drain current between current conducing electrodes of drain and sorces.
U_g–voltage at control electrode (gate).

Another classification criterion for transistors as toggles is voltage and current characteristics (VCC). VCC exist in two fundamentally different types: normally closed (NC) and normally open (NO). The transistors of various operational

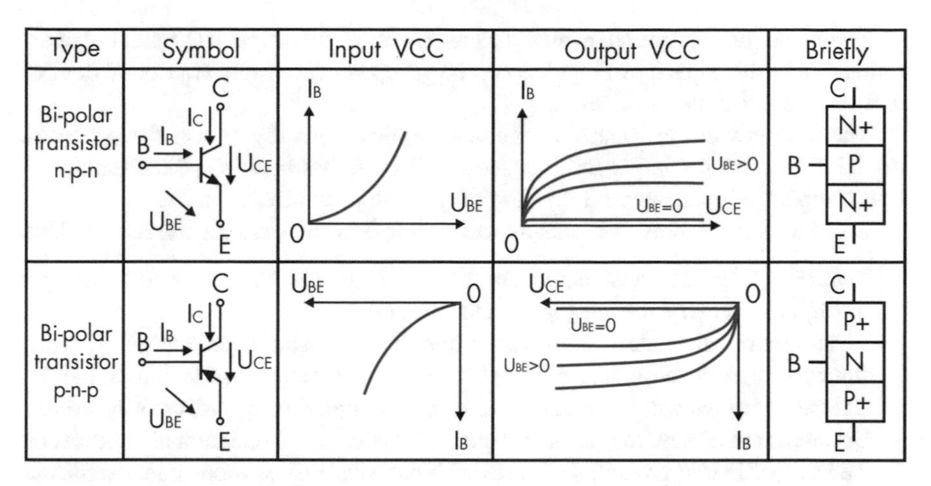

Fig. 1.20 VCC of bipolar transistors

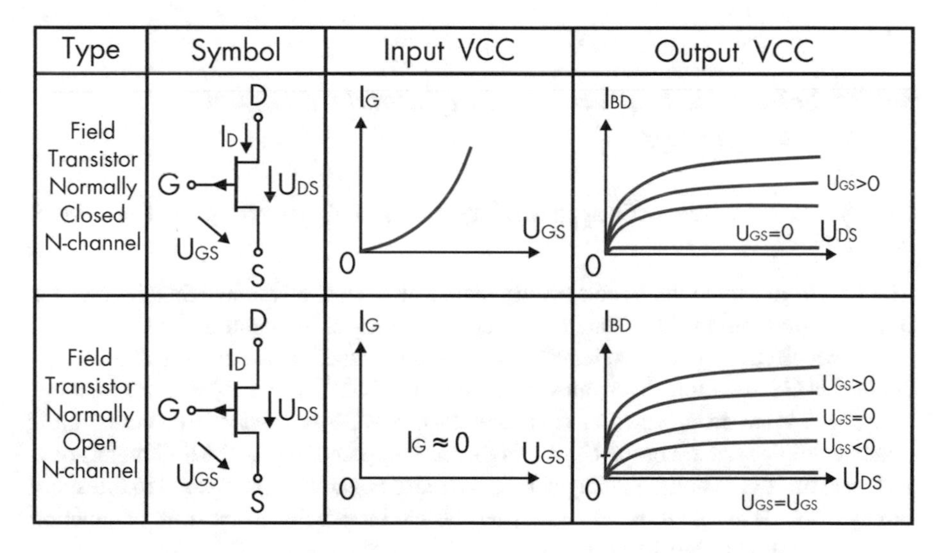

Fig. 1.21 VCC of field transistors with control junction

mechanism and various structural types, which are closed when the voltage at control electrode is 0, are normally closed. Standard VCC of transistors are shown in Figs. 1.20 and 1.21.

Bipolar transistors, field transistors with control p-n junction, MOS transistors with induced channel, and Schottky field transistors are classified as normally closed (NC) transistors. Field transistors with built-in channel (Fig. 1.21) pertain to normally open type.

The consequence of fundamental differences in NC VCC and NO VCC is that the transistors in the capacity of toggles may be grouped into current controlled devices and voltage controlled devices.

The progress in the sphere of microelectronics generally and in the sphere of digital devices in particular for over recent 50 years has been propelled mainly by improvement of transistors on the base of technological advancement.

Improvement of transistor designs seems to follow the strategic trends as below:

- Reduction of geometrical features of "classical" transistors with standard designs through elementary scaling (linear and vertical).
- Formation of full dielectric insulation to reduce parasitic capacitances of switch structures and development of new schematic and design and layout and architectural solutions within the frameworks of traditional mechanisms of operation.
- Development of new toggles using quantum-mechanical operational concepts on the basis of heterojunction structures with nanoscale dimensions under semiconductor technology framework.

1.3.2 Schematic Implementation of Standard Digital Microcircuits

CMOS The most frequently applied standard LCs of digital microcircuit are CMOS logic cells.

The field transistors which are currently in use have such advantage that their gate is oxide isolated and in static state no current is passing at the input. In Fig. 1.22 there are shown circuits of toggles, junction, and output characteristics of n-channel and *p*-channel MOS field transistor; they are also called N-MOS and P-MOS transistors.

As it is clear from junction characteristics, NMOS transistor is switched on if positive voltages gate-source U_{gs} are higher as compared to U_{th}. P-MOS transistor is switched on if gate-source voltages U_{gs}, *are* more negative as compared to threshold voltage U_{th}.. It is obvious however that the characteristic in the area of junction OFF/ON state is not of linear nature.

Drain current of N-MOS transistors is positive, whereas with P-MOS transistors, it is negative. Hence those transistors are linked into circuit joining the drains of both transistors in the manner shown in Fig. 1.23. The gates of the both transistors are interfaced in such a way $U_{gs(NMOS)} = U_e$ and $U_{gs(PMOS)} = U_e$ - V_{DD}. Due to correct selection U_{th} and U_{CC} there are ensured steady states when one of transistors is switched off and the other one is switched on. The circuit thus obtained is functioning as an inverter because at $U_I = 0$ V, N-MOS transistor is switched off and PMOS transistor is a conductor. Therefore, $U_O = 0$ *V*. For instance, at $U_I = U_{CC}$, P-MOS transistor is switched off, and N-MOS transistor is a conductor, and $U_O \approx 0$ V.

Standard CMOS LCs are formed from an inverter through completing each of N-MOS and P-MOS transistors with one more transistor either in serial or in parallel

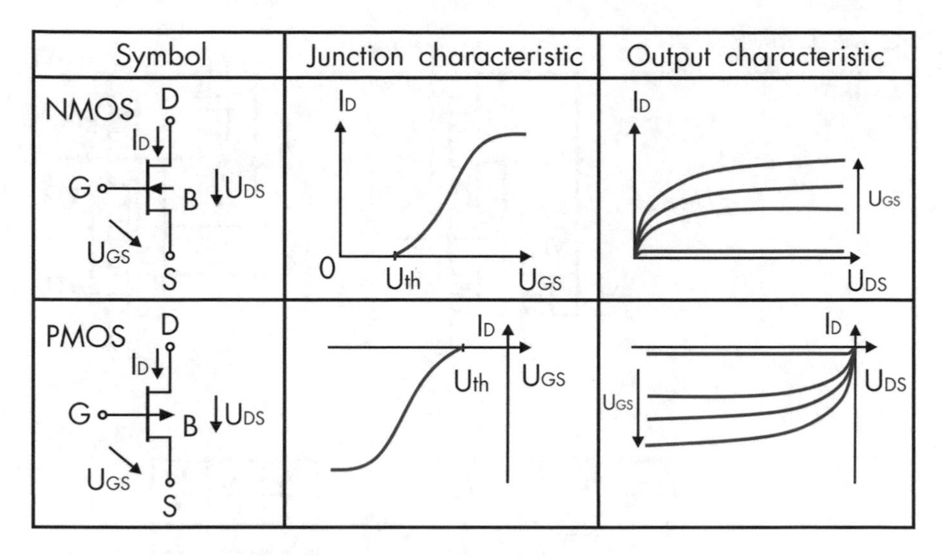

Fig. 1.22 Characteristics of N-MOS and P-MOS transistors

Fig. 1.23 CMOS inverter

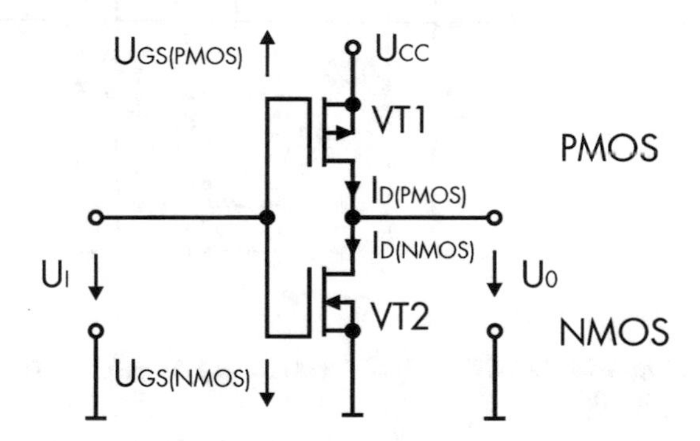

mode. Thus, there is obtained circuit “2 NAND” or “2 NOR” (Fig. 1.24a, b). In circuit “2 NOR,” provided in Fig. 1.24, an output signal corresponds to level U_{OL} any time when one of the inputs corresponds to level U_{IH}, because then at least one of *n*-channel field transistors is closed. On the contrary in circuit “2 NAND” Fig. 1.24b *y* is located at level U_{OL} only when both inputs are at H level. In this case n-channel field transistors are conductors, and p-channel field transistors are in closed state.

TTL

The prevailing method of logic elements implementation has been transistor-transistor logic (TTL) (Fig. 1.25). The functions of TTL LC are based on application of multi-emitter transistor VT1 at the input. If at all inputs the voltages U_I are close to supply voltage U_{CC}, then input transistor collector is operating as emitter, and the transistor is working in inverse active mode. Then next transistor VT2 is activated, and as a result, the voltage at the output corresponds to level U_{OL}. In order to minimize input current gain ratio of transistor VT1, inverse current should be close

Fig. 1.24 Electrical circuits CMOS LC "2 NOR" (**a**) and "2 NAND" (**b**)

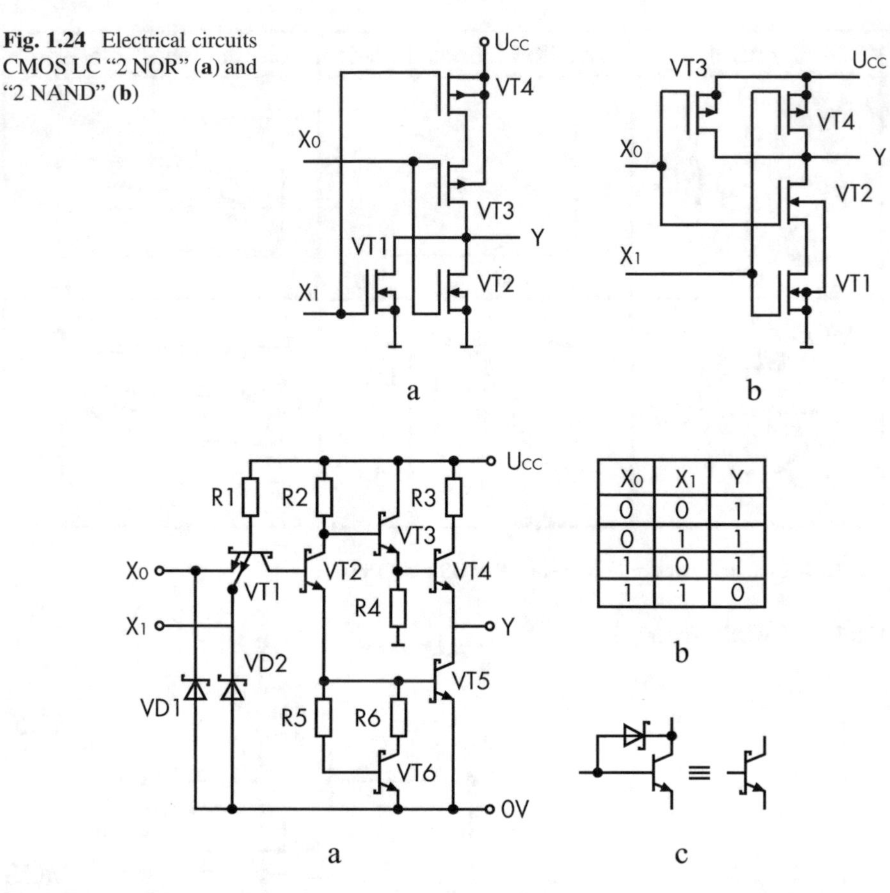

Xo	X1	Y
0	0	1
0	1	1
1	0	1
1	1	0

Fig. 1.25 Circuit TTLS LC "2NAND" (**a**) designation of transistors with Schottky diodes (**c**) truth table (**b**)

to 1. Should at one of the inputs the voltage corresponds to level 0 V, then input transistor VT1 is operating in active normal mode (current is passing). Voltage collector-emitter goes down to the minimum residual voltage, and next transistor VT2 is closed. Output voltage corresponds to U_{OH}.

Thus circuit TTL LC shown in Fig. 1.25 performs function of "NAND" (Fig. 1.25b). Function "2 NOR" shall be implemented by parallel connection to transistor VT2 of circuit similar to VT1,VT2 R_1.

As far as semiconductor technology had been improved and non-saturated bipolar transistor with Schottky diodes and good high-speed performance had been created (Fig. 1.25c), their application in TTL integral circuit has induced dramatic increase of LC high-speed performance and reduction of power consumption.

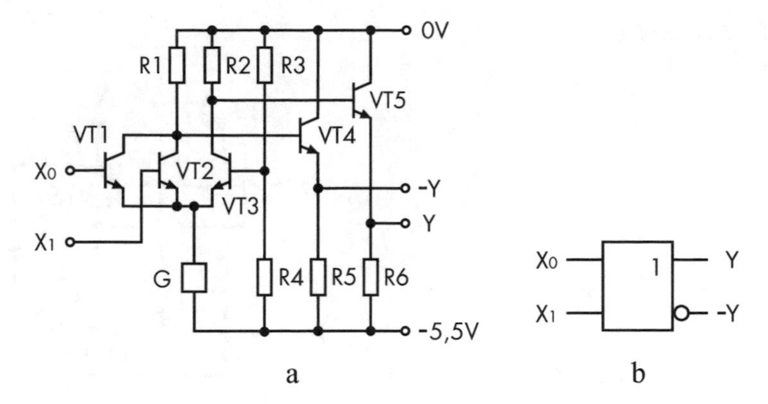

Fig. 1.26 Standard ECL based LC "2NOR": (a) схема; (b) logic symbol for positive logic

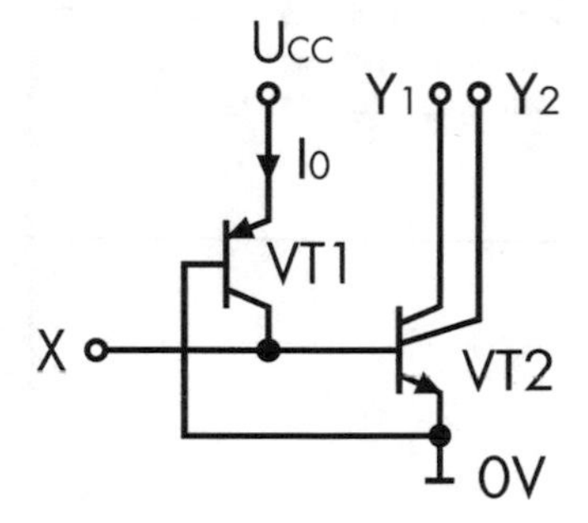

Fig. 1.27 Electrical circuit of I^2L inverter

Emitter Coupled Logic In emitter coupled logic (ECL) circuits, there are applied differential amplifiers where the transistors are not put into saturation state (Fig. 1.26). In this way these circuits feature the enhanced high-speed performance.

In the input differential amplifier input signals x_0 and x_1 voltage circuits are compared with the reference signal. If x_0 and x_1 have value, 5.5 V, then transistors VT1 and VT2 are closed, and transistor VT3 is open. If conversely, x_0 *and* x_1 have values OV, then VT1 or VT2 is open, and VT3 is closed. Output signal *-y* has value U_{OH}. Consequently circuit performs the function of gate "2 NOR." Switching threshold may be set by the voltage at base VT_3, provided by means of current generator G.

Integral Injection Logic Integral injection logic (I2L) is particularly suitable for making bipolar circuits because its standard LC need very small area on the die surface.

If Fig. 1.27 there is shown a standard inverter. Transistor VT1 operates as source of direct current I_0. Should input signal x be located at level ($U_I \approx U^{VT1}{}_{BE}$), the whole current shall flow into base VT2 which gets open. Outputs y_1 and y_2 in this case are at level ($U_{OL} \approx 0$ V). Should the input match the level (OV), then current I_0 flow into the previous LC, and the output signals correspond to level ($U_{OH} \approx U_{BE}$).

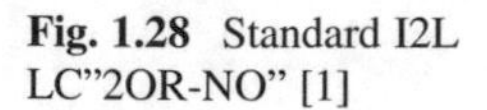

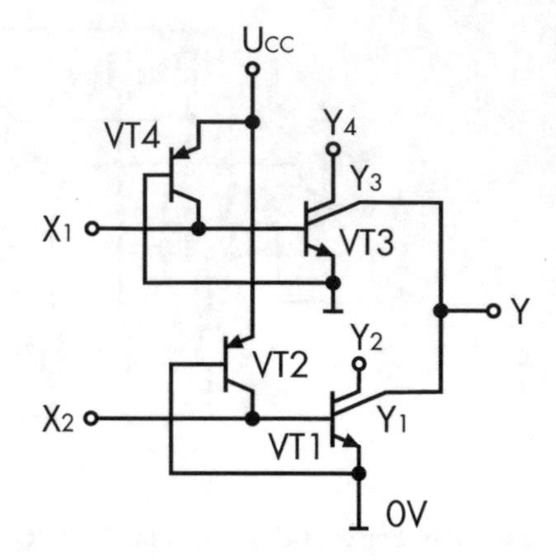

Fig. 1.28 Standard I2L LC"2OR-NO" [1]

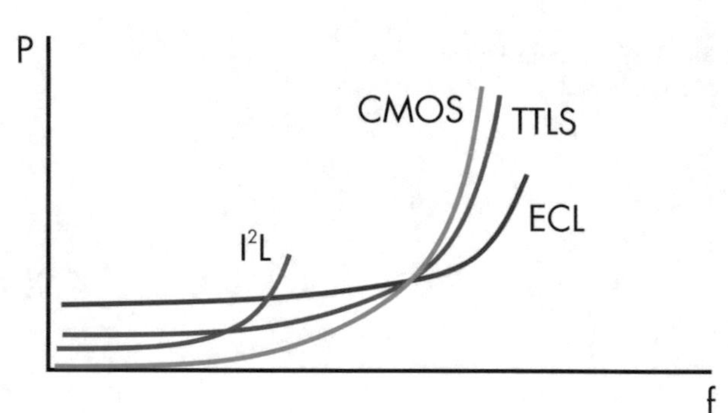

Fig. 1.29 Dependence of various LC power on switching frequency [1]

Logic functions in MICROCIRCUIT I2L are generated by joining the collectors of switching n-n transistors (Fig. 1.28). Both open collectors Y2 and Y4 may be used for implementation of other logic functions. I2L LC may be used at supply voltages U_{CC} up to 0.8 V and have high-speed performance programmable in dependence of current I_O which constitutes a relevant advantage.

Power Dissipated and Characteristics of Various LC Switching (Fig. 1.29)

1.3.3 Techniques of Digital Microcircuits Element Base Selection

Selection of digital microcircuit element base appears to be quite a sophisticated task and is determined in the first turn by the conditions of the microcircuit application which imposes certain requirements on electrical parameters of microcircuits. As is clear from Sect. 1.2, contemporary digital microcircuits have a great diversity of performance parameters; however, when picking out element base of digital micro-circuits, there may be used parameters of standard logic cells and parameters of matching components. In this case the integrate index of cell quality may be set forth as in [1]:

$$E = E_{LC} * E_{MC};$$

where:

E_{LC}–index of standard LC quality.
E_{MC}–index of IC quality.
$E_{LC,MC}$ may be shown in the format:

$$E_{LC,IC} = Z[\mathrm{a}_1 * 0_1, \ldots .\mathrm{a}_\mathrm{n} * 0_\mathrm{n}],$$

where: Z–evaluation function.

$$E_{ЛЭ,ЭС} = 1 + \frac{\sum_{1}^{n} a_i \cdot \eta_i}{n}$$

Incidentally the following criteria may be used in the capacity of standard ones:

η_1–high-speed performance criterion,
η_2–energy criterion,
η_3–functional criterion,
η_4–load-carrying capacity,
a_i–weight factors defined by application conditions

$$\eta_i = \left| \frac{P_i - P_Б}{P_{MAX}} \right|$$

where: P_i, $P_Б$, P_{MAX}–values of indices of standard element, element under comparison, and the maximum index.

1.4 Impact of Destabilizing Factors on Serviceability of Digital Microcircuits

Serviceability of digital microcircuits to a considerable extent is determined by the ambient conditions where the microcircuits is running.

As a rule the destabilizing factors are understood as all kinds of impacts any microcircuit is subjected to in the process of operation being part of the digital systems. The major factors are ambient (operation) temperature, penetrating ionizing radiation (electron, proton, neutron, hard γ-radiation), mechanic impacts, electromagnetic emission, and also unauthorized occurrences of voltages and currents at pins of microcircuits. If methods of protection against penetrating ionizing radiation and electromagnetic fields are depicted in depth and they are based mainly on some technological practices, then the mechanisms and schematic methods of protection against unauthorized voltage and current signals at the pins of digital microcircuits are worth of special consideration. Among the latter, the mechanisms of electrical static discharge, latch-up, high-frequency and low-frequency interferences, and electrical overloads at the pins of microcircuits need to be addressed.

1.4.1 Immunity of Digital Microcircuits to Electrostatic Discharge

Mechanism of Electrostatic Discharge Impact on Microcircuits If two samples of material are in contact with each other, where one is an insulator, and rubbed together on the surface, there may be accumulated charges resulting in, even rather high, potentials to occur. Such charges obtained through transfer of electrons from one surface to another one, which is very similar to the electrons, have a value dependent upon dimensions, nature, and physical state of surfaces in contact and also upon energy (normally mechanical energy) applied to these surfaces. Such process is called electrostatic charge which may be positive (loss of electrons) and negative (excess of electrons) on the surface of the materials which are in contact. Polarity and amplitude of such charges are determined by the density of electrons for each couple of materials. While electrical contact between the materials is maintained, no problem of static electricity exists. However, if the areas are separated, the difference of potential between the surfaces of materials increases. If the materials are current-conducting, then the charges flow up to the point of contacting and then disappear. If one of the materials is an insulator, then migration of charges will never happen, and the electrostatic charge is held by the material. When one of the materials is an insulator, then migration of the charges on its surface is not possible, and the electrostatic charge is held by the material. If in contact there are two materials with different levels of electrostatic charge levels, and at formation of conducting path between them under impact of the emerging difference of electrostatic potentials, there occurs drainage of electrostatic electricity of the same polarity from one material to another. Such process is called electrostatic discharge (electrostatic discharge, ESD).

The principal electrostatic characteristics of electrostatic discharge are charge potential and the value of power (or current) of charge. In the vast majority of cases, electrostatic discharge appears as a random phenomenon because the mechanism of its display is substantially dependent on the form, physical state, and nature of the contact between interacting contacting surfaces and time. Besides the nature of electrostatic discharge depends on the ambient environment (temperature, humidity, presence of electromagnetic fields), where quite frequently it generates such secondary phenomenon as corona discharge and point discharges.

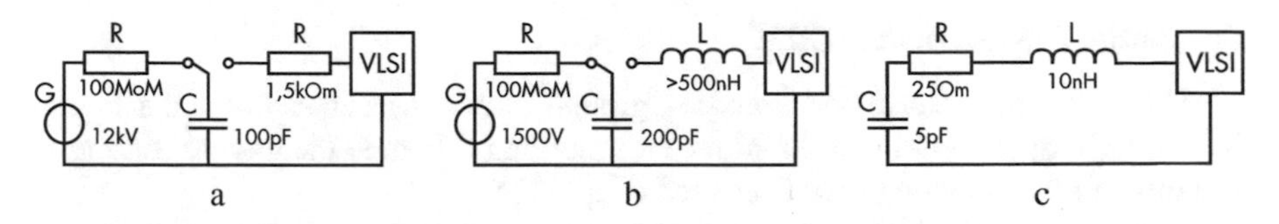

Fig. 1.30 Test circuit of models: (**a**) human body; (**b**) machine model; (**c**) charged device models

Presently there are considered three principal models of electrostatic discharge (ESD) [1]:

1. **Human Body Model (HBM).**

A man being a carrier of charge gets in contact with the device (microcircuit), and through this device the current flows toward the ground (Fig. 1.30a). In this circuit the capacitor ($C = 100$ pF) is being charged though high Ohmic resistors (r = 100mOhm) with voltage ±2 kV and then discharged through 1.5 kOhm resistor in the device under testing.

The capacitor simulates the capacitance of the human body which actually may vary within 500 pF. The resistance of the human body also may vary within rather broad ranges–from few tens Ohm to hundreds of Ohm depending upon the conditions. Discharge voltage may amount to 4 kV and even higher.

One of the most crucial parameters in the test is current rise time during the discharge. It should be around tenths of nanoseconds. Here however it is relevant to emphasize that the discharge current does not immediately spread along the conducting area. Therefore, at the beginning, there is a danger of overloading of protection (guard) circuit. On the other hand, such case perhaps is not so dangerous because as a rule electrostatic discharge happens not inside the circuit itself but somewhere inside the package or on the conductor not linked with the contact, which in its turn features rather high inductance affording the guard circuit some time for complete switching on.

2. **Machine Model.**

In this model, electrostatic discharge (machine model, MM) happens as a result of various mechanic impacts always observed in the equipment for IC production. The case and gears of such equipment are made of metal but inevitably incorporate some plastic parts greatly differing in dimensions and shapes. When these parts are in motion, electrostatic charge may be generated which is then followed by its discharge.

This test circuit (Fig. 1.30b) is similar to the previous one with the only difference that the resistance of the metal parts of the device is low which results in noticeably higher peak current in the device under testing. But on the plus side, as a result of high inductance occurring in this very case, no overloading of the guard circuit takes place, and the range and time of current rise is limited. The power observed in this model is higher than HBM power due to low resistance. Therefore in order to avoid damaging of the device under testing, the voltage needs be reduced to ±200 V.

3. **Charge Device Model (CDM).**

The device being insulated from the "ground" works as a capacitor and accumulates the charge in relation to its surface discharging it then to the ground as long as the discharge conducting pass is created (Fig.1.30c).

At occurrence of electrostatic discharge in the device itself, ESD peak current is higher than in any of the cases addressed, and it enters the device under testing almost without delay (rise time is less than 200 ns) which simplifies the designing of the guard circuits for this model.

It should be emphasized that there is no correlation between HBM and charged device, i.e., the components remaining functional in the first case do not have to work in CDM test. There are also known the models of charged cable (charged cable model, CCM) and model of transfer line pulse model; however, in practice, they are not frequently used [7].

Damages caused by electrostatic discharge may be divided into two groups: the ones defined by excessive power consumption and by availability of excessive voltage gradient. In the first group, electrostatic discharge results in passing of high current, and this current (its effect may be amplified due to non-uniformity of the circuit geometrical parameters or non-uniformity or defects of the microcircuits emerging at its fabrication) normally tends to take the most beneficial path. In the meantime there is observed considerable heat emission which may cause failure of the microcircuits (fatal failure) or degradation of operational characteristics of microcircuits (parametric failure). The phenomenon as above is called heat breakdown.

In the second group, considerable potential gradient may cause breakdown even in sufficiently thick oxide layer in MOS structure, and it contributes to formation of short circuits and leakages between the insulated regions of microcircuits further on in the process of operation.

All types of damages of digital microcircuits induced by electrostatic discharge may be divided into the following groups:

- Deterioration of connections and rupture of microcircuit element structure at the result of heat emission. In the channel of electrostatic discharge, there occur current non-uniformities resulting in development of hot point–channel resistance R_{CH} is increased, temperature inside it may reach 1000°C, and after that R_{CH} drastically decreases (current density increases, and there is observed temperature rise up to 1450 °C) and start of local silicon melting and further increase of channel temperature up to 2500–3000°C.
- Oxide destruction.
- Deterioration of Si and Al contacts (up to change of junction structure).
- Melting of metal connections, diffusion and polysilicon resistors of microcircuits, and burning out of connections between die pads IC package.
- Electric current migration in metal and polysilicon alloys.
- N Latent damages (local melts, cracks, hillocks not causing immediate failure, but affecting the reliability of microcircuits).

Methods of Digital Microcircuits Protection Against Impact of Electrostatic Discharge As it appears from reviewing the mechanism of electrostatic discharge, risk of damage gets higher along with increase of microcircuit integration due to inevitable effects of reducing the thickness of insulating and active regions of transistors and thickness of IC interconnections conductors which definitely needs introduction of protection tools against electrostatic charge into structure of digital microcircuits.

There is known the entire range of protection tools against electrostatic discharge having proper characteristics, but for the perfect IC protection, they should be selected and introduced in the best possible way.

There exist two possible ways of developing total protection of the device: random way method and sample-based method.

In random way method, ESD current takes its own random path between contact pads. Weak link of protection circuit in the first turn gets to break down voltage. These elements differ depending on various circumstances. In a few iterations, such links shall be selected and rejected. Random way method is not a universal flexible method, and it is specific not only for technology but for option of circuit implementation on "silicon."

Another method consists in directing ESD current along the specified path. The selected path has the lowest possible resistance as if compared with other parasitic circuits. In case of this method debugging of protection circuit does not take so much time because it is easier to search and eliminate the weakest elements. Method of the sample-based way is more prone to streamlining and classification.

It is important to observe that the devices applied in protection circuits may be divided into two groups: breakdown devices (BD) and non-breakdown devices (NBD) [7]. The devices of the first type run in the mode between the first and the second (heat) breakdown, and their operation is heavily dependent on the layout implementation and technological processes; subsequently their designing is quite a sophisticated process. It is much easier to use non-breakdown devices, i.e., devices which run in normal mode. It is easier to simulate such circuits, and its results are more predictable. Also it is relevant to note that when there is a trend to increase the number of elements and respectively the IC size, there is increase in intrinsic capacitance of the device. In order to keep up with accurate values of supply voltages, it is appropriate to use bypass (decoupling) capacitors in the ICs. Such capacitors amidst decrease of supply voltage value and current increase may be helpful at reducing ESD voltage for the circuit, *Vcc-Vss*, protecting thus the main device. Also reduction of layout features shall result in increase of the operational frequency. As a consequences rise time of ESD pulse for the model of the charged device is comparable with the time of protection circuit operation which affords faster response to this pulse.

Let's have a look at operation concept of protection device against ESD. The circuit, incorporating ESD source and protection device are shown in Fig. 1.31 where *R1* and *C1* – are equivalents of ESD source; *Z1* – parallel element protection unit which removes ESD (in compliance with full resistances) from the element

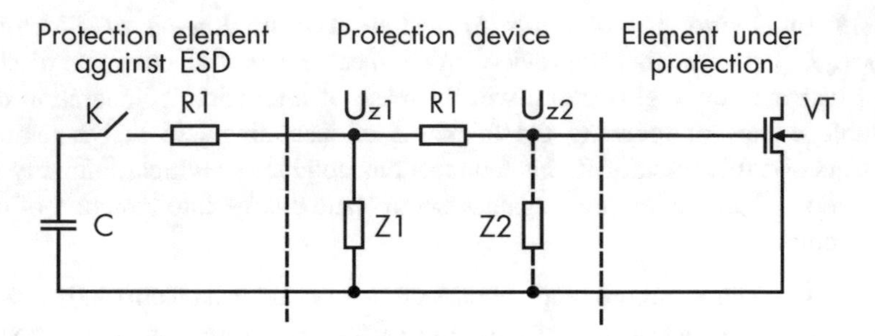

Fig. 1.31 General block diagram of digital microcircuit pin protection device

under protection. To increase the level of current rejected through protection element Z1, the device is completed with serial protection element R1 which increases the impedance of the component under protection. At protection of MOS transistors the device is equipped with an additional parallel protection element Z2. The purpose of this element is to decrease the clamping voltage *UZ1*, which is observed when ESD takes place at protection element Z1.

Basic requirements applied to ESD protection circuits:

(a) Clamping voltage $U_{Z1}(U_{Z2})$ should be below maximum tolerable voltage of microcircuit elements.
(b) Parallel impedance of protection device for the range of operational voltages of microcircuit pin should be indefinitely large.
(c) Serial resistance and capacity of protection device for the range of operational voltages of microcircuit pins should be indefinitely small.
(d) Time of protection device operation should be rather short (shorter than the time of element local overheating).
(e) Delay time of operational signal at microcircuit pin should be rather short (significantly less than switch delay of the element under protection).

Besides, ESD may occur between any microcircuits pins; therefore, protection device should efficiently remove (reject) ESC irrespective of the combination of microcircuits pins where ESC arrives.

ESD Testing Modes in Dependence on the Options of Its Occurrence on the Contact Pads Most frequently ESD occurs between contact pad of input signal and contact pad of one of the supply lines. ESD may be of positive or negative polarity with respect to potentials of supply lines. Thus, there are four modes of ESD occurrence at the contact pad of input/output signal resulting in various paths of stress current flowing [1]. Options of ESD measurements for these modes are shown in Fig. 1.32a.

For the first mode (PS mode), electrostatic discharge positive in relation to ground potential (V_{DD}) is provided to one of the contact pads of input signals, supply line (V_{CC}) is not connected, and stress current flows out of the device through ground

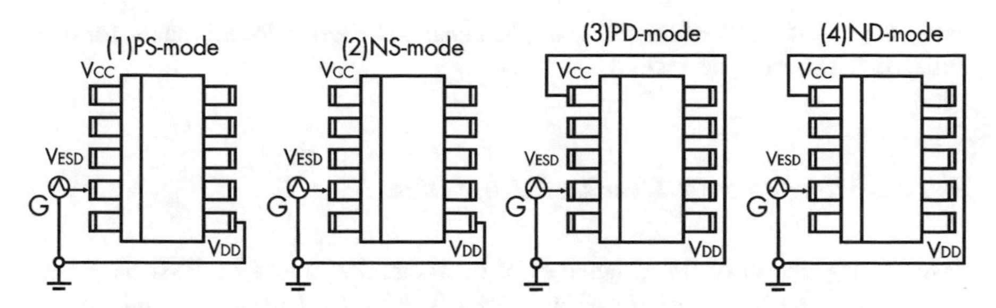

Fig. 1.32 Check of device immunity against ESD for four modes of discharge occurrence between contact pad of input/output signal and contact pad of one of supply lines

line. In the second mode (NS-mode), ESD is negative with reference to ground potential (V_{DD}) of contact pad, supply line (V_{CC}) is not connected, and stress current flows out of the device trough ground line (V_{DD}). For the third mode (PD-mode), ground line is ruptured (V_{DD}), and ESD is positive with reference to supply line potential through which stress current flows out of the device. For the fourth mode (ND-mode), ground line is ruptured (V_{DD}), and ESD stress is negative with reference to supply line (V_{CC}), through which stress current flows out of the device. In all the four cases, the rest of input-output contact pads are not connected. These stress voltages may damage both n-MOS and p-MOS devices of input or output buffer. For example, if there is no ESD protection circuit between contact pad and supply line (V_{CC}), the IC will be damaged.

In ND-mode stress current initially shall flow through protection circuit from contact pad to ground line (V_{CC}) and then through protection circuit between supply lines (if such is available) or through cells of internal circuit to supply line V_{CC}. Also any CMOS IC shall have parasitic capacitances and resistances of lines where voltage drops. It results in damage of IC internal cells in spite of the protection available. ESD ND-mode results in formation of stress voltage between supply lines (V_{CC}). If ESD current is not removed fast and efficiently from the IC, then there may occur damages in the protection circuit V_{CC} и V_{DD} and precisely ruptures of field oxide of parasitic structures between two n + diffusion regions [6]. Thus, it is essential to ensure ESD protection between contact pad and both supply lines to ensure safety of the device against damage in case of ESD occurrence in any of the four modes.

But even having ensured the complete protection against the four above listed ESD modes, one cannot be absolutely confident about the IC being perfectly protected. There is probability of ESD emerging between contact pads of input and output signals or between contact pads of supply lines. The current passes from one contact pad to another along some path through the internal device cells. In order to identify protection of the device, some additional measurement modes are introduced.

For these modes it is difficult to detect IC damages simply by tracking leakage current. For this purpose, that is, to detect failure, complete functional check is needed. For sufficient ESD protection, IC needs to be tested for all modes for three models: HBM, MM, and CDM.

1.4.2 Microcircuits Overload Tolerance

Over a long period of time, damages of microcircuits caused by ESD have been attributed to the impact of electrical overloads. This term means the influence of abnormally high currents and (or) voltages over various time spaces, wherein their values and duration are deemed sufficient to cause failure (more often than not manifested in melting of metal conducting layer or in breakdown of insulating layer). The cause of such phenomena was normally deemed deviation from rated voltages, supplies, or erroneous activities performed with the said device or with the present microcircuit.

However, presently, it is thought more correctly to differentiate the failures caused by ESD and electrical overloads, because the respective phenomena need various protection circuits and have various nature: in particular, for the first process, the parameter changes, or microcircuit fails due to the discharges taking place because of accumulation of triboelectric (electrostatic) charges, and the second process emerges when the relevant conditions of microcircuit operation are not complied with. Unfortunately, very occasionally, one manages to differentiate by the nature of failures those caused by ESD from the failures induced by electrical overload because the more damaged and destructed the microcircuit cells are, the more identical the damages incurred seem to be.

General types of electrical overloads as below are differentiated:

- Short time increase of voltage values (currents) provided to pins (inputs and/or outputs) of microcircuits.
- Unauthorized mode of short-time circuit of output (or group of inputs) of microcircuits to common line.
- Unauthorized short time getting to open output (logic “0” state) of voltages close to supply voltage U_{CC}.
- Short-time exceedance of maximum tolerable value of supply voltage on the part of supply source.

Overload modes may occur at fabrication operations, mounting, and debugging of digital systems and surely in the process of operation. As long as in real it is impossible to avoid all emergencies or unforeseeable situations, the majority of up-to-date digital microcircuits is equipped with built-in guard circuits protecting against electrical overload. For this purpose, MC of digital microcircuits (input or output components) are completed with special guard circuit which at occurrence of overload conditions prevents from deterioration of digital microcircuit components. Two options of guard circuit protection are commonly known [11]: Guard circuit is connected in parallel with the IC circuit under protection; GC is connected in parallel with IC circuit under protection.

Guard circuit may be non-controllable and perform restriction functions (in the first case, for current; in the second case, voltage) or controllable. In this case, guard circuit under control analyzes signal levels at output, compares them with reference, and in case of overload mode restricts signal levels at pins or disconnect IC circuit from the pin, thus preventing its destruction.

In case of supply interruption U_{CC} and its exceedance, digital chips may be augmented with special guard circuits which at occurrence of overload mode generate OVERLOAD signal at MP pin and may disconnect internal supply circuits of digital microcircuits from supply pin.

1.4.3 Dependence of Electrical Characteristics of Digital Microcircuits Upon Operational Modes

At development of digital systems, it is appropriate to take into consideration the changes of general parameters of digital microcircuits along with temperature, supply voltage, and other external influencing factors. Here the calculation should be performed for the worst case of the integrated effect of these factors which results in total change of parameters. The system should be designed in such a way that its serviceability is ensured for the case of its incorporating any sample of the digital microcircuits of this type.

Let's address temperature's impact T_A and supply voltage U_{CC} at high-speed performance, interference immunity, dissipation power, and load capacity of digital microcircuits.

Enhancement of ambient operational temperature induces change of the key electrical parameters of digital microcircuits due to the impact of known physical mechanisms.

Thus, for bipolar microcircuits, there takes place enhancement of gain ratios for all transistors, increase of lifetime of minority carriers, and increase of transistor saturation degree. The rated values of diffusion or implanted resistors of digital microcircuits normally also increase along with temperature rise which results in decrease of values and currents being toggled in the digital microcircuits. The value of logic swing is decreased. All these factors cumulatively are influencing on the microcircuit high-speed performance. Normally, the influence of the temperature on the dynamic parameters of microcircuits is studied separately–times of switching ON/OFF.

Here switch-ON time t_{pHL} of bipolar digital microcircuits along with temperature rise is slightly reduced, whereas the most sensitive switch-off time t_{pLH} increases. Thus, for TTLS microcircuits, the standard specific value of such change amounts to 0.04 ns/C° [5].

Increase of supply voltage otherwise things being equal contributes to high-speed performance increase. For bipolar digital microcircuits, faster toggling of transistors is determined by increase of their control base currents, while for CMOS microcircuits, increase of high-speed performance is linked with re-charging time decrease of loading and parasitic capacitances of MOS transistors.

Dissipation power P_{CC} is defined by microcircuit consumption current value which is in linear dependence upon supply voltage U_{CC}; therefore, the power reviewed along with supply voltage increases in compliance with standard square law. While F transistors toggling frequency rises, microcircuit power consumption increases. While the temperature varies, there happens a comparatively small change of consumption current; however, for practical purposes, calculation of dissipated power dependence upon the ambient temperature needs to be performed being guided by the temperature dependences available in the performance specifications on specific types of digital microcircuits in the section of reference data and dependences.

Load-carrying capacity of digital microcircuits also depends upon the temperature because its critical transistor gain ratio, resistors, and output voltages are temperature-dependent. However, one should have in mind that the currents by which load-carrying capacity and fan-out by output are calculated, in reference books normally, are provided with account of temperature changes over the entire operational range; therefore, in the calculations, changes of actual current values may be ignored.

Decrease of supply voltage within the tolerable voltage change range may result in reduction of output avalanche current. At high output level, decrease of supply voltage in bipolar microcircuits may induce such decrease of supply voltage of logic unit U_{OH}, that for digital microcircuits the required value of output current will not be ensured. Therefore, at calculation of destabilizing impact of supply voltage at load-carrying capacity, it is feasible to explore the impact of supply voltage at microcircuit interference immunity.

Reserves of ΔU_T^+, ΔU_T^-, in general case, are defined by inter-relations between output voltage levels of control microcircuits and input threshold voltages U_{TH}, U_{TL} of the controlled microcircuit. Voltage of logic zero U_{OL} of bipolar microcircuit is defined by saturation voltage of standard output transistor. At increase of T_A temperature, this voltage may slightly rise. As with temperature T_A rise, output voltage of logic "1" U_{OH} also increases, and static interference immunity by low level ΔU_T^+ will decrease and by high level ΔU_T^- will increase. Along with temperature T_A, there is subjected to change the dynamic interference immunity, that is, interference immunity against positive pulses tends to decrease at low-level signal, and interference immunity against negative pulses at high level tends to increase.

Considering that input threshold U_T of bipolar microcircuits toggling shall be defined by voltage drop at serially connected transistor junctions, then threshold U_T is in minor dependence of supply voltage. Changes of supply voltage U_{CC} do not have noticeable impact at output level of logic zero U_{OL} as well. However, output voltage of logic "1" U_{OH} is defined by voltage drops at junctions counted from supply voltage level; therefore, any changes of output voltage shall produce an effect on level of logic "1" U_{OH} at output. For instance, if the supply voltage value is reduced by 0.4V, then the value ΔU will be diminished by 0.4V.

Reserves of interference immunity by low-level ΔU_T^+ do not depend upon change of supply voltage U_{CC}. While reviewing the dynamic interference immunity, it is possible to phrase a general conclusion: the higher the microcircuit's high-speed performance is, the lower its dynamic interference immunity is. Rise of supply voltage U_{CC} may result in minor decrease of dynamic interference immunity which for level logic "1" U_{OH} at output will be compensated by increase of static reserve of interference immunity.

1.4.4 Immunity of Digital Microcircuits to the Impact Produced by Interferences

Consistent increase of high-speed performance of digital microcircuits and their application in digital computing and control systems have demonstrated that interference immunity of digital microcircuits is one of the characteristics instrumental for reliability of system operation. In relation to high-speed performance of digital microcircuits, the problem of interference immunity is getting even more complicated since high-speed performance of microcircuit elements is getting comparable with time spaces of signal propagation in the system communication lines. As a consequence, the noise signal in the line may be perceived by the microcircuit as true signal, and as a result the serviceability of the system may suffer. In the most comprehensive manner, the issues of interference immunity are elaborated in [8–10]; however, the issues of internal noise generation within microcircuit circuits (chains) and schematic techniques of their immunity increase are not completely covered.

There are known two parameters of interference immunity of digital microcircuits:

- Interference immunity of input matching components, output matching components, and supply circuits of digital microcircuits to the impact of external noises.
- Interference immunity of input interfacing elements, output interfacing elements internal logic cells, and their supply circuits to the influence produced by internal interferences and also their ability to their generation.

The relevance of interference immunity issue of digital microcircuits against internal interferences relates to not very high logic swing of the voltages in LC circuits and ability of high-speed performance logic cells and input and output matching components to generate high-level interferences at parasitic capacitances and inductive components of microcircuits.

In a particularly strong way, the effects of noises generation are displayed in high-speed performance output matching elements which control higher capacitive loads and commuting high output current.

The following are the main types of noises:

- Noises generated in signal lines due to their crossover parasitic interaction.
- Noises in signal lines related to mismatching between cell outputs and load, availability of L,C components in load.
- Noises in signal lines caused by signal "competition".

The types of noises (interferences) listed above have various mechanisms of generation, and in various ways they are affecting the microcircuits; however, their existence in microcircuit chains may have an influence on the stability of microcircuit operation and characteristics.

Interference Immunity of Digital Microcircuits Against Impacts Produced by External Noises.

Stability of Input Matching Components Against the Impact Produced by External Noises.

"Interference immunity" of input matching element is understood as its ability to sustain its steady-state condition under interference impact. Static reserve of interference immunity is defined by the following parameters:

$$\Delta U_T^+ = U_{TL} - U_{IL}; \ldots \Delta U_T^- = U_{IH} - U_{TH};$$

These parameters are illustrated in Fig. 1.33a where it shows transfer characteristic of input inverting interfacing element (graph 1) controlled by a device with transfer characteristic of similar type (graph 2). The values of threshold voltages of high and low levels U_{TH}, U_{TL} are defined on the transfer characteristic in points A and B where $dU_0/dU_I = -1$. The values of input voltages of high and low levels U_{IH}, U_{IL}, correspond to points *C* and *D*, where $U_{IH}=U_{OH}$ and $U_{IL} = U_{OL}$ *of control device.*

However the availability of steady-state reserve of interference immunity and input matching component appears to be essential but insufficient requirement to ensure noise immunity of MP microcircuits. Therewith depending upon noise signal duration and own switching delay of input matching component, there are cases when input matching component shall not change its state even if steady-state reserve of noise immunity is exceeded by noise signal. Therefore, matching components of digital microcircuits are defined by pulse noise immunity t_n, the range of which is equal to logic swing of voltages at input $\Delta U_T = U_{IH}$-U_{TL}.

The Influence of External Noises on the Inputs of Digital Microcircuits.

The influence of external interferences on the inputs of digital microcircuits involves their influence on the inputs of input interfacing elements and change of logic state of input matching components. Thus, herewith at the output of input interfacing element, there will be generated a false short signal which when propagating along the chains of digital microcircuits is likely to change the states of functional blocks of digital microcircuits, including the memory cells, and to generate false (phantom) signals at outputs of digital microcircuits which is equivalent to the failure of digital microcircuits.

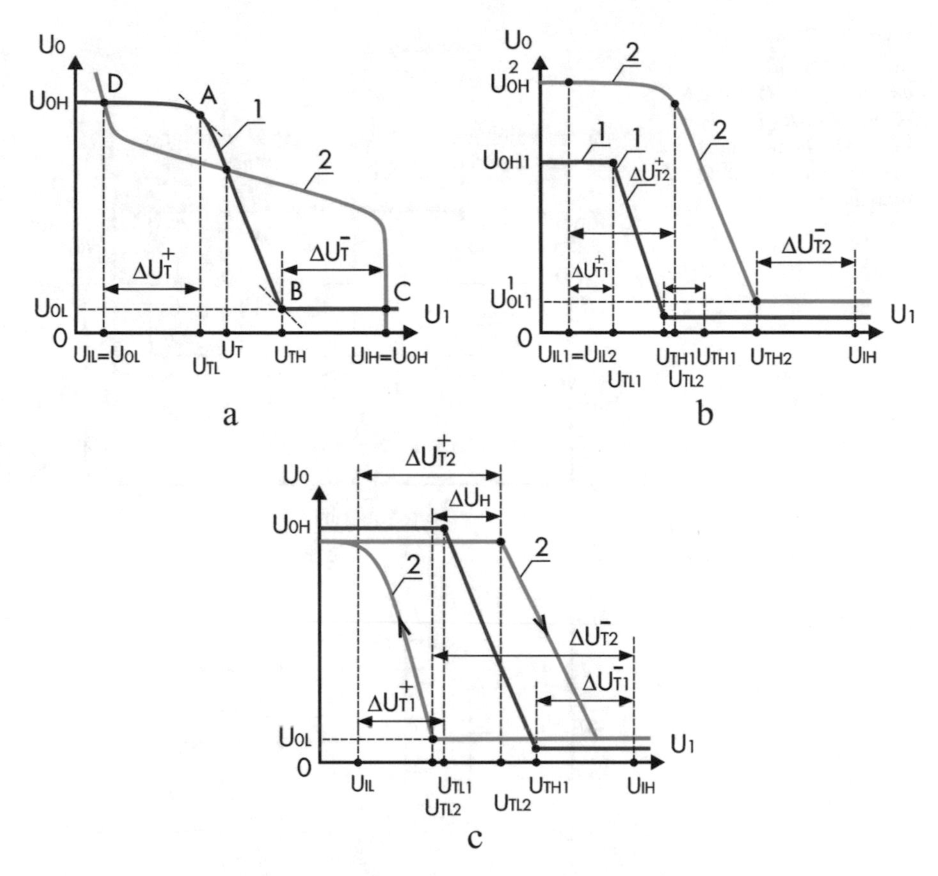

Fig. 1.33 Transfer characteristics of input interfacing elements of digital microcircuits: inverting (**a**), with enhanced toggling threshold (**b**) hysteresis type (**c**)

Standard Methods of Preventing External Noises at the Inputs of Digital Microcircuits.

Enhancement of noise immunity of digital microcircuits at inputs is attained due to increase of both static and dynamic noise immunity of input matching components. Among the methods of improving steady-state noise immunity of input matching components, the methods as follows may be singled out:

- Increase of input threshold toggling voltages U_{TH}, U_{TL} and logic swing of input voltages $\Delta U_T = U_{OH}$-U_{OL} (Fig. 1.33b, graph 2). The disadvantage of this method is the need for increase in the supply voltage U_{CC} and deterioration in high-speed performance of input interfacing element.
- Generation of transfer characteristic in input matching component symmetrical to logic swing of input voltages LU_T.

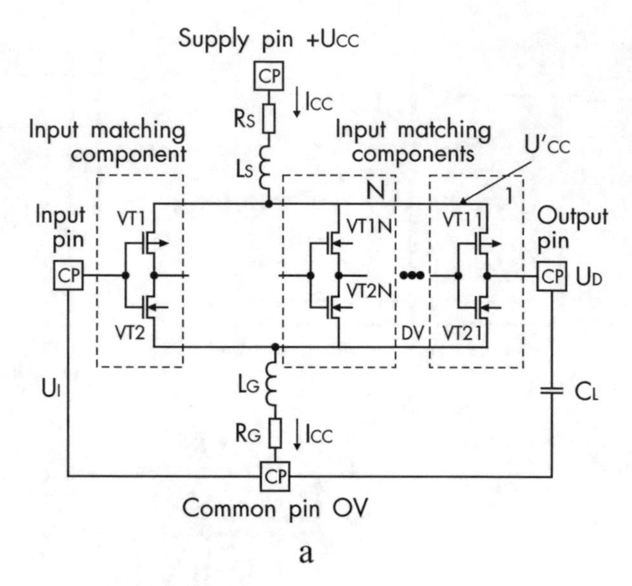

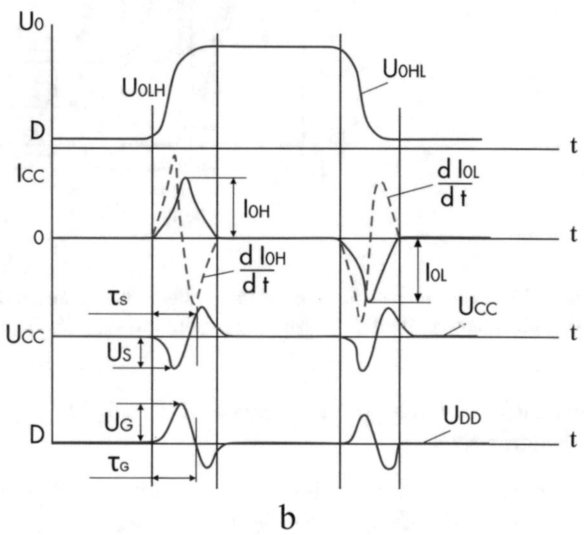

Fig. 1.34 Diagram of digital microcircuit explaining the mechanism of interferences generation in the supply circuits of microcircuits (**a**) and signal time diagrams (**b**)

– Use of internal feedbacks in the input matching component to generate a transfer characteristic of hysteresis type with loop width ΔU_H (graph *2* in Fig. 1.34c).

The following methods of dynamic interference enhancement may be distinguished:

– Limiting minimum toggling delay time and durations of input MC fronts which decrease the sensitivity of input MC to duration of dynamic noises.
– Use of special built-in limiting (clipper) circuits to limit dynamic noises (capacitance circuits).

– Use of paraphrase input interfacing elements in microcircuits for two-line communication lines receiving direct and inverse input signals and insensitive to dynamic noise, generated simultaneously at direct and inverse inputs.

Robustness of supply circuits of digital microcircuits against impact produced by external noises. Apart from the inputs of digital microcircuits, external noises may occur in the supply circuits of digital microcircuits and have an influence on their operation.

Stability of supply circuits of digital microcircuit against impact produced by external noises may be defined by tolerable deviation of supply voltage of digital microcircuits from rated $\pm\Delta U_{CC}$, where operational stability and dynamic and electrical parameters of digital microcircuits do not vary. In literature there is no reference to the parameter defining the immunity of digital microcircuits to the impact of dynamic noises along supply circuits because dynamic noise immunity to the impact produced by this type of noises is determined to significant degree both by functional composition of microcircuits and modes of its linking up (supply source, load, etc.).

The Influence of the External Interferences Over the Supply Circuits of Digital Microcircuits The influence of the external interferences on the supply circuits of digital microcircuits involves their impact upon the supply circuits of the MP elements of microcircuits and change of logic levels at the outputs of microcircuit logic elements. At the influence over the combinational circuits, such change of supply voltage not reducing voltage levels at the output which are lower than the threshold levels shall mainly alter the rate of signal advancing along the logical circuits without changing however their integrity. At reduction of logic voltages level at the element output below the threshold level it is possible to change the logical state of memory cells, invalid clocking of cell memories and loss of data transmission validity along microcircuit circuits.

General Methods of Tackling the External Interferences in Supply Circuits of Digital Microcircuits.

The main methods of protection against the external interferences in supply circuits of digital microcircuits are methods of designing interconnections of microcircuit supply circuits in the device, choosing the most preferable and efficient power values and output resistance of microcircuit supply sources, and also electrical decoupling of supply circuits. The other group of methods incorporates special interferences limiting or smoothing circuits (capacitive chains, etc.) which can be both external for microcircuits and can be embedded into microcircuits.

Stability of Output Matching Components Against External Interferences The influence of external interferences over the output matching components resides mainly in transition of the matching components into the state close to overload which may cause failure of matching components. The influence created by the interference of this type on operation of digital microcircuits is rather weak if there is no connection between the output and microcircuit internal components, and connections are implemented through supply circuits of matching elements.

Interference Immunity of Digital Microcircuits to the Influence of the Internal Interferences

Interferences Generated in the Supply Lines of Digital Microcircuits

The mechanism of interferences generation in supply lines is explained on the example of CMOS microcircuits, the simplified diagram of which is shown in Fig. 1.34a. In the diagram, U_{CC} and 0 V supply pins and common line of MICRO CIRCUITS the external supply source is connected to; U'_{CC} and 0 V' are internal supply lines and common line of microcircuits.

L_S, R_S–parasitic (stray) inductances and resistance of internal supply line (pin inductance and resistance including).

L_G, R_G,–parasitic inductance and resistance of common line (pin inductance and resistance including).

External load capacitance C_L is connected between output pin U_O and pin 0 V. Assume that the step-down output transistor VT21 of output matching component DO1 is in closed state and step-up output transistor VT11 in open state. Then load capacitance C_L is charged up to high-level voltage *U*OH. At unlocking (enabling) of step-down transistor VT21 capacitance C_L is discharged, and through the open transistor VT21 time-variant current is flowing (Fig. 1.34b).

$$i_{OL} = C_L dU_O^{HL}/dt,$$

Which at parasitic resistances R_G and inductance L_G of common line 0 V generates voltage swing:

$$\Delta U_G = L_G di_{OL}/dt + i_{OL} R_G$$

This voltage at the time of switching affects the common line 0 V' and changes its potential in relation to the potential of the external output 0 V, as shown in Fig. 1.35a, b.

At switching step-up output transistor VT11 ON and switching step-down transistor VT21 *OFF* load capacitance C_L is charged by drain current of P-MOS transistor VT11:

$$i_{OH} = C_L dU_O^{LH}/dt$$

The said current, while flowing through stray inductance L_S and resistance R_S of supply line U_{CC}, shall generate voltage swing at them:

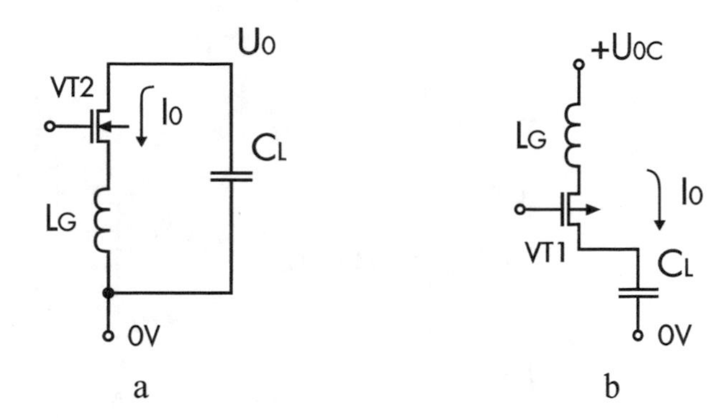

Fig. 1.35a Equivalent circuits of charging (**a**) and discharging (**b**) of load capacitance on the basis of MOS transistor

$$\Delta U_S = L_S di_{OH}/dt + i_{OH}R_S$$

Voltage ΔU_S at the time of switching has an effect on the internal supply line U'_{CC} and changes its potential in relation to the potential of the external pin U_{CC} (Fig. 1.34b).

For example, at load capacitance $C_L = 50$ pF and output voltage swing $L_O = 5$ V over 2 ns overcharge current value =50x5/2 = 125 mA.

Assume that line inductance (supply line or common line) makes.

$L_{G,S} = 10$ nH and resistance $R_{G,S} = 2$ Ohm. The value of peak voltage shall be:

$$\Delta U_{G,S} = 10 \times 125/2 = 125 \times 2 = 0,875 \text{ V},$$

i.e., it is comparable with microcircuit supply voltage $U_{CC} = 5$ V.

At simultaneous toggling of a few output interfacing elements, the value of interferences voltage ΔU_S, ΔU_G is increased in proportion to the number of interfacing elements which are simultaneously being toggled into the same state.

Actually the value of interference voltage ΔU_S, ΔU_G, is higher than the shown one. The reason is the presence of let-through consumption currents in the output matching components which are generated at toggling into the states when both MOS transistors VT11 and VT21 are in open state. These currents being added together with overcharge currents of load capacitance i_{OH}, i_{OL}, significantly increase the currents passing through stray inductances L_G,L_S AND resistances R_G, R_S, and increase the interference voltage ΔU_G, ΔU_S.

Estimation of Interference Voltage For estimation of interferences voltage, equivalent circuits of discharge (charge) of load capacitance on the basis of MOS transistors are used shown in Fig. 1.35a where the influence of let-through currents is not taken into account.

Assuming that at discharging of load capacitance the voltage at transistor gate VT2 (Fig. 1.35a, *a*) is varying over the time in such a way that the discharge current I_{OL} of load preliminary charged up to U_{OH} is of linear type (Fig. 1.35b, *c*), then

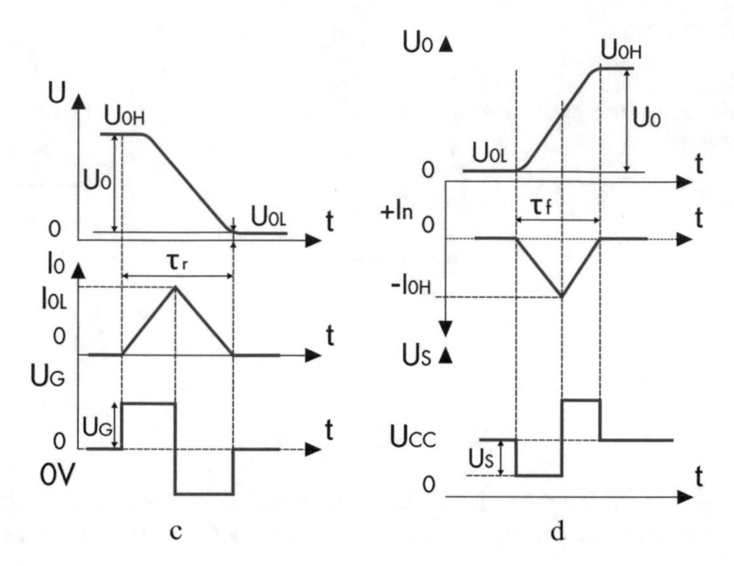

Fig. 1.35b Time diagrams of signals (*c, d*) to define the interference level in the common line 0 V in supply line U_{CC}

interference voltage ΔU_G has a constant value. The charge accumulated by load capacitance C_L:

$$Q = C_L U_O,$$

where $U_O = U_{OH}$-U_{OL}–logic voltage swing at the output.

This charge may be discharged over the time τ_r by output current:

$$I_{OL} = 2Q/\tau_r = 2C_L U_{OL}/\tau_r,$$

Then with account of the linear dependence of discharge current i_{OL} interference voltage in common line shall be:

$$\Delta U_O = L_O di_{OL}/dt = L_G 2I_{OL}\tau_r = \Delta L_G C_L U_O/\tau_r^2$$

Considering the fact that simultaneously there may be toggled into the same state *N* output matching components, the result is:

$$\Delta U_G = 4N\Delta L_G C_L U_O/\tau_r^2$$

If the voltage at transistor gate VT1 (Fig. 1.35a, *b*) is changing over the time τ_f in such a way that the charging current i_{OH} of load capacitance C_L is of linear type, interference voltage ΔUS in supply line will also have a constant value (Fig. 1.35b, *d*).

Similarly at charging of load capacitance the charge accumulated in the load capacitance C_L at voltage U_O shall be written:

$$Q = C_L U_O$$

This charge may be generated over the time τ_f by output current:

$$I_{OH} = 2Q/\tau_r 2C_L U_O/\tau_f,$$

Then with account of linear dependence upon the charge current I_{OH}, supply voltage in supply line shall be:

$$\Delta U_S = L_S di_{OH}/dt = L_S 2I_{OH}/\tau_r = 4L_S C_L U_O/\tau_r^2$$

Considering the fact that simultaneously there may be toggled into the same state N output interfacing elements, the result is:

$$\Delta U_S = 4N\Delta L_S C_L U_O/\tau_r^2$$

As it appears from the formula obtained, the greatest influence upon the value of interference voltages in supply line and common line is made by capacitance charging (discharging) durations τ_f, τ_r, i.e., durations of output signal front.

Influence of Interferences in Supply Lines on Microcircuit Inputs Let's explore the influence of interferences in supply lines on microcircuit inputs on the example of CMOS microcircuits: the equivalent circuit of input interfacing element is shown in Fig. 1.35a. Input voltage U_I provided to microcircuit inputs normally is referred to the common external pin. Here input interfacing element as a rule has input threshold voltages of high U_{TH} and low U_{TL} levels, when microcircuit retains its state. The difference between input voltages and input threshold voltages defines the reserve of interference immunity ΔU_T^+ of input matching component of low-level $\Delta U_T^+ = U_{TL} - U_{OL}$ and high-level ΔU_T^- U_{OH}-U_{TH} .

Emergence of interference dC/$_c$ in common line results in increase of the potential of microcircuits internal common line *0 V*′, and as a consequence to change of threshold voltages and reserves of interference immunity:

$$\Delta U_T^+ = U_{TL} + \Delta U_G - U_{IL} = \Delta U_{TO} + \Delta U_G,$$

$$\Delta U_T^- = U_{IH} - U_{TH} - \Delta U_G = \Delta U_{TO} - \Delta U_G.$$

As it appears from the obtained formula, an interference in common line U_G impairs interference immunity at high-level signal at input U_{IH}. At interference voltage in common line $U_G > U_{IH}$-U_{TH} and its duration T_G, comparable with toggling delay of input interfacing element, there may occur false operation of input interfacing element and microcircuits in general.

Occurrence of interference U_S in supply line shall result in potential decrease of internal supply line U'_{CC} of microcircuits and consequently to change of threshold voltage:

$$\Delta U''_T = U_{TL} - U_S, \qquad \Delta U_{TH} = U_{TL} - U_S$$

and to corresponding change of interference immunity reserve.

$$\Delta U^+_T = U_{TL} - \Delta U_S - U_{IL} = \Delta U^+_{TO} - \Delta U_S,$$
$$\Delta U^-_T = U_{IH} - \Delta U_S - U_{TH} = \Delta U^-_{TO} - \Delta U_S.$$

As it appears from the formula obtained, the interference in supply line U_S impairs the interference immunity of microcircuits at low-level signal at input U_{IL}.

At interference voltage in supply line $U_S > U_{TL}\text{-}U_{IL}$ and its duration τ_S, comparable with switching delay of input matching component its faulty actuation and failure of microcircuit operation are possible.

Influence of Interference in Supply Lines on Microcircuit Outputs The influence of interferences in supply lines on microcircuits outputs consists in direct change of the levels of output voltages of low U_{OL} and high U_{OH} levels.

Such change of voltage, while affecting the inputs of load microcircuits at specified duration and range of interferences, may cause their false actuation. Besides, such interferences, while affecting the internal cells of output matching component, may result in triggering (enablement) of insulating junctions, interaction of cells with each other, and their fatal failures ("latching," etc.). Another mechanism affecting the microcircuit inputs pertains to reduction of supply voltage at the internal microcircuits lines U'_{CC}, *0 V* (at negative interference ΔU_S, positive interference ΔU_G) and reduction of load capacitance of output matching components as a consequence, i.e., output currents of low I_{OL} and high I_{OH} levels. Reduction of output currents in its turn results in increase of charging (discharging) durations of load capacitance and extending the fronts of output signals when the interferences appear.

General Methods of Tackling the Interferences in Microcircuit Supply Lines To minimize the influence of interference in supply lines on serviceability of digital microcircuits, appropriate steps and measures are applied:

(a) Separation of supply lines of input 1 and output 3 matching components, internal 2 functional blocks of microcircuits, and their connection to individual electrically isolated pins (Fig. 1.36). Such approach at building microcircuit supply chains does not help eliminate interference generation in supply lines of matching component at their toggling; interferences, however, are being diminished due to minimizing stray inductance of supply lines and eliminating their influence on the inputs and internal functional blocks of microcircuits.

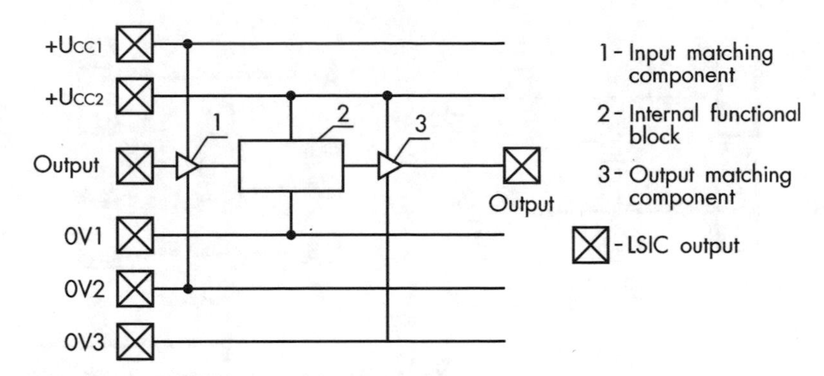

Fig. 1.36 Diagram of arranging supply circuits of digital microcircuits with reduced interferences level

(b) Use of the circuit of output voltage rise rate control which is built in the control block of output matching component. Such circuit permits to efficiently control the output element di_o/dt relation and to reduce the interferences in microcircuit supply lines.
(c) Use of output generators switching ON/OFF delay circuit embedded into control block of output interfacing element. In output element at toggling, there is a state when both output transistors VT11 and VT21 of generators are open (Fig. 1.35a); at such a state in supply circuit, there flows high let-through consumption current generating the interferences. For the purpose of lowering interference level at low–/high -state toggling, the delay circuit inhibits connection of step-up transistor VT11 till the step-down transistor VT21 gets into the close state. At toggling from high level to low level, the delay circuit inhibits switching the step-down transistor VT21 on till the step-up transistor VT11 gets into close state. Thus, in such a way, consumption of let-through currents is reduced.

Interferences Generated in Signal Lines Due to Cross-Coupling Mechanism of interference generation. Cross-coupling interferences are induced by mutual influence of adjacent signal lines located in immediate proximity to each other provided there is capacitance coupling between the outputs of two output matching components.

There are two known types of cross-coupling interferences:

[1]: Capacitance interference and inductive interference, where capacitance interference is rather substantial at huge swings of output voltages and high capacitances. Inductive interference is typical for huge swings of output current in signal lines with high inductances and low capacitances. For MP due to rather low values of commutated currents and microcircuits, the most typical are capacitance interferences in signal lines. The mechanism of their emergence is explained by the circuit shown in Fig. 1.37a and b, where D1 and D2- are output interfacing elements of microcircuits

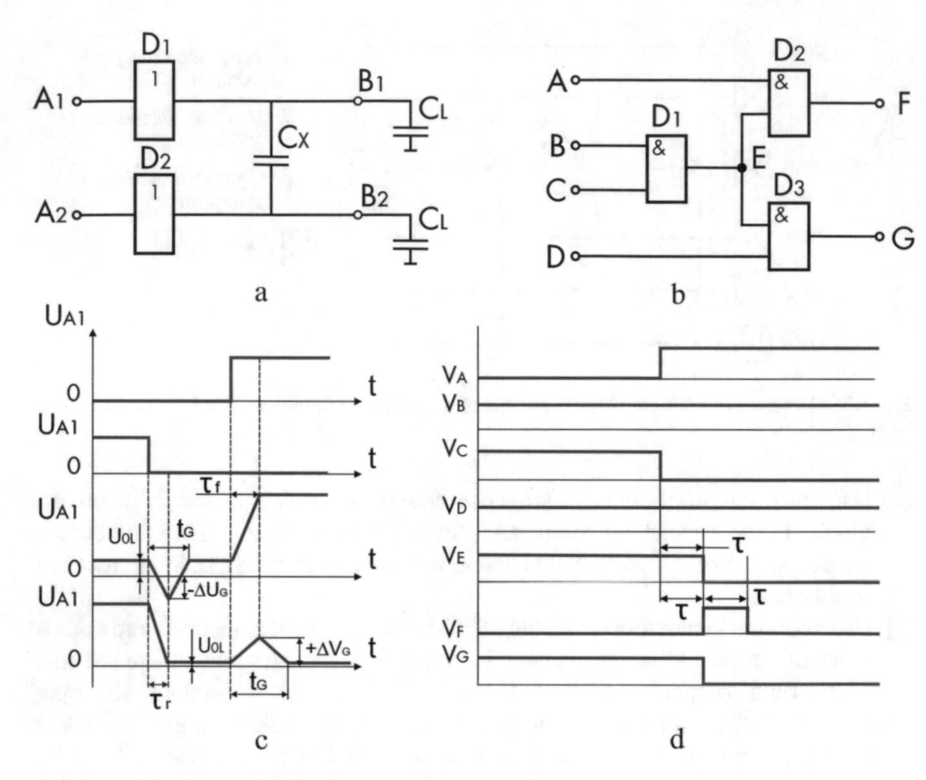

Fig. 1.37 Diagrams explaining the mechanism of cross-interference generation (**a**), caused by signal "competitions" (**c**) and signal time diagrams (**b** and **c**)

and Cx- is coupling capacitor of two system lines. At negative voltage swing in one activated line, for example, in B2 in the other D1 found in passive state of low level, there is generated a negative interference.– ΔU_G. At positive voltage swing in line B1, in the other B2 found in passive state of low level, there is generated positive interference $+\Delta U_G$.

Calculation of Interference Parameters Calculation of the parameters of cross-coupling interferences may be done by the algorithm set forth in [1]. Voltage of interference:

$$\Delta U_G = \left|\frac{\tau A}{\tau_{f,r}}\right| \left[1 - \exp\frac{\tau_{f,r}}{\tau}\right],$$

where:

$A = C_X \Delta U/C$–a certain constant value.
Cx–coupling capacitance.
C–$C_X + C_O + C_L$.
ΔU–differential of voltage swing between lines.

Co–output capacitance of matching component.
C_L–load capacitance.
$\tau = CR_O R_L/(R_O + R_L)$,
Ro–output resistance of matching component.
R_L–load capacitance.

Interference duration [7].

$$t_G = \tau_{f,r} + \tau \ \ln \frac{|\tau A/\tau_{f,r} - U_{OL}|}{U_{OL}} \left[1 - \exp \frac{\tau_{f,r}}{\tau}\right]$$

Influence of Cross-Interferences Upon Microcircuit Pins The influence of cross interferences resides in direct changing of microcircuit output signal pins which affect the inputs of load microcircuits and at the specified range and duration can cause their false actuation. Besides these interferences, the negative ones in the first turn, through affecting the components of output interfacing element, may result in enabling the insulating junctions of components and fatal failures of microcircuits.

Standard Methods of Tackling Cross-Interferences Standard method of crosstalk attenuation involves reduction of parasitic cross-capacitance and inductance couplings in the hardware where digital microcircuits are applied. However in order to improve the reliability of digital microcircuits, output matching components may be supplemented with special circuits of interferences limiting which reduce the interferences at the outputs to levels not affecting in any way the insulation of the microcircuits components.

The interferences in the signal lines due to mismatching with the load, availability of some components in load L,C which cause presence of multiple reflections between output and load, deteriorating the form of output signal. Generation of the interferences of such type is, in fact, entirely determined by application conditions; therefore, the mechanism of their generation and interference parameter calculations is shown in [12] and is not elaborated in this book.

The influence of misalignment interferences on the inputs and outputs of microcircuits is similar to the impact produced by cross-interferences. The methods of tackling the misalignment interferences incorporates the techniques of interconnection designing, mounting, and arranging the digital microcircuits inside the hardware that ensures minor signal reflections from mismatched loads and non-uniformities and minor signal attenuations in the lines. The methods of interconnections design and mounting are based on reasonable wiring of PCBs with account of noise immunity of the ICs applied. Another group of techniques is related to introduction of special circuits into the output matching component which track the output signal at toggling, making comparison with the reference signals and modifying the gain ratio of output matching component, if differential is observed. In such a way, the dynamic alignment between matching element and load is ensured.

The third group of techniques relates to inserting the embedded circuits of interference limitations at interfacing elements' outputs not affecting in any way the reliability of operations of digital microcircuits and load circuits.

Interferences in the Signal Lines Caused by Signal "Competitions" Mechanism of interferences generation. In actual microcircuits delay time of logic cells' signals propagation has finite value depending upon application conditions (load capacitance, supply voltage). Here the values of toggling delay times are of statistic nature and may vary for various microcircuits. The finite values of toggling delay times and their technological variation result in the situation when the signals along various circuits may propagate with various rate and trigger "competitions" of signals which results in interferences in signal lines. The mechanism of generation of the specified type of interferences in the digital microcircuits is explained in Fig. 1.37c, d where one can see the fragment of functional circuit of digital microcircuits in randomly selected logic and time diagrams of operation. Assume that inputs [A, B, C, D] were in the initial state [0111]. On change of the initial state of inputs at [A, B, C, D] = [0111] in compliance with the functional circuit, the state of outputs [F, G] should undergo change from [01] to [00]. However, considering that logic elements have finite value, direct modification of signal at input A shall induce transition of element output into high-level state [1] with delay 1τ.

Change of signal at C input shall result in emergence of signal delayed at 1 T at the second input of logic cell D2 and switching back of its output F into low-level state [0] after time 2τ, i.e., due to different signal transmission times along the circuits: input A-output F and input C-output F and finite-time values of logic element delay in output in place of retention of low-level signal; there will be generated short (with duration $\approx 1\tau$) signal of low level or glitch. This signal, while propagating along the circuits, may induce emerging of false signals at outputs of digital microcircuits and loss of data in memory cells (at emerging of such interferences in clocking circuits).

Generation of interference signals and their duration depend upon the length of logic circuits and delay values. The interferences of this type may also occur due to the impact produced by other factors: external electromagnetic influences or ionizing radiation and may be positive - as generation of high level signal at the background of low level signal $\ll 0 \gg$, and negative as low level signal at the background of high level signal $\ll 1 \gg$.

Methods of Tackling with Glitches in the Signal Lines In order to eliminate this type of interferences at designing of digital microcircuits, some special practices are applied:

- Simulation of digital microcircuits operation with account of signal propagation delays in the circuits, taking into account the finite value of logic cell delay time, modes of their loading, and other external factors.
- Strobing of signal transmission circuits.

– Application of structured methods permitting detection and correction of error at the output of the device by threefold and many-fold task solution with further comparison by majority rule, and also by excessive coding of input signals and further data correction.

These methods relate mainly to logic design of digital microcircuits and may be covered separately.

1.5 Parasitic Elements and Effects in Digital Microcircuits

Availability of common semiconductor substrate (total volume of semiconductor) for components of digital microcircuits determines the presence of various parasitic couplings between microcircuit components and also parasitic effects. These effects under some certain conditions may affect the serviceability and parameters of microcircuit components and impair the serviceability of microcircuits in general. Therefore in case of digital microcircuits, there should be undertaken some special measures (schematic or design and technological ones) to mitigate parasitic effects and elements.

1.5.1 Parasitic Transistor Elements Inside Digital Microcircuit Dice

Parasitic (stray) transistor structures stem from the presence inside digital microcircuit dice, semiconductor areas of various types, and also multilayered structures of "metal insulator conductor" which in their totality may form parasitic transistor structures between the components.

There exist parasitic transistor structures of two types:

– Bipolar structures which are formed in the microcircuits between the adjacent semiconductor areas of various types within microcircuit die and which may be both of vertical and horizontal type.
– Field structures which are formed by metalized or polysilicon interconnections on sacrificing insulator between two semiconductor areas.

For normal operation case parasitic transistor structures may be closed, which ensures full insulation of components in microcircuits.

Considering the fact that there are differences in insulation and generation of bipolar components and CMOS microcircuits, let's explore parasitic transistor structures for MOS and Bi-CMOS microcircuits separately.

Parasitic Transistor Structures in MOS Digital Microcircuits. Field Structures Parasitic n-MOS transistor (its structure), formed between two active

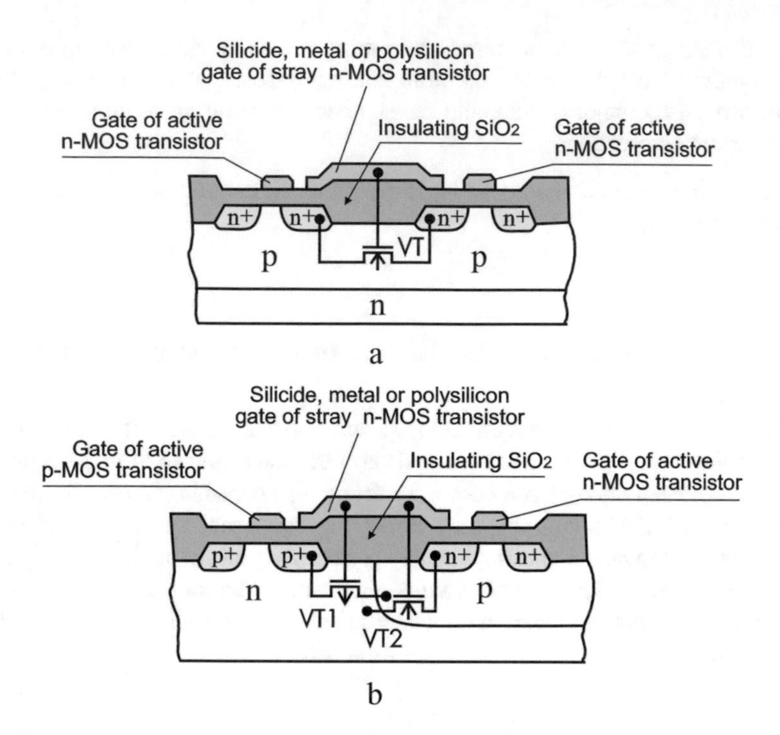

Fig. 1.38 Structure of parasitic n-MOS transistor of CMOS microcircuits (**a**) and parasitic (stray) MOS transistors with various types of conductance of CMOS microcircuits (**b**)

n_mos components [1] in diagram form, is presented in Fig. 1.38a. The functions of source and drain of this transistor are performed by regions of n^+ type of active transistors, and the gate is formed by metal (or polysilicon) interconnection which is located on insulating dielectric between two active n-MOS transistors. Considering the fact that drain and source of parasitic n-MOS transistor are interchangeable, such transistor structure is a symmetrical one: The influence of parasitic transistor manifests itself in formation of conducting channel between two active n-MOS transistors at emergence of positive potential at the gate of stray transistor that exceeds its threshold voltage $U^n{}_T$. As a consequence, there are possible changes of potential in-circuit nodes and loss of logic state in microcircuit logic cells. On average, in the microcircuits along with functional failure, this effect is displayed as emerging high static consumption current $I^C{}_{CC}$. So long as this structure cannot be actually removed, the main method of mitigating its impact turns out to be an increase of the threshold voltage $U^n{}_T$ up to the level which is considerably exceeding the maximum tolerable levels of voltage in the microcircuits. It is attained through increasing the thickness of insulating oxide and enhancing the concentration of p^+-type dopants under insulating oxide.

Similar parasitic structure (only of p type) exists between two active p-MOS transistors and differs in region conductivity type. However, due to positive cumulative charge density in the insulating oxide and consequently high negative values of threshold voltages $U^{p}{}_{T}$, it is much more difficult to have these transistors switched on as compared to the parasitic n-MOS transistors.

Therefore the insulation of active p-MOS transistors, as a rule, is better than that of active n-MOS transistors.

In CMOS microcircuit parasitic MOS transistors may be formed between active MOS transistors of various types of conductance. The parasitic structures of such type are shown in Fig. 1.38b. In this case the source of parasitic p-MOS transistor is formed by p" + region of active p-MOS transistor and its drain by p-well of active n-MIS transistor. The source of parasitic n-MOS transistor is formed by n + region of active n-MOS transistor and its drain by n-substrate of circuit. The gates of parasitic transistors are formed by metalized or polysilicon interconnection on the isolation oxide SiO_2 between active n and p-MOS transistors. Whereas the dopant concentration in the regions of such parasitic transistors differs greatly, such structures are asymmetrical ones. The influence produced by such structures manifests itself in formation of the conductance channels between the regions of active p and n NOS transistors. As the p^+ and n^+ areas are the drains appropriately of the p^- and n^- N-MOS transistors during activation of these transistors there may be incorrect functioning of the element. Assuming that p + and n + regions are drains of p-n MOS transistors of logic cells, such channels at switching the active transistors ON shall vary the voltage levels at active transistors which will be equivalent to the loss of logic state of logic cells and failure of the microcircuits in general. Apart from functional failure, this type of parasitic structure will manifest itself in huge static consumption current $I^{C}{}_{CC}$ of microcircuits. Methods of mitigating the impact of parasitic structure of this type are similar to the methods of mitigating symmetrical transistor structures.

Bipolar Structures Presence of MOS transistors of two conductance types in CMOS microcircuits induces emerging of two types of parasitic bipolar transistors in the structure of die: p-n-p и n-p-n. Schematic cross- section of die structure of CMOS microcircuit shown in Fig. 1.39 demonstrates the options of parasitic bipolar structures. The first one out of n-p-n type VT1 has the form of symmetrical horizontal structure generated by drain, source, and p-well n-MOS active transistor which is connected in parallel to the active n-MOS transistor. The second one being also of n-p-n VT2 type represents an asymmetrical vertical structure generated by source of n-MOS active transistor, its p-well region and n-substrate of die.

Similarly, there are generated the parasitic structures of p-n-p type. The first one being asymmetrical horizontal structure VT1 is generated by source, drain, and substrate of p-MOS active transistor and connected in parallel to the active p-MOS transistor. The second one being an asymmetric horizontal structure VT4 is generated by the source p- MOS active transistor, n-substrate, and p-region of well n-MOS active transistor. In isolated option the influence of stray bipolar transistors is not relevant as their junction base emitters are short-circuited by low Ohmic resistors *Rg*.

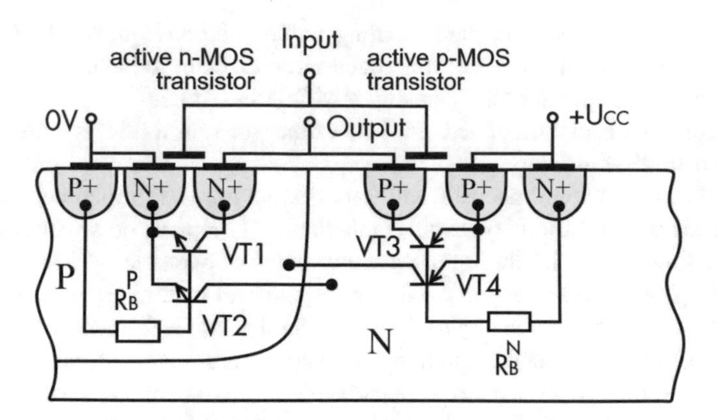

Fig. 1.39 Structure of bipolar parasitic transistor of CMOS microcircuits

Cumulatively, however, p-n-p transistor VT4 and n-p-n transistor VT2 form parasitic thyristor combined structure. At occurrence of the conditions within the structure of die (capacitance currents, influence of ionizing radiation, etc.), there's possible direct shift of base-emitter junction of parasitic p-n-p transistor VT4 which results in its switching ON and emerging of the collector in its chain. Such current, when getting into the base of parasitic n-p-n transistor VT2, shall result in its enabling and emitter current avalanche rise of parasitic p-n-p transistor VT3, i.e., connection of parasitic thyristor.

At exceeding limiting tolerable current density, heat destruction of structure components and failure of microcircuits in general may take place. Whereas the said effect is particularly important for CMOS microcircuits, the mechanism of its formation and methods of counteracting are elaborated in Sect. 1.4.

Parasitic Transistor Structures in BiCMOS Digital Microcircuit

Field Structures.

An example p-type parasitic field structure generation is shown in Fig. 1.40 where there is shown a fragment of cross section of microcircuit die with resistors (1 isolation). P-region of R1 resistor connected to supply line U_{CC} generates the source of stray p-MOS transistor, p-region of R2 resistor generates the respective drain, and gate 3 is formed by metalized or polysilicon line located between the resistors on sacrificing dielectric 4. At change of potential on line below the threshold potential, switching of stray p-MOS transistor ON is possible; hence, between resistors R1 and R2 conducting channel is formed.

This channel may contribute to re-distribution of currents between the resistors, changes in the parameters of the circuits which are using these resistors, and failure of the microcircuits.

By the similar concept in microcircuit dice, there may occur parasitic field n-type structures as well.

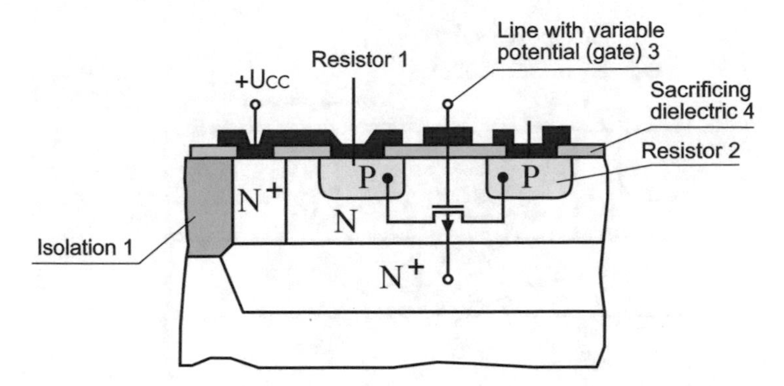

Fig. 1.40 Structure of parasitic p-MOIS transistor of Bi-CMOS microcircuits

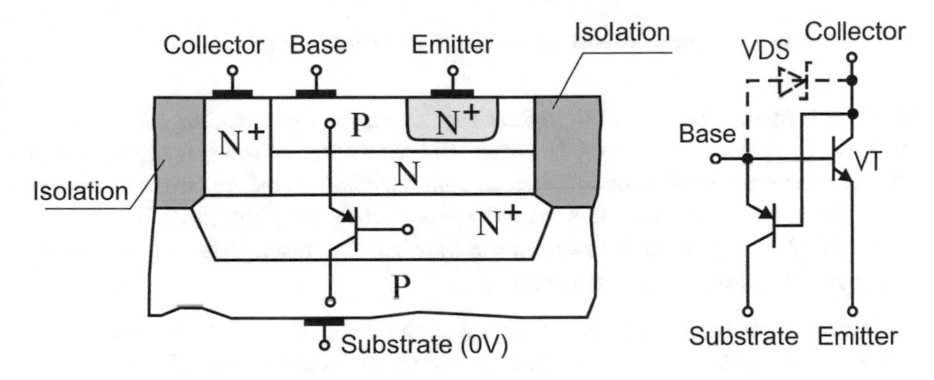

Fig. 1.41 Structure of parasitic p-n-p transistor of bipolar microcircuits

Methods of field structures mitigation (attenuation) in the dice of bipolar microcircuits are similar to the methods of inhibiting parasitic field structures in CMOS microcircuits.

Bipolar Structure In the dice of bipolar and Bi-CMOS digital microcircuits, creation of parasitic bipolar structures is associated mainly with semiconductor substrate. Formation of the first type, that is, vertical asymmetrical p-n-p structure, is shown in Fig. 1.41 where there is displayed a cross section of active bipolar n-p-n transistor. In this structure parasitic p-n-p transistor is formed by p-base of n-p-n transistor, n-, n + region of the collector, and p-substrate. At switching active n-p-n transistor ON and at direct biasing of its collector junction, there takes place direct biasing of emitter-base junction and switching parasitic p-n-p transistor ON. As potential is provided to p-substrate, some portion of active n-p-n transistor through the switched parasitic n-p-n transistor is shunted into the substrate and may vary the operational mode of the active n-p-n transistor. Besides, that current, when flowing along the voluminous resistance of substrate, may result in direct bias of isolation junctions and damaging the isolation of microcircuit components. The standard method of attenuating the impact of this parasitic structure is generation of n + layer in the base

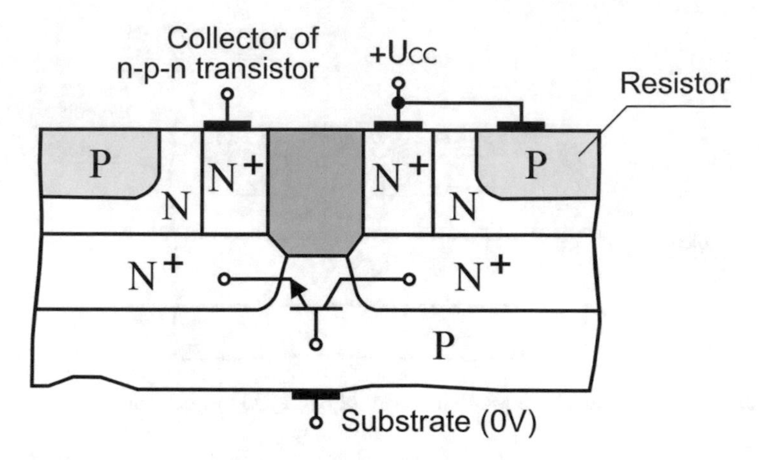

Fig. 1.42 Structure of parasitic n-p-n transistor of bipolar microcircuits

of p-n-p parasitic transistor with high dopant concentration, reducing current transmission ratio or p-n-p transistor. The most efficient method however is application of shunting Schottky diodes connected in parallel to collector-base junction of active-p-n transistor. Simultaneously this Schottky diode when complying with the requirements $U_{ПР}^{VDS} < U_{БЭ}^{VT}$ keeps parasitic p-n-p transistor in closed state and completely eliminates the current in the substrate.

Formation of horizontal n-p-n structure in the die of Bi-CMOS microcircuits is shown in Fig. 1.42. This structure is a symmetrical one, and is formed between two n$^+$-regions, where there are formed microcircuit cells, and one n$^+$- region represents an emitter, the second region is made as a collector, and p-substrate is made as a base. The influence of this parasitic transistor may be studied on the example of Fig. 1.42, where in the adjacent regions of ra- type, there are formed an active n-p-n and p-resistor. The region of emitter in parasitic n-p-n transistor is formed by n + region of collector in active n-p-n transistor, where at active n-p-n transistor being in switched ON state, there may be observed a minor voltage $U_C \approx 0.1$ V. There is provided voltage $+U_{CC}$ for reverse bias of the well to the well region with resistors forming the collector of parasitic n-p-n transistor. As a consequence between collector and emitter of parasitic n-p-n transistor, there is applied voltage $U_{VEЭ} \approx U_{CC}$, and on its base, there is the substrate potential $U_{SUB} = 0$, and in static state parasitic n-p-n transistor is closed. However, at occurrence of dynamic or static currents, and when voltage drop is observed at the voluminous resistance, there may happen direct bias of emitter junction of parasitic n-p-n transistor, and its switching ON. Here between the collector of active n-p-n transistor and n-well with resistor, there is formed the conducting channel, and an auxiliary current will be flowing into the collector of active n-p-n transistor which is quite able to modify the operational mode of active n-p-n transistor and to disturb the operation of the microcircuits. Attenuation of such transistor effect may be attained through increase of dopant concentration in the substrate between two n + regions with microcircuits components or introducing a supplementary guard ring between them, where there is provided close to zero voltage.

In bipolar microcircuits there may exist some other options of parasitic transistor structures depending upon some specific physical structure of microcircuit die.

1.5.2 Miller's Effect

Parasitic Miller's effect [1, 7, 10] manifests itself both in bipolar and MOS microcircuits. The essence of this effect is increase of transistor output capacitance at its switching OFF due to availability of feedback between its output and input. Let's review the said effect by the example of bipolar n-p-n transistor shown in Fig.1.43a.

Output transistor capacitance:

$$C_O = dQ_O/dU_O - i_O dt/dU_O,$$

where

$i_o = i_K + i_c$, U_o–output current and voltage,
t–time,
i_K–current flowing into transistor collector,
i_c–current flowing through collector capacitance C_K.

Assuming that the whole i_c gets into the transistor base and ignoring input transistor capacitance, we'll obtain:

$$i_O = i_C\beta + i_C = i_C(\beta + 1),$$

where: β–gain ratio n-p-n transistor base current

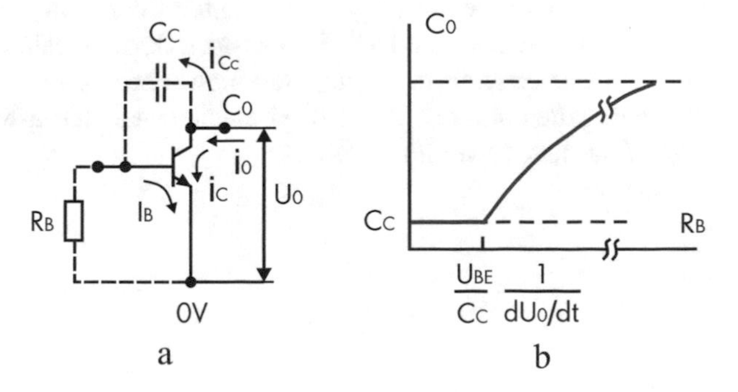

Fig. 1.43 Circuit for calculation of Miller capacity (**a**) and dependence of output capacity upon base resistance R_B (**b**)

$$i_C = C_K \frac{dU_O}{dt} \approx C_K \frac{dU_O}{dt},$$

then

$$C_O = C_K \frac{dU_O}{dt}(\beta + 1) / \left(\frac{dU_O}{dt}\right) = C_K(\beta + 1),$$

i.e., transistor output capacitance is approximately β times higher than collector capacitance. Assuming that β=30÷150, it's apparent that such effect is significantly increasing output capacitance C_O of transistor worsening meanwhile its dynamic properties. The standard method of attenuating this effect is creation of the circuits in n-p-n- transistor shunting capacity current of feedback i_c in output 0 V (Fig. 1.43a, resistor R_B).

Here:

$$i_O = i_C + i_D = \left(i_C - \frac{U_{BE}^{VT}}{R_B}\right)\beta + i_D = i_D(\beta + 1) - \frac{\beta U_{BE}^{VT}}{RB},$$

Then:

$$C_O = (\beta + 1)C_K - \beta \frac{U_{BE}}{R_B} \frac{1}{dU_O/dt},$$

Assuming that $R_B \to \infty$ (mode of broken base when the whole current i_c gets into transistor base), we present $C_O \approx (\beta + 1)\ C_C$.

At:

$$R_И \approx \frac{UBE}{C_C} \frac{1}{dU_O/dt},$$

We obtain $C_O = C_C$.

Within the range 0 < R_B < R_{BO}, the transistor is switched OFF; therefore, the whole current i_d is shunted to line 0 V, and output capacitance и $C_O = C_C$ is actually not subjected to changes (Fig. 1.43b). In the similar way, one can obtain the formula for output capacitance and MOS transistors. Specific values of base circuit resistance need to be selected from the required high-speed performance because its decrease, improving the output capacitance simultaneously, has affect adversely on the duration of transistor switching ON.

1.5.3 Latch-Up Effect

One of the possible causes for electrical overloads in CMOS microcircuits is the thyristor effect referenced in the literature as "latch-up" effect. If at any moment of time to the input(output) of the running microcircuit there is provided a voltage, exceeding the value of the voltage at supply U_{CC} or less that at common line 0 V, then CMOS microcircuit may enter the latch-up mode which is peculiar for development of low-resistance channel between pins U_{CC}, 0 V and drastic increase (peak)of consumption current. Here the microcircuit has no response to the control activities (it does not obey the commands) even if the destabilizing factors which have caused the shift of the microcircuits into this mode are eliminated. The essence of this effect [1, 5] is the operation of thyristor n-p-n structure always available in the design of any CMOS microcircuit, but revealing itself as a rule under specific conditions. In Fig. 1.44a, b, there is provided standard circuit of basic CMOS cell and sketch of its semiconductor structure, explaining the concept of thyristor effect formation. In Fig. 1.44c, there is depicted the structure diagram equivalent to the one in Fig. 1.44b, which is simulated by two bipolar transistors (VT_p and VT_n), shunting resistors R_S and R_W, and capacitor C_S. Here VT_p x specified the sideways transistor of p-n-p type formed by p + –region (emitter), n^- -substrate (base), and p^-well (collector), and VT_n specified the vertical n-p-n thyristor formed by n + –region (emitter), p^-- well (base), and n^--substrate (collector). Resistor R_S specifies

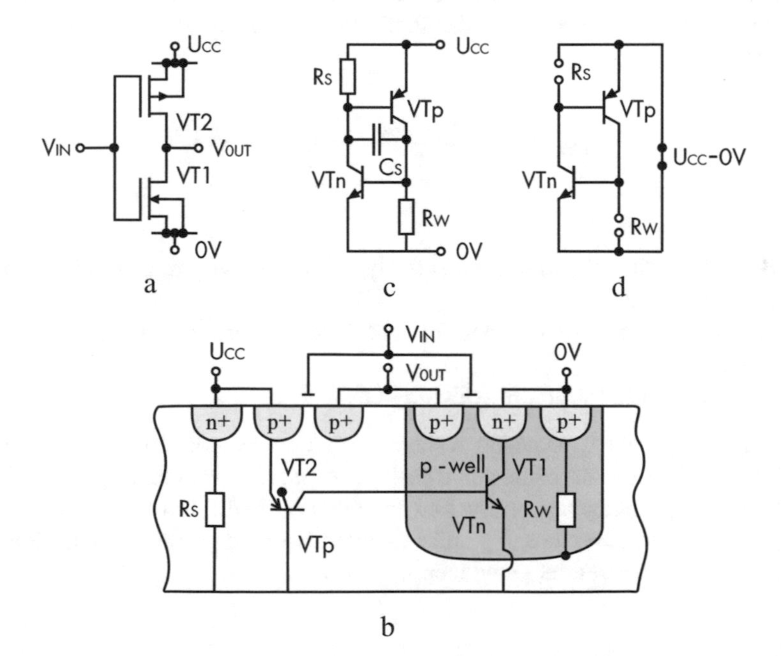

Fig. 1.44 Circuit explaining the causes for emerging thyristor effect: (**a**) CMOS inverter, (**b**) cross-section of active structure; (**c**) equivalent electrical circuit of inverter; (**d**) equivalent electrical circuit of thyristors

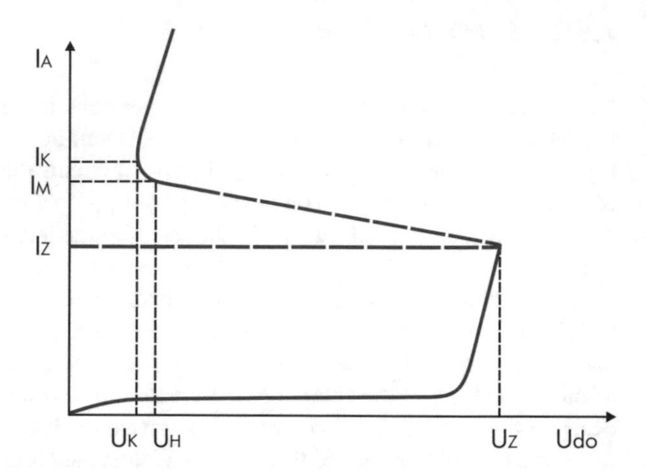

Fig. 1.45 Voltage current characteristic of thyristor structure

resistance n¯substrate on the range between contact n + type on planar side of chip and contact from the bottom side and R_W, resistance of p¯-well between p + −contact and low boundary with substrate region. C_S specifies the capacity of depleted layer of p-well-substrate junction.

If pins U_{CC} and U_{SS} (short circuit mode by supply) are connected and to break the shunting equivalent resistors R_S and R_W, as it is shown in Fig. 1.44d, then conventional thyristor structure is obtained as one of the circuit diversities with feedback. The condition of connecting similar thyristor structure may be expressed by the formula:

$$A_{pnp} \cdot A_{npn} = 1 + \frac{\left(A_{npn} U_{EBP}/R_W + A_{pnp} U_{EBN}/R_S\right)}{I_Z},$$

where:

A_{npn}, A_{pnp}–gain ratio by current, respectively, n-p-n and p-n-p transistor in the circuit with common base.

U_{EBP}, U_{EBN}–voltage emitter base in forward direction, respectively, n-p-n and p-n-p transistor.

I_Z–current of thyristor structure switching ON.

In Fig. 1.45, there is provided voltage current characteristic (VCC) of thyristor structure with the region of negative resistance peculiar for latch-up effect. In the figure U_Z and I_z- voltage and current switch ON;; U_C and I_C- voltage and current in the figure of graph curvature, U_H and I_H – retention voltage and current. Retention current (one of the essential parameters):

$$I_H = \frac{A_{pnp}(U_{E\,pnp} + U_{CC})/R_S + A_{npn}(U_{E\,npn} - U_W)/R_W}{A_{pnp} + A_{npn} - 1},$$

where:
U_S and U_W–voltages on n-substrate and p-well

Whereas U_{CC}, A_{pnp}, A_{npn}, R_S, and R_W after being switched ON have values different from those at switching mode, the holding current I_H exceeds connection current I_Z.

Conditions for inserting the parasitic thyristor structure emerge if in one of the base regions (n-substrate or p-well), or in the area of bulk charge (spatial charge) of junction n-substrate-p-well (area of spatial charge of bipolar transistor parasitic collector), there are generated minority charge carriers, and thus, switching ON current I_Z. is exceeded.

Even if then the hold- on current is not attained at increasing of I_Z, operation of the circuit is being impaired, because the supply voltage U_{CC} of CMOS structure falls to the level that is less than the level required to ensure the serviceability.

The physical fundamentals of CMOS microcircuit operation are rather sophisticated, but such knowledge is essential to ensure protection against latch-up phenomenon. Fig. 1.46 provides a complete model of CMOS element with lumped parameters [1] which ensures close proximity of rated and experimental data. In Fig. 1.46a the CMOS structure is represented with the parasitic elements, where the resistors R_p^+ and R_n^+ are equivalent to the p^+ and n^+ of drain / source, and the resistors R_{cp} and R_{cn} are equivalent to the optic resistances of collectors of the parasitic p-n-p and n-p-n transistors. In Fig. 1.46b there is shown its equivalent electrical circuit with lumped elements. The circuit in Fig. 1.46c makes it possible to calculate the thyristor effect for the case demonstrated in Fig. 1.47a on the basis of feedback mechanism at corresponding short circuit mode of supplies U_{SS}, U_{DD} and additional collector and emitter resistors. By means of such circuit on the base of known mathematical mechanism, it is possible both to quantify and qualify the influence of each individual parasitic resistance on the "latch-up" effect parameters of CMOS structure.

The results of such qualitative evaluation are available in Fig. 1.47a-e where there is shown the influence on VCC parameters of CMOS element produced by thyristor effect and also standard parameters of cell design: a) R_S; b) R_W; c) R_p^+; d) R_n^+; and e) R_{Cn}. By varying design and technological parameters, it is possible to avoid "latch-up" effect within the frameworks of external stresses known in advance. For example, one of the triggering events for latch-up effect of CMOS microcircuits may be written in the form of $R_S R_W \geq R_p^+ + R_n^+$. Introduction of epitaxial layers into the microcircuit design shall reduce R_{CN} as well as the probability of the effect occurrence.

The addressed thyristor mechanisms of parasitic effects make their appearance not only in the basic CMOS elements (such example may be conveniently used for reviewing the essence of the effect) but also in the safety circuits for inputs and input and output interfacing elements of the microcircuits and also between the adjacent

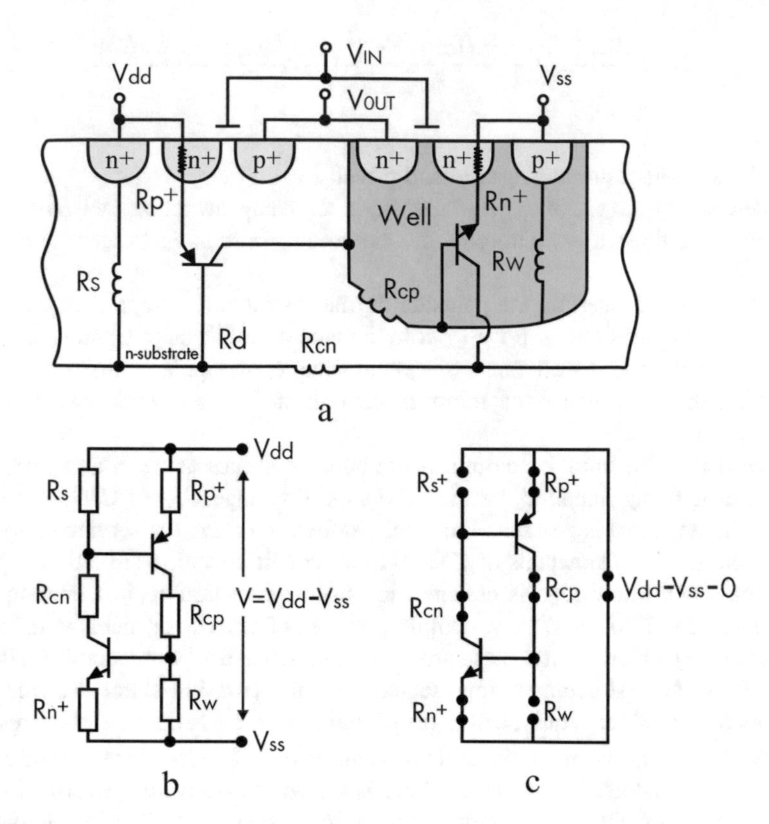

Fig. 1.46 Full model of thyristor structure of CMOS microcircuit

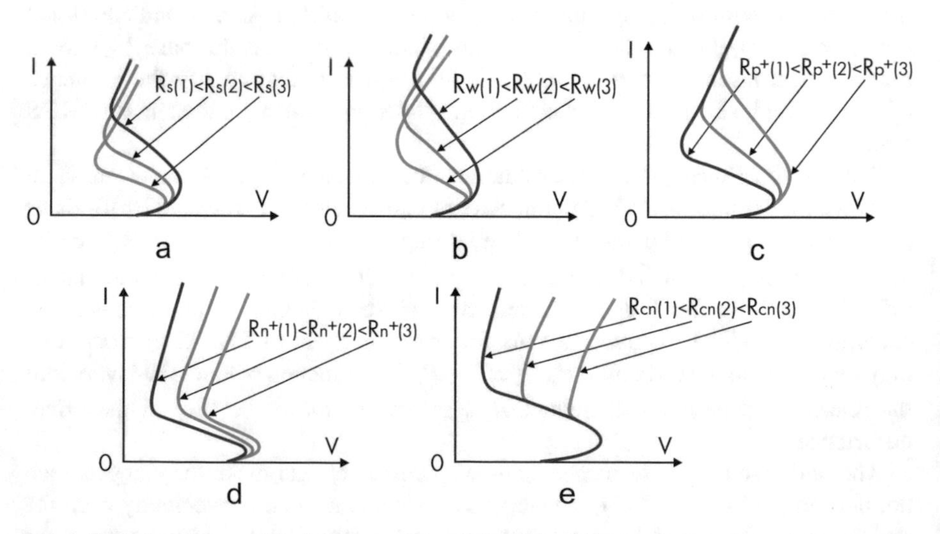

Fig. 1.47 Dependences of the influence on the VCC produced by the thyristor structure parameters

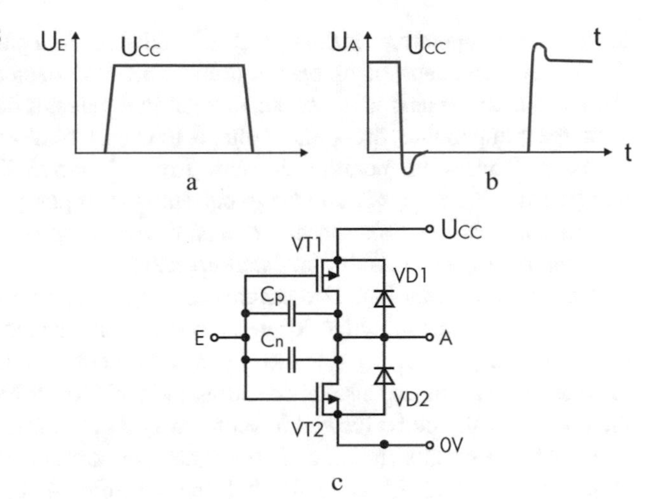

Fig. 1.48 Circuit of CMOS inverter (**a**) and form of actual input and output pulses (**b**)

CMOS elements of microcircuits. Apart from those above explored, there may exist other causes enabling parasitic thyristor effects:

1. Exposure of the regions to radiation or gamma radiation (energy of alpha-radiation is not sufficient for thyristor enabling).
2. Injection of minor carriers. Should in the process of microcircuit operation here is passing a substantial current through the equivalent resistances R_S or R_W (Fig. 1.46a) t which results in putting one of the bipolar transistors of thyristor couple into saturation mode, then either of the emitters starts injecting the carriers and the thyristor is switched on. Similar results may take place at power-up U_{CC} (in case of low rate of voltage rise) due to the influence capacitor charge current C_S (Fig. 1.44), and also at peak rises (surges) of voltage U_{CC}.
3. Injection from drain regions (control thyristor electrodes) into the base regions of bipolar transistors. Such effect may be observed at short-time peak exceedances of output voltage CMOS elements for 0.7 V over voltage U_{DD} or for 0.7 V below U_{SS}. Such surges of output voltages may be attributed to both external pulses noises and parasitic influence of MOS transistor drain-gate capacitor in high-speed performance standard CMOS elements. In Fig. 1.48a–c, there are shown graphs of input U_E and output U_A pulses of CMOS element with parasitic drain-gate capacitors C_p, C_n using diodes VD1, VD2 at the output for protection against enabling of thyristor structure. The steeper input front U_{EI} is the higher the surge at front U_A. will be.
4. Avalanche breakdown of junction p-n junction between substrate and p-well.
5. Electrical fields directed perpendicularly to CMOS microcircuits surface also may result in enabling the thyristor, particularly in case of microcircuits with the epitaxial layers.

In designing and application of digital microcircuits, it is essential to acknowledge and take into consideration temperature dependence of the "latch-up" effect. As long as the temperature of the operating environment T_p is decreasing, resistances R_W, R_S and gain ratio of bipolar transistors are also decreasing, and generation of carriers due to impact ionization and breakdown voltages decreases. Basically along with the temperature drop, the voltage (current) of thyristor structure enabling increases. Thus at temperature decrease from $T_p = 375$ K to $T_p = 77$ K, hold-up current is multiplied by 3, and firing current is multiplied by 5.

There are differentiated technological, design, topological, and schematic activities and measures to inhibit the latch-up effect .

Design and technological measures as below need be set out: The contacts of wells and substrates should be located in such a way that the resistances R_S и R_W tend to their minimum values; gain ratios by the current of an oblong p-n-p transistor need to be reduced at the expense of extending the width of its base; the regions acting as the collectors should be located in such a way that the thyristor's breakover current and holding currents increase. If necessary so, around the critical regions, there should be put guard rings and also to bring to minimum drain-gate capacitors C_S. In the first turn, it is relevant for microcircuit input and output interfacing elements to apply transistor ring structures with drain located in the center.

The most suitable and effective technological and topological designing is possible only with recourse to extremely expensive two- and three-dimensional models and methods of computation.

The standard schematic measures and activities are as follows: application of special circuits of substrate bias voltage generators and isolating wells, which are placed in the dice; application of Schottky diodes to prevent injection of charge carriers through drain p-n junctions at surges of output voltages U_A of CMOS-elements; application of input and output protection circuits proof against thyristor effect; enabling of diodes at inputs and outputs of integrated microcircuits with pass voltages below 0.7 V; and support of some minor rise rate of voltage U_{DD} at switching ON (use of filter capacitor in supply network).

In conclusion, it should be noted that the conditions of thryristor effect occurrence in MOS microcircuits are influenced by a plethora of factors which quite frequently defy tracking or metering. Even at entirely identical values of static and dynamic parameters of CMOS microcircuits, the values of their thyristor characteristics and, hence, the environment of latch-up effect occurrence may be substantially different. It may be true for one and the same type of digital microcircuits of various manufacturers or for the microcircuits by the same manufacturer but fabricated over various periods of time.

References

1. Belous, A. I., Emelyanov, V. A., & Tourtsevich, A. S. (2012). *Fundamentals of schematics in microelectronics*. M.: Technosphere, (p. 472) (in Russian).
2. Aleksenko, A. G. (2002). *Fundamentals of Microschematics*. M.:Physmatizdat.
3. Ugrumov, E. P. (2002). *Digital Schematics*. SPb:BVH-St.Petersburgh.

4. Belous, A. I., & Yarzhembitsky, V. B. (2001). *Schematics of digital microcircuits for data processing and transmission systems*. Minsk: Unitary Enterprise "Technoprint", p.116 ISBN985–464–064-7 (in Russian).
5. Belous, A. I., Ovchinnikov, V. E., & Tourtsevich, A. S. (2015). *Aspects of microelectronic devices designing for space application*. Ministry of Education, Republic of Belarus, Homel State University named after F.Skoryna, (p. 301), in two volumes, Volume 1. (in Russian).
6. Belous, A. I., Saladukha, V. A., & Shvedau, S. V. (2015). Space electronics. Technosphere Publishin House, in two volumes, (p. 1184).
7. Belous, A., Saladukha, V., & Shvedau, S. (2017). *Space microelectronics volume 2: Integrated circuit design for space applications* (pp. 720). London, Artech House, ISBN: 9781630812591.
8. Waykerly Дж, J.. (2002) *Design of digital devices*. Postmarket.
9. Myrova, L., & Chepyzhenko, A. (1983). Ensuring radiation stability of communication equipment. *Radio and communication*.
10. Korshuniv, F. N, Bogatyrev, Y. V., & Vavilov, V. A. (1986). Radiation impact on integral microcircuits. *Minsk, Science and Engineering*.
11. Mkrtchyan, S. O., Melkonyan, C. R., Abgaryan R.A. (1990) Schematics of digital ICs protection against overloads. *Electronic Engineering*. Microelectronic devices, Issue 4(82) Series 10: 30–34.
12. Scarlette, J. (1974). *Transistor –transistor ICs*. And their application: Translation from English language. Edit. B. E. Ermolaeva. M/, Mir.

Chapter 2
Schematic Solutions of Digital CMOS Microcircuits

This chapter addresses in great depth and consistent manner any and all niceties of schematic solutions for digital CMOS microcircuits (both classified in literature sources, like publications or patents, and designed by the authors and well proven then in off-the-shelf microcircuits): standard logic cells and their modifications, static and dynamic logic cells, and memory cells (both controlled by sync signal level and clocked by sync signal front).

2.1 Standard Logic Cells of Digital CMOS Microcircuits

Due to low static consumption power, high interference immunity, and packing density, the complementary MOS schematics have grown to be a predominating tool in the digital microcircuits. However, the high-speed performance of CMOS circuits is still less compared to the bipolar ones, and the area occupied on a die by complex logic cells is rather large as compared to the logic cells on MOS transistors of the same conductivity type. Two major approaches to how to increase both high-speed performance and packing density of CMOS microcircuit are popular [1].

The first approach relates to development of new technologies with scale decrease of linear dimensions of microcircuit components and determined by the level of the technological equipment and materials. The status of the problem as well as the prospects of these high technologies will be covered more in depth in the final chapter of this book.

The second approach has to do with the development of some new schematic solutions for CMOS microcircuits, the most relevant of which are addressed below and seem to be the most attention worthy for the following reasons:

(a) In complex LSIIC miscellaneous requirements may be applied to various functional blocks. For instance, for arithmetic and control devices, an accurate support of timing relationship is essential, whereas for register blocks of

A. Belous, V. Saladukha, *The Art and Science of Microelectronic Circuit Design*,
https://doi.org/10.1007/978-3-030-89854-0_2

microcircuits, high packing density is a must. Therefore, application of various types of LC circuits contributes to attaining better parameters of digital microcircuits.

(b) Employing some new schematic solutions affords obtaining higher packing density and higher speed performance and extending application scope of CMOS circuits.

2.1.1 Static CMOS Logic Cells

This section delves into both convenient logic cell and some of the most popular modifications of logic cells.

Static CMOS LC with the standard architecture. The basis of standard static CMOS LC is formed by the electrical circuit shown in Fig.2.1a, containing two MOS transistors of various conductivity types and performing inversion function. The principal characteristics of such circuit are detailed in [1]. Output voltage levels of CMOS LC may be estimated by a well-known formula for drain current I_C^n, I_C^p of MOS transistors.

$$U_{OL} \approx \left(U_{CC} - U_T^n\right) - \sqrt{\left[0,5_{kn}\left(U_{CC} - U_T^n\right)^2 - I_{OL}\right]/0,5_{kn}}, \text{at} U_{OL} < U_{CC} - U_T^n \quad (2.1)$$

$$U_{OL} \approx U_T^n + \sqrt{\left[0,5_{kp}\left(U_{CC} - \left|U_T^n\right|\right)^2 - \left|I_{OH}\right|\right]/0,5_{kn}}; \text{at } U_{OH} < U_T^p, \quad (2.2)$$

where: U_T^p, U_T^n—threshold p-MOS and n-MOS transistors VT1, VT2

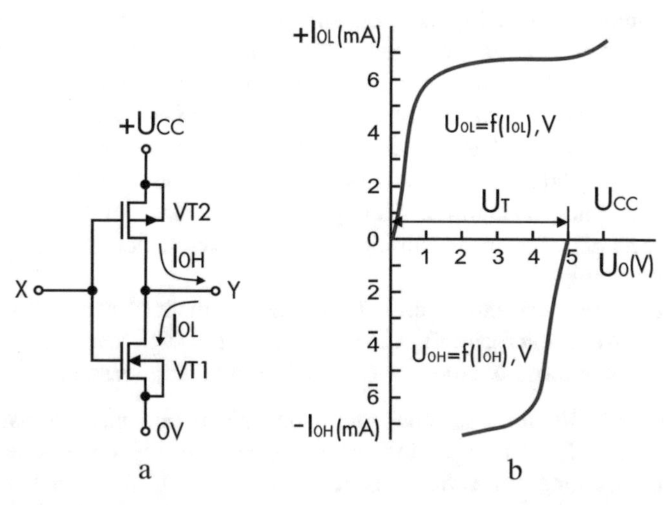

Fig. 2.1 Diagram of (**a**) standard and output characteristics (**b**) of static CMOS LC inverter

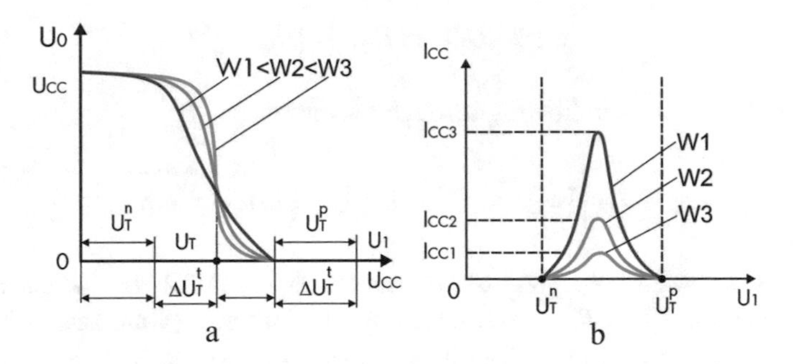

Fig. 2.2 Transfer (**a**) and current (**b**) characteristics of static CMOS LC in transient mode.

K_p, K_n–transconductance of p-МОП and n-МОП транзисторов VT1, VT2

Output characteristics of standard CMOS LC corresponding to the formula provided are shown in Fig. 2.1b. Whereas input CMOS LC is of capacitance nature, its input characteristic has no practical relevance. Here, meanwhile, the capacitance nature of CMOS LC input is responsible for actually zero input currents $I_{IH}{\approx}0$, $I_{IL}{\approx}0$; therefore, in the chain of serially connected CMOS LC in static mode, there is:

$$U_{OL} \approx 0; U_{OH} \approx U_{CC}; \tag{2.3}$$

At making use in the capacity of load standard LC of similar type with input switching threshold $U_T{\approx}U_{cc}/2$, the values of the IC input threshold:

$$U_T = \left[U_T^n + \sqrt{K_P/K_n}(U_{CC} - |U_T^p|)\right]/\left(1 + \sqrt{K_p/K_n}\right); \tag{2.4}$$

When equality of transconductance $K_p{=}K_n$, is obtained through selection of n- and p-MOS -transistors *W, L*, then $U_T = \left[U_T^n + (U_{CC} - |U_T^p|)\right]/2$, and in case of equal threshold voltages of MOS transistors $U_T^n = | U_T^p |$ (we receive $U_T{=}U_{cc}/2$), then reserve of interference stability of CMOS LC shall make (Fig. 2.2a):

$$\Delta U_T^+ = U_T - U_{OL} \approx U_{CC/2}; \Delta U_T^- = U_{OH} - U_T \approx U_{CC}/2; \tag{2.5}$$

Whereas in each of two logic states, one of the transistors VT1 and VT2 is open and the second one is closed, CMOS standard logic cell is characterized by low static consumption $P_{CCL}^C \approx P_{CCH}^C \approx 0$.

Supply current I_{CC} and power P_{CC} are consumed at switching-over only when (Fig. 2.2b):

$$U_T^n < U_I < U_{CC} - |U_T^p|; \tag{2.6}$$

and let-through current is flowing through the circuit:

$$I_{CC} = K_P\left[(U_{CC} - U_T^n - |U_T^p|)^2/\left(1 + \sqrt{K_p/K_n}\right)\right]; \tag{2.7}$$

When the width W_p, W_n of transistors VT1, VT2 is increasing, their transconductance K_p^{VT2}, K_n^{VT1} and consumption current I_{CC}. are increasing. Since the standard circuit of logic cell is consuming the current at switching-over only, then its value depends upon switching frequency F; therefore, for inverter [1]:

$$I_{CC}^{Д} = C_L U_{CC}^2 F; \tag{2.8}$$

High-speed performance of standard CMOS logic cell to a large extent is defined by load capacity, and it can be estimated rather accurately by the formula as below:

$$t_{PHL} = \frac{C_L}{K_n'\left(\frac{W_n}{L_n}\right)(U_{CC} - U_T^n)} - \left[\frac{U_T^n}{(U_{CC} - U_T^n)} + \frac{1}{2}\ln\left(\frac{3U_{CC} - 4U_T^n}{U_{CC}}\right)\right]; \tag{2.9}$$

$$t_{PLH} = \frac{C_L}{K_p'\left(\frac{W_p}{L_p}\right)(U_{CC} - U_T^p)} - \left[\frac{U_T^p}{(U_{CC} - [U_T^p])} + \frac{1}{2}\ln\left(\frac{3U_{CC} - 4U_T^n}{U_{CC}}\right)\right]; \tag{2.10}$$

where K_p',K_n'–specific transconductance of p- and n-MOS transistors

W_p, W_n–width of p and n MOS transistors
L_p, L_n–gate length *of p and n MOS* transistors

Generation of logic functions on the base of CMOS standard logic cell is done by parallel (serial) connection of p- and n-MOS transistors. An example of electrical circuit of CMOS LC with function 2 NOR is shown in Fig. 2.3. The methods of generation of CMOS LC circuits performing sophisticated logic functions are elaborated in work [2]. As it is obvious from the figure, in standard CMOS LC logic function is generated twice: in block D1 of n-MOS transistors VT1 and VT2 and block D2 of p-MOS transistors VT3–VT4. It is responsible for the doubled quantity of components in the circuit of logic cell, which results in increase of input capacitors, degrading of high-speed performance, and increased area on the die. Besides, serial (parallel) connection of MOS transistors in LC circuit shall result in the dependence of output resistance of logic cells upon combination of input signals X and type of function performed by the LC. Hence, for elimination of these faults, there has been suggested a range of electrical circuits of CMOS logic cells with the structure not similar to the standard one.

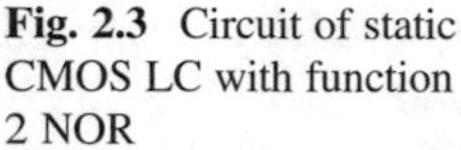

Fig. 2.3 Circuit of static CMOS LC with function 2 NOR

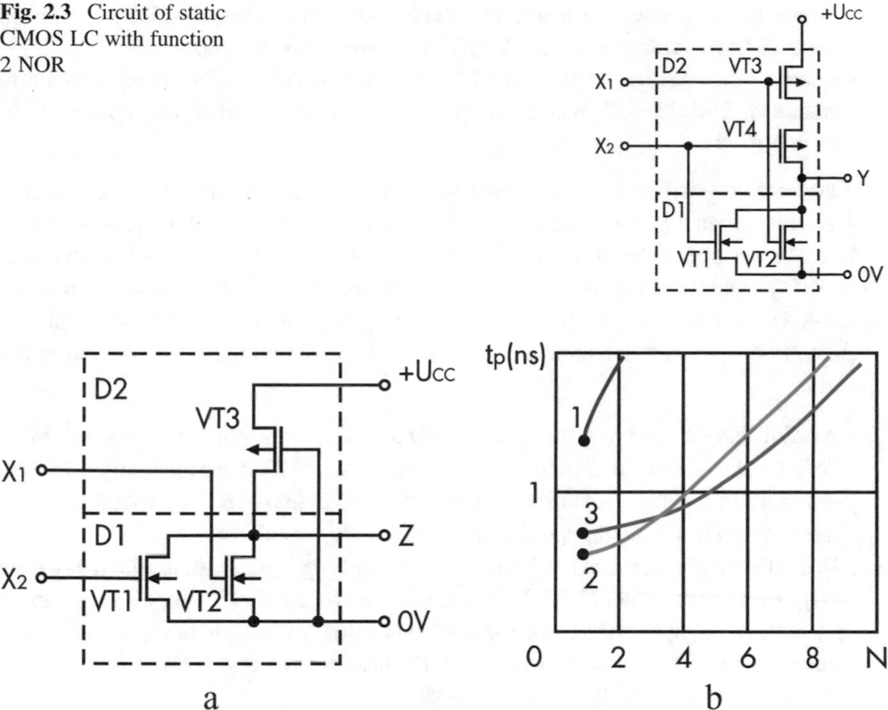

Fig. 2.4 Circuit (**a**) of static CMOS LC of pseudo n-MOS and comparison dependences of the average delay time of static CMOS logic cells of various types upon number of loads, N (**b**)

Static CMOS Logic Cells of Pseudo-MOS Type Emergence of CMOS LC of this type has to do with need to reduce the number of components in LC through avoiding duplication of generation circuit of logic function.

The circuit of such CMOS standard logic cell [1] is shown in Fig. 2.4a and called "pseudo" n-MOS. In this logic cell, the logic function is formed by one block (*D1* on n-MOП transistors VT1–VT2), and the second block D2 containing one p-MOS transistor only VT3 with "grounded" gate performs the function of current supply of load capacity charge. Such technique has made it possible to reduce the number of components in the logic cell to N+1 in N-input logic cell with function NOR and to reduce by half the input capacitances of the LC.

Delay time t_p of signal propagation for such LC is lower as compared to CMOS LC of standard architecture, which is shown in the graph describing the dependence of the average delay time t_p upon number of loads N, as in Fig. 2.4b where *1* is dependence of the average delay time on the load for standard CMOS LC of 3NOR type; 2 dependence of the average delay time upon the load for pseudo p-MOS LC of type 3 NOR; and 3 dependence of the average delay time upon the load for symmetrical CMOS LC of type 2 NOR. Such decrease may be explained by two reasons:

(1) Single loading transistor has much less capacity than a kit of serially connected p-MOS transistors of standard CMOS LC with NOR function.
(2) Loading component of the said LC is permanently in ON state, whereas in standard CMOS LC block of p-MOS transistors block is operating in switch mode.

Output voltage of high-level $U_{OH}{\approx}U_{CC}$. However, as transistor VT3 is constantly is ON state, then LC has enhanced output voltage of low-level U_{OL}. In order to assure normal operation of such LCs circuits, normally, one is guided by the rule: $U_{OL} < U_T^n$, which is supported by appropriate selection of dimensions of n-MOS and p-MOS transistors VT1-VT3: $(W/L)_p/(W/L)_n=A$, where A is a constant value.

In comparison with the standard CMOS LC, such circuit manifests the following defects:

(a) Availability of static consumption power P_{CCL}^C in static state of low level. Such defect stems from the fact that p-MOS transistor VT3 is permanently switched on; therefore, if high-level signal is provided to at least one of the outputs X, in the supply circuit, static consumption current I_{CCL}^C will flow.
(b) Reduced toggle threshold of logic cell U_T, resulting in skew of toggle delay time t_{PLH}, t_{PHL}. This defect stems from the fact that to support the required values of the voltage U_{OL} dimension of p-MOП transistor VT3 W_p is made much less as compared to dimension W_n of n-MOS transistors VT1 and VT2, which contributes to decreasing of U_T; toggle threshold.
(c) Dependence of LC high-speed performance upon combination of input signals. Such defect is appertaining to the fact that permanently active load P-MOS transistor VT3 has permanent output I_{OH}. Therefore, output resistance of logic cell in open state shall depend on the number of powered n-MOS transistors VT1 and VT2, and its value is less with the more number of powered n-MOS transistors.

As a consequence, toggle delay time t_p will be decreasing in proportion to the number of simultaneously toggled transistors.

Static CMOS Logic Cells with Symmetrical Structure One of the attempts to reduce the dependence of CMOS LC high-speed performance on the combination of input signals is the development of CMOS LC symmetrical structure [1]. The standard electrical circuit of such LC performing NOR function is shown in Fig. 2.5a, where it is obvious that the load block D2 of p-MOS transistors VT3 and VT4 has electrical circuit symmetrical to block *D1* of n-MOS transistors VT1 and VT2. Therefore, each n-MOS transistor generating the logic function corresponds to one p-MOS load transistor. As a consequence, at toggling one n-MOS transistor of D1 block, recharge of output capacitor will be supported by on p-MOS transistor. At toggling two n-MOS transistors, the recharge of output capacitor will be done by double current at the expense of two load p-MOS transistors, etc.

Therefore one can make a conclusion concerning higher high-speed performance of "symmetrical" CMOS LC as compared to "pseudo" n-MOS LC. Within such

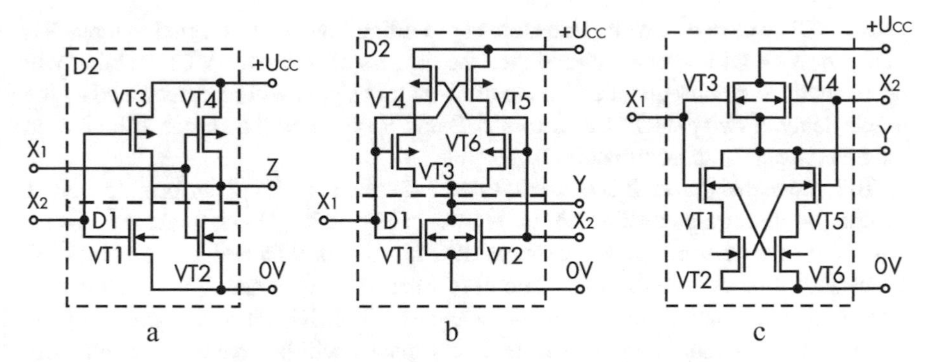

Fig. 2.5 Electrical circuits of symmetrical CMOS LC:
(**a**) standard
(**b**) with controlled charge (discharge) circuits
(**c**) with accelerating circuits

logic cell, it is essential also to assure the respective dimensions of p and n-MOS transistors to support the relevant output voltage of low level U_{OL}:

$$(W/L)_p/(W/L)_n = A/(N\text{-}1); U_{OL} < U_T^n; \tag{2.11}$$

where N is the number of n and p-MOS transistors connected in parallel way in the logic cell.

In [1] however, it was found that the structure of symmetrical type does not provide noticeable advantage in terms of high-speed performance as compared to pseudo n-MOS LC, which is shown in Fig. 2.4b (graph 3). Besides, if at least at two LC output combinations of opposite signal levels are observed in the feed circuit, also the static consumption current I_{CCL}^C will be passing, which is not found in standard CMOS logic cells. Due to the causes mentioned herewith, the logic cells of such type have not gained practical application. A more promising method of eliminating the dependence of LC high-speed performance upon the combination of the input signals, not increasing meanwhile the static consumption power P_{CC}^C of CMOS LC, turns out to be generation of the additional controlled charge (discharge) circuits of load capacity, converting the standard circuit of CMOS LC into the structure of symmetrical type.

Let's consider the work of a standard CMOS LC with 2 NOR function, the diagram of which is shown in Fig. 2.3. Let's assume that low-level signal is provided to input X1 and to input X2 high-level signal is provided. Here low level of signal will be set at Y output. As transistor VT4 is open, then at transistor VT3 drains there will be set potential $U_H^{VT3} \approx U_{CC.}$ Therefore, low-level signal is provided to input X2, and LC is toggled into high-level state, and output signal will be changed after time t_{pO}, determined by toggle delay one transistor VT3 only. In case of opposite combinations of input signals (X1, high level; X2, low level), transistor VT4 will be closed; consequently, notwithstanding the low level of voltage at transistor VT3 gate, it will be closed as well. Therefore at providing low level of signal to input X1, transistor VT4 will be opened, and then the process of transistor VT3 enabling will be initiated. As a consequence, change of LC output signal will be determined by the toggle delay of two p-MOS transistors VT3 and VT4 and will be twice as high as in the first case, i.e., approximately $2t_{pO}$.

To remove this defect, it is possible to use LC electrical circuit performing 2 NOR function with extra-controlled circuit shown in Fig. 2.5b [1]. The circuit is distinguished by the presence of two extra p-MOS transistors VT5 and VT6 in block D2 connected in parallel to the main circuit of transistors VT3 and VT4. As a consequence at combination of the signals at inputs XI=L,X2=H and toggling of the signal at input X2 into low level state, load capacity will be charged through more high-speed transistor VT5 and VT6 circuit because due to low-level voltage at gate, transistor VT 5 from X1 input is preliminarily open. At combination of the signals at inputs *X1=L, X2=H* and signal toggle at input X1` in low-level state, load capacity will be charged along another high-speed transistor chain VT3 and VT4 where transistor VT4 will be preliminarily open due to low level of voltage at input X2. Thus LC turning-off delay time will be the same for various combinations of input signals.

Similar circuit with the accelerating circuits may be generated for the logic cells with function 2 NAND. The electrical circuit of such LC is shown in Fig. 2.5c [2]. If logic function is changing, the number of additional parallel chains is increased in proportion to the number of MOS transistors serially connected in the charging (discharging) circuit. For example, for logic cells with 3 NAND block D1 is completed with two parallel circuits comprising n-MOS transistors 4NOR – three circuits and so on [3].

CMOS Static LC with Differential Structure One of the well-known and popular schematic techniques enabling to extend the functional potentialities of CMOS LC and to improve the dynamic characteristics is using CMOS LC of differential architecture. Complementary CMOS LC of such type have been denominated *Complementary Differential Cascade Voltage Switch (DCVS)* Logic [4], and apart from the advantages and merits listed above, they are characterized with intrinsic ability to self-testing. Such circuits may be easily designed by means of the procedures based on Karnaugh map or table methods. A summary structure of DCVS LC is shown in Fig. 2.6 and incorporates two key blocks: Differential Binary Cascade Voltage Solution circuit and Load Circuit.

DCVS block (Fig. 2.7) should be in conformity with the conditions as follows: if the input vector *X=(X1...Xn)* is true for toggle function *Q(X)*, then output Q is disconnected from node G, and output Q is linked with node G; if the input vector switched *X=(X1...Xn)* – is false, then a reverse case is observed.

Electrical circuit of the block DCVS performing the adder function is shown in Fig. 2.7 and may be designed with Karnaugh maps. Several options may be used as load capacity. In Fig. 2.8 there is shown a standard circuit of LC DCVS where in the capacity of load there is used an elementary circuit of trigger cell on transistors CT3 and VT4. Depending upon state of inputs (*X1...Xn, X1...Xn*), block DCVS is setting

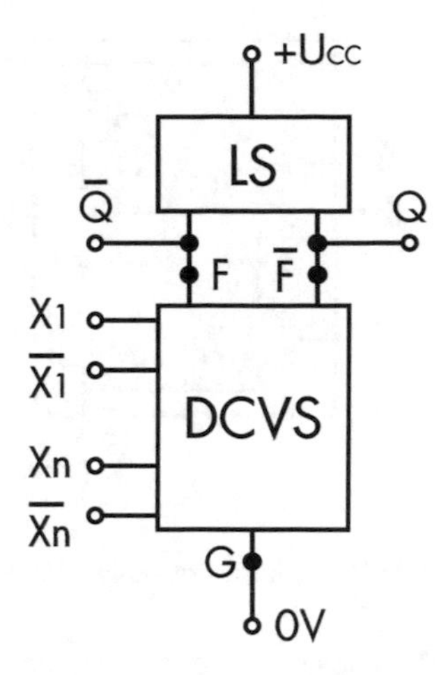

Fig. 2.6 Summary structure of static CMOS LC with DCVS architecture

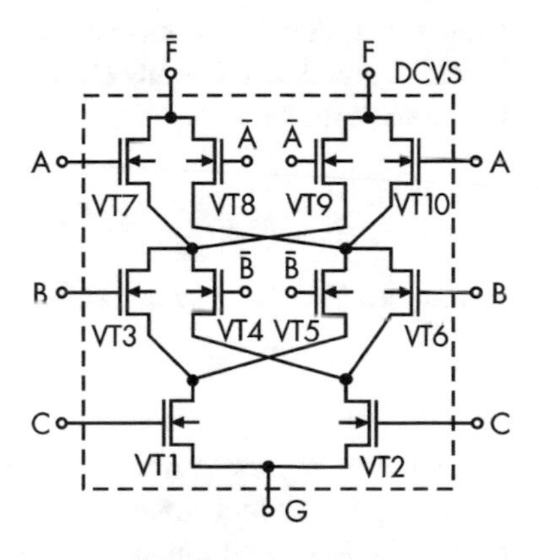

Fig. 2.7 Block diagram of DCVS performing the adder's function

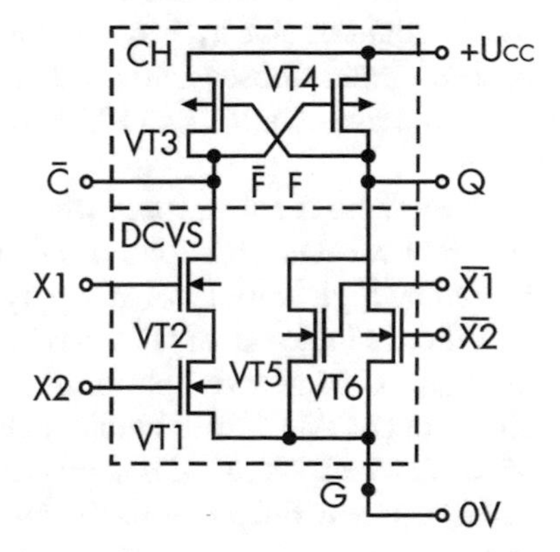

Fig. 2.8 Standard electrical circuits of static CMOS LC with differential structure

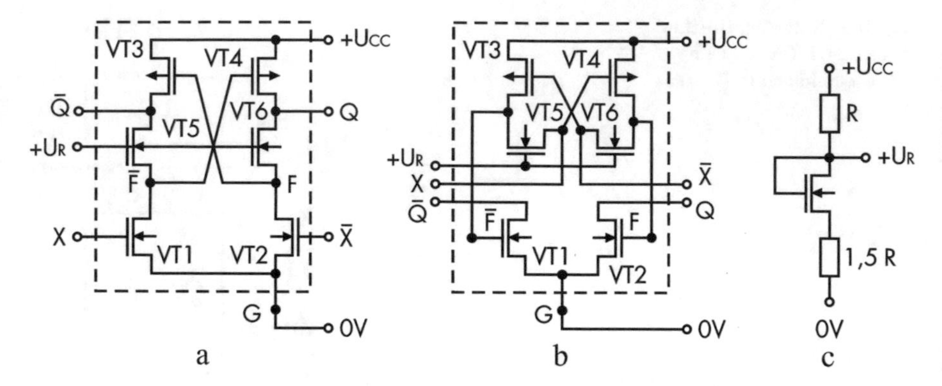

Fig. 2.9 Circuits of static CMOS LC with differential split level logic (DSL)

low level of voltage on one of the outputs Q or $\overline{Q}$. Under influence of this level trigger cell on p-MOS transistors VT3 and Vt4 in regenerative manner sets voltage levels U_{CC} and *0,* at outputs *Q,* $\overline{Q}$ and reversely. As a consequence the output voltage levels shall make:

$$U_{OL}^{Q,\overline{Q}} \approx 0V; U_{OH}^{Q,\overline{Q}} \approx U_{CC}, \tag{2.12}$$

And logic voltage swing at the output

$$\Delta U_T \approx U_{CC}. \tag{2.13}$$

The consumption current of the logic element I_{cc} passes only at the moment of its switching over. Since toggling of load circuit in the logic cell is performed by DCVS block with low-level signal, then for stable LC toggling, n-MOS transistors of DCVS block should be able to commutate drain current of p-MOS transistors VT3 and VT4 of load circuit. Consequently, one shall need an appropriate choice of dimensions of p-MOS transistors VT3 and VT4 of load circuit and n-MOS transistors of DCVS block.

Since there is a time space when some of p-MOS transistors of load circuit and a circuit of n-MOS transistors of DCVS block are open, such logic cells are characterized by high noise level in supply circuit U_{CC}.

Whereas logic cell of DCV has huge logic swing of voltages ΔU_T at output that is limiting LC high-speed performance, there was suggested an improved electrical circuit of DCVS LC with limit of output levels [5, 6]. This circuit was named CMOS of differential logic with split levels—CMOS Differential Split Level Logic (DSL)—and it differs from the standard circuit by load. The standard electrical circuit explaining the operational concept of DSL LC is shown in Fig. 2.9, and it incorporates in the load circuit two additional n-MOS transistors VT5 and VT6 with the gates connected to the reference supply $+U_R$. The values of the reference supply potential supporting the most preferable high-speed performance:

$$U_R \approx 0,5U_{CC}/2 + U_T^n. \tag{2.14}$$

Gates of p-MOS load transistors VT3, VT4 are connected not to the inputs as it is the case in the standard circuit to DCVS but to the sources of additional n-MOS transistors VT5 and VT6.

Assuming that to input X of logic cell there is provided low level of signal and to $\overline{X}$ high level, then in node $\overline{F}$, there will be set high level of potential and in the node F low level. As with high-level voltage in the nodes F $\overline{F}$ is limited by additional transistors VT5 and VT6, then the potential in the node F: $U_F \approx 0.5U_{cc}$. However, since such level of U_F does not support the closed state of load p-MOS transistor VT4, in node F, there will be set an intermediate level of voltage $U_{\overline{F}} \approx$ 100mV, and its open state shall result in emergence of static consumption I_{CC}^C.

High voltage level in node F shall result in closing of n-MOS transistor VT5, and low level in node $\overline{F}$ shall result in enabling load p-MOS transistor VT3; therefore, at output $\overline{Q}$, there will be set an output high level voltage $U_{OH} \approx U_{CC}$.

High level of voltage on node F shall result in enabling the additional transistor VT6, and at output Q there is set a low level of output voltage U_{OL} with the value defined by the dimensions of transistors VT2, VT4, and VT6. Maximum logic swing of the voltages:

In nodes F and $\overline{F} \Delta U_T^{F,\overline{F}} \approx 0,5U_{CC}$;
At output Q И $\overline{Q}$ $\Delta U_T^{Q,\overline{Q}} \approx U_{CC}$;

It makes possible to reduce the width of the gates of n-MOS transistors VT1 and VT2 of block DCVS and to reduce the influence of "hot" electron effect upon functioning of LC. Besides, in order to reduce the signal delays at the interconnections, it is preferable to use these nodes F, $\overline{F}$ as inputs and outputs. Circuit of DSL LC reconfigured in this way is shown in Fig. 2.9b. In the capacity of LC inputs X, $\overline{X}$, there are used sources of n-MOS transistors VT5 and VT6, limiting the signal levels in load. Outputs of Logic cells are generated by open drains of n-MOS transistors VT1 and VT2 of block DCVS. The levels of internal signals in this circuit (in nodes A, $\overline{A}$) correspond to the signal levels in nodes F and $\overline{F}$ of circuit in Fig. 2.9a. In the capacity of LC reference voltage generator U_R, it is possible to use the circuit shown in Fig. 2.9c which supports the appropriate value of U_R. Comparison-rated characteristics of CMOS LC with the differential structure at an example of full single-bit adder circuit are shown in Table 2.1; however, the real advantages of these circuits are less significant which are noted in [5]. This is because LC DCVS and DSL in the capacity of load regenerative have trigger circuit at p-MOS transistors VT3 and VT4. Therefore maximum high-speed performance of this circuit is attained at minor width of p-MOS transistors VT3 and VT4. And in its turn, it affects the switch-off time tp_{LH} of LC. Besides, the presence of static consumption power P_{CC}^C constitutes the defect of LC DSL.

With a view to eliminating such defects, there has been suggested a new type of differential CMOS LC: **Complementary Pass Transistor Logic (CPL)** [7]. The fundamental concept of this LC resides in arranging the logic circuit at n-MOS pass

Table 2.1 Comparison characteristics of the full adders at CMOS LC with differential architecture [7]

Type of logic cell	Capacity		Number of p-MOS/n-MOS transistors	Max signal delay time, ns	Power dissipation at maximum frequency mW	Standardized quality ratio
	Input Φ	Output Φ				
Standard static CMOS	155	155	15/15	20	0.58	1.0
Static DCVS	85	85	4/18	22	1.01	2.11
Static DCL	85	85	4/22	14	1.35	1.63

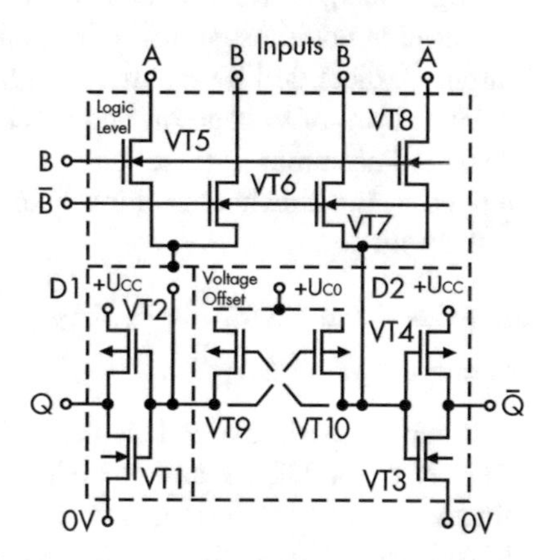

Fig. 2.10 Circuit of static differential CMOS logic circuit at pass transistors (CPL)

transistor and removal of trigger load circuit as it is shown in Fig. 2.10. Logic cell CPL incorporates complementary inputs A, $\overline{A}$, B, $\overline{B}$, logic device (LD) at "pass" transistors VT5–VT8 with paraphrase outputs and output circuit (OC0 (inverters D1 and D2 at transistors VT1–VT4). Here "pass" transistors VT5–VT8 perform the function of components in output capacity charge/discharge of logic circuit which makes it possible to eliminate trigger load of load circuit in the output circuit and makes use in full the advantages of differential structure. Whereas the high-level voltage U_{OH} at outputs of pass transistor VT5-VT8 circuit is reduced for the value of the threshold voltage U_T^n of pass transistors, level offset is needed which is provided by output inverters *D1* and *D2*.

Load circuit at p-MOS transistors VT9 and VT10 makes it possible to reduce consumption power *Pcc*. In this case the width of p-MOS transistors VT9 and VT10 may be minimal. Logic functions in logic device may be generated by a combination of four standard logic modules: AND (NAND), OR (NOR), Excluding OR (excluding NOR), and Mounting AND (NAND), electrical circuit of which are shown in Fig. 2.11a–d. Thus in the circuit provided in Fig. 2.10, logic cell performs the function AND (NAND) from two variables A and B. LC CPL is distinguished by availability of elementary advantageous double transistor circuits D1 and D2.

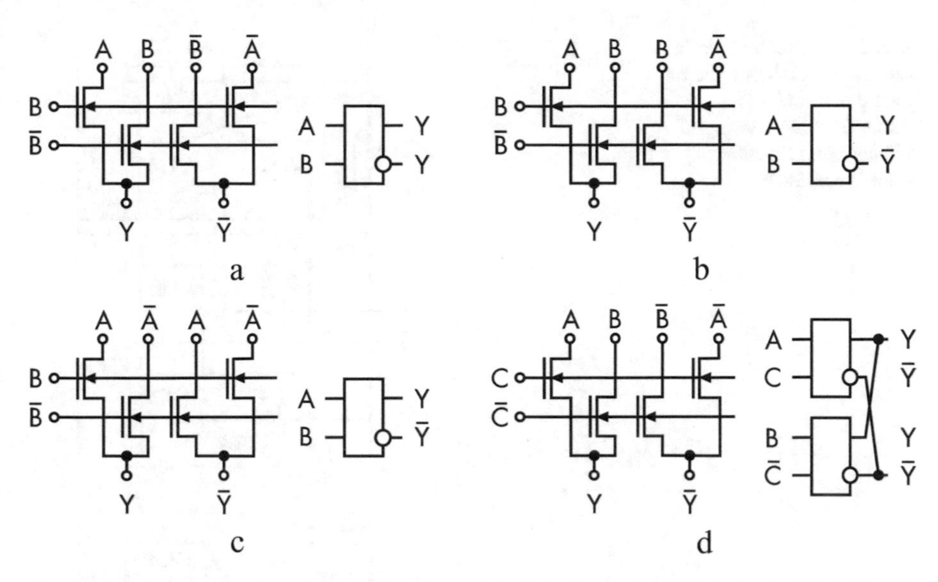

Fig. 2.11 Electrical circuits of standard logic modules LC CPL: (**a**) 0 AND/NAND, (**b**) OR/NOR, (**c**) excluding OR, (**d**) mounting AND/AND-NO

Table 2.2 Comparison characteristics of full adder on CMOS LC at pass transistors [7]

Type of LC	Total number of transistors	Area on die, um2	Signal delay time, ns	Power for на 100 МГц, мВт
Standard static CMOS	40	4730	0.63	1.2
CPL	28	4218	0.26	0.86

Whereas inverters *D1 and D2* are not always necessary for logic cell, they may be eliminated from the circuit, and logic devices may be connected in serial mode, reducing thus the length of the circuit and improving high-speed performance. Comparison-rated characteristics of LC CPL when they are used in the circuit of full single-bit adder are provided in Table 2.2.

Significant defect manifested by of CPL LC is restriction by the number of serially connected pass n-MOS transistors which has to do with reduced output voltage of high-level logic device output. Such restriction may be removed by using the improved modification of CPL LOC, which has been named **Complementary Reduced Swing CMOS logic (CRSL)** [8]. Functioning of such logic cell is explained by the electrical circuits shown in Fig. 2.12 and performing the function of three pass adder. The circuit is distinguished by availability of input buffer incorporating the inverters, performed with a view to increase the high-speed performance at n-MOS transistors VT3–VT16 and updated output circuit incorporating two additional n-MOS transistors VT1 and VT2. If high-level signal is provided to input A and low-level signal to input A, respectively, high-level signal is set at node F $U_F \approx U_{CC} - U_T^n$ and at node $\overline{F}$ low -level signal $U_F \approx 0$.

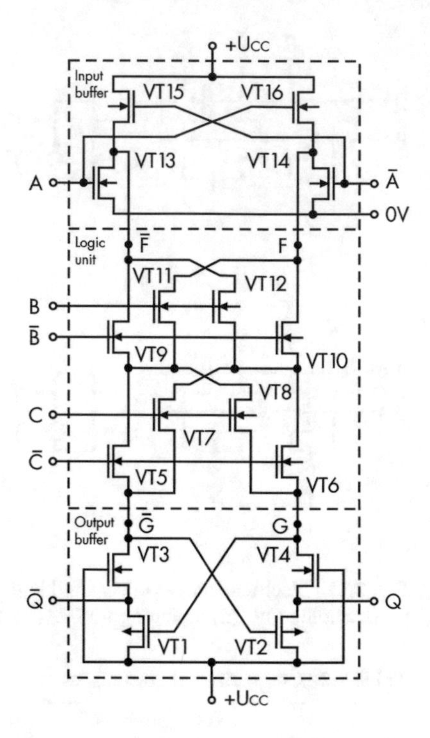

Fig. 2.12 Circuit of static differential CMOS logic on pass transistors with diminished logic swing of voltages and function of three input adder

At appropriate combination of signals at inputs $B, \overline{B}, C, \overline{C}$, which is opening serial circuits of pass transistors in mode $\overline{G}$, there should be set high-level signal different from signal level in node $\overline{F}$:$U_{\overline{G}} = U_F - NU_T^n$, where N is the number of serially connected pass transistors.

However, due to additional n-MOS транзистора VT1 and open p-MOS transistor VT3, connected in parallel to the circuit of pass transistors, in node $\overline{G}$, there will be set level $U_{\overline{G}} \approx U_{CC} - U_T^n$. Here, such voltage level in node $U_{\overline{G}}$ may result in enabling the second p-MOS transistor VT4 and emergence of static consumption current I_{CC}^C. Consequently, there is a need for appropriate choice of the transistor dimensions in the output circuit VT1–VT4. However, provided $U_T^n < U_T^p$, there may be a closed mode of p-MOS transistor T2 and to reduce the static consumption current I_{CC}^C.Use of such configuration in logic cell is particularly efficient in serial high-speed functional devices, such as multipliers and adders.

Multi-output Static CMOS Logic Cells The promising method used in schematic of CMOS microcircuits is application of technical solutions of high-speed bipolar schematics. One of such solutions is the one suggested in [1].

CMOS LC Multi-drain Logic—The CMOS Multi-drain Logic (MDL) Structure is taken from I2L schematics. Such logic cell incorporates two MOS transistors where the first one (multi-drain) is used to generate the logic functions and the second one is used as the current source to charge the input-output capacity of logic

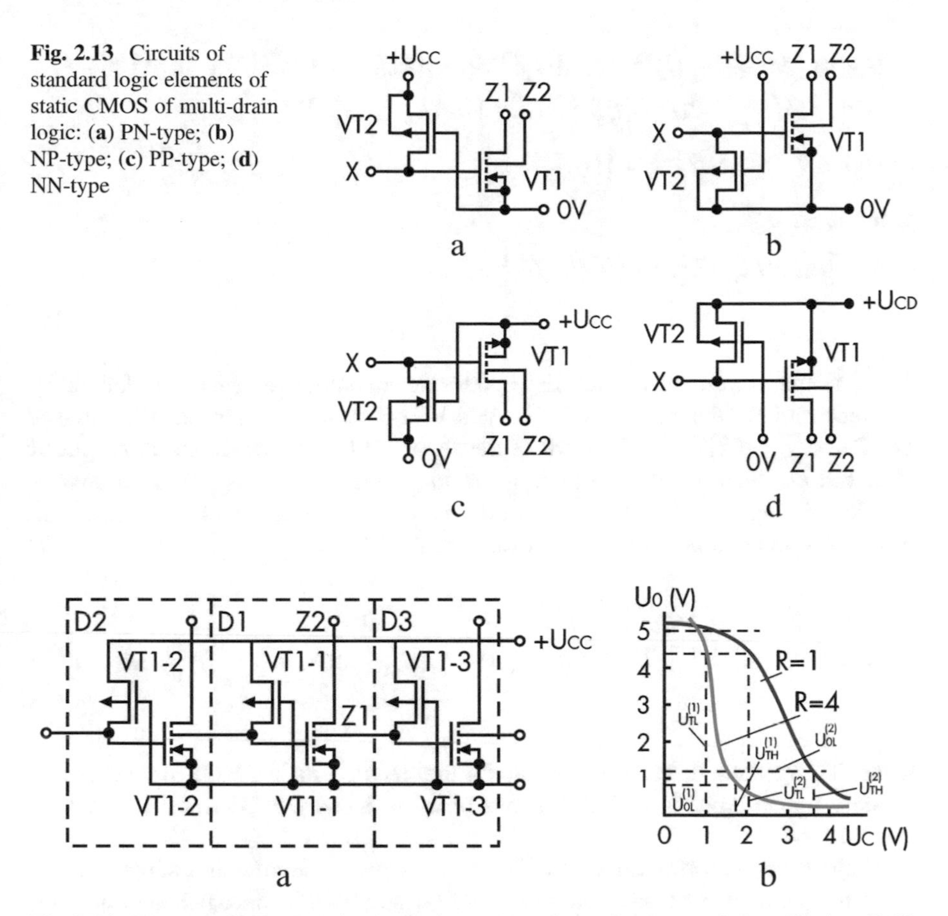

Fig. 2.13 Circuits of standard logic elements of static CMOS of multi-drain logic: (**a**) PN-type; (**b**) NP-type; (**c**) PP-type; (**d**) NN-type

Fig. 2.14 Diagram of MDL LC circuit of pn-type (**a**) and transfer characteristics of logic cell (**b**)

cell and to support the required levels of input/output signals. Here, whereas in CMOS schematics there are applied two types of MOS transistors, there exist four standard configurations of electrical circuits of MDL LC shown in Fig. 2.1 a–d and designated as PN, NP, PP, and NN type. Let's consider LC generation of such type on an example of electrical circuit incorporating three LC D1–D3 of PN type and shown in Fig. 2.14 a. Should multi-drain transistor VT1-2 preceding LC D2 is powered, then the potential at input LC D1 (gate of transistor VT1-1) is generated by p-MOS transistor VT2-1 .At $U_{IH} = U_{cc} > U_T^{VT1-1}$, transistor VT1-1 is switched ON, and at output *LC* D1, low-level output voltage is generated (Fig. 2.13).

$$\begin{aligned}
&U_{OL} \approx \{R(U_{CC} - U_T^{VT1-1}) - |U_T^{VT2-3}| - [(RU_{CC} - RU_T^{VT1-1} - |U_T^{VT2-3}|)^2 - \\
&-(R-1)(U_{CC}^2 - 2U_{CC}|U_T^{VT2-3}|)]^{1/2}\}, at U_{OL} > U_T^{VT2-3}; \\
&U_{OL} \approx U_{CC} - U_T^{VT1-1} - \left[(U_{CC} - U_T^{VT1-1})^2 - (U_{CC} - U_T^{VT2-3})/R\right]^{1/2}, \\
&\text{at}\, U_{OL} < U_T^{VT2-3}, \\
&R = [\mu_n (W/L)_n C_3^n] / \left[\mu_p (W/L)_p C_3^p\right] -
\end{aligned} \tag{2.15}$$

As it appears from the formula listed, the value of output voltage U_{OL} is defined by the dimensions of the transistors VT1-1 and VT2-3. At low level of signal at input of LC D1, $U_{IL}^{D1} = U_{OL}^{D2}\text{-}U_T^{VT2-1}$, n-MOS transistor VT1-1 is closed, and at output of logic cell D1, there is set output voltage of high level: $U_{OH} \approx U_{CC}$. Transfer characteristic of logic cell D1 for various values of constant R is shown in Fig. 2.14b. Interference immunity of logic cell:

$$\begin{aligned}
&\Delta U_T^+ \approx U_T^{VT1-1} + (U_{CC} - U_T^{VT2-3})/(R^2 + R)^{1/2} - U_{OL}^{D2}; \\
&\Delta U_T^- = U_{CC} - U_T^{VT1-1} - |U_T^{VT2-3}|/R - (2R-1)(U_{CC}^2 - 2U_{CC}|U_T^{VT2-3}|)/\left[R(3R-1)^{1/2}\right];
\end{aligned} \tag{2.16}$$

where it is apparent that it is lower than for the standard static CMOS logic cells and depends upon correlation of dimensions of n-MOS and p-MOS transistors VT1-1 and VT2-3.

High-speed performance of LC MDL in a great measure is defined by the structural parameters of MOS transistors VT1-1 and VT1-2. Since p-MOS transistor is used as the current supply, then to accelerate its activation, its dimensions should be greater. However, in this case, the value of the output voltage of low-level U_{OL} is increased, and interference immunity of ΔU_T^{+-} is degrading. Simultaneously, the increased drain current of p-MOS transistor VT2-3 affects the duration of n- MOS VT1-1 transistor switching on. Therefore, high-speed performance of LC and appropriate correlation of p-MOS and n-MOS transistors shall be defined by means of improvement methods. Since p-MOS transistor is permanently switched on, then logic cells of such type at low signal level at the input of logic cell has static consumption current P_{CCH}^C, which constitutes a significant defect in comparison with the standard static CMOS logic cell.

Generation of logic functions by means of MDL LC is done by installation combination of LC outputs. An example of circuit generation which performs the logic functions OR (NOR) is shown in Fig. 2.15. As it appears from the figure, the configuration LC NDL is similar to LC structure of pseudo n-MOS type where n-MOS transistor VT1 is a single-drain transistor. However, due to multi-output architecture LC MDL possess considerably higher functional abilities and contribute to implementation of the ICs with better high density and high-speed performance.

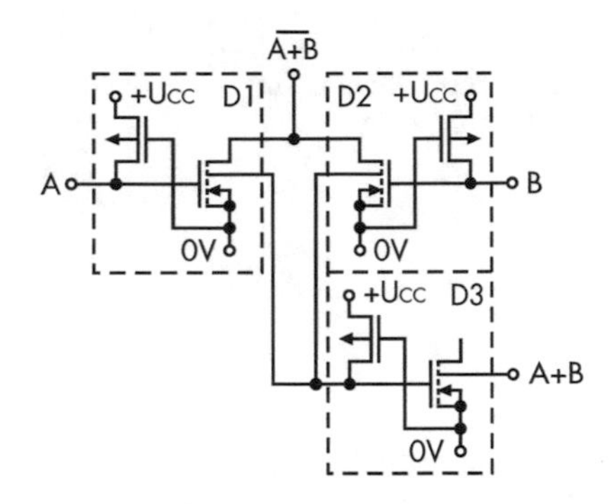

Fig. 2.15 Example of MDL LC -based circuit performing the functions of OR/NOR

Table 2.3 Comparison characteristics of LC by functional capabilities

LC type	Standardized area of LC			
	Inverter	2 NAND	2NOR	Excluding OR
Standard	1.0	1.0	1.0	1.0
PN-MDL	1.42	0.62	0.84	0.57
Pseudo n-MOS	1.42	0.85	0.84	0.60

Table 2.4 Comparison characteristics of static CMOS LC by high-speed performance

Type of LC	Standardized delay time		
	Inverter	3 NAND	3 NOR
Standard CMOS	1.59	2.44	3.47
PN-КМЛ	1.0	1.0	1.0
«псевдо» n-МОП	1.0	2.35	0.97

Table 2.5 Standardized switch-over delay time of various CVTL LC

C_S,,pF	Inverter	2NOR	3NOR	2NAND	3NAND
0	0.76	0.52	0.44	1.03	1.25
0.2	0.46	0.31	0.26	0.73	1.06
0.25	0.44	0.29	0.24	0.69	1.00

The comparison characteristics of MD004C Digital CMOS microcircuits: multi-drain logic (MDL)

LC in terms of functional potentialities and electrical parameters are shown in Tables 2.3–2.5.

When making a comparison between options of electrical circuits of standard MDL LC shown in Fig. 2.13a–d, one can make an inference that LC of PN type allow serial connection into logic circuit as well as logic cells of NP type. However, load of LC of PN type with elements of NP type and conversely is not permitted because in this case, there are no charge/discharge circuits for LC power supplies. For operation of LC of PP and NN type in the sequencing circuit, for the same reason, it is required interchange of the LC of PN and NP type. Multi-drain LC logic allows direct control of standard CMOS LC which makes it possible to arrange them on one and the same die complying at the same time with contradicting requirements of various functional blocks of microcircuits. For the purpose of such interfacing and

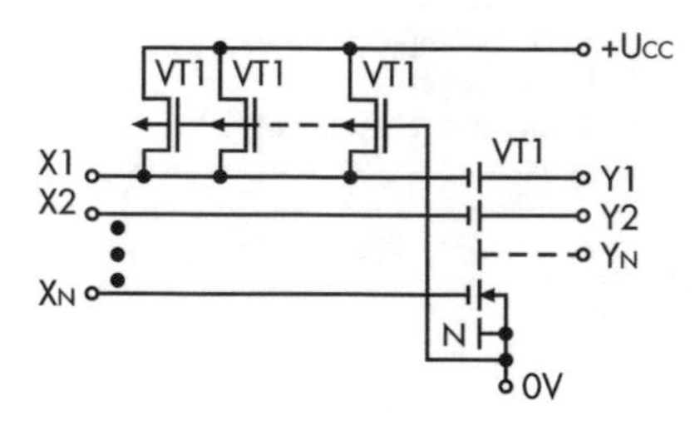

Fig. 2.16 Circuit of multi-input static CMOS LC of multi-drain logic

aligning each of output nodes of MDL LC need to be equipped with p-MOS- or n-MOS-based supply source (depending upon LC type).

Unlike standard LCs, MDL LC appear to be single-input structure which induces its less favorable functional abilities. In [1] there was suggested new type of MDL LC where development of functional abilities is attained through creation of a few extra electrically isolated inputs. Electrical circuit of such PN type LC is shown in Fig. 2.16 and incorporates multi-gate multi-drain n-MOS transistor VT1 and a few p-MOS transistors VT2-VTN, acting in the capacity of voltage generator of transistors VT1 gate offset. Assuming that to inputs $X1...X_N$ there is provided low level of signal, then outputs $Y1...Y_N$ will be closed, and load-determined high-level signal will be set at them. In case high-level signal is provided to at least one of the inputs $X1 ...X_N$ through one of the transistors VT2-VTN between source and one of drains corresponding to gate $X1 ...X_N$, conductive channel is formed.

However, since other channels controlled by other gates of transistors VT1 will be closed, then at outputs $Y1...Y_N$, high level of signal will be maintained. All the channels in n-MOS transistor VT1 will be open at high level of signal at all the inputs $X1...X_N$, and consequently at the outputs $Y1...Y_N$, low level of signal will be set. Thus, this logic cell along with multiplication of the output signal with the inversion at isolated outputs $Y1 ...Y_N$ performs the function I from input variables $Y1...Y_N$:

$$Y1 = \overline{X_1X_2\ldots X_n};\ Y2 = \overline{X_1X_2\ldots X_n};\ Y_N = \overline{X_1X_2\ldots X_N};$$

Functional capabilities may be developed in a similar way through generation of multi-input structure due to combining essential MOS transistors into cascade or stack circuit[1]. Such logic cells were denominated *Cascade Complementary Multui-Drain Logic Cells (C^2MDL)*, and an example of n-MOS transistor-based LC electrical circuit is shown in Fig. 2.17a and p-MOS transistors in Fig. 2.17b. In this case, logic cell performs the function AND of input variables with further inversion and multiplication of output signal. However, an additional characteristic of C^2 MDL LC is that it makes possible to generate transient functions in serial circuit of MOS transistors. For example, in the circuit shown in Fig. 2.17a, the major function being executed is as follows, $\overline{(A+B)\bullet B\bullet C\bullet D}$, and a transient additional function is $\overline{(A+B)\bullet BC}$. Now it is possible to reduce the number of microcircuit components and to enhance their packing density. In order to maintain the same value of LC output resistance and high-speed performance that MDL has, the

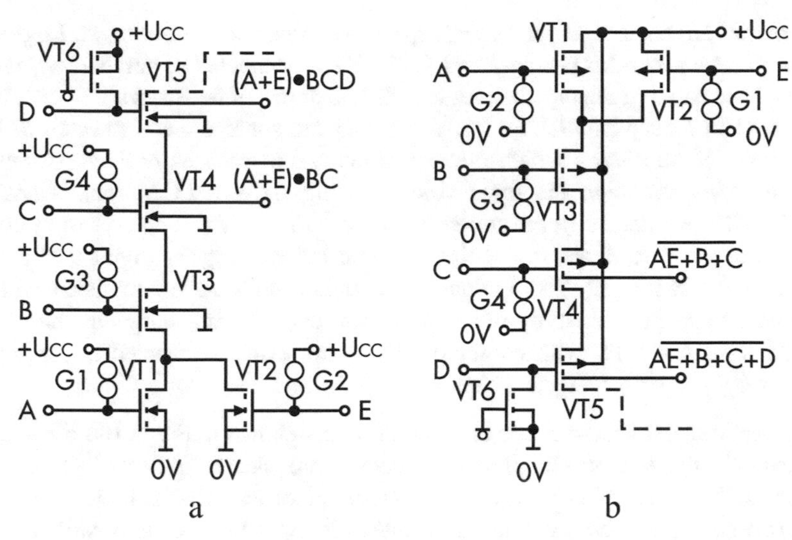

Fig. 2.17 Circuits of static CMOS LC of stack multi-drain logic on the base of n-MOS (**a**) and p-MOS (**b**) transistors

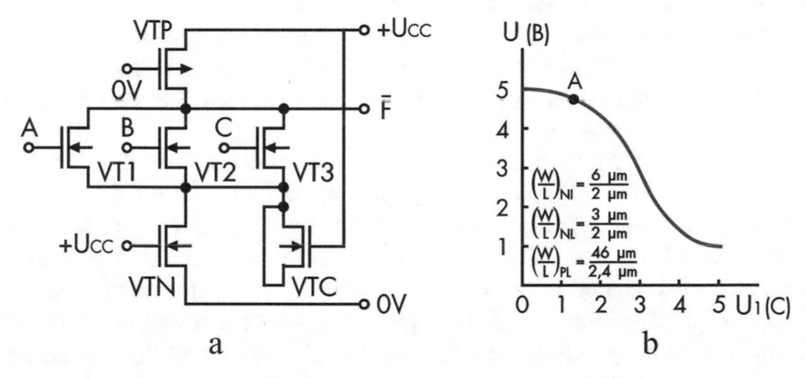

Fig. 2.18 Standard electrical circuit of static CMOS LC non-threshold transistor logic (**a**) and its transfer characteristic (**b**)

dimensions of MOS transistors need to be increased in proportion to the length of the chain formed by the transistors connected.

Complementing MOS LCs of Non-threshold Transistor Logic Microcircuits It is known that application of non-threshold concepts of functioning in bipolar schematics makes it possible to attain the maximum high-speed performance of the circuits. In [1], there was elaborated the possibility to use non-threshold concepts of operation in CMOS LCs. The standard electrical circuit of CMOS LC of *non-threshold transistor logic called CNTL* is shown in Fig. 2.18a, and by its configuration, it is similar to bipolar analogue. In CNTL schematics function NAND may be implemented through serial connection of a few n-MOS transistors; however, it presents no advantages as far as high-speed performance is concerned. Logic cell incorporates a few n-MOS transistors VT1÷VT3, connected in parallel way where the gates are forming the inputs of logic cell. The drains of transistors VT1÷VT3 are connected to load p-MOS VTP transistor, and the sources are connected to load n-MOS VTN transistors which are permanently powered as dividers of supply voltage $+U_{CC}$, reducing the logic voltage swing ΔU_T at LC output. Transistor n-MOS VTN connected in the source circuit of VT1 ÷ VT3 shall result in occurrence of negative feedback in the circuit which is impairing the high-speed performance of the logic cell. To eliminate such effect, node A is completed with an additional capacitor, shunting VTN transistor at high frequency. In the circuit n-MOS transistor VTC with combined drain and source connected to the supply source $+U_{CC}$ is performing as a capacitor.

Typical transfer characteristic of CNTL inverter shown in Fig. 2.18b has smooth character in the transient region, and there is no clearly indicated switch-over threshold. At overlapping transient characteristics of two CNTL logic cells, there exist two points of crossing which guarantee the availability of two various stable states and correct logic operation; therefore, such logic cells are called "non-threshold." Logic swing of output voltages ΔU_T is within the limits 3.5 V at U_{CC}=5 V and assures high interference immunity of CNTL circuits.

Formulae for calculation of CNTL LC switching-over delay time t_{pLH}, t_{pHL} are of very complicated format and inconvenient for practical application; therefore, in the vast majority of cases, high-speed performance is calculated at computers using some popular established models.

If parameters of MOS transistors models of standard CMOS LC are used, computation of CNTL LC high-speed performance shown in [1] provides the results summarized in Table 2.3. As it appears from the table, if LC of NAND type, application of CVTL basis does not grant any advantages in terms of high-speed performance as compared to the standard CMOS LC even at shunting capacitance $C_S^{VTC} \approx 0, 25$pF. Therefore, in CNTL logic cell in the capacity of basic on there is used function NOR, granting considerable advantage as far as high-speed performance is concerned.

Circuits CNTL LC are compatible by the levels with standard CMOS LC, which allows their combining on the same IC die and using the advantages of both types of the LC.

As load transistors, VTN and VTP have fixed dimensions, and total conductance of the block, consisting of the transistors VTI connected in parallel, may vary depending upon the code combination at LC inputs; then the logic swing CNTL LC varies in dependence upon the input signal combination. In order to make stable signal output levels and logic swing [1], it is suggested to implement load transistor VTN, VTP using controlled levels of LC output signal.

An example of electrical circuit of such CNTL LC is shown in Fig. 2.19, and it is distinguished in such a way that the gate of load transistor VTN is connected not to the supply source but to the output $\overline{F}$. The circuits of VT1 transistors drains are completed with stabilization Z block of signal output levels. In such a circuit change

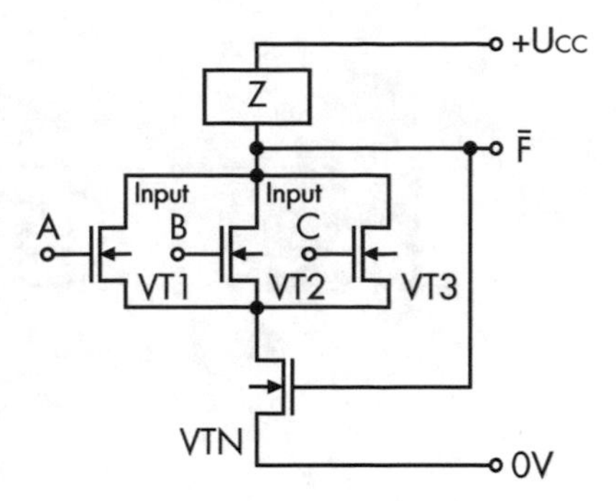

Fig. 2.19 Circuit of static CMOS LC NTL with stabilized levels of output signal

of output levels will result in change of VTN transistor conductance blocking, stabilizing output levels.

In comparison with CMOS LC standards, the vast majority of the LCs investigated have much greater high-speed performance. However, their default is revealed in presence of static consumption current that predetermined extensive use in the digital microcircuits.

Application of the static logic cells as above is based upon the efficient selection of the parameters in their totality: power on operational frequency/high-speed performance and in the first turn at high frequencies where power of static CMOS LC is comparable with consumption power of bipolar BiCMOS LC.

2.1.2 *Standard LC of Dynamic CMOS Logic*

Standard Dynamic CMOS LC Emergence of dynamic CMOS LC was dictated by the need in reducing the area taken on the die and improvement of time characteristics of standard static CMOS LCs. For this purpose, there was removed data redundancy of static CMOS LC due to the fact that inside of them output signal is generated in parallel along two complementary circuits: along circuit of n-MOS transistors and along the circuit of p-MOS transistors (Fig. 2.3).

In Fig. 2.20a, b, there are shown standard structures of dynamic CMOS LC: the first structure is on the base of logic circuit of n-MOS transistors (Fig. 2.20a), and the second one is on the base of logic circuit of p-NOS transistors (Fig. 2.20b). It follows from the figure that the standard dynamic CMOS LC have one logic circuit of transistors only and two additional clocking transistors VT1 and VT2. As a consequence, the number of transistors in N-input LC is reduced from 2N static CMOS LC to N+2 in dynamic LC, and the area taken by the LC on the IC die is reduced. Simultaneously input capacities of logic cell are reduced by half which improves the dynamic characteristics of the logic cell. For the dynamic logic cells, there should be distinguished two conceptually different phases of work: (example in Fig. 2.20a):

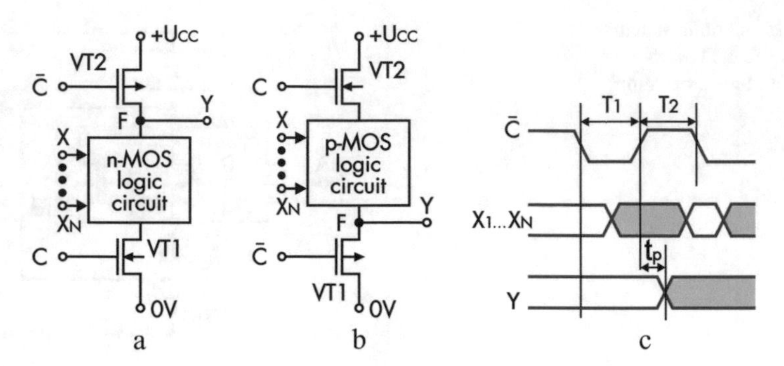

Fig. 2.20 Standard structure of dynamic CMOS LC of n-type (**a**) and p-type (**b**) and time diagram of operation of dynamic CMOS LC of n-type

Fig. 2.21 Examples of dynamic CMOS LC circuits of n-type (**a**) and p-type (**b**) performing function 2NAND

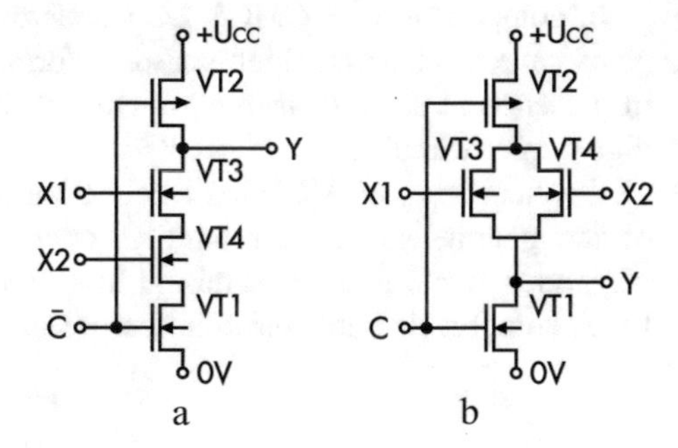

Phase T1 *(C = "L")*—phase of preliminary accumulation of charge or pre-charge (Fig. 2.20c). For this time space p-MOS transistor VT2 is a conductor, and n-MOS VT1 transistor is closed, and as a consequence node Y is charged up to U_{OH} close to $+U_{CC}$ irrespective of the state of n-MOS logic circuit.

Phase T2 *(C = "H")*—estimation phase (Fig. 2.20c). During this period, p-MOS transistor VT2 is closed, and n-MOS transistor VT1 is open. Depending upon signal levels at input $XI–X_N$, logic circuit of n-MOS transistors is either open or closed. As a consequence, this node *F* for the time t_p is discharged to low level of voltage U_{OL}, close to zero *0V*, or maintain the state of high-level voltage U_{OH}. By a similar concept, operation of dynamic CMOS LCs is effected, which uses logic circuit at p-MOS transistors (Fig. 2.20b). Examples of electrical circuit of dynamic CMOS logic cells implementing the circuit 2NOR and using the logic circuits at transistors of n-MOS and p-MOS type are shown in Fig. 2.21a, b.

High-speed dynamic CMOS LC is determined by a few factors:

(a) Load capacity which is twice as low as the standard CMOS LC
(b) Layout configuration which is formed by serial/parallel circuit of transistors of the same conductivity type and clock p-MOS transistor VT2 and n-MOS transistor VT1 connected in addition

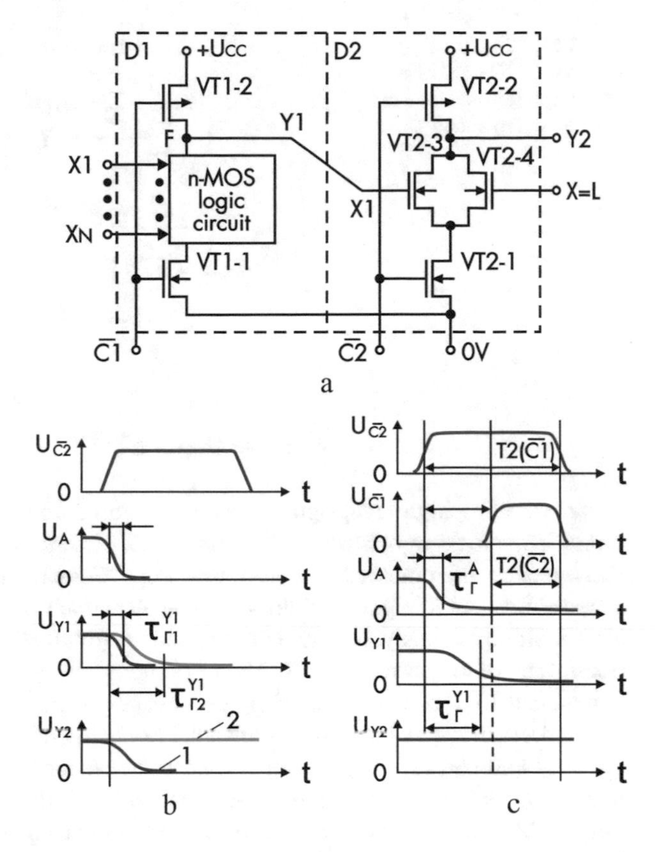

Fig. 2.22 Diagram of serial circuits of dynamic CMOS LC (**a**) and time diagrams of signals: in the nodes the circuit (**b**) and in the nodes of circuit when two clock signals $\overline{C1}$ И $\overline{C2}$ (**c**) are used.

(c) Threshold switch-over voltage which in dynamic CMOS LC is approximately equal to the threshold voltage of n-MOS (p-MOS) transistors whereas in static CMOS LCs is approaching $U_{CC}/2$

Dynamic CMOS LCs in the stable states of phases T1 or T2 as well as static CMOS LCs do not consume power. This is because the clock transistors VT1 and VT2 are operating in antiphase, i.e., one of them is closed, and the other one is open, the other way round.

As a consequence in static states, there is no circuit of consumption current flowing, and the consumption power P_{CC}^{C} of LC is insufficient. Notwithstanding obvious advantages, application of the dynamic CMOS LCs is limited by a few factors.

"Race conditions" of signal front. In Fig. 2.22a and b, there are shown serial circuits of their two D1, D2 dynamic CMOS LCs and time signal diagrams in points of circuit. Assume that within the phase of preliminary charge T1, nodes Y1 and Y2 were charged till the level $U_{OH} \approx U_{CC}$. Also let's assume that n-MOS logic circuit D1 is completely open, i.e., $X1 \div X_N = H$. High signal level at Y1 output of circuit D1 when getting to the gate of transistor VT2-3 shall result in its enabling, and consequently in node A voltage level as below is set:

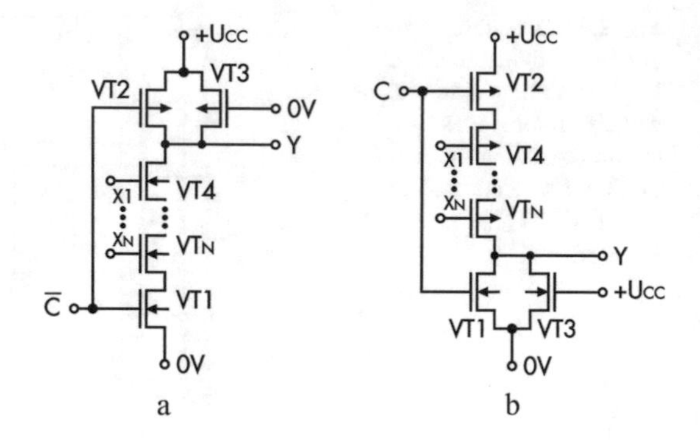

Fig. 2.23 Circuits of dynamic CMOS LC of n- and p-type for operation at low clock frequency

$$U_A \approx U_{CC} - U_T^{VT2-3} \quad (2.17)$$

At providing high-level signal to inputs $\overline{C1}$ and $\overline{C2}$ and transfer of the circuit into phase T2, clock transistors VT1-1 and VT2-1 pass into open state, and as a consequence, the potential at their drains drops. Potential drop at the drain of *VT1-1* transistor results in potential drop in node *Y1*, which is striving to close transistor *VT2-3*. Simultaneously, potential drop in node A (drain of *VT1-2* transistor) supports open state of transistor VT2-3. Value of transistor VT2-3 conductivity will be determined by the difference of signal voltages in nodes *Y1* and *A*, which in its turn is determined by the duration of signal fronts in nodes *Y1* и *A*. In the state when $\tau_\Gamma^{Y1} > \tau_\Gamma^A$ and $U_{Y1-A} > U_T^{VT2-3}$, open state of transistor *VT2-3* will lead to that some portion of the charge from the capacity of node Y2 draining into line 0V, and at output Y2 there will be set pseudo-level of output signal (graph 1, Fig. 2.22a). Therefore, for normal operation of dynamic logic cell, it is essential that $U_{Y1\text{-}A} < U_T^{VT2-3}$; capacity discharge duration of node *Y1* is less than capacity discharge duration of node *A*: $\tau_\Gamma^{Y1} << \tau_\Gamma^{A1}$.

Such state of things shall support the closed state of transistor VT2-3 in phase T2, maintaining voltage U_A (graph 2) of true signal at output Y2 (Fig. 2.22b). The most straightforward solution of this problem is application of the additional clock signals. In this case the process of transition into phase *T2* of circuit *D2* by clock signal *C2* is detained for some time *T0* during which signals *X1* are being set at input *D2* of the circuit (Fig. 2.22c). After signals *X1* are set to clock input *C2* of *D2* circuit, there is provided an additional clock pulse transferring it to phase *T2*.

However, some defects seem to be intrinsic to this method, such as need for generation of additional clock signals and restriction of signal front "racing" depth. Therefore, to solve this problem, some schematics depicted below are used.

Charge Retention Issue in accumulation nodes. At the output Y (Fig. 2.23 a,b) at the low frequency of switching there may be a loss of information (failure) because of influence of the parasitic leakage currents. Such situation may take place if at clock output C high level of signal is provided and to inputs *X1*–X_N low level. Here node Y passes into the third state, and the potential on it is determined by the charge accumulated at its parasitic capacity. Therefore, for low operational frequency of the

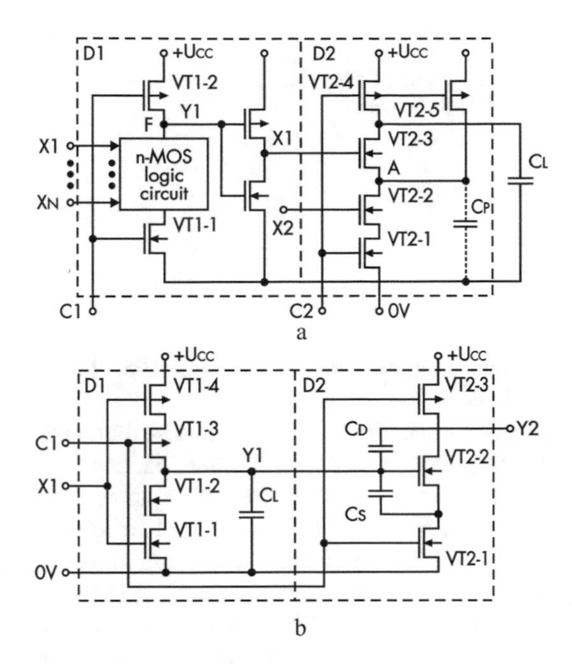

Fig. 2.24 Circuit of dynamic CMOS explaining the problem if charge re-distribution among the parasitic capacities of n-MOS transistors (**a**) and parasitic capacities of n-MOS transistors and coupling capacity (**b**)

logic cell node Y incorporates permanently open transistor with high output resistance (p-MOS VT3 for logic cell of n-type and n-MOS VT3 for logic cells of p-type) supporting the potential level in the node.

Re-distribution Problem Issue. This problem is appertaining to charge re-distribution between parasitic capacities in n-MOS (or p-MOS) transistors of logic circuit. The essence of the problem may be explained by means of the circuit in Fig. 2.23a. Assuming that within evaluation phase voltage low level is provided to input X2 and transistor VT2 is closed, voltage at input X1 shall vary its value from low level to high one. Here transistor VT2-3 is being opened, VT2-4 is being closed, and the charge accumulated at load capacity C_L is being distributed between capacity C_L and parasitic capacity C_p of node *A*. Consequently, this voltage at Y2 is being reduced to the level, $U_{OH}^{Y2} \approx U_{CC}C_L/(C_p + C_L)$, which affects the operation of logic cell circuit.

Potential solution for this problem is connection of additional charge transistors (transistor VT2-6) to intermediate nodes of serial circuit of n-MOS transistors. However, it is increasing the number of components in LC and the area occupied by LC on a die. The best possible solution is the use of separate control of charged transistor VT2-5 and discharge transistor VT2-1. While for the purpose of state evaluation discharge transistor VT2-1 performs full voltage swing from 0V, U_{CC} charged VT2-5 performs a partial swing. Consequently in the phase T1 of preliminary charge charged transistor VT2-5 is completely powered; in the evaluation phase T2, it is partially powered. Thus there may be provided an additional current to

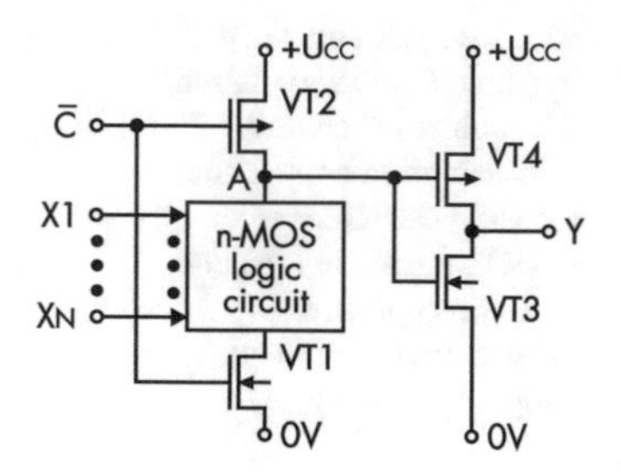

Fig. 2.25 Standard structural circuit of dynamic CMOS logic "domino"

support voltage level in nodeY2 (Fig. 2.24a) without application of additional transistors.

Another option of charge re-distribution is relating to charge re-distribution between couple capacity C_L and parasitic capacities of drain C_D source C_S of n-MOS transistors VT2-2 and is explained in Fig. 2.24b. Assuming that in phase *T1* nodes *Y1* and *Y2* were charged to level U_{OH}~U_{CC} at transition to phaseT2, transistor VT2-2 of circuit *D2* is going into open state. Here the charge accumulated at capacity C_L and is being re-distributed between capacity c C_L and capacities C_D, C_S, which results in reduction of voltage level on node Y1, degradation of interference immunity, and possible operation disruption.

Considering the presence of major defects in the standard structure of dynamic CMOS LC, there have been suggested a range of new schematic solutions for dynamic CMOS LC where no said defects and faults are found.

Dynamic CMOS LC of "Domino" Type Dynamic CMOS LCs of "domino" type represent an improved updated type of two-phase dynamic CMOS logic cells being clocked by one clocking sequence of signals [1]. Application of such logic cells simplifies MP microcircuit clocking and contributing to more comprehensive application of dynamic properties of logic cells. Major LC electrical circuit of "domino" type is shown in Fig. 2.25 and is distinguished from standard circuit of dynamic LC by availability of output static inverter at transistors VT3 and VT4. As a consequence, the number of the components for N-input logic cell is increased up to N = 4. As LC output there may be used inverter Y only, and output A in n-MOS logic cell is connected to inverter input only. The high voltage level at the point AN-MOS of the logic circuit at the time of the phase T1 ensures emergence at the output Y of the low level of the logic signal.

Finally, over the phase T1, all the LC inputs connected to LC outputs "domino" are in the state of low level; therefore, the transistors of load n-MOS logic circuits of LC are switched off. Therefore, at transition to T2 phase, there is prevented undesirable discharge of output nodes A and generation of false levels of output voltages in the LC. All the nodes of LC in phase T2 perform one transition only and maintain the set level until the next phase T1 of preliminary charge occurs. The behavior of LC circuit reminding, in fact, the behavior of domino line has defined the name of this LC type.

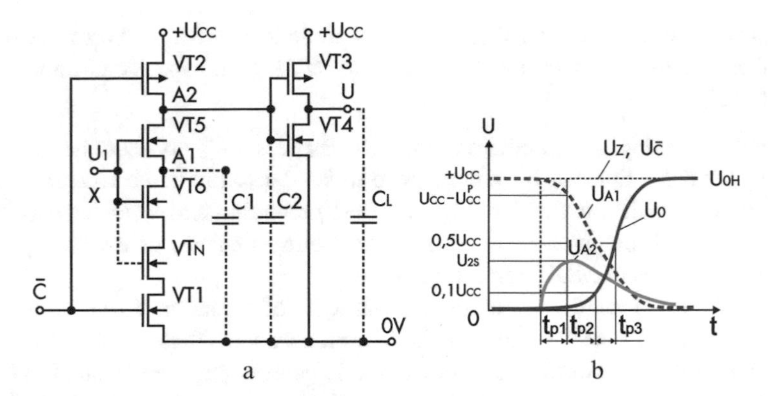

Fig. 2.26 Dynamic circuit of CMOS LC "domino" for the worst case of operation (**a**) and time voltage diagrams in the circuit nodes (**b**)

The function of the standard circuit of dynamic CMOS LC "domino" is explained in Fig. 2.26 where there is shown LC circuit for the worst case of operational mode and time diagrams (b) of LC operation in T2 phase.

In the phase of preliminary charge T1 at low voltage level at clock input C parasitic capacity C1 of mode A1 is charged up to level $U_A \approx U_{CC}$ through open p-MOS transistor VT2. Here the voltage at LC output drops to level $U^Y_{OL} \approx 0$. Assuming that the LC under study is operating in the serial circuit of similar LC, at input X, there will be observed low voltage level closing n-MOS transistors VT5-VTN. At providing high voltage level to clock input C, the LC goes into phase T2. Assuming here that the voltage at input X of LC also goes to high level.

The circuit of serially connected transistors VT5-VTN is open, and parasitic capacity C1 start discharging through this circuit. At the start of discharge (diffusion) mode, upper transistor VT5 operates in saturation mode, whereas the rest operates in the linear area until the voltage U_A attains the value below $+U_{CC}$. In the area located higher than this point, all the transistors VT5-VTN shall be operating in the linear region. Here, there occurs charge re-distribution between parasitic capacities C1 and C2 which attains saturation state (U_{2S}). This level shall be maintained until all the transistors VT5-VTN adhere to linear mode.

Transistor Vt5 works as constant current supply discharging C1, and voltage U_{A1} varies linearly with time. Delay t_p of output signal U_O in relation to input U_I at level $0.5U_{CC}$ is normally represented as the one consisting of three:

$$t_p = t_{p1} + t_{p2} + t_{p3}; \tag{2.18}$$

where *tpl*–time over which p-MOS transistor, *VT3*, shall attain the conductance state.

This time depends upon the primary charge re-distribution between C1 and C2.

t_{p2}–time over which output voltage U_O shall attain level $0.1U_{CC}$ after attaining the state of VT3transistor conductance. This time shall be defined by primary charge re-distribution between C1 and C2 and also by voltage change rate in A node.

t_{p3}–time over which output voltage U_O shall attain level $0.5U_{CC}$. This time shall be defined by p-MOS parameters of transistor VT3, load capacity C_L, and voltage value U_1.

Notwithstanding the popularity enjoined by this LC, no accurate model for defining the delay time has been developed so far. Therefore in the process of digital microcircuit simulation for defining the most preferable and efficient parameters, there are applied computer-aided tools for computing the delay time with application of popular models of MOS transistors [1].

Whereas in the logic cell, there exist one type of transition, from the low level state to high level state, in either of the LC circuit nodes, there are no pulse signal punching (cutting through); therefore, all the logic cells may be switched over from the phase of preliminary charge T1 into phase T2 by one and the same clock signal C front. It simplifies clocking of LC in the microcircuits and contributes to better and more comprehensive use of LC dynamic properties (in the circuit there is no dead time between the output state of one LC and operating of the next LC).

As the phase *T2* and unlocking of the logic circuit of the n-MOS transistors become evident with the unlocking of the lower synchronizing transistor VT1, the logic element has the switching threshold, close to the threshold voltage of the transistor VT1. As at the LE input the static inverter is activated, then the LE has the output levels of signals, similar to the static CMOS LEs. Besides, this inverter isolates the output of the logic circuit (node *A1*) from the output *Y* and can ensure the large value load currents. Similar to the regular dynamic CMOS LEs, the "domino"--type circuits have the static consumption power P^C_{CC} missing.

The main deficiency of the dynamic LEs of the "domino" type is that they all are non-inverting and cannot ensure formation of the entire system of logic functions. However, as the LEs of the "domino" type are compatible with the standard static CMOS LEs, then during formation of the complex logic circuits, this deficiency is eliminated by introduction of the additional inverter.

In some cases, in the digital integrated circuits, there may be evident the static or low power functioning of the dynamic "domino" LEs. Meanwhile, there takes place the level loss of potential in the "floating" nodes of the LE circuit (such situation arises in node *A1*, when to the synchroinput $\overline{C}$ the high level of the signal is applied and to the inputs $X1$-X_N the low level). In this case, in node *A1*, they introduce the permanently activated p-MOS transistor VT10, fixing its potential (Fig. 2.27a). The version of the LE circuit "domino" is possible, in which the charging transistor VT2 is excluded while keeping the permanently activated transistor VT10. In such a circuit, the dual-phase operational mode is ensured by the lower n-MOS transistor VT1.

The deficiency of the CMOS LE "domino," as in the other logic with the non-complementary circuit structure, is in the feature that it can generate only one output function (while the group functions are quite often realized in the shape of the logic tree). For elimination of this deficiency, they use the CMOS LE circuit of the type "domino" [1] with multiple outputs. The essence of the structure is explained on the circuit of the signal transfer formation for the adder (Fig. 2.27b). The basic

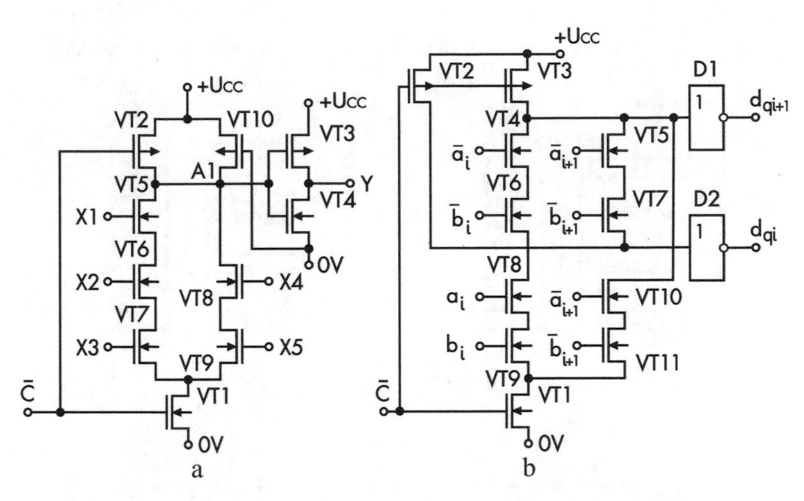

Fig. 2.27 Improved electric diagrams of the dynamic CMOS LEs "domino" for functioning at the low synchronization frequency (**a**) and with multiple outputs (**b**)

concept of LE is that the intermediate functions are used of the logic tree of the n-MOS transistors, thus protecting the circuit from splitting. As the total economy in the number of the components depends on reduction of the number of splittings, then the actual advantage of this LE "domino" depends on the recursiveness of the realized logic functions and especially evidently in the circuits of ALU and adders. As the intermediate nodes of the logic tree are additionally charged (p-MOS transistor VT2 Fig. 2.27b), then such circuits in comparison with the regular LEs "domino" possess the improved dynamic characteristics.

Dynamic CMOS LEs with the Preliminary Charge In the dynamic CMOS blocks, there is no problem of the signal "race," if during the phases each from the LE inputs maintains a stable state. For this at the outputs of the dynamic LEs, there may be set the dynamic memory elements. The logic elements of such type are called the dynamic CMOS logic with the preliminary charge [1]. The LE circuits of this type during application of the logic blocks of the n- and p-type are indicated in Fig. 2.28a, b. During application of the low-level signal to the synchroinput of the LE n-type circuit, the transistor VT1 is closed, VT2 is open, and at the output *A* of the n-MOS logic circuit via the transistor VT2 independently on the logic state of the inputs $X1$-X_N, the signal high level is set. Therefore the transistor VT3 gets open, and the transistor VT5 remains closed.

However, as at the gate of the synchronizing transistor VT4 of the memory element the low level signal is in evidence, the gate is closed, and the output Y of LE is in the third state. Therefore, at it during the phase of the preliminary charge *T1*, the signal level is maintained, earlier stored in its parasitic capacitance. During transition to phase *T2* (Fig. 2.28b) and application of the high-level signal to input *C* (Fig. 2.28a), the transistor VT1 gets open, and VT2 gets closed, and in dependence on the levels of signals at inputs $X1$-X_N, the n-MOS logic circuit either gets open or

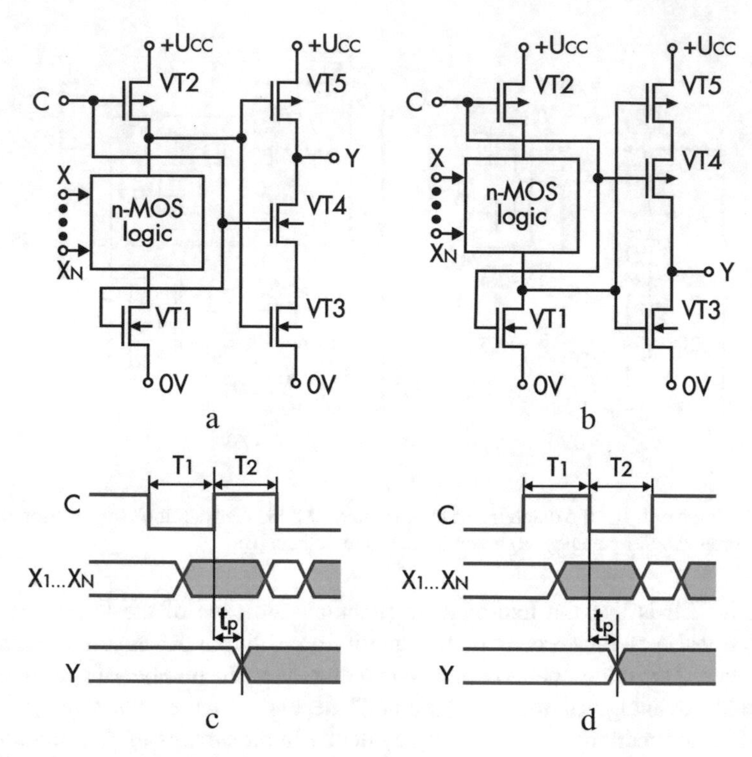

Fig. 2.28 Diagrams of the CMOS LE of n-type (**a**) and p-type (**b**) with the preliminary charge and the time charts of their operation (**b**)

maintains its closed state. During its unlocking, the voltage in node A drops, opens the transistor VT5 of the memory element, and closes the transistor VT3. As during this the synchronizing transistor VT4 of the memory element gets open, then at the LE output, the high level of the signal is set. With the closed n-MOS logic circuit, the high level of the signal at the gate of the synchronizing transistor VT4 opens it, and at the output Y of LE, the low level of the signal is set. While applying the new low-level synchropulse, the memory element switches over to the storage mode of the transferred information to the output Y and the logic block to the phase of the preliminary charge $T1$. By the similar principle also functions the p-type LE (Fig. 2.28b). The time chart of its operation is indicated in Fig. 2.28c and differs only by the feature that the phase of the preliminary charge $T1$ takes place with the high level of the signal at the input C.

Dynamic CMOS LEs with the Differential Structure

It is possible to expand the potentialities of the dynamic CMOS LEs by means of application of their differential-type structure [1]. The diagram of LE of such type (Fig. 2.29a) includes CMOS LE and differential stage potential circuit (DSPC) with complementary inputs $X1...X_N$, $\overline{X1}\ldots\overline{X_N}$ and outputs Y, $\overline{Y}$. The levels of potentials in the nodes Y, $\overline{Y}$ are firmed by means of the synchronizing transistors VT1, VT2,

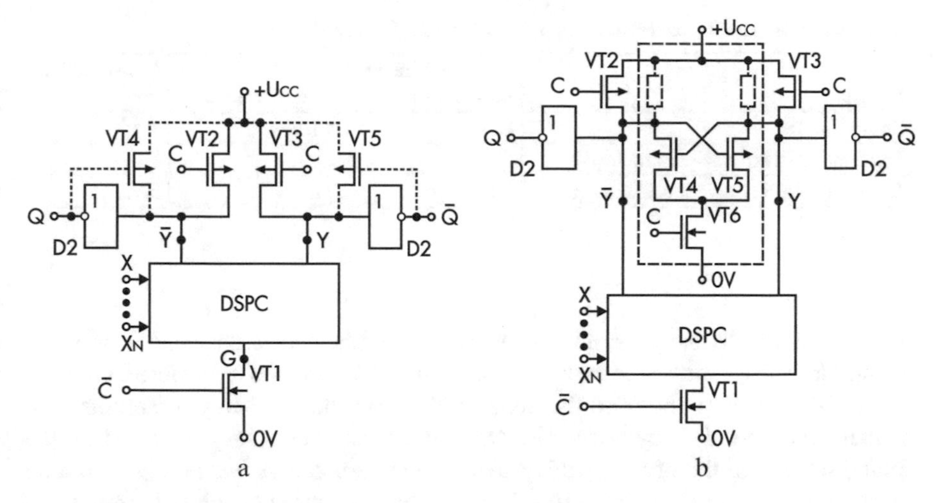

Fig. 2.29 The layout diagram of the dynamic CMOS LE of the differential type

and VT3 by the synchrosignal C and are transferred to the outputs Q, $\overline{Q}$ by means of the inverters $D1$ and $D2$. The functioning of such circuit is similar to the dual-phase dynamic CMOS LE "domino"; however, there are two output signals formed in it: the direct Q and inverse $\overline{Q}$. For prevention of the loss of levels in the "floating" nodes Y, $\overline{Y}$ with the low-frequency functionality, the circuit has two transistor clamps introduced of the p-MOS type of the transistor VT4 and VT5.

One of the substantial deficiencies of such circuit is a large voltage drop in nodes Y, $\overline{Y}$. This determines the great time of discharging the capacitances of nodes Y, $\overline{Y}$, while in order to switch over the output voltage inverters the voltages in these nodes should be reduced below the level of $U_{CC}/2$. One of the approaches making it possible to improve performance of LE of such type is in bringing down the voltage drop in the nodes Y, $\overline{Y}$. This approach is determined by the electric diagram of LE, indicated in Fig. 2.29b, and received the identification of the CMOS differential logic with setting by the reference standard (sample-set differential logic SSDL) [18]. The logic element contains the differential stage potential diagram; three standard transistors VT2, VT3, and VT6; the setting transistor VT1; the additional discharge circuit of the flip-flop type on the transistors VT4 and VT5; and two buffer inverters $D1$ and $D2$. Functioning of LE takes place in two stages: the reference sample $T1$ and the setting of $T2$. In phase $T1$ the synchrosignal $\overline{C}$ is in the high level C; in the high level, transistors VT1, VT2, and VT3 are activated and VT6 deactivated. Meanwhile, the capacitance discharge circuit of one of the nodes Y, $\overline{Y}$ is formed by one of the circuits of the differential stage potential circuit. As a result, one from the nodes Y, $\overline{Y}$ will charge to the voltage level, close to $+U_{CC}$, and the second of the smaller $+U_{CC}$, determined by the dimensions of the transistors. With transition to phase $T2$, transistor VT1 is deactivated, and VT6 gets activated, which results in activation of the discharge circuit on the transistors VT4 and VT5. With the activated transistors VT2 and VT3,

Table 2.6 Comparative characteristics of LE of the SSDL type

LE type	Number of rated nominal values	Performance	Consumption power
Static CMOS	28	10	1.75
Dynamic with differential structure	17	12	1.67
Dynamic differential of the *SSDL* type	22	8	3.72

activation of the discharge circuit ensues discharge of the node capacitance Y, ($\overline{Y}$) with the smaller voltage via one from transistors VT4 and VT5 to the level close to zero.

The high level in one from the nodes Y, ($\overline{Y}$), and the low level in the other one switch over the output inverters *D1, D2* into the complementary states. Thus, the final discharge of the node capacitance Y, ($\overline{Y}$) to zero occurs not through the long sequential circuit of the n-MOS transistors of the differential circuit but through the discharge circuit of VT4 and VT5. The dimensions of these transistors can be performed as large, as required, which improves durations of the junctions in the nodes Y, $\overline{Y}$ and enhances performance of LE. The comparative characteristics as per the performance of the different dynamic LEs with the function $Q = \overline{A} \bullet \overline{B} \bullet \overline{C} \bullet \overline{D} + \overline{A} \bullet (B + C + D)$ are provided in Table 2.6.

Opposite to other types of LE in the considered circuit, performance does not depend on the number of the consequently connected transistors. This is related to the fact that the difference of levels in the nodes Y, $\overline{Y}$ constitutes 300–400 mV. However, availability of the additional discharge circuit augments approximately by two times the LE consumption power. As at the LE outputs are set the static CMOS inverts D1,D2, then dynamic CMOS LES with the different structure are compatible with the standard CMOS LES as per the levels.

The performance enhancement at the expense of reduction of the logic voltage swing ΔU_T with maintenance of the noise immunity is possible by means of the LE circuit, described in [2] and called the dynamic differential CMOS logic on the "pass" transistors—CDPTL (CMOS differential pass-transistor logic). An example of the LE electric diagram of such type, performing the function of multiplexor 4 for 1, is represented in Fig. 2.30. The circuit is formed by several parallel circuits sequentially connected n-MOS transistors, and its functioning is based on passing two symmetrical circuits of the n-MOS transistors of the differential input variables $Y0 \div Y3$, $\overline{Y}0 \div \overline{Y}3$ in dependence on the control signals $X0$, $\overline{X0}$, $X1$, and $\overline{X1}$. Staging CDPTL LEs is possible by means of the sequential connection of several circuits; however, for the degradation reduction of the levels of signals along the sequential circuit and minimization of the quadratic dependence of the delay time of the signal passing on the number of the connected in sequence n-MOS transistors, it is necessary to activate via the regular intervals of the buffer circuits.

Apart from the degradation reduction of the signals in the circuit and besides the increasing of the LE output currents by means of the output buffer circuits, the LE synchronization comes into effect. A number is known also of other dynamic CMOS

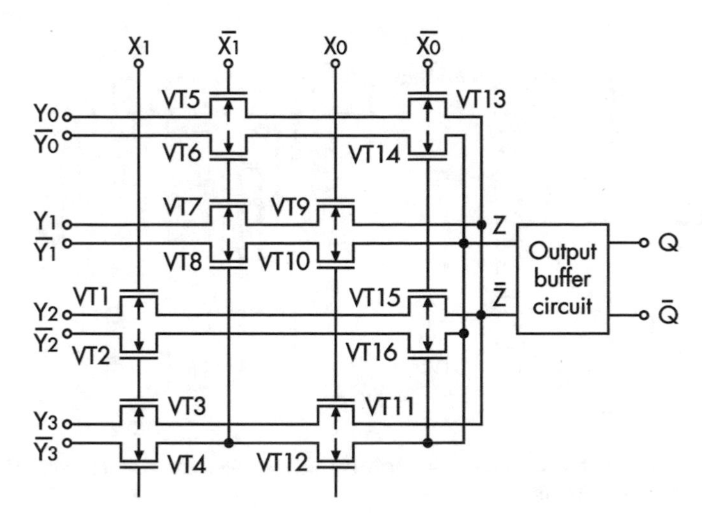

Fig. 2.30 Diagram of the dynamic CMOS logic of the differential type on the "through-type" transistors

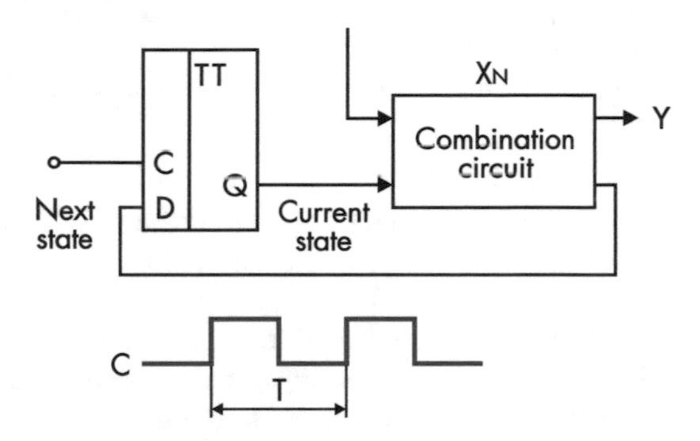

Fig. 2.31 Diagram of the dynamic CMOS circuit with the single-phase clocking

LEs, such as the dynamic CMOS LEs of the multi-drain (DCML) and cascaded multi-drain logic (DC^2ML) [1] dynamic CMPS LEs without the "races" as per the type NORA [2] and others [3]; however, they are hardly practically applied and are rather of a theoretical interest.

A simpler system, occupying a minimum area on the die and simplifying synchronization of the integrated circuits, is a single-phase system, applied both in the static and dynamic LEs [8]. A most elementary circuit diagram with the single-phase clocking is indicated in Fig. 2.31 and contains the combination circuit (logic) and the memory element (D_t-flip-flop). By the positive edge of the synchrosignal *C*, the following state of the combination circuit becomes the present state and generates the new next state. As the D_t-flip-flop is clocked by the edge and is "non-translucent" for the incoming information *D*, then the signal of the state cannot pass through the recombination circuit more than 1 time for a period of *T*. Thus, the data shift to such an extent is controlled by the clocked memory element. In Fig. 2.32a is presented the diagram of the dynamic edge synchronized D_t-flip-flop, with application of nine

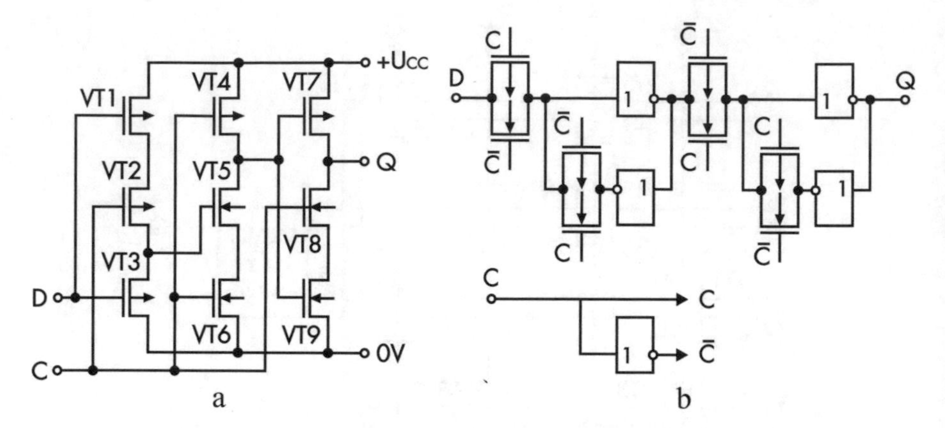

Fig. 2.32 Diagram of the dynamic (**a**) and static (**b**) memory elements, edge-clocked, of the system with the single-phase clocking

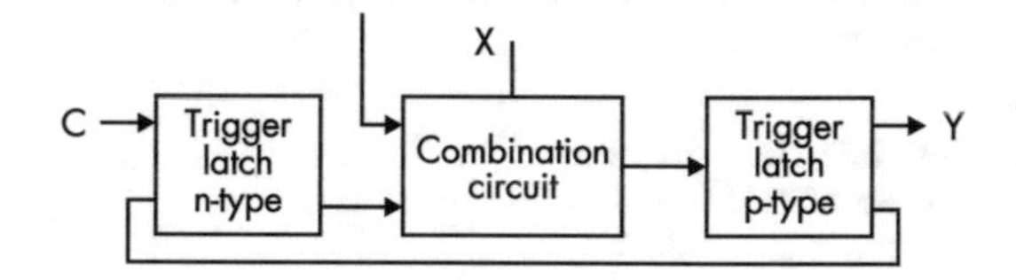

Fig. 2.33 Diagram of the dynamic CMOS system with the single-phase clocking on the basis of the dynamic D-flip-flops of the "latch" type

transistors and one line of synchronization. With the necessity of application of the static D_t-flip-flop, one can use the traditional CMOS circuit of the D_t-flip-flop on the basis of the signal transmission switches, indicated in Fig. 2.32b; however, such a circuit has a large number of components. In application, instead of one D_t-flip-flop, edge-clocked, two D-flip-flops of the "latch" type, it is possible to attain a higher performance of the circuit and to cut down the number of components (Fig. 2.33). However, during application of the traditional CMOS circuits-"latches" the single-phase clocking does not enhance the rate of performance, and such a circuit requires the multiphase clocking. The reason for this is in the necessity of creating two various points in each cycle of synchronization for storage of the signal. Therefore, for prevention of the "races" in each flip-flop-"latch," there should be formed a period of securing the signal. However, this task can be solved by creation of two flip-flop- "latches," which are "translucent" for the input signal as per the various parts of the synchrosignal, i.e., as per the high and low levels. In Fig. 2.34a the diagram is indicated of such D-flipflop, "translucent" for the output signal D as per the high level of the synchrosignal, in Fig. 2.34b by the low level.

This task can be solved also with application of the dynamic LEs with the preliminary charge. The diagram of such circuit is indicated in Fig. 2.35 and is based on two LEs: the n- and p-type. Thus, the n-block is in the phase of the preliminary charge, when the synchrosignal C is at the low level, and it evaluates the state of its stable inputs X_n; when the synchrosignal C is at the high level, the p-block is in the phase of the preliminary charge at the high level of the

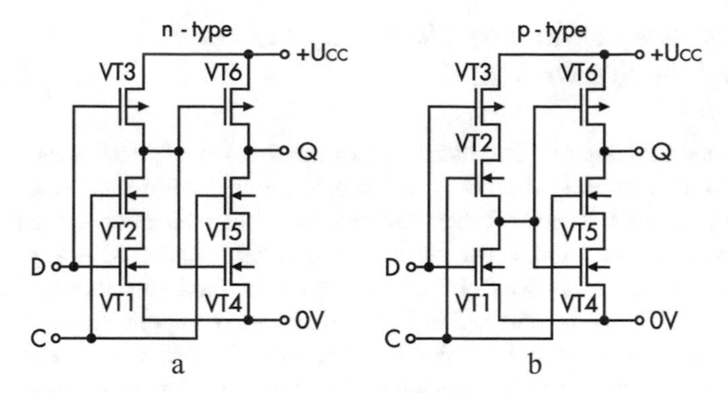

Fig. 2.34 Diagrams of the dynamic D-flip-flops of the "latch" type

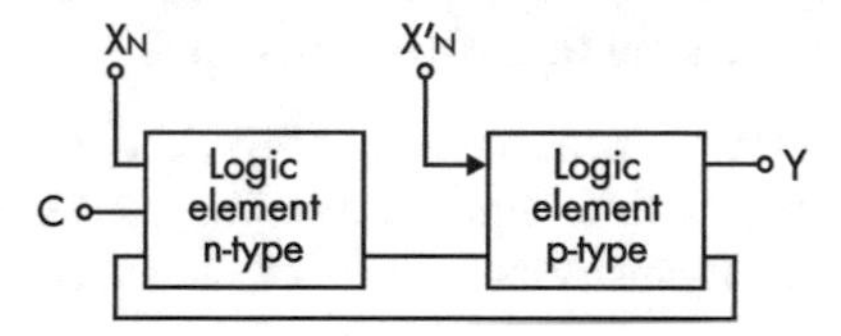

Fig. 2.35 Diagram of the dynamic CMOS system with the single-phase clocking on the basis of the dynamic CMOS LEs with the preliminary charge of the p- and n-type

synchrosignal *C* and in the evaluation phase at the low level of the synchrosignal *C*. Therefore, both n- and p-blocks can jointly use one synchrosignal *C*. Meanwhile, in the evaluation phase, the p-block monitors the state of the n-block and alters the state of the system's outputs, irrespective of the state of the n-block inputs. The synchronization principles of the mixed structures are indicated in [1].

2.2 Memory Elements of the Digital CMOS Integrated Circuits

In the CMOS integrated circuits, they use the memory elements (ME) on the basis both of the most elementary bistable cells and the more sophisticated flip-flops of the D-type. However, application of the CMOS element base with the most low level of the static consumption power and the schematic versions of the CMOS ME are more varied because of the possibility of application in MEs of both the static and dynamic principles of information storage.

2.2.1 Memory Elements, Clocked by the Level of the Synchrosignal

Static Memory Elements. The basic type of the memory element of the given class in the digital integrated circuits is the synchronous JD-flip-flop, clocked by the synchrosignal level. The most elementary method of realization of such memory element is its construction as per the known diagrams, indicated in [1] with application of the static basic CMOS LEs. The diagram of the D-flip-flop on the basis of LE of the NAND type is indicated in Fig. 2.36. However, application of ME of such type in the digital integrated circuits is of low efficiency due to a great number of components and the large area, occupied on the die. Therefore, in a number of designs for application in the microprocessor of the high complexity integrated circuits, there were proposed the improved versions of the electric diagrams of the D-flip-flops, clocked by the level of the synchrosignal.

The diagram of the memory element of type 1 (Fig. 2.37) is constructed on the basis of two dynamic switches (the first, input one on the transistors VT1–VT4 and, the second one, of feedback, on transistors VT5–VT8) and uses the dual-phase signal of synchronization $C, \overline{C}$.

Suppose to the input C was applied the high level of signal, to the input $\overline{C}$ the low level. Meanwhile transistors VT2 and VT3 are open, and the first switch functions as a regular inverter, and the input signal from the input $\overline{D}$ is transferred via node A and LE D to the output $\overline{Q}$ in the direct form. In this mode, transistors VT6 and VT7 are closed and isolate transistors VT5 and VT8 of the feedback switch from node A.

During the phase variation of the synchrosignals (C to the low level, $\overline{C}$ to the high level), transistors VT2 and VT3 get closed and disconnect the input switch from the input D, and in node A on the parasitic capacitance, the last level of signal is maintained. Meanwhile, transistors VT6 and VT7 are activated, and the feedback switch together with LE D forms the bistable static cell, in which the level is stored in node A, and ME switches over to the storage mode. With the purpose of cutting down the number of components in the circuit of ME in the function of the feedback inverter, it is possible to apply the static inverter (transistors VT5 and VT6; Fig. 2.37b) [6].

However, in such diagram, for switching from the low-level state to the high-level state, it is necessary that transistors VT1 and VT2 should be capable of switching over the current, released by the activated transistor VT6, and, vice versa, during transition from the high-level state to the low-level state, the current, released by

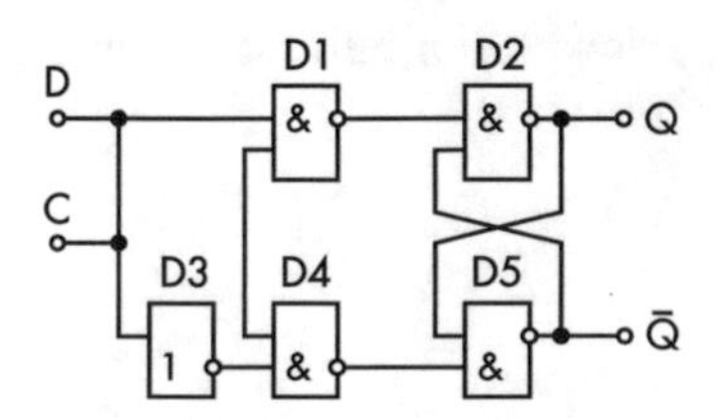

Fig. 2.36 Diagram of ME on the basis of LE of the NAND type

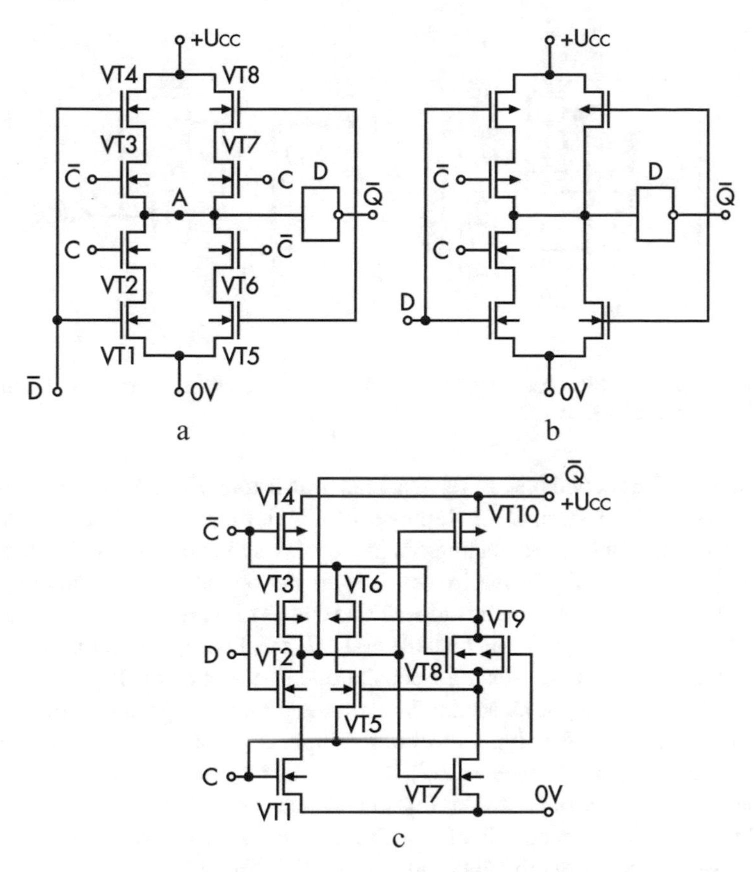

Fig. 2.37 Diagrams of MEs on the basis of the dynamic switches: (**a**) basic; (**b**) with application of one static inverter; (**c**) with the reduced consumption power

transistors VT3 and VT4, should be sufficient for switching over the enabled transistor VT5 to the saturation area. For this purpose, transistors VT1–VT4 have larger sizes, than from VT5 and VT6.

However, such a method of controlling ME is characterized by the enhanced values of the consumption current I_{CC} in the switch-over mode and consumption power *Pcc*.

The circuit of CMOS ME with the reduced consumption power is indicated in Fig. 2.37b [7]. For this purpose, in the output inverter, an additional "pass-through" switch is introduced on transistors VT8 and VT9. With application of the low-level signal to input $\overline{C}$ and the high level to input C (the mode of the through signal transfer to the output Q), this switch isolates the transistors VT10 and VT7 of the output inverter, disrupts the feedback circuits of the bistable cell of ME, and thus reduces the current, consumed by ME.

The ME diagram of type 2 [5], indicated in Fig. 2.38a, is constructed on the basis of two switched "pass-through" switches (the first one, in the input one on transistors

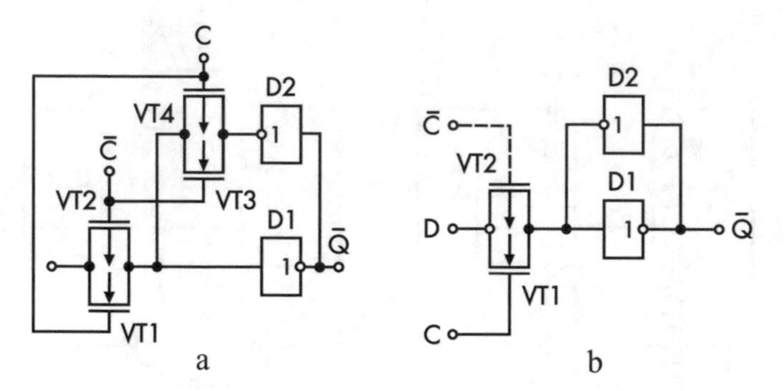

Fig. 2.38 Diagrams of MEs, based on the "pass-through" switches: (**a**) basic; (**b**) with application of one "pass-through" switch

VT1 and VT2; the second one, of feedback, on transistors VT3 and VT4 and uses the dual-phase synchronization $C, \overline{C}$). Suppose the high level of signal is applied to input C, while to input $\overline{C}$ the low-level one. Then the input switch is open and transfers the signal from the input D via the inverter $D1$ to the output $\overline{Q}$ in the inverse form. Meanwhile, the feedback switch is closed and cuts off inverter $D2$ from node A and thus disrupts the feedback in the bistable cell $D1$, $D2$. During variation of the phase of the synchrosignals C, $\overline{C}$ for the opposite one, the input switch gets closed and isolates node A from input D. Meanwhile, on the parasitic capacitance of node A is stored the last value of the signal level of the input D. Simultaneously the feedback switch gets open, and the inverters $D1$, $D2$ form the bistable static cell, in which the signal level is stored of node A. The flip-flop switches over to the storage mode. It is possible to simplify the circuit of the D-flip-flop by means of exclusion of the feedback switch [8] as per the diagram, indicated in Fig. 2.38b.

However, for switch-over of such diagram, it is required that the load capacitance of the circuit, controlling the input D, as well as of the transistors of the switch *VT1*, *VT2*, should be sufficient for commutation of the output currents of the low I_{OL} and high I_{OH} levels of the inverter $D2$. For this purpose, the sizes of transistors VT1, VT2, as well as those of the circuit, controlling input D, are made with the dimensions, exceeding the sizes of the transistors of the inverters *Dl*, *D2*. Sometimes in the input switch for simplification of synchronization, they exclude the p-MOS transistor *VT2* (the dotted line indicated in the figure), and synchronization is performed by one synchrosignal C. However, as the signal level in node A at output D is reduced for the value U_T^{VT1}, the switching threshold U_T^{D1} of the inverter $D1$ for the reliable switching is made reduced:

$$U_T^{D1} < U_{CC}/2; \tag{2.19}$$

The ME diagram of type 3, having gained the widest popularity in the literature, is indicated in Fig. 2.39a and represents the ME combination of two types. At the input, circuit D contains "the pass-through" switch on transistors VT1 and VT2, and in the feedback circuit the dynamic switch on transistors VT3 and VT6. Suppose to the

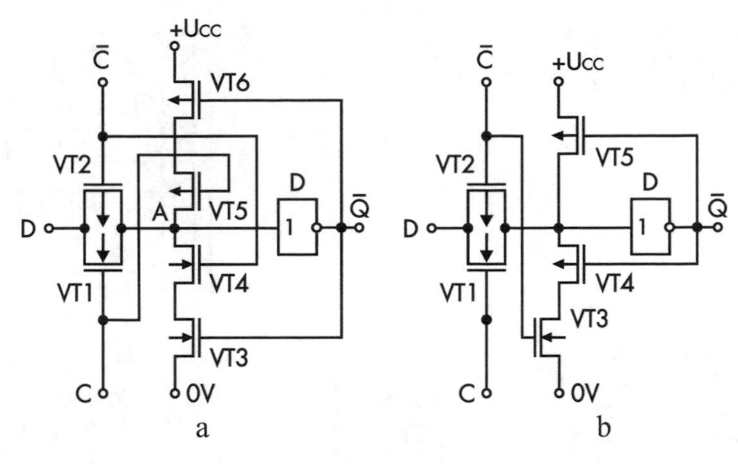

Fig. 2.39 ME diagram of the combined type: (**a**) basic; (**b**) simplified version

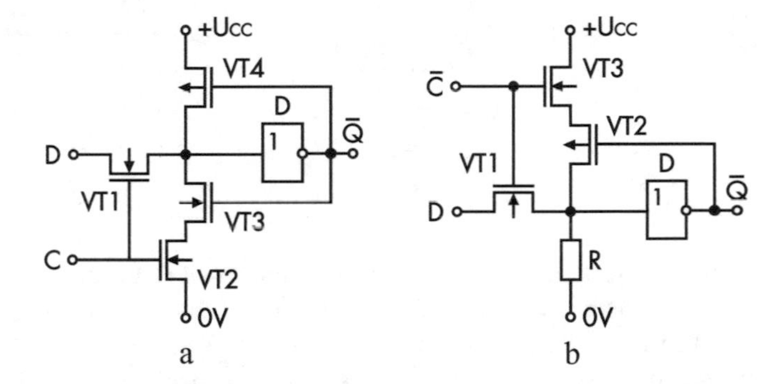

Fig. 2.40 The ME electric diagrams with one forward (**a**) and inverse (**b**) synchrosignal

input C is applied the high-level signal and to the input $\overline{C}$ the low level one. Then the input switch is open, and the signal from the input D in the inverse form through the inverter $D1$ is transferred to the output $\overline{Q}$. Meanwhile, transistors VT4 and VT5 are deactivated and isolate transistors VT3 and VT6, the feedback switch from node A. During change of phase of the synchrosignals C, C for the opposite one, the input switch gets closed and isolates node A from input D. During this time on the parasitic capacitance of node A, the last signal level is stored of the input D. Simultaneously, transistors VT4 and VT5 get open, and the activated feedback switch together with the inverter $D1$ form the bistable static memory cell. Then the D-flip-flop switches over to the storage mode. Sometimes with the purpose of simplifying the circuit, they exclude one from the synchronizing transistors (Fig. 2.39b) [1].

It is possible to simplify synchronization of the integrated circuit blocks by means of the D-flip-flops, using one synchroinput C ($\overline{C}$). The diagrams of such type, the first one from them, is synchronized by the forward signal C, and the second one by the inverse $\overline{C}$, which are indicated in Fig. 2.40a and b. In the diagram [30] in

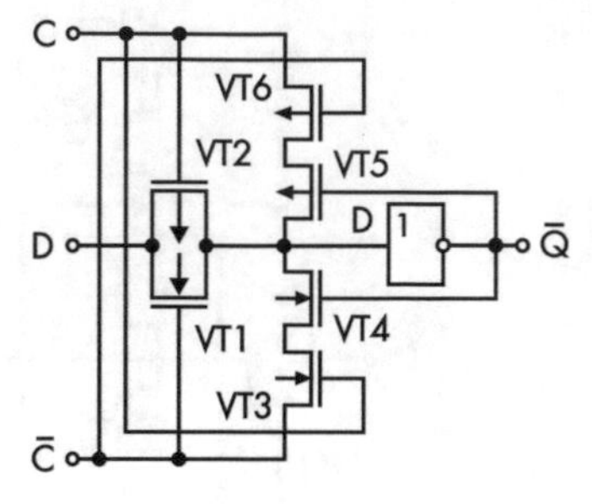

Fig. 2.41 The ME diagram of the combined type with the reduced consumption power

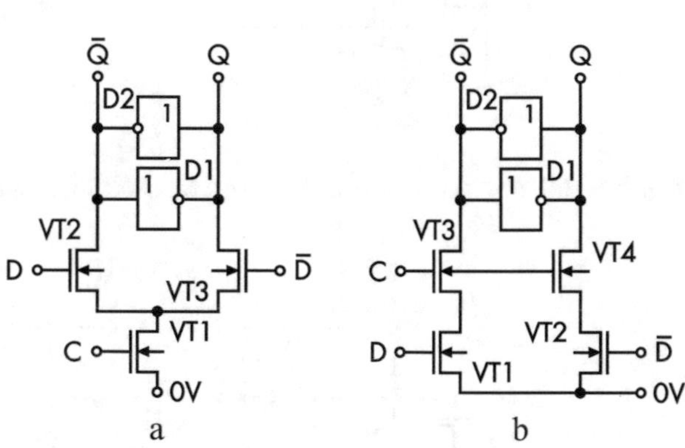

Fig. 2.42 The basic ME diagram with the differential input stage (**a**) and altered synchronization diagram (**b**)

Fig. 2.40a during application of the high-level synchrosignal *C*, the transistor VT1 is open and ensures transfer of the signal from the input *D* to the output $\overline{Q}$ via the inverter *D*. Transistor VT2 gets closed and cuts off VT3. The transistor VT4 promotes the regenerative activation of LE *D* and enhances the voltage level in node A up to the level U_{CC}, lowered by the input transistor VT1. With application of the low-level synchrosignal *C*, the transistor VT1 gets closed, locks the signal transfer from the input *D*, and maintains on the parasitic capacitance of node A the last value of the signal. Meanwhile, the transistor VT2 is activated, and the pair of transistors VT3, VT4 together with the inverter *D1* form the static bistable cell, maintaining the last state of node A. Also the ME diagram functions similarly, indicated in Fig. 2.40b, during application of the synchrosignal $\overline{C}$ of the opposite polarity [1].

Figure 2.41 shows the electric ME diagram of type 3, possessing the reduced power consumption level. This is attained by cutting off the input signal circuit in the transfer mode: synchroinput $\overline{C}$ (*C*)—synchronizing transistor of the feedback inverter VT5(VT8)—the information input *D*. For this the source of the transistor VT6 is connected to the synchroinput *C* and the source of the transistor VT3 to the synchroinput $\overline{C}$ [2].

The electric diagram of the D-flip-flop of type 4 has become quite popular in the ME systems of the "pipeline" type, as ensuring the maximum performance (Fig. 2.42a) [3]. The circuit contains the bistable cell on the inverters *D1, D2* and the synchronized differential input stage on the transistors VT1–VT3. With the high

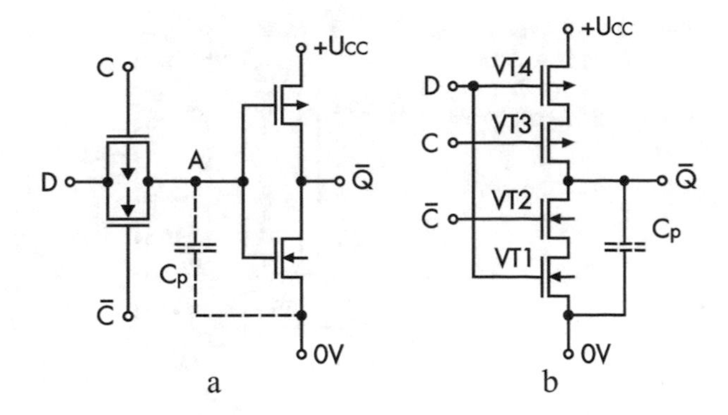

Fig. 2.43 Basic electric circuits of the dynamic MEs on the basis of the "pass-through" (**a**) and dynamic (**b**) switches

signal level at the synchroinput *C*, the synchronizing transistor VT1 is open, and the signals from the inputs $D, \overline{D}$ are transferred to the outputs $Q, \overline{Q}$. In parallel with the transfer the signal is written into the bistable cell *D1, D2*. For the stable ME operation the transistor sizes of the inverters *D1, D2* should be smaller than the sizes of the transistors VT1–VT3. With application of the low-level signal to the input *C*, the transistor VT1 closes the input transistors VT1, VT2 and locks the signal transfer from the inputs $D, \overline{D}$. The flip-flop switches over to the storage mode and to the outputs $Q, \overline{Q}$ arrive the levels of signals, written in the bistable cell *D1, D2*. As per the similar principle functions also the electric diagram of the flip-flop, indicated in Fig. 2.42b [34] and differing with the altered synchronization circuit.

Dynamic Memory Elements. The limit low leakage currents of the CMOS integrated circuit components make it possible to use in them the memory elements, constructed on the principle of the dynamic information storage (dynamic MEs). Such MEs have a minimum number of components, and they are especially suitable for application in the microprocessors of the integrated circuits of the "pipeline" type.

The versions of the D-flip-flops, level-clocked, are indicated in Fig. 2.43 a, b) and are described in detail [5]. The diagram in Fig. 2.43a is designed on the basis of the static inverter on the transistors VT3, VT4 and the "pass-through" switch on the transistors VT1, VT2. With the low level of the synchrosignal at the input *C*, and the high level at the input $\overline{C}$, the "pass-through" switch is opened, and the informational signal from the input *D* in the inverse form arrives to the output. With the phase alteration of the signals $C, \overline{C}$ for the opposite ones, the switch gets closed, and in node *A* on the parasitic capacitance of the C_p inverter, the last level of the signal is stored for some time, holding up the state of the converter, which forms the output signal $\overline{Q}$. With the voltage level alteration on the capacitance *C* (at the expense of charge/discharge by the leakage currents), the second-level regeneration is required.

The version of the diagram in Fig. 2.43b differs with the fact that the flip-flop is constructed on the basis of the dynamic switch, and, as the element, storing the level, the load capacitance is used of the C_p flip-flop. The basic deficiency of the indicated circuits of the dynamic D-flip-flops is the necessity of having two synchrosignals:

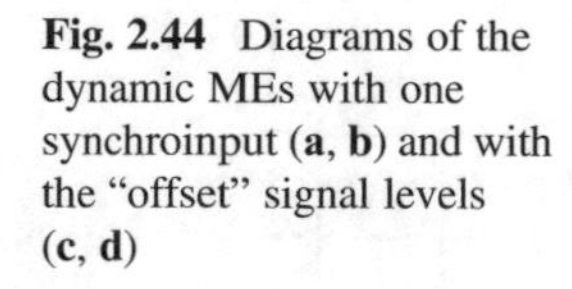

Fig. 2.44 Diagrams of the dynamic MEs with one synchroinput (**a**, **b**) and with the “offset” signal levels (**c**, **d**)

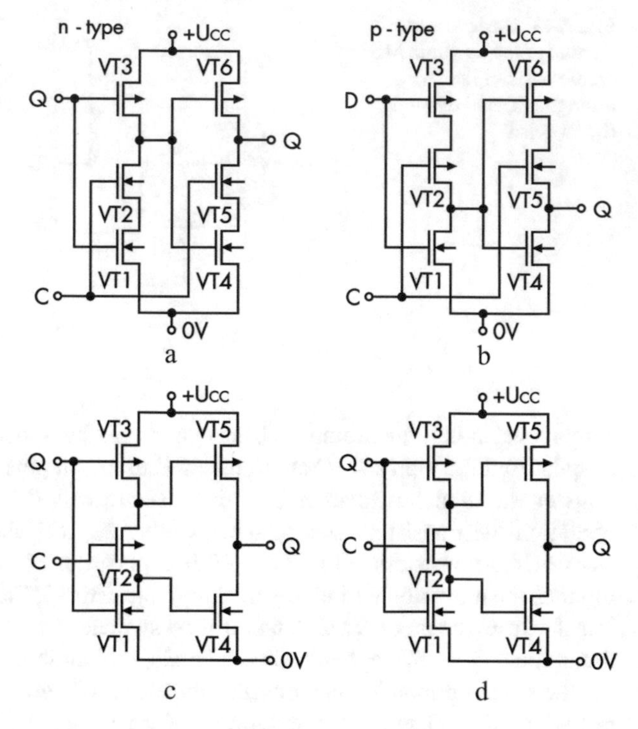

the forward C and the inverse $\overline{C}$. For simplification of synchronization, one can use the dynamic D-flip-flops, clocked by one synchrosignal C. The versions of diagrams of such type are reviewed in [6], and their electric diagrams are indicated in Fig. 2.44a, b. The first from the diagrams is constructed on the basis of the combined in series two dynamic inverters of n-type. With the high level of the synchrosignal at the input C, the synchronizing transistors VT2, VT5 of the inverters are open, and the information signal from the input D is transferred in the direct form to the output Q. With the phase alteration of the synchrosignal for the opposite one, the synchronizing transistors VT2, VT5 are closed, and in the flip-flop, the level of the last transferred signal is maintained. The second diagram is constructed on the basis of the dynamic inverters of p-type and functions similarly to the first one.

For simplification of the considered circuits in the second inverters, it is possible to exclude the synchronizing transistor VT5, resulting in the output inverter becoming static.

Such MEs are known to be dynamic with the “offset” levels [6], and the versions of the n- and p-type electric diagrams are indicated in Fig. 2.44c, d; the functionality is similar to the above-reviewed diagrams. In such MEs, the capacitance is twice reduced at the input, thus enhancing their performance.

2.2.2 Memory Elements, Clocked by the Synchrosignal Edge

The basic type of ME of the given class from the digital integrated circuits is the D_t-flip-flop, clocked by the synchrosignal edge. The most convenient form of realization of such ME is the known diagram of the D_t-flip-flop, having the structure of "three flip-flops," and making use of the static CMOS LEs. The diagram of such a flip-flop with application of LE of the NAND type is indicated in Fig. 2.45, and its functionality and parameters are considered in [1].

However, because of a large number of components, application of ME of such type is also low effective in ME of the high complexity integrated circuits. Therefore, the most widely used ME, clocked by the edge, is the structure of the type "M-S" (master/slave). This structure presupposes the serial connection of two edge-clocked D-flip-flops. Meanwhile, the clocking phase of the first flip-flop (M) is opposite the clocking phase of the second flip-flop (S). The first from these circuits is indicated in Fig. 2.46a. With the low level of the signal at the input *C*, the high $\overline{C}$ at the input, the master flip-flop *T1* functions in the mode of the signal transfer of input-output $\overline{Q}$, the slave one *T2* in the storage mode of the previous information. With the phase alteration of the synchrosignal for the opposite one, the master D-flip-flop switches over to the information storage mode, and the slave one T2 transfers information from the master flip-flop to the output *Q*.

The circuits of the D_t-flip-flop, edge-clocked and using the single-phase synchronization, are indicated in Fig. 2.46b. In this circuit for the purpose of ensuring the

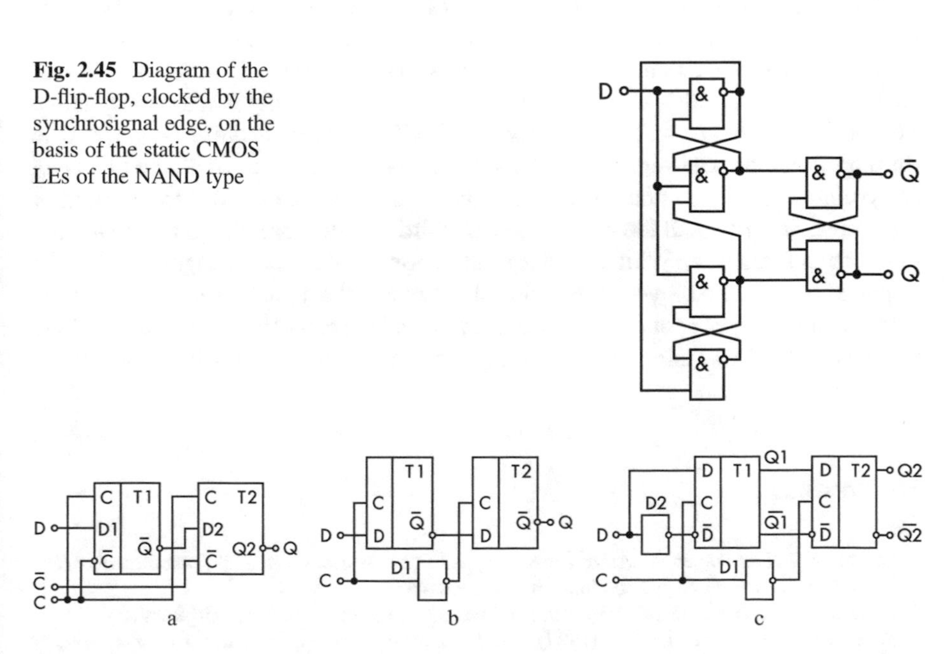

Fig. 2.45 Diagram of the D-flip-flop, clocked by the synchrosignal edge, on the basis of the static CMOS LEs of the NAND type

Fig. 2.46 Basic diagrams of D-flip-flop, edge-clocked of the M-S type: (**a**) with dual-phase synchronization C, $\overline{C}$; (**b–c**) single-phase synchronization C; (**c**) with application of the D-flip-flops

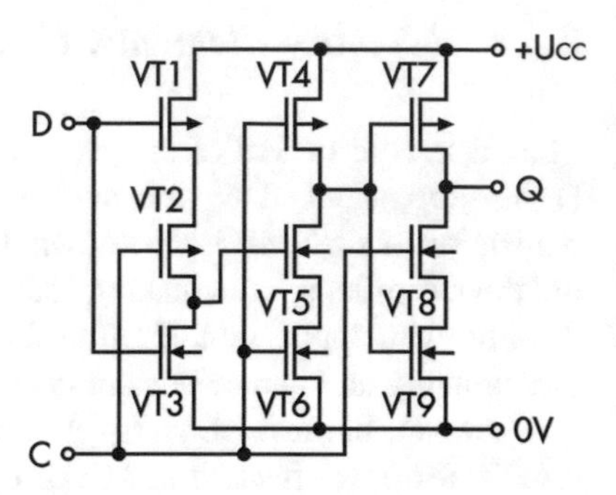

Fig. 2.47 Electric diagram of D-flip-flop, edge-clocked on the basis of the dynamic CMOS LEs

counter-phase synchronization of the master T1 and the slave T2 flip-flops, an additional inverter D1 is introduced into the synchronization circuit. When used as the master and the slave flip-flops, having the differential inputs $D, \overline{D}$, it is necessary to introduce an additional inverter D2 also at the input *D* ME (Fig. 2.46c). As the master and slave flip-flops T1, T2 of the considered MEs, one can apply the above-viewed circuits of the static D-flip-flops. When using the microprocessor structures of the “pipeline” type, it is possible to use the dynamic circuits of the level clocked D-flip-flops. However, taking into consideration that fact that the master D-flip-flop T1 is used for the temporary information storage, and the slave flip-flop for the static storage, it is possible to use the structures of the mixed type with the appropriate synchronization. In such structure, the master D-flip-flop is made dynamic and the slave one static.

With application of the dynamic CMOS LE, it is possible to construct the edge-clocked D_t-flip-flops of the more elementary structure, more responsive and containing a smaller number of components. The diagram of the D_t-flip-flop of such type is indicated in Fig. 2.47 and uses three sequentially connected dynamic inverters (the first one, on the transistors VT1–VT3 of p-type and the second and the third one, on the transistors VT4–VT6 and VT7–VT9 of n-type). With the low level of synchrosignal *C*, the first inverter is open and transfers information to the input of the second inverter, and the second and the third inverters are closed and retain the previous information. With the phase alteration of the synchrosignal *C* for the opposite one, the first inverter gets closed; however, the signal level at its output is retained at the expense of the input capacitance of the second inverter. This level via the open second and third inverters is applied to the output *Q* of ME.

References

1. Belous, A. I., Emelyanov, V. A., & Turtsevich, A. S. (2012). *The schematics fundamentals of the microwave devices* (472 pp). Technosphere (in Russian).
2. Alekseenko, A. G. (2002). *Fundamentals of microschematics.* Fismatizdat (in Russian).
3. Kazennov, G., & Kremlev, V. (1987). *Semiconductor integrated circuits. Under edition of Ya. Koledov.* Higher Education.
4. Frike, K. (2003). *Introductory course of digital electronics.* Technosphere (in Russian).

5. Ugryumov, E. P. (2002). *Digital schematics*. BVKh-Petersburg (in Russian).
6. Waykerly, J. (2002). *Designing digital devices*. Post Market.
7. Mkrtchyan, S. O., Melkonyan, S. R., & Abgaryan, R. A. (1990). Schematics for protection of digital LSIs from overloads. *Electronic Machinery Microelectronic Devices, 4*(82 Series 10), 30–34.
8. Belous, A. I., & Yarzhembitsky, V. B. (2001). *Schematics of the digital integrated circuits for the information processing and transfer systems* (116 pp). UE Technoprint. ISBN 985-464-064-7 (in Russian).

Chapter 3
Schematic Technical Solutions of the Bipolar Integrated Circuits

This voluminous chapter is dedicated to elaborate analysis of schematic solutions of bipolar digital microcircuits (Schottky transistor-transistor logic, ECL, I2L, etc.) Such close attention given in this book to the bipolar microcircuits may be explained by the fact that these microcircuits along with CMOS microcircuits are extensively applied in modern smart systems to control defense systems and military and space hardware exhibiting generally higher tolerance to diverse ionizing radiation and high loading capacity.

3.1 Digital Integrated Circuits on the Bipolar Transistors with the Schottky Diodes

As it is known, the bipolar and field transistors perform similar functions: they operate in the electric circuits or as a linear amplifier or as an electronic switch.

The advantages and the shortcomings of the bipolar and field transistors flow from the physical phenomena, lying in the basis of their operation. Thus, the operation of the bipolar transistors is based on the phenomenon of injection of the minor charge carriers into the base via the forward-biased emitter transition: with the alteration of the input control current the flow of the injected charge carriers, which results in the alteration of the output current. Thus, *the bipolar transistor is current controlled,* its input resistance is small, and the output current is ensured by the carriers of both signs (holes and electrons).

The work of the field transistors is based on the *phenomenon of the field effect.* Under alteration influence of the voltage, applied to the gate, conductivity of the channel changes, which results in the value alteration of the output current. Thus, *the field transistor is controlled by voltage*, and its input resistance is large, as the input circuit is isolated from the output one with dielectric. The output current of the field

A. Belous, V. Saladukha, *The Art and Science of Microelectronic Circuit Design*,
https://doi.org/10.1007/978-3-030-89854-0_3

transistors is ensured with similar sign carriers (or holes or electrons in dependence on the type of conductivity of a channel).

However, the MOS-transistors are not perfect. First, due to a rather high resistance of the channel in the closed state, the voltage drop on the open MDS-transistor is noticeably greater than the voltage drop on the saturated bipolar transistor. This shortcoming is aggravated by the fact that the temperature dependence of the channel resistance is stronger, than dependence of the saturation voltage of the Bipolar transistor on temperature (channel resistance of the open MDS-transistor within the temperature range of 25–150 °C is increased 2 times, the bipolar transistor saturation voltage approximately 1.5 times). Second, the MOS-transistors have the smaller value of the limit temperature, equal to 150 °C, and as for the silicon bipolar transistors, the limit operating temperature may reach 200 °C. This fact limits application of the MDS-transistors in the operational modes with the stepped-up temperature of the ambient medium. The bipolar integrated circuits are also more resistant to influence various ionizing effects (radiation); therefore, they are widely used in the weaponry systems and military hardware.

3.1.1 Basic TTLS Logic Elements of Digital Integrated Circuits

As the basis logic elements of the transistor-transistor logic with the Schottky diodes (TTLS) in the integrated circuits are used the various versions of the TTL circuits with a simple inverter. The simplest electric circuit of TTLS of a basic logic element is indicated in Fig. 3.1 and includes two transistors with the Schottky diode (one of which is a multi-emitter transistor (MET)) and three resistors. Analysis of functioning and static and dynamic parameters of such logic element is indicated [1, 2]. A specific feature of the given circuit is a low switching threshold voltage:

$$U_{\mathrm{T}} = U_{BE}^{VT1} - U_{CE}^{VT2} = 0,75\text{-}0,2 = 0,55\mathrm{A}$$

This determines the high response and the low noise resistance of the logic elements and requires during application in the integrated circuits of the additional elements of matching with the external devices. The input characteristic of a logic element is indicated in Fig. 3.2a, and the logic element is characterized with the low-level input current:

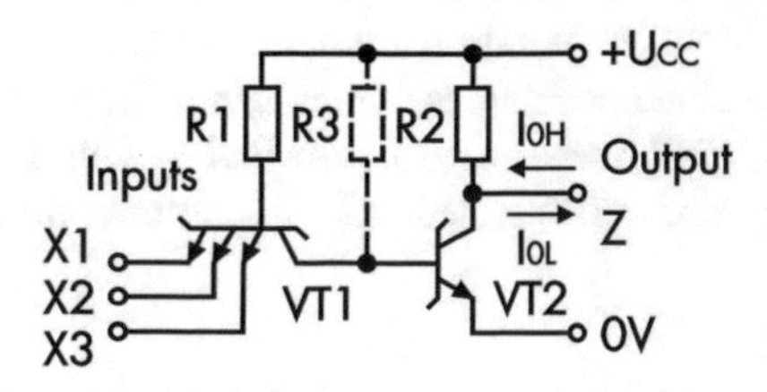

Fig. 3.1 Circuit of the basic TTLS logic elements

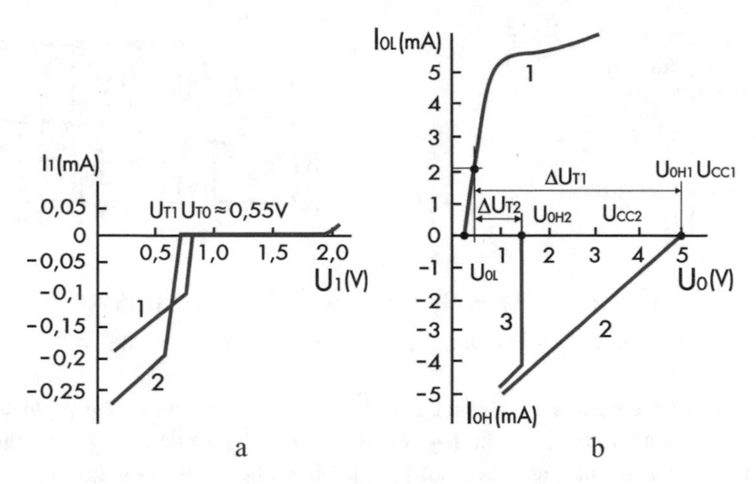

Fig. 3.2 Input (**a**), output (**b**) static characteristics of the basic TTLS logic element

$$I_{IL} = \frac{U_{CC} - U_{BE}^{VT1} - U_I}{R_1}.$$

The high-level input current I_{IH} is equal to the leakage current of the collector-emitter junction of the VT1 transistor with the Schottky diode.

The output characteristics are listed in Fig. 3.2b, and in the open state, they correspond to the curve 1 and in the closed state to the curve 3. In the open state, the voltage U_{OL} at the output of the logic element is as follows:

$$U_{OL} = U_{CE}^{VT2} + \left[\frac{U_{CC} - U_{CE}^{VT2}}{R_2} + I_{OL}\right] \cdot r_K \approx 0,25 \div 0,3B;$$

where: U_{CE}^{VT2}–residual collector-emitter of the open output transistor VT2 at I_C=0

r_C–output resistance (collector resistance) of the open output transistor VT2

In the closed state, the output voltage U_{OH} of the logic element is determined by the value of the supply voltage U_{CC}:

$$U_{OH} = U_{CC} - I_{OH} \cdot R_2;$$

and at $U_{CC} = 5B$; $U_{OH} = 5B$.

Comparing the values of the output voltages U_{OH}, U_{OL} with the value of the input threshold voltage U_T, it follows that the logic element has the substantially varying margins of noise resistance to the positive and negative noises:

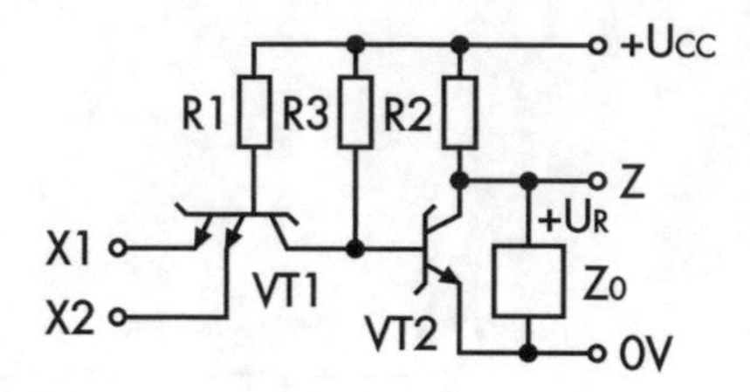

Fig. 3.3 TTLS LE schematic diagram with limitation of the output levels

$$\Delta U_T^+ = U_T - U_{IL} = U_T - U_{OL} = 0.55 - 0,3 = 0.25\ \text{A};$$
$$\Delta U_T^- = U_{IH} - U_T = U_{OH} - U_T = 5.0 - 0,55 = 4.45\ \text{A}.$$

This difference results in the fact that the junction duration of the input voltage from the high state to the low one t_{HL} is substantially higher than the swing duration t_{LH} from the low to the high state. In [2, 3] it is shown that the delay times of the TTLS logic element in the approximated shape can be represented as follows:

$$t_{PHL} \approx \left[C_C^{VT2} + C_S^{VT2} + (C_L + C_P)/\beta_N^{VT2}\right] \cdot \frac{U_{OH} - U_I}{I_Б^{VT2}};$$
$$t_{PLH} \approx \left[C_C^{VT2} + C_S^{VT2} + C_L + C_P\right] \cdot \frac{U_I - U_{OL}}{U_{CC}/R_2 + I_{OL}}$$

where:C_C^{VT2} –capacitance of the collector junction of the VT2 transistor

C_S^{VT2} –capacitance of the VT2 Schottky transistor junction
C_P –isolation parasitic capacitance of the VT2 transistor and R2 resistor
C_L –load capacitance

From these formulae it follows that the time periods of t_{PLH}, t_{PHL} depend on the voltage swings at the input of the logic element (LE); therefore, owing to the great differences between ΔU_T^+ and ΔU_T^-, the values t_{PLH}, t_{PHL} will substantially differ. The simplest means for reduction of this difference is bringing down the supply voltage U_{CC}, making it possible to reduce the output voltage $U_{OH.}$ However, with absence of the resistor *R3*, the minimum values of the voltage U_{CC} are limited with the following level:

$$U_{CC} > U_{BE}^{VT2} + U_{CB}^{VT1}.$$

Introduction of the resistor R3 (Fig. 3.1) makes it possible to reach reduction of the supply voltage value down to the level of $U_{CC\mu\iota\nu} = U_{BE}^{VT2}$ and to cut down difference in the values of the time delays of t_{PLH}, t_{PHL}. However, during this, there is a substantial increase of the low-level input current I_{IL}. Another quite popular means of limiting the high-level output voltage U_{OH} and ensuring the symmetrical relative to the threshold switching over U_T of the input influences U_{IL}, U_{IH} is application at the output of the logic element of the limitation circuit Z_O (Fig. 3.3). The versions of the limitations circuits can be different, and they are chosen on the

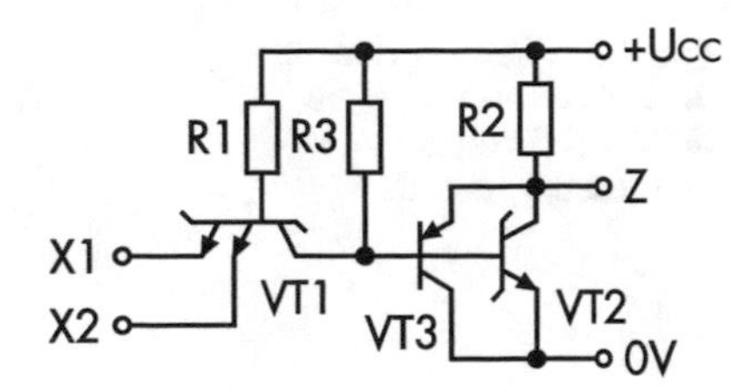

Fig. 3.4 Schematic diagram of TTLS logic element with the limitation circuit on the basis of the p-n-p transistor

basis of the required values of the input voltage U_{OH}. Introduction of the limitation circuit Z_O bears practically no significance on the output characteristic in the turned on state but results in the considerable alteration of the output characteristic of the logic element in the turned off state (the curve 3 in Fig. 3.2b).

From the figure, it follows that during alteration of the output current I_{OH}, the high-level output voltage U_{OH2} practically does not change, and its value is determined by the limitation voltage U_R. With the output current:

$$I_{OH}\rangle \frac{U_{CC} - U_R}{R2}$$

the limitation circuit is disabled, and the output voltage U_{OH2} diminishes.

The effective means of limiting the high-level output voltage U_{OH} is application of the p-n-p transistor VT3, included as per the schematic diagram in Fig. 3.4 [5]. With the high voltage level U_{IH} at the input of the logic element, the transistor VT1 is closed, and VT2 is opened.

Meanwhile, on the collector-emitter junction of the transistor VT2, the following voltage is set:

$$U_{CB}^{VT2} = U_{BE}^{VT3} \approx U_P^{VDS} \approx 0,55B;$$

therefore, the p-n-p transistor VT3 is closed and has no effect on performance of the circuit. With application of the low-level input voltage U_{IL}, the transistor VT1 is opened, and in the base of the p-n-p transistor VT3 and the n-p-n transistor VT2, the low-level voltage is set:

$$U_B^{VT3} \approx U_{IL} + U_{CE}^{VT1};$$

therefore, the transistor VT2 is closed, and VT3 is open. Because of this, at the output Z the circuit «output-junction» «base – emitter VT3» – junction «collector – emitter» «VT1 – input» the high level output voltage is set:

$$U_{OH} \approx U_{IL} + U_{CE}^{VT1} + U_{CE}^{VT3} \approx 1,25B;$$

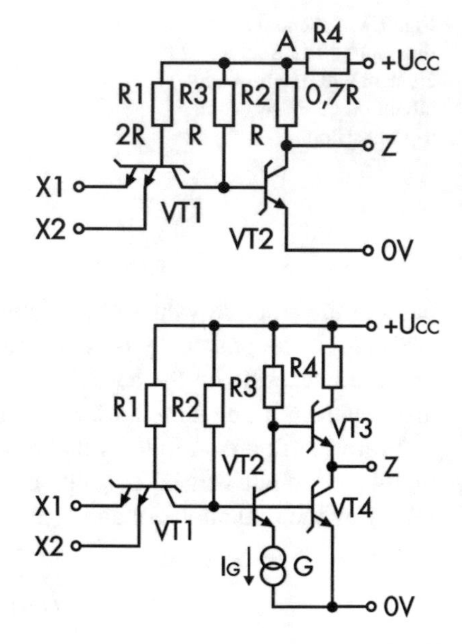

Fig. 3.5 TTLS Diagram of the logic element with the limitation circuit in the power supply circuit

Fig. 3.6 TTLS Diagram of the logic element with the active output

Meanwhile the noise resistance margins at the input are as follows:

$$\Delta U_T^+ = 0,55 - 0,25 = 0,3B$$
$$\Delta U_T^- = 1,25 - 0,55 \approx 0,65B.$$

It is possible to limit the high-level output voltage at the expense of introduction of the limitation circuits into the supply voltage of the logic element. The electric circuit, indicated in Fig. 3.5 [4], illustrates the logic element of such type. As a limitation element, reducing the high-level output voltage, the resistor R4 is used. At the expense of the currents, consumed by the logic element and passing through the resisters R1–R3 in the switched-off state, on the resistor R4, the voltage drop is created, reducing voltage in the node A and reducing the high-level output voltage U_{OH}. The voltage value in the node A at $U_{CC} = 1.5$ V, and the nominal rated values of the resistors R1–R4, indicated in the circuit (Fig. 3.5), constitute about 1 V.

Application at the output of the logic element of the resistor as a load increases the time t LH and the consumption power value in the open state (Fig. 3.1) determines the stepped up edge duration of switching off the output signal t_{LH} and the increased consumption power in the open state P_{CCL}.

Therefore, a number of the TTLS circuits of the logic element are purposed for elimination of this discrepancy by means of application at the output of two controlled active components—transistors. Thus, in the electric diagram of the logic element, indicated in Fig. 3.6, [1], the load capacitance discharge is ensured by the transistor VT4 and the charge by the transistor VT3. The transistor VT3 is controlled by means of the circuit, containing the transistor VT2 and the current generator G, connected across the output transistor VT4. During application of the high-level input voltage U_{IH}, the current, set by the transistor VT1 and resistors Rl

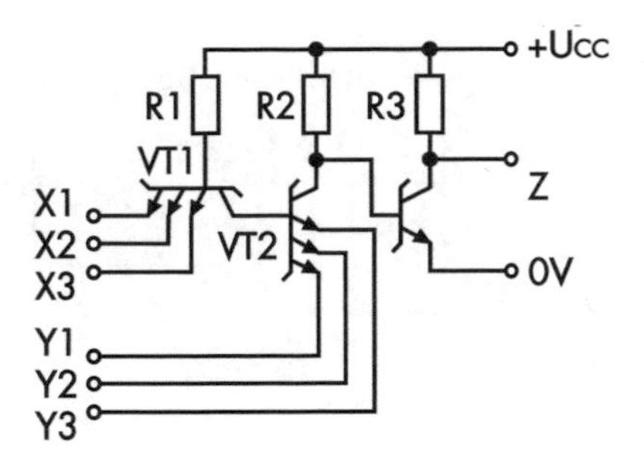

Fig. 3.7 TTLS diagram of the logic element with the expanded functional features

and R2, arrives to the bases of the transistors VT2 and VT4 and opens them. The open transistor VT4 discharges the load capacitance, and the open transistor VT2 passes the generator current, which steps down the voltage in the base of the transistor VT3 and closes it. For reliable locking, the transistor VT3 is required to perform the following conditions:

$$U_{CC} - I_G \cdot R3 < U_{BE}^{VT3}..$$

Meanwhile, in comparison with the circuit of the logic element (|Fig. 3.1) at the expense of the closed transistor VT3 reduction is ensured of the consumption power P_{CCL}. When applying the low-level input voltage U_{IL}, the transistors VT2 and VT3 switch over to the closed state, and the current source G switches off. Meanwhile, the transistor VT3 gets open, charging the load capacitance. The charge current value can be limited with the resistor R4, having resistance several times lower than of the resistor R2 (Fig. 3.1).

The circuit of the logic element, indicated in Fig. 3.1, as known, makes it possible to perform the function "AND-NOT" of the input variables by means of the multi-emitter transistor VT1 and the function "Installing OR" by combining the outputs of several logic elements. The electric circuit, indicated in Fig. 3.7, is illustrated by the example of the logic element with the expanded functional features, thus being different with the fact that apart from formation of the function of formation of the function "AND" by means of the multi-emitter transistor VT1, it forms an additional function "AND" as per the emitter of the transistors VT2. For formation of the required output signal levels, the transistor VT3 is used, simultaneously performing the function of inversion.

In this case, the logic element performs the following function:

$$Z = X1 \cdot X2 \cdot X3\left(\overline{Y1} \cdot \overline{Y2} \cdot \overline{Y3}\right);$$

As it was noted, TTLS of the logic element has a low input threshold of switching over U_T. One of the methods, enhancing its noise protection, is formation of its transfer characteristic of the "hysteresis" type. An example of the electric diagram of such logic element is shown in Fig. 3.8 [4]. In the initial state at the low-level input voltage U_{IL}, the transistor VT2 is closed and at the output the high level output

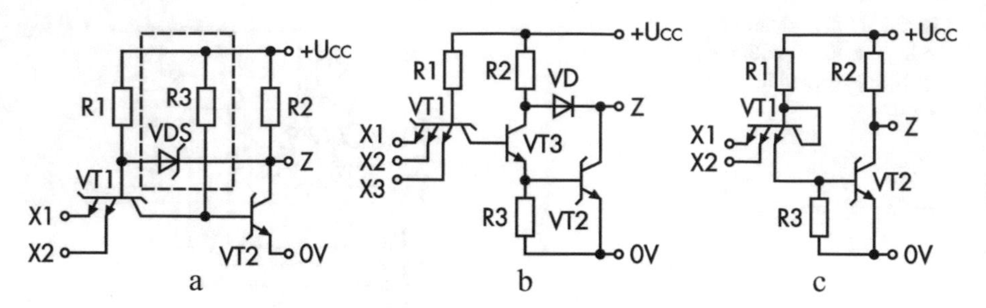

Fig. 3.8 TTLS diagram of the enhanced noise resistance logic element:
(**a**) With the transfer characteristic of the "hysteresis" type;
(**b**) With the stepped-up input switching threshold with the help of the transistor VT3;
(**c**) With the stepped-up input switching threshold at the expense of a diode

voltage U_{OH}. This voltage maintains the feedback diode VDS in the closed state; therefore, when stepping up the input voltage U_{IL}, switching over of the logic element will occur at the following level:

$$U_{TL} = U_{BE}^{VT2} + U_{CE}^{VT1} \approx 0,55B;$$

With the unlocking of the transistor VT2 at the output of the logic element, the low-level voltage U_{OL} will set up, which will result in the unlocking of the diode VDS and the base current interception of the input transistor VT1, switching over to the closed state. However, the open state of the output transistor VT2 will be meanwhile maintained by the current, passing through the resistor R3, and at the base of the transistor VT1 voltage will set up:

$$U_{B}^{VT1} = U_{CE}^{VT2} + U_{P}^{VDS};$$

therefore, with reduction of the input voltage U_{IH}, deactivation of the transistor VT1 (and, consequently, and deactivation of the transistor VT2) will take place with the input voltage:

$$U_{T2} = U_{B}^{VT1} - U_{BE}^{VT1} \approx 0,1B;$$

Another known method of enhancing the threshold voltage of switching U_T logic element is introduction of the schematic components between the collector of the input transistor VT1 and the base of the output transistor VT2. Thus, in the diagram, indicated in Fig. 3.8b as a component, stepping up the switching threshold, the transistor VT3 is used.

In the diagram, indicated in Fig. 3.8c [4] as such a component, a diode is used, whose function is performed by one from the junctions «base-emitter» of the transistor VT1. In the logic element of such type, the switching threshold is stepped up to the level:

$$U_T = U_{BE}^{VT2} + U_{BE}^{VT3(VT1)} - U_{CE}^{VT1} \approx 1,3 \div 1,4B;$$

Quite known also is the number of other logic elements of the TTLS type of the applied in the digital integrated circuits and described in the academic articles [1].

3.1.2 Basic Logic Elements of Schottky Transistor Logic

The basic electric diagram of the basic logic element of the Schottky transistor logic (STL) [1] is indicated in Fig. 3.9 and is essentially a simplest inverter on the key transistor VT1 with the "bypassing" Schottky diode VDS and a set of the "isolating" Schottky diodes VDS1–VDS3 in its collector. Such design, containing a minimum number of the schematic components, determines a small area of the logic element, occupied on the chip, and the low values of the parasitic capacitances of the logic elements and realizes the function of the "multiplied" inversion; meanwhile, each from the outputs, formed by the anodes of the Schottky diodes VDS1–VDS3, is electrically isolated. This permits the "mount" combination of outputs of the various logic elements and formation in such a way the function "NOT – AND". The static characteristics of the STL logic element are considered in [4]. A peculiar feature of the STL logic element is a small logic swing in the output voltages U_T, related to availability of two types of the Schottky diodes: "bypassing" and "decoupling" VDS1÷VDS3:

$$\Delta U_T = U_{OH} - U_{OL} = U_{BE}^{VT1} - U_{CE}^{VT1} + U_P^{VDS1(VDS2,VDS3)}$$
$$= U_P^{VDS0} - U_P^{VDS1(VDS2,VDS3)};$$

Hence it follows that in order to ensure serviceability, it is necessary to perform the conditions:

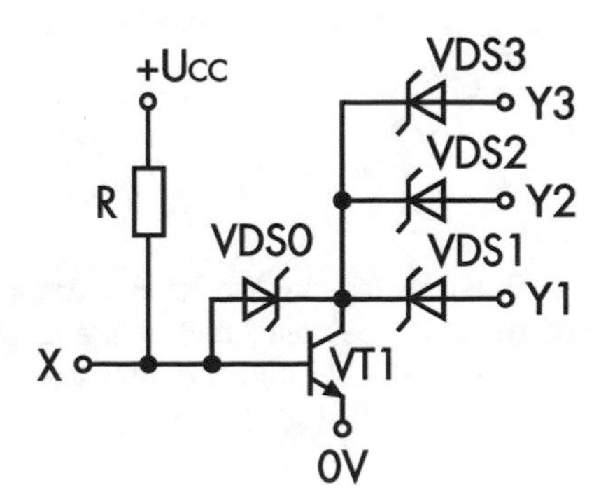

Fig. 3.9 Diagram of the base STL logic element

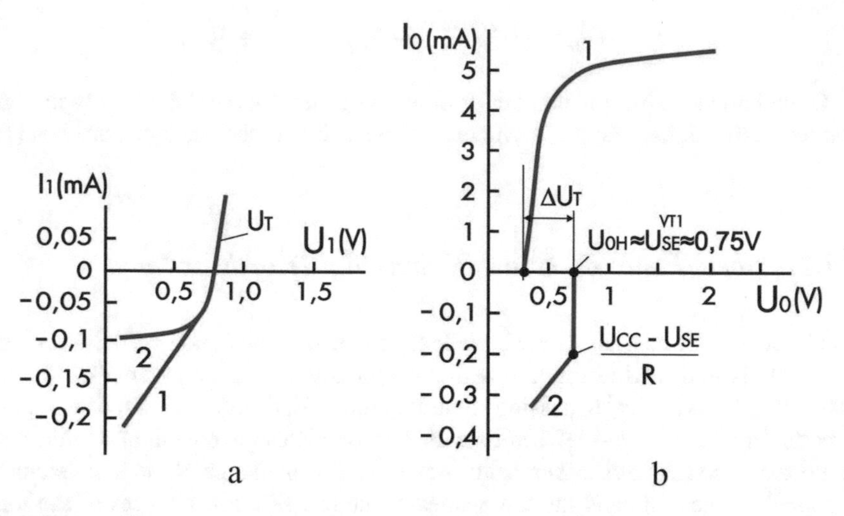

Fig. 3.10 Input (**a**) and output (**b**) static characteristics of the basic STL logic element

$$U_P^{VDS0} > U_P^{VDS1(VDS2,VDS3)};$$

On the basis of the optimum requirements to the performance and noise immunity, this difference is selected in the range:

$$U_P^{VDS0} - U_P^{VDS1-VDS3} \approx 0,15 \div 0,25B;$$

which is attained by application of the Schottky diodes on the basis of the various contact systems: VDS0 on the basis of the system PtS i-Si and VDS1–VDS3 on the basis of the system TiW-Si.

The output characteristic of the logic element in the open state is indicated in Fig. 3.10b (the curve 1) and represents the output characteristic of the open bipolar n-p-n transistor, shifted as per the voltage axis for the value $U_P^{VDS1(VDS2,VDS3)}$. The output characteristic of the logic element in the closed state (when used as a load a logic element of the similar type) is indicated in Fig. 3.10b (the curve 2). From the last characteristic, it follows that the value *UOH* is practically constant and is equal to U_{BE}^{VT1} up to the level:

$$I_{OH} < \frac{U_{CC} - U_{BE}^{VT1}}{R};$$

When surpassing this level, current is intercepted from the load transistor's base of the logic element, and at the expense of the voltage drop on the resistor R (Fig. 3.9), the high-level output voltage U_{OH} drops. From Fig. 3.10a (the curve 1) it is obvious that at the expense of application at the input of the resistor's logic

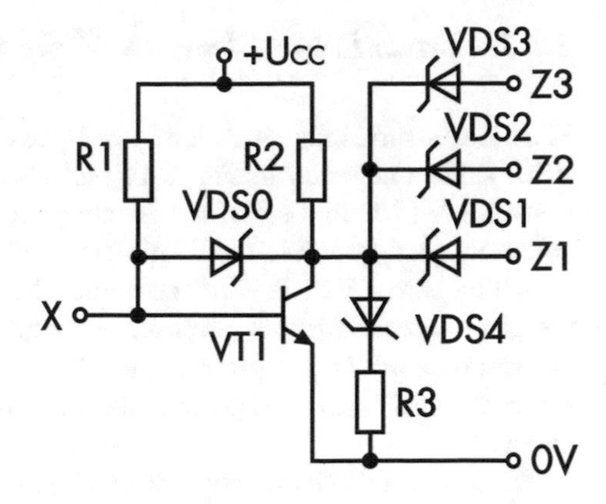

Fig. 3.11 Diagram of STL logic element with fixation of the collector's potential of the switch transistor

element, the input current I_I depends on the input voltage U_I, and the switching threshold of the logic element constitutes:

$$U_T \approx U_{BE}^{VT1};$$

In order to stabilize the input current (the curve 2 in Fig. 3.10a) in the logic element as a source of the base current of the transistor VT1, one may use the p-n-p transistor [3].

The average switching time of the STL logic element is determined by the parasitic capacitances of the switch transistor VT1, the logic swing ΔU_T, and the consumption current I_{CC} [1]:

$$t_P = \frac{\left[(3 + K1 + K2)C_{CB}^{VT1} + (1 + K2)C_{CP}^{VT1} + C_{BE}^{VT1}\right]}{2I_{CC}} \cdot \Delta U_T;$$

where C_{CB}^{VT1}, C_{CP}^{VT1}, C_{BE}^{VT1}—capacitances of the junctions base-collector, collector-substrate, and base-emitter of the transistor VT1

K1, K2—certain dimensionless constants, depending on the design, when $0<K1$, $K2<1$

In the diagram, indicated in Fig. 3.9, the collector of the transistor VT1 in the switched-off state has a “floating” potential, which impairs the response of the logic element. In order to eliminate this shortcoming, one uses the various circuits for fixing the collector's potential of the transistor VT1, one from which is indicated in Fig. 3.11. The fixation diagram includes two resistors R2 and R3 and the Schottky diode VDS4, connected to the output and the supply source and fixes the collector's potential of the transistor VT1 at the level, approximately equal to 1–1.5 V.

3.1.3 Basic Logic Elements of the Integrated Schottky Logic

The basic electric diagram of the basic logic element of the integrated Schottky logic (ISL) [1] is integrated in Fig. 3.12 and also represents a simplest inverter on the transistor VT1 with a set of the "decoupling" Schottky diodes VDS1–VDS3 in its collector. As opposed to the STL logic element in which the saturation state of the switch transistor VT1 is eliminated with the "bypassing" Schottky diode VDS0, in the given circuit, for this purpose, one uses the structural superposed with n-p-n-транзистором VT4 p-n-p-transistor VT2. With the low-level input voltage, the n-p-n-transistor VT1 is closed, and its collector junction is shifted in the reverse direction.

Because of this, the junction of base-emitter of the p-n-p-transistor VT2 is closed and has no effect on operation of the logic element. With the high-level input voltage, the transistor VT1 is opened, and on its collector, the low-level voltage is established. Meanwhile, the collector junction of the transistor VT1, and, consequently, and the junction of base-emitter of the transistor VT2 are shifted in the forward voltage. The transistor VT2 is opened and intercepts a portion of the base current of the transistor VT1, thus reducing its degree of saturation. ISL logic element also realizes the function of the "multiplied" inversion of the input signal by means of a set of the output Schottky diodes VDS1–VDS3, and realization of the logic functions is performed by means of the "mount" integration of outputs of the various logic elements, which makes it possible to avail the function "NOT-AND" at output.

The static characteristics of the logic element of such type are considered in detail in [1]. A specific feature of the given circuit is a small logic swing in voltages ΔU_T, related to availability of the output Schottky diodes VDS1–VDS3 and making it possible to attain the prompt response of the logic element:

$$\Delta U_T = U_{OH} - U_{OL} = U_{BE}^{VT1} - U_{CE}^{VT1} + U_P^{VDS1(VDS2,VDS3)} = 0,25 \div 0,35B;$$

Meanwhile, in the circuit of the logic element, there may be used the Schottky diodes, applied in TTLS integrated circuit (for instance, on the basis of PtSi) with:

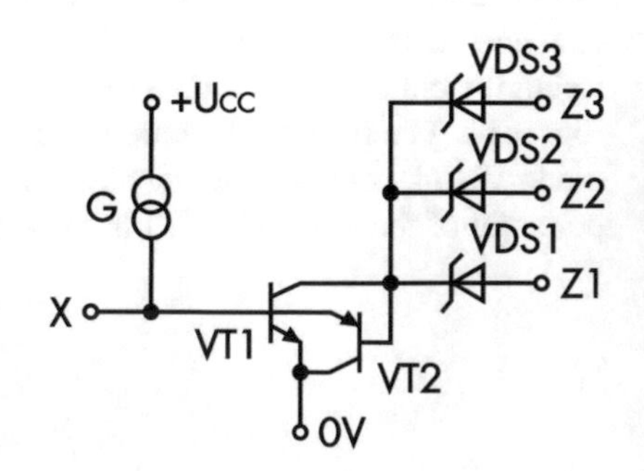

Fig. 3.12 Diagram of the basic STL logic element

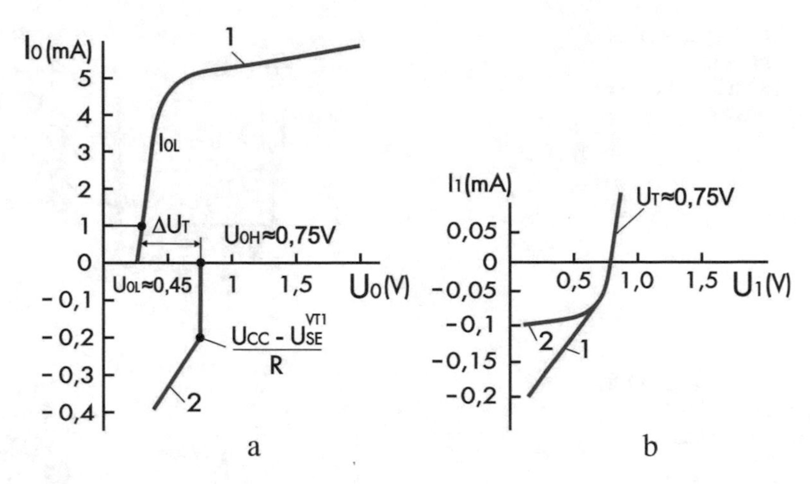

Fig. 3.13 Output (**a**) and input (**b**) static characteristics of the basic ISL logic element

$$U_P^{VDS1(VDS2,VDS3)} = 0,40 \div 0,50B.$$

The output characteristic of the logic element in the open state is indicated in Fig. 3.13a (the curve 1) and represents the output characteristic of the open n-p-n-transistor, biased as per axis of voltages for a value of the direct voltage drop of the Schottky diodes VDS1–VDS3. The output characteristic of the logic element in the closed state (during application as a load of the logic element of the similar type) is indicated in Fig. 3.13a (the curve 2). From the latter one, it is obvious that the value of the high-level output voltage U_{OH} is practically constant and is equal to U_{BE}^{VT1} of the output current level:

$$I_{OH} < \frac{U_{CC} - U_{BE}^{VT1}}{R};$$

When exceeding this value I_{OH}, current is intercepted of the generator G of the load logic element, and the output voltage U_{OH} drops.

The input characteristic of the ICL logic element in Fig. 3.13b is similar to STL logic element, and the logic element itself permits application both of the resistor (the curve 1 in Fig. 3.13b) and of the p-n-p transistor (the curve 2) as the current source G. The switching threshold of the logic element constitutes:

$$U_T \approx U_{BE}^{VT1};$$

The dynamic parameters of the ISL logic element are considered in detail in [1]. The average delay time of switching ISL logic element is several times more than of STL logic element, which is related to the p-n-p transistor VT2:

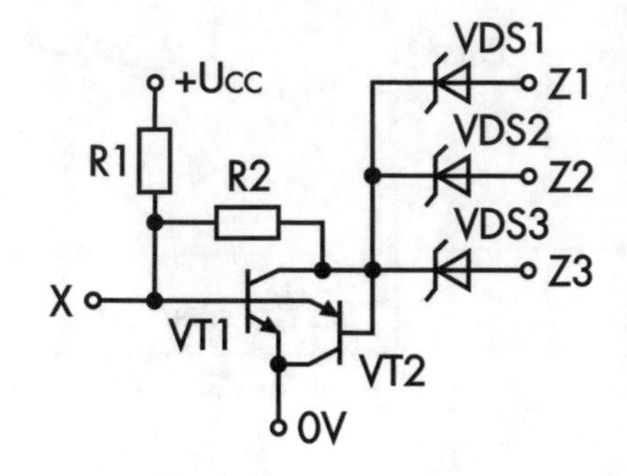

Fig. 3.14 Diagrams of ISL logic element with saturation limitation of the p-n-p transistor

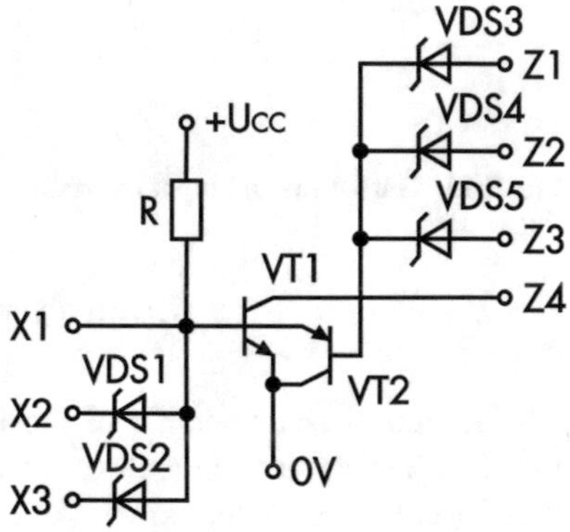

Fig. 3.15 Diagrams of ISL logic element with the expanded functional potentialities

$$t_P = \left[\frac{\left[(3 + K1 + K2)C_{CB}^{VT1} + (1 + K2)C_{CP}^{VT1} + C_{BE}^{VT1}\right]}{2I_{CC}} \cdot \Delta U_T\right] + \frac{\alpha^* \cdot \tau_P^*}{2};$$

where: α^*—effective ratio of the emitter current transfer of the p-n-p transistor VT2

τ^*_p—effective time of the charge carriers transfer in base of the p-n-p transistor VT2

In the diagram of the logic element, indicated in Fig. 3.12, on the collector of the switch transistor VT1, similar to STL logic element, in the switched-off state is the "floating" potential, aggravating the performance of the logic element. In order to eliminate this state, one also uses the various methods of the collector potential fixation of the n-p-n-transistor VT1. In one of such circuits of the logic element (Fig. 3.14) [4] for the potential fixation, one uses the bias circuit, connected to the supply source, containing the current generator G2, resistor R, and diode VDS4 (Fig. 3.11). In the other circuit [5], indicated in Fig. 3.15, the collector potential is fixed by means of the resistor R2, connected between the collector and the base of the switch transistor VT1.

ISL logic elements (also STL) possess smaller functional potentialities, as compared with TTLS, because in them, the logic functions are formed only by integration of the logic element outputs.

In [1] is also suggested the electric diagram of the ISL logic element, possessing the enhanced functional potentialities. In such diagram (Fig. 3.15), the logic

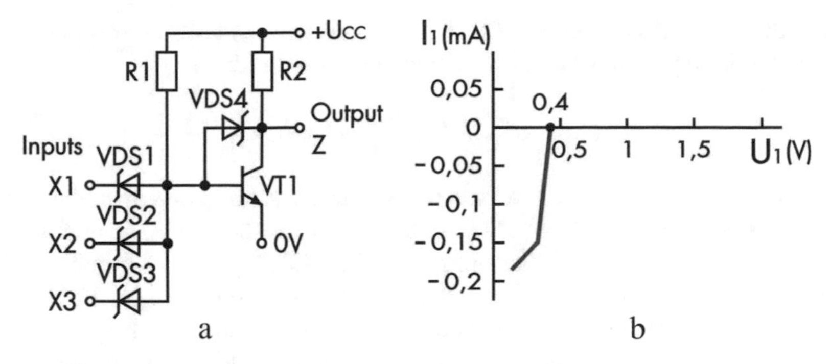

Fig. 3.16 Diagram (**a**) and input characteristic (**b**) of the base DTLS LE

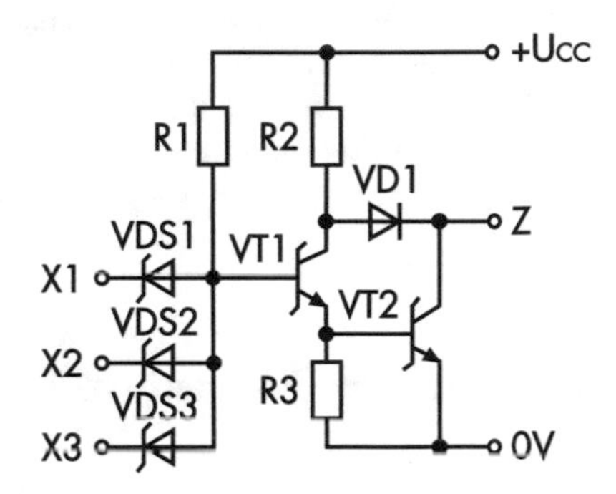

Fig. 3.17 Diagram of the base LE DTLS with the enhanced switching threshold

functions are formed by means of a set of the Schottky diodes, both at the input (VDS1, VDS2) and at the output (VDS3, VDS4). Meanwhile, if the logic function "AND" is formed at the inputs *X2*, *X3*, then for its pin ISL logic element, they use the "clean" output *Z4*. During formation of the logic function "AND" at the output, the outputs *Z1–Z3* ISL LE (logic element) are connected to the "clean" input *X1* of the load ISL LE.

One can also attain expansion of the functional potentialities of ISL LE by means of the cascading connection of the switch transistors VT1–VT3 (Fig. 3.16b) [1]. For the purpose of limitation of the saturation degree of the transistors VT1–VT3, one matched multi-collector p-n-p-transistor VT4 is used.

3.1.4 Base Logic Elements of the Diode-Transistor Logic with Schottky Diodes

The base electric diagram of the logic element of the diode-transistor logic with the Schottky diodes (DTLS) [1] (Fig. 3.17) is similar to the electric diagram of TTLS LE, in which at the input instead of MET a set of the "low barrier" Schottky diodes VDS1–VDS3 is included. Analysis of the static characteristics of such logic element is provided in [4]. As opposed to TTLS LE, the given circuit owing to application of the input Schottky diodes has a lower switching threshold:

$$U_T = U_{BE}^{VT} - U_P^{VDS1-VDS3} \approx 0,4B;$$

which also requires application in the integrated circuits the special matching circuits with the external devices (Fig. 3.17). As DTLS LE has the output, similar to TTLS LE, then its output characteristics are identical to TTLS LE. Taking into consideration the conclusion, stated in [3], that the delay times t_{PLH} and t_{PIHL} of TTLS LE are determined by the parameters of the output transistor VT2 and weakly depend on the input transistor VT1, for computation of the time values of the switch delays t_{PLH} and t_{PHL} DTLS LE, it is possible to use the formulae, indicated in Sect. 3.1.1. for TTLS LE, and for enhancing its characteristics practically all circuits are suitable, described for TTLS LE.

In order to enhance the noise immunity of DTLS LE, it is possible also to introduce into the base of the output transistor VT the additional components (as, for instance, the transistor VT1 in Fig. 3.17), stepping up the LE switching threshold.

3.2 Memory Elements of TTLS Integrated Circuits

In TTLS LSI as the static memory elements (ME), the complex elements are used both of the type of synchronous flip-flops and the simplest bistable cells (BC) with the control circuits. In the majority of cases, ME are realized on the basic LE, on the basis of other functional blocks of integrated circuits as per the known methods and techniques, described in [2, 5]. However, the TTLS schematics provides a possibility of applying the below indicated non-traditional methods and techniques of constructing ME, making it possible to cut down the number of the circuit elements and enhance the characteristics of the integrated circuits.

3.2.1 Memory Elements, Synchrosignal Edge Cycled

When using the base logic elements as the ME basis, it is possible to use the flip-flops of Dt-type, built as per the pattern "master-slave" (M-S). Example of the electric diagram of such flip-flop, using TTLS base LEs, is provided in Fig. 3.18. The flip-flop contains two BE on the transistors VT3, VT4, VT7, VT8 и VT11, VT12, and VT15; four record switches on the transistors VT1, VT2, VT5, VT6, VT9, VT11, VT13, and VT14; and two buffer elements on the transistors VT17 and VT18. The first two switches control writing into the master R-S flip-flop and the second one into the slave one. Application in the synchronization circuit of the master R-S control flip-flop as per the emitter of the transistors VT1 and VT5, and in

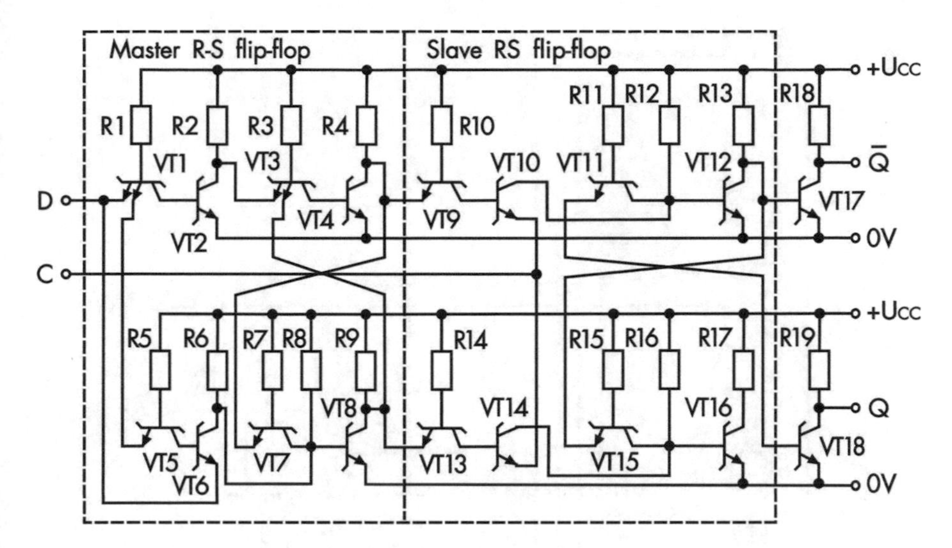

Fig. 3.18 Diagram of the D-flip-flop of the type M-S on the basis of TTLS LE

the slave one control by the emitters of the transistors VT10 and VT14 ensures formation of the paraphrase synchrosignals for the master and slave R-S flip-flops without application of the additional inverter. The output buffer components VT17, VT18 isolate the feedback circuits from the outputs Q, $\overline{Q}$ and enhance the noise immunity of the Dt flip-flop.

Criticality of the synchronization circuits to the levels of input voltages, as well as a great number of the circuit elements (up to 10), determine a wider application in the integrated circuits of the Dt-flip-flops, built as per the pattern of "three flip-flops." Example of the electric diagram of a such flip-flop on the basis of TTLS LE is indicated in Fig. 3.19.

When applying the multi-output basic LE of the type STL or ISL, the electric diagram of a similar Dt-flip-flop will have a similar layout. However, a shortcoming of such circuits is a necessity of a greater number of components to control the flip-flops. Therefore, in the TTLS schematics, a greater popularity is attributed to the Dt flip-flops of the type M-S with application of the threshold control principles. Example of the electric diagram of such flip-flop is indicated in Fig. 3.20. The circuit contains the master R-S flip-flop on the transistors VT11 and VT12, the slave R-S flip-flop on the transistors VT1–VT10, and the input threshold circuit on the transistors VT13 and VT14.

The circuit functions in the following way: at the high signal level at the synchroinput C, the transistors VT11 and VT12 are closed; at the inputs $R2$, $S2$ of the slave flip-flop are set the high-level signals and the slave R-S flip-flop in the storage time. When applying the high signal to the input D or that of low levels, the input circuit forms at the outputs Y, $\overline{Y}$ two paraphrase signals with the levels:

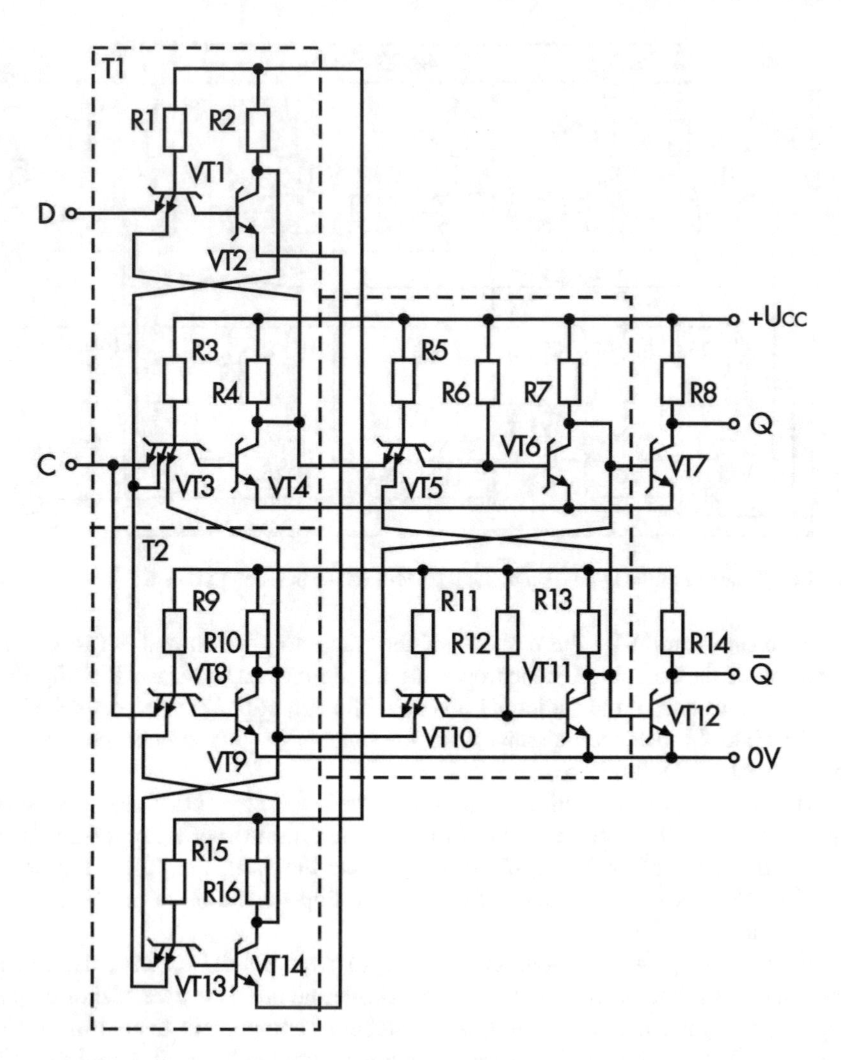

Fig. 3.19 Diagram of the Dt-flip-flop, built as per the pattern of the "three flip-flops" on the basis of TTLS LE

$$U_{OH} = U_{CC};$$

$$U_{OL} = U_{CE}^{VT1} + 2U_{P}^{VD};$$

arriving to inputs of the master R-S flip-flop. As the transistors VT11 and VT12 are closed, writing to the master R-S does not take place. When reducing the voltage at the synchroinput *C* below the level:

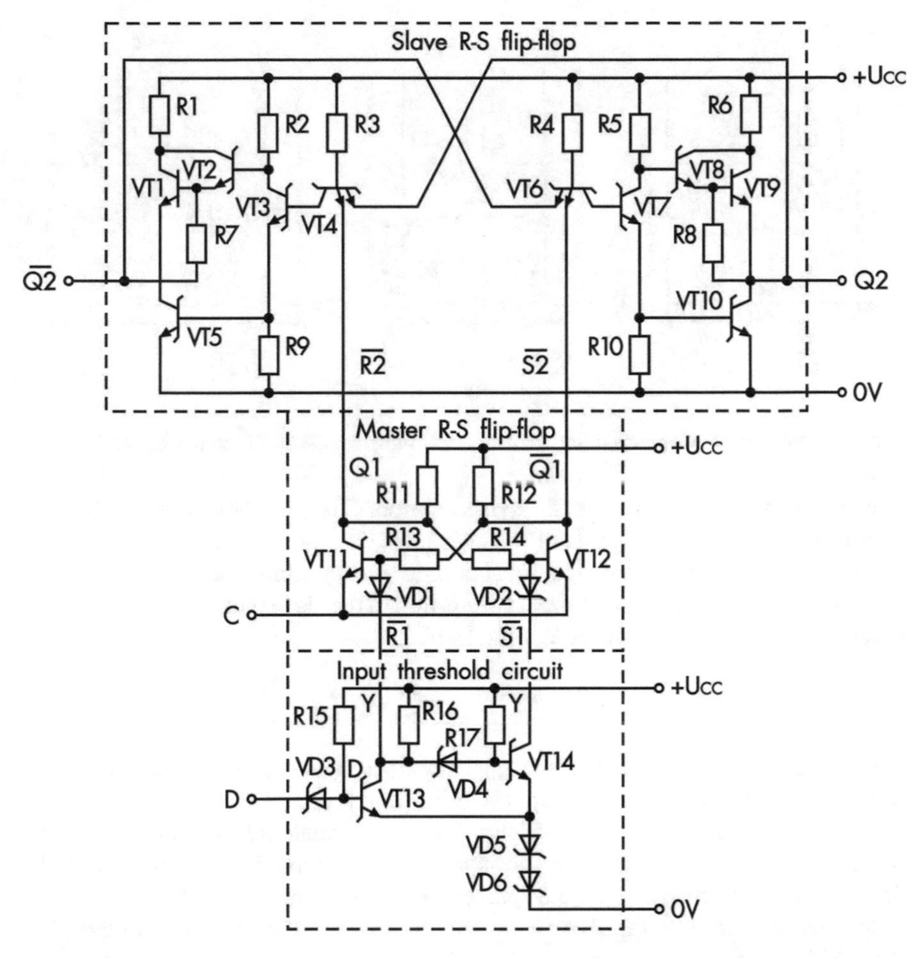

Fig. 3.20 Diagram of Dt-flip-flop, built as per the pattern M-S with application of the threshold control principles

$$U_{T1} \approx U_{CC} - U_B^{VT11(VT12)};$$

unlocking takes place of one from the transistors VT11 and VT12, on the base of which from the input circuit, the signal of the high-level U_{OH} arrives. The signal of the low-level U_{OL} from the input threshold circuit, arriving on the base of another from the transistors VTU, VT12, keeps it in the closed state. Thus, at the outputs $Q1$, $\overline{Q}1$ of the master R-S flip-flop appear the combination of the signals of the low and high levels:

$$U_{T2} = 2U_{BE}^{VT} - 2U_{CE}^{VT};$$

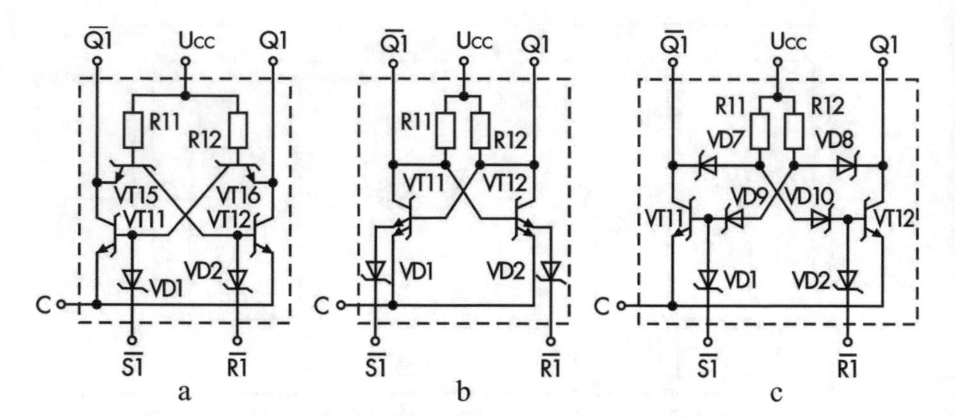

Fig. 3.21 Versions of diagrams of the master R-S flip-flop for the Dt-flip-flop (Fig. 3.20)

which to the voltage level U_{T2} at the synchrooutput C is sensed by the slave R-S flip-flop as a combination of signals of the high level.

With the subsequent voltage reduction at the synchroinput below U_{T2}, the information is rewritten from the master flip-flop to the slave one. During the voltage reduction at the synchroinput below the level:

$$U_{T3} = 3\mathrm{U}_P^{\mathrm{VDS}} + U_{CE}^{VT} - U_{BE}^{VT};$$

both signals at the outputs $Y, \overline{Y}$ of the input circuit are again sensed by the inputs $\overline{R}, \overline{S1}$ of the master R-S flip-flop as the signals of the high level.

Meanwhile, in the master R-S flip-flop, the written information is retained irrespective of the information at the input D. The information is written into the flip-flop of the given type by the negative edge of the synchrosignal C.

As the master R-S flip-flop, it is possible to use the most elementary ME (Fig. 3.21a, b, c).

As opposed to the circuit, indicated in Fig. 3.20, the circuit of the master R-S flip-flop, indicated in Fig. 3.22, has a reduced consumption power [1]. This accounts to the fact that in this circuit, the master R-S flip-flop is dynamically powered, which is set to the required logic state and consumes power only at the moment of switching the synchrosignal from the high level to the low one, and in the static states, it is disconnected from power supply. Such a specific feature of the flip-flop is related to the following: with the high level of the synchrosignal C, both transistors VT5 and VT6 of the master R-S flip-flop are closed, and at its outputs $Q1$, $\overline{Q}1$, high-level signals are set, and the slave R-S flip-flop is in the storage mode.

As the voltage drops as per the circuits VT2-VD1, or VT3-VD1 of the slave, R-S flip-flop is lower than via the circuits VDS1-R6-VT8-VDS2, or via VDS2-R5-VDS5-VT7-VD2 of the master R-S flip-flop; then the consumption current of the master R-S flip-flop is missing.

During reduction of the voltage level at the synchroinput C, a short-time unlocking takes place of one from the transistors VT5 and VT6 (in dependence on

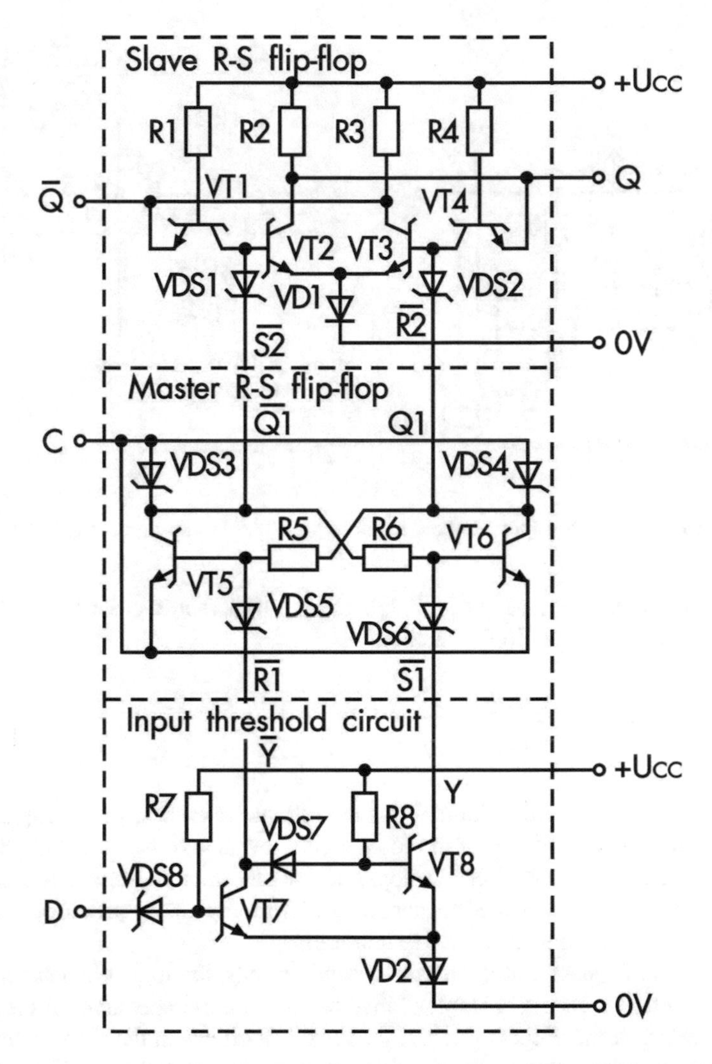

Fig. 3.22 Diagram of the base TTLS Dt-flip-flop of the type M-S with the reduced power consumption

the fact, which from the transistors VT7 and VT8 of the input circuit is open), and on one from the outputs $Q1$, $\overline{Q}1$ of the slave R-S flip-flop, a short low-level pulse is formed, setting the slave R-S flip-flop to the required state. After transition of the synchrosignal C into the static state of the low level, the transistors VT5 and VT6 get closed, and the consumption current of the master R-S flip-flop is missing. As the input threshold circuit of the flip-flop, one can use both the DTLS circuit (Fig. 3.22) and the TTLS (Fig. 3.23a, b); however, the diode circuit requires a more thorough selection of the element parameters.

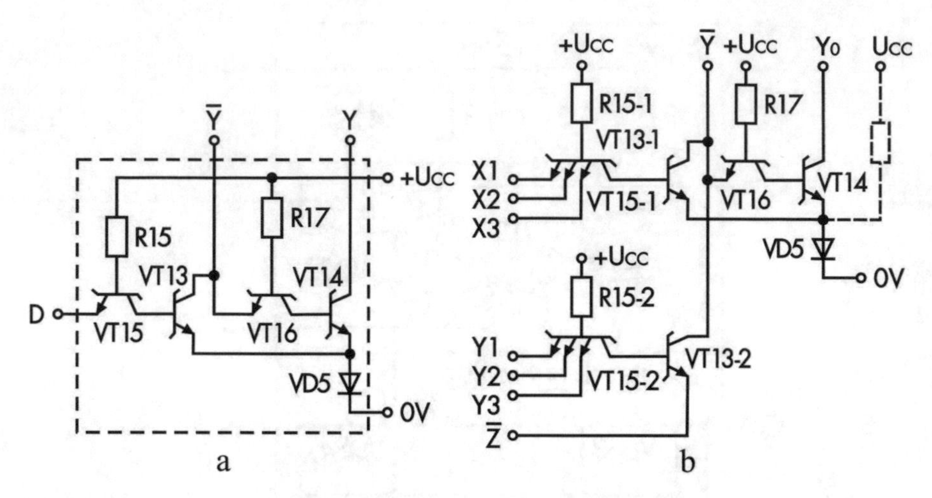

Fig. 3.23 Diagrams of the input threshold buffer of the TTLS Dt-flip-flop of the M-S type

In the diagram, indicated in Fig. 3.23b, the input threshold circuit, apart from the write control, realizes the logic functions:

$$\overline{Y} = \overline{X1 \cdot X2 \cdot X3 + Y1 \cdot Y2 \cdot Y3 \cdot \overline{Z}};$$
$$Y = X1 \cdot X2 \cdot X3 + Y1 \cdot Y2 \cdot Y3 \cdot \overline{Z};$$

When using the inverting input Z for the reliable operation of the flip-flop, one requires a thorough matching of the input signal levels and the potential fixation of the VD5 diode. As a slave R-S flip-flop, one uses the more complex circuits of the R-S flip-flops, which is related to necessity of forming the required output signal levels, the load capacitance and noise immunity.

Thus, for performance of the first two requirements, the logic elements are made as per the pattern with the "complex" inverter, and for enhancement of the level of noise immunity at the outputs $Q1, \overline{Q}1$, the feedback effects in the slave R-S flip-flop are realized from the collectors of the phase separating transistors VT3 and VT16 (Fig. 3.24a).

However, the margin of the noise immunity in the feedback circuits is reduced down to the level:

$$\Delta U_T^+ = U_{BE}^{VT} - 2U_{CB}^{VT};$$

which makes such a flip-flop so sensitive to the noises as per the power supply circuits.

For sensitivity reduction of the flip-flop to this type of noises, it is possible to use the circuit of the R-S flip-flop, indicated in Fig. 3.24b [2]. In this circuit at the inputs of the slave R-S flip-flop, the diodes VD1 and VD2 are introduced, enhancing the threshold of switching the logic element, and the feedback circuits are formed by

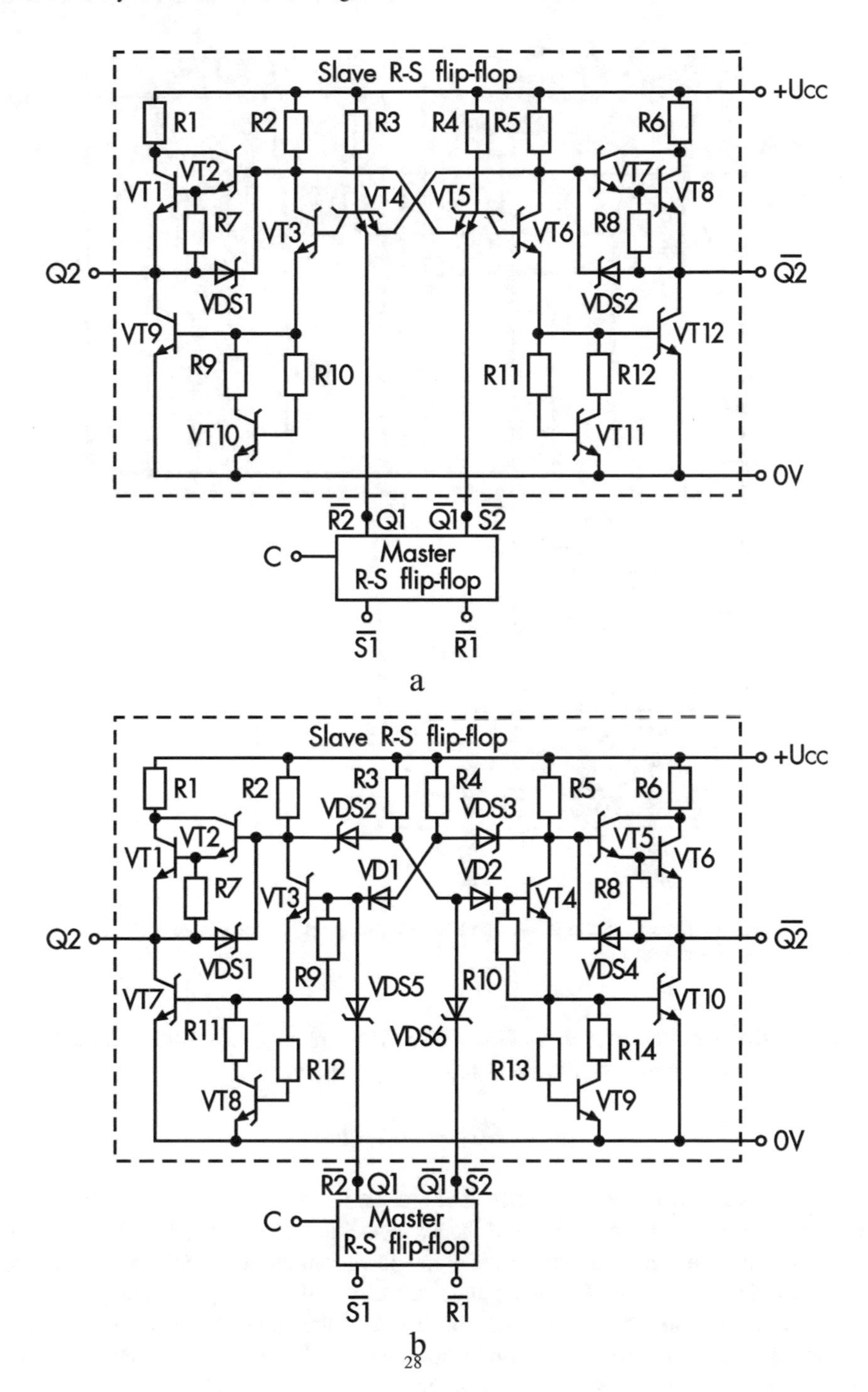

28

Fig. 3.24 Diagrams of the slave R-S flip-flop on the basis of the logic element with the transistor (**a**) and diode (**b**) inputs

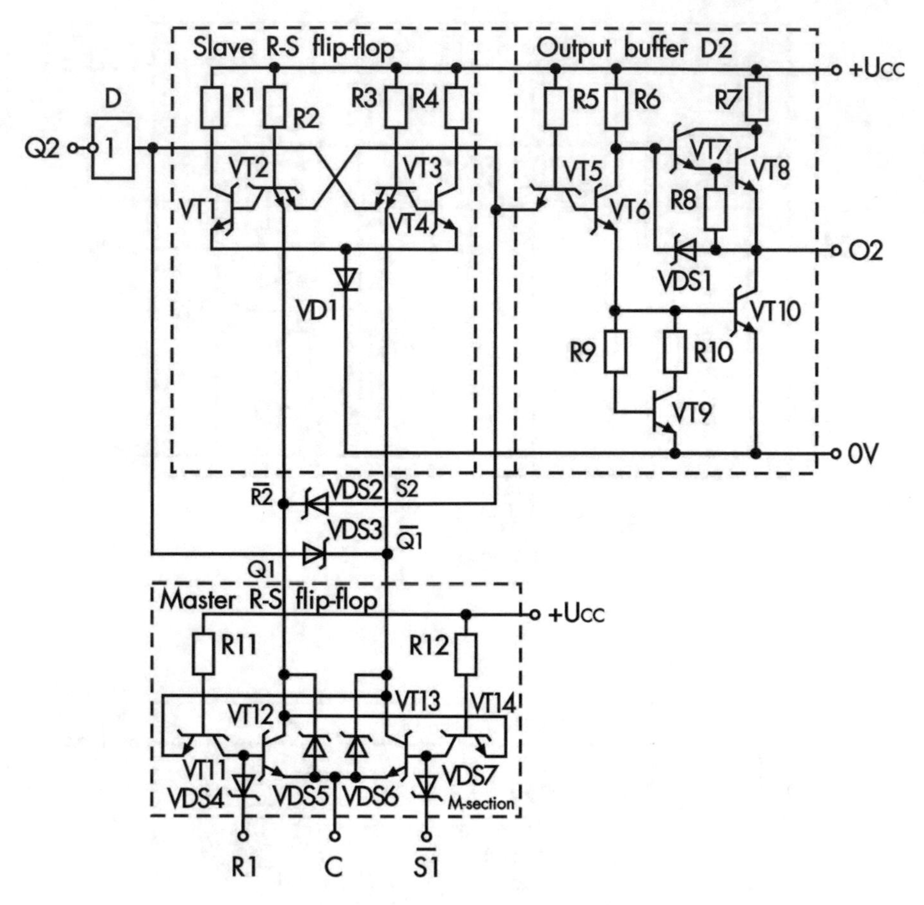

Fig. 3.25 Version of the diagram of the slave R-S- flip-flop on the basis of low-power TTLS LE with the power output buffer

means of the Schottky diodes VDS2, VDS3. This enhances the noise immunity as per the feedback circuits of the slave flip-flop to the level:

$$\Delta U_T^+ = U_{BE}^{VT};$$

In other way, ensuring enhancement of resistance of the Dt flip-flop to the noises at the outputs Q, $\overline{Q}$ is construction of the slave R-S flip-flop on the elementary logic elements with connection to its outputs of the power output buffer $D2$, as indicated in Fig. 3.25 [1]. This buffer forms the required levels of the output signals, the load capacitance, and isolates the feedback influence of the slave R-S flip-flop from the outputs Q, $\overline{Q}$. In order to enhance the response rate of the Dt flip-flop, they introduce the "accelerating" components, for instance, the diodes VDS2, VDS3 [1]. These diodes are switched to the state of the high-level simultaneously and the output

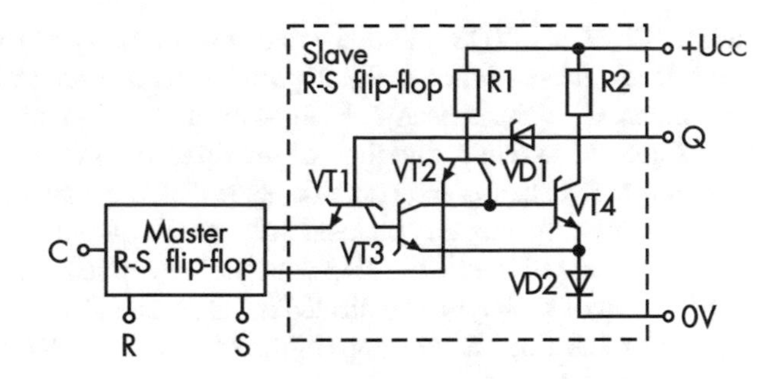

Fig. 3.26 Diagram of the TTLS Dt-flip-flop with the asymmetrical structure

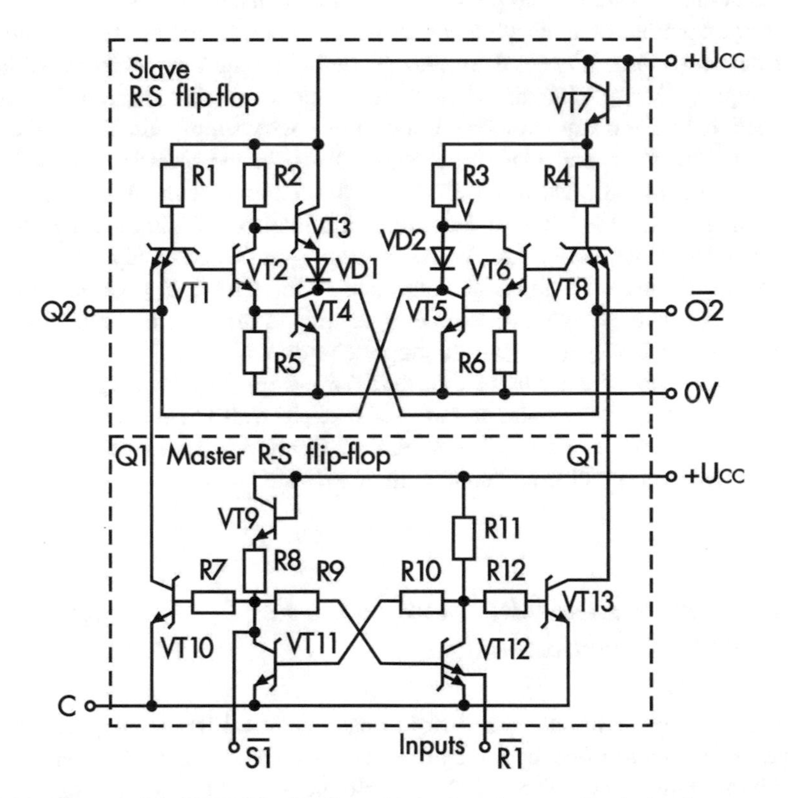

Fig. 3.27 Diagram of the TTLLS Dt-flip-flop with the forced state setting during the power supply

buffer, thus enhancing its rate of response. A peculiar feature of the memory element circuits, using such kind "accelerating" bonds, is their enhanced noise immunity.

This is related to the fact that setting the slave R-S flip-flop to the required logic state occurs after switching over the power output buffer, generating the great noises in the power supply buses. In the Dt flip-flops of such type, in order to enhance its rate of response, it is also possible to use the "accelerating" bonds of the other type, introduced between the synchronization circuit and the slave R-S flip-flop. In the circuit, indicated in Fig. 3.27, these bonds are formed by means of the Schottky diodes VDS6 and VDS7. With the transition of the synchrosignal C to the state of the high level, these diodes make it possible to accelerate the charge of the output capacitances of the master R-S flip-flop and the input capacitances of the slave R-S flip-flop, to reduce the time of rewriting the information from the master R-S flip-flop and enhance the response speed of the memory element. When during application of the memory element only one its output is required (Q or $\overline{Q}$), efficient are the circuits of the Dt flip-flop with the "asymmetrical" structure. One version of the flip-flop of such type is indicated in Fig. 3.26 [4], and its functioning is similar to the earlier described Dt flip-flops of the M-S type. Such a circuit has a fewer number of the schematic elements and possesses a smaller consumption power.

In certain options of application of the memory elements in the integrated circuits, it is required to forcefully set them into the definite logic state when switching on power supply. One of the main schematic technical methods of tackling this task is introduction of the delay components into the power supply circuits of the logic element of the slave and master flip-flops. Thus, in the electric diagram of the memory element, indicated in Fig. 3.27 [3], the function of the delay components is performed by the transistors in the diode introduction: VT1 in the slave R-S flip-flop and VT8 in the master one. When switching on power supply U_{CC}, the logic element of the slave R-S flip-flop, connected to the circuit U_{CC} via the transistor VT1, will get enabled with a certain delay relative to the logic element of this flip-flop, connected directly to the power supply circuit.

This will result to the fact that the slave R-S flip-flop will get set to the state of low level at the output $Q2$. A similar process will take place also in the slave R-S flip-flop by means of the transistor VT8. The master R-S flip-flop will get set to the same state as the slave one and will keep maintaining it set state.

3.2.2 Memory Elements, Cycled by the Level of the Synchrosignal

Wide applications in the integrated circuit have acquired the memory elements on the basis of the Dt flip-flop, cycled by the level of synchrosignal [1]. When applying as the basic multi-input TTLS LE, the electric diagram of the Dt flip-flop will have the layout, indicated in Fig. 3.28.

The flip-flop contains the basic cell on the transistors VT3, VT4, VT8, and VT9; two input write switches on the transistors VT1, VT2, VT6, and VT7; and two output buffer components on the transistors VT5 and VT10. The switch on the transistors VT1 and VT2 performs the inverse transfer of information from the input D into the basic cell, and application in the second switch of write control as per the transistor emitter VT7 ensures the direct information transfer from the input D to the basic cell. The output buffer components VT5 and VT10 isolate the feedback circuits from the outputs Q, $\overline{Q}$ and enhance the noise immunity of the Dt flip-flop. During application as the basic cell of the multi-input LE of the STL type or ISL, the electric

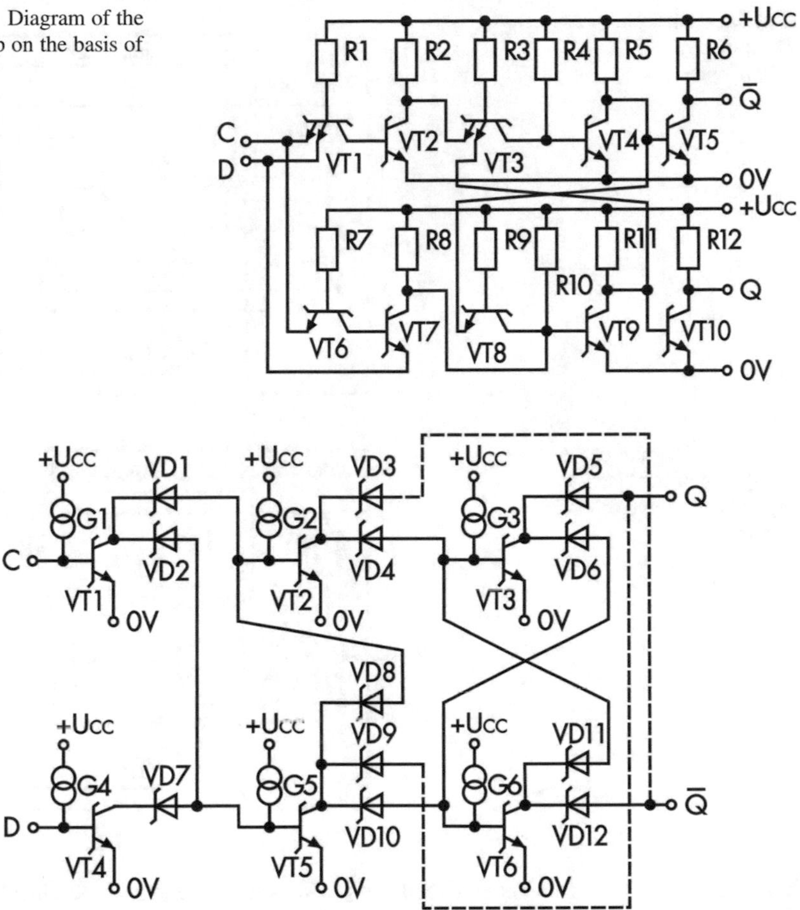

Fig. 3.28 Diagram of the D-flip-flop on the basis of TTLS LE

Fig. 3.29 Diagram of the D-flip-flop on the basis of STL LE

diagram of such D-flip-flop will have a layout, indicated in Fig. 3.29. Performance of such flip-flop is similar to the TTLS flip-flop. In this flip-flop it is also possible to form the “accelerating” bonds (in the Figure they are depicted with a dotted line), enhancing the rate of response of the flip-flop. For isolation of the input *D* from the circuit of the synchrosignalization at the output of the flip-flop, an additional logic element is introduced on the transistor VT4. The shortcoming of such memory elements during their application in the integrated circuits is a large number of the schematic components, enhanced power consumption, and a long time of transferring information from the input *D* to the outputs $Q$and $\overline{Q}$.

As in the memory elements, cycled by the synchrosignal levelquite often, only one output is used (Q or $\overline{Q}$); widely popular are the Dt flip-flops of the “latch” type. The diagram, indicated in Fig. 3.30 [3], contains the write key (wk) on the transistors

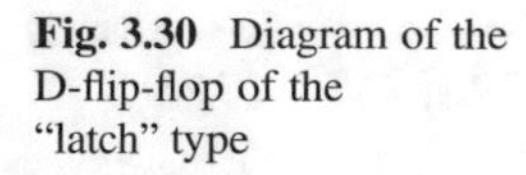

Fig. 3.30 Diagram of the D-flip-flop of the "latch" type

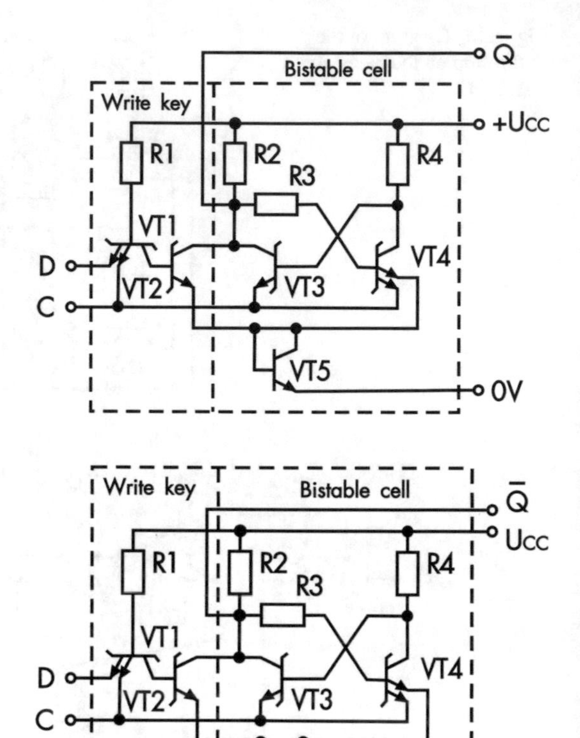

Fig. 3.31 Diagram of the D-flip-flop of the "latch" type with the enhanced write stability

VT1 and VT2 and the bistable cell (bc) with the switched-off feedback on the transistors VT3 and VT4, connected across the output $\overline{Q}$.

The transistor VT5 is required for enhancement of the switching thresholds at the inputs *D* and *C* and to ensure the noise immunity of the D-flipflop. With the high-level signal at the input *C*, the transistor VT3 is closed, owing to which the feedback to the base cell is disabled.

The write key transfers to the output $\overline{Q}$ the information from the input *D* in the inverse form, bypassing the basic cell and thus ensuring the minimum time of transferring the information t_{PHL}, t_{PLH}. Meanwhile the transistor VT4 gets set to the state, corresponding to the signal at the input D. With application of the low-level signal to the input *C*, the write switch (key) closes up and locks the information transfer from the input *D*. The emitter potential of the transistors VT3 and VT4 sinks, which results in disabling the feedback in the basic cell. Meanwhile, in the basic cell, the signal level of the transistor VT4 is retained. The informational level of the basic cell, plotted from the transistor collector VT2, arrives to the output. As per the similar principle, the flip-flop functions, whose electric diagram is indicated in Fig. 3.31 and which differs with the realization diagram of the basic cell.

3.3 Schematics of the Input Matching Elements of the TTLS Integrated Circuits

3.3.1 Input Matching TTLS Elements of Integrated Circuits with the standard TTL input levels

The electric diagram of the input matching element of the digital TTLS integrated circuits with the standard TTL input levels consists of the gain circuit, to the input of which the level matching circuit is connected, ensuring recognition of the TTL signal levels.

As a gain circuit, one usually uses a push-pull gain stage, whose electric diagram is indicated in Fig. 3.32. From the input level matching circuits, most popular are the following: type 1, diode circuit (Fig. 3.33a); type 2, transistor circuit (Fig. 3.33b); and type 3, diode-transistor circuit (Fig. 3.33c). Comparison of the input characteristics of circuits indicate that during application of the transistor circuit of type 2, the matching element switching threshold:

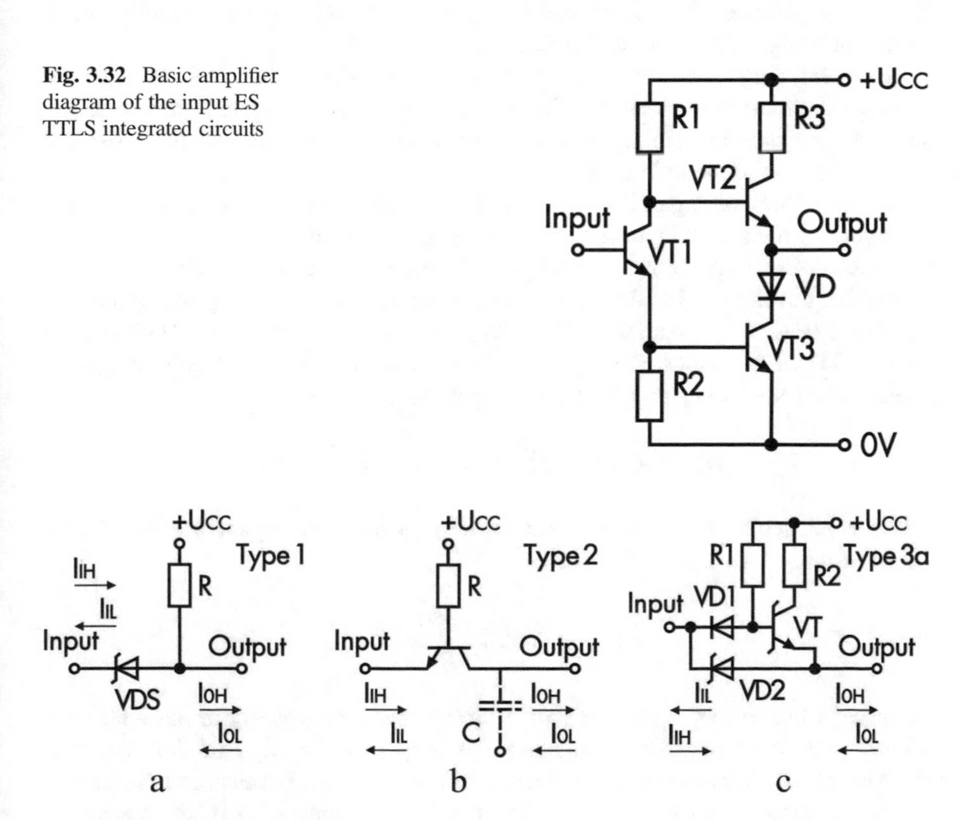

Fig. 3.32 Basic amplifier diagram of the input ES TTLS integrated circuits

Fig. 3.33 Diode (**a**), transistor (**b**), and diode-transistor (**c**) of the matching diagram of the integrated circuit ES TTLS levels

$$U_{T2} = U_{BE}^{VT1} + U_{BE}^{VT2} - U_{CE}^{VT} \approx 1,3B;$$

i.e., this circuit makes it possible to obtain a higher switching threshold than that of the diode circuit of type 1, for which:

$$U_{T1} \approx U_{BE}^{VT1} + U_{BE}^{VT3} - U_{P}^{VD4} \approx 1,0B;$$

With the similar resistance values of the resistors R, the maximum values of the low-level input currents I_{IL} circuit of type 1 and 2 differ insignificantly and are as follows:

$$I_{IL1} = \frac{U_{CC} - U_{P}^{VDS} - U_{IL}}{R};$$
$$I_{IL2} = \frac{U_{CC} - U_{BE}^{VT} - U_{IL}}{R};$$

The diode circuit possesses a higher rate of response, as the barrier capacitance of the Schottky diode VDS contributes to the charge transfer to the matching circuit output or to the charge dissipation from the output.

However, the junction capacitance of collector-substrate of the input transistor of the type 2 circuit is connected to the input, which reduces its rate of response and prompts necessity in stepping up the output current I_{OH} by means of the resistance value reduction of the resistor R.

Another shortcoming of the type 2 circuit is in the fact that with the similar input voltages, the high-level input current I_{IN} is higher in it. And, finally, the type 2 circuit has a lower input breakdown voltage, which determines its lower reliability.

Further development of the type 1 and 2 circuits resulted in emergence of the type 3a circuit (Fig. 3.33c) with the additional gain stage on the transistor VT1 and the diode VD1 at the input. Owing to application of the additional gain stage, the matching element switching threshold rose up to the level:

$$U_{T3} \approx U_{BE}^{VT1} + U_{BE}^{VT3} + U_{BE}^{VT} - U_{P}^{VD} \approx 1,5B;$$

which enhanced the matching element noise by the low-level signal. Meanwhile, the input current of the low-level circuit:

$$I_{IL3} \approx \frac{U_{CC} - U_{P}^{VD1} - U_{IL}}{R}$$

is close to the current I_{IL} of the type 1 circuit. However, owing to the additional resistor R2, it is possible to increase the output current I_{OH} of the circuit to compensate the total time delay reduction of enabling t_{PHL} because of introduction of the additional gain stage. The delay time of disabling t_{PLH} is not practically influenced by introduction of the additional stage, as the Schottky diode VDS

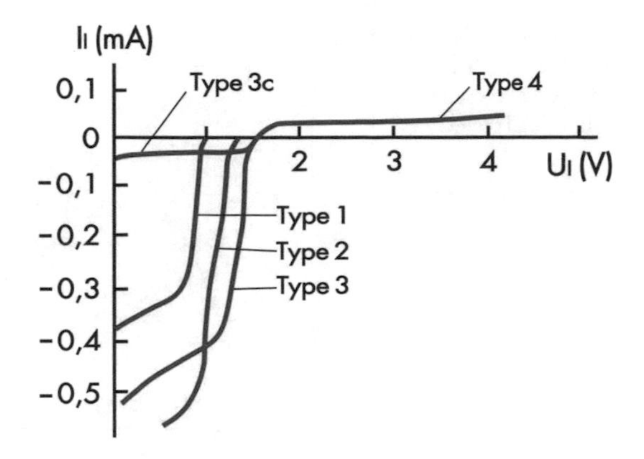

Fig. 3.34 Input characteristics of the matching element input TTLS of the integrated circuits

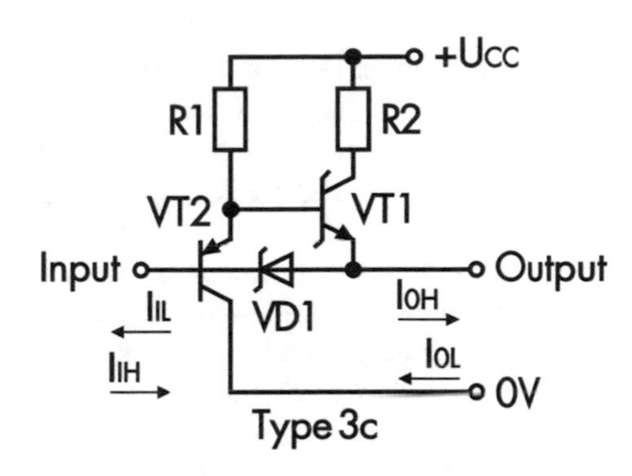

Fig. 3.35 Matching diagram of the input ES TTLS of the integrated circuits with the input p-n-p-transistor

performs the same discharge function and ensures the same value of the low-level output current I_{OL}, as the diode VDS in the type 1 circuit. With the purpose of enhancing the matching element reliability, the diode VD1 is made on the junction basis of collector-base, having the maximum breakdown voltage in the structure of the integrated circuits.

The average switching delays of all three types of the circuits are approximately similar, but the type 3a matching element has a somewhat greater consumption power. Nonetheless, this shortcoming is successfully compensated by a higher noise immunity of the circuit and reliability of its performance.

Input characteristics of the matching element are indicated in Fig. 3.34.

The important improvement of the input circuit of the type 3 consists of replacement of the input diode VD for the p-n –p transistor VT2 (Fig. 3.35).

Such a circuit makes it possible to abruptly reduce the low-level input current to the value:

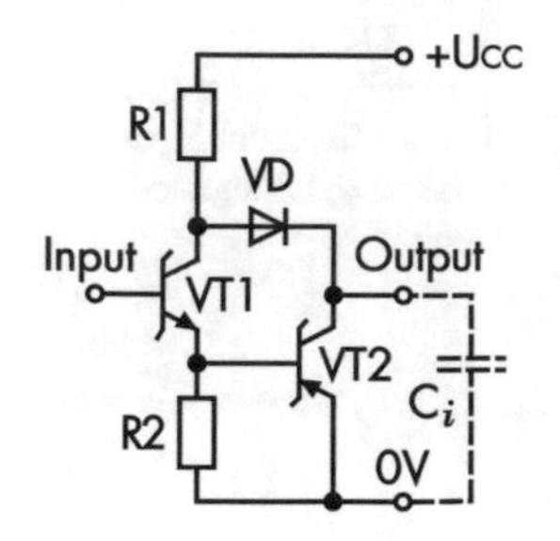

Fig. 3.36 Gain diagrams of the input ES TTLS integrated circuits with a small load capacitance

$$I_{IL3} = \frac{U_{cc} - U^{VT0} - U_{IL}}{\left(\beta_N^{VT2} + 1\right) \cdot R1};$$

where: β_N^{VT2}—gain ratio of the base current of the transistor VT2

Meanwhile, as the p-n-p-transistor VT2 functions in the circuit with the "common collector," the rate of response of the matching element in comparison with the diode circuit of type 3 practically does not change, and the switching threshold U_T remains the same.

One of the important requirements to the input matching elements of the digital integrated circuits is a minimum number of the schematic elements; therefore, there may be the simplest modifications of the gain circuits of the matching element input signals, the majority of which is described in [1]. In Fig. 3.36a, a simplified gain circuit is indicated, having a small load capacitance. This is related to the fact that load capacitance charge C_L ЭС happens through the diode VD and the resistor R1, having high resistance. As during the circuit activation at the expense of the feedback diode VD there happens the collector current interception of the transistor VT1 (and, appropriately, of the base current of the output transistor VT2), then with the purpose of preventing the duration deterioration of the activation edge t_{HL} of the output signal, the area of the emitter junction of the transistor VT2 should exceed the area of the emitter VD.

In the inner circuits of the TTLS integrated circuits are widely popular the high-speed logic elements with the threshold switching voltages of $UT{\approx}0.5{\div}1.3$ V, which causes the necessity of reducing the logic change of voltages at the output of the input matching elements. Thus, in Fig. 3.36, the gain diagram is indicated of the input matching element, in which because of application of the additional circuit limitation is ensured of the high-level output voltage:

$$U_{OH} = U_R - U_{BE}^{VT3} - U_P^{VDS} \approx 3U_P^{VDS}, U_{OH} = U_R \approx 3U_P^{VDS} + U_P^{VD6};$$

One of the known methods, making it possible to enhance the integration extent of the bipolar integrated circuits and their performance is in reduction of the power supply voltage of the internal logic elements to the level of $UCC \approx 1.5{\div}3.0$ V. However, such levels of the power supply integrated circuits do not permit to ensure power supply of the input matching elements with the standard TTL-levels. This

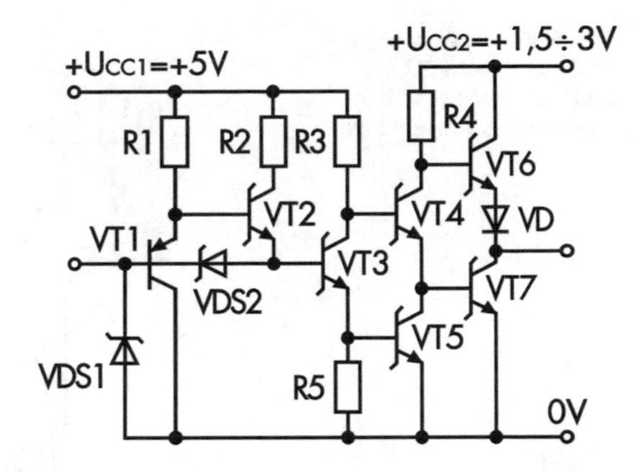

Fig. 3.37 The power supply diagram of the input ME TTLS integrated circuits with the various supply voltage values of the input matching elements and the internal functional blocks

causes the necessity of two values of supply voltage: the standard one of 5 V for the input matching elements and the low voltage one for the input matching elements and the low voltage one of 1.5÷3.0 V for the internal logic elements. However, during transfer of the signal from the output of the matching element to the input of the logic element there will happen deterioration of the rate of response due to the great logic voltage swing at the output of the matching element (at the inputs of the logic element). Influence of this shortcoming can be minimized during power supply of the matching element gain output circuit from the low voltage power supply of 1.5÷3 V and of the input matching element matching circuit from the standard one of +5 V, as indicated in Fig. 3.37 [4].

3.3.2 Input ME TTLS of Integrated Circuits with the Enhanced Load Capacitance

In some cases of application of the internal matching elements, for instance, in the synchronization circuits, it is required that the matching element should handle a large number of loads. In such cases, they use the special input matching elements, differing with the output gain circuit and possessing the increased output currents I_{OL}, I_{OH}. The required values of these currents are ensured basically by selection of the resistor values and the structural design of the output transistors. However, a change in the values of resistors results in the build-up of both the static and dynamic consumption currents, which, in its turn, determines a high level of noises in the power supply bus U_{CC} and the common bus *0V* that can disrupt the performance of the integrated circuits. In order to reduce the noise level, they use various schematic methods.

As an example, let's consider the gain circuit of the input matching element with the enhanced load capacitance (Fig. 3.38). The circuit differs from the standard one with availability of the additional Schottky diodes VDS1–VDS3, absence of the resistor in the circuit of the output transistor VT3, and application of the Darlington

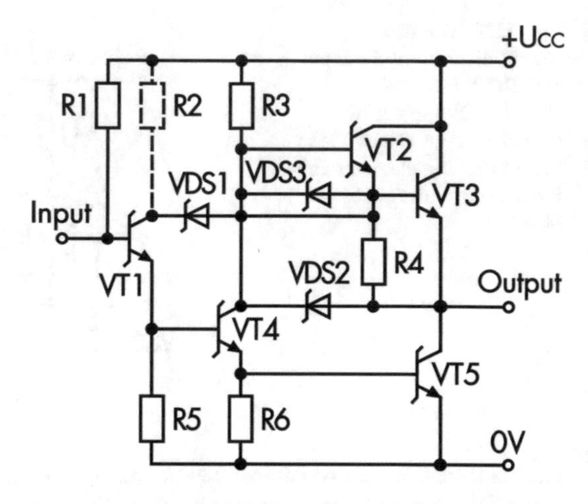

Fig. 3.38 The gain diagram of the input ME TTLS integrated circuits with the enhanced load capacitance

transistors in the output matching element circuit. Availability of the Shottky diode VDS2 between the matching element output and the collector of the phase separating transistor VT4 ensures reduction of the output resistance of the transistor VT5 during turning power on and improvement of the edge duration in activation of t_{HL}. This is related to the fact that during the load capacitance discharge a part of the output current I_{OL} through the diode VDS2 and the phase separating transistor VT4 and gets in the base of the output transistor VT5 and cuts down its output resistance. The Schottky diodes VDS1, VDS3 during activation of the matching element ensure the preceding deactivation of the output transistor VT3 with respect to activation of the output transistor VT2, thus reducing the noise level in the common bus *0V* and the power supply bus U_{CC}.

Application of a couple of transistors VT2 and VT3, as per the Darlington circuit and exclusion of the resistor in the circuit of their collectors, makes it possible to step up the charge output current I_{OH} of the matching element and to improve the edge duration of the matching element deactivation.

3.3.3 Input ME TTLS of the Integrated Circuits with the Paraphrase Outputs

In the majority of the application cases, the input ME TTLS of the integrated circuits do not bear the functional load and perform the matching functions and those of protection and form at the output the inverse or direct signal from the input one. The functional potentialities of the input matching element can be stepped up by introduction of two outputs into the matching element: the "repeating" input signal and the "inverting," i.e., paraphrase ones. With the necessity of forming such signals, one can use the serial connection of the input ME inverter and the internal logic element, also with the function of inversion. The other realization version presupposes the

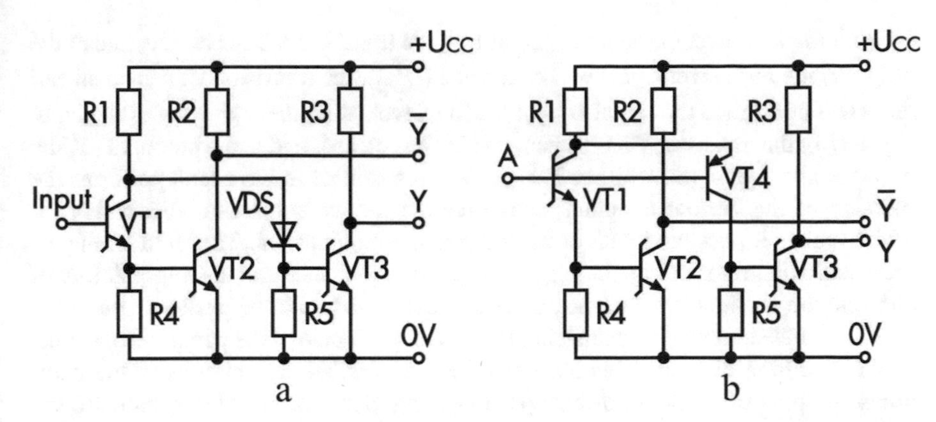

Fig. 3.39 Gain diagrams of the input matching elements with the paraphrase outputs on the basis of transistors of one (**a**) and different (**b**) types of conductivity

parallel connection of two input matching elements: with the inversion function and the repetition function. However, for the paraphrase signals, the important parameter is the phase synchronism of the output signals, i.e., difference in time of the formation delays of the output forward Y and inverse $\overline{Y}$ signals Dt_c.

It is known that for the reliable operation of the digital devices, it is required that this difference should be as minimum as possible.

From the first considered version, it follows that the difference in delays of the output signals will be equal to the switching time delay of the internal logic element. The shortcoming of the second version is the two times reduction of the load capacitance because of integration of two matching element inputs.

For elimination of these shortcomings, they use the special input matching elements with the paraphrase outputs. In the circuits of such type, the difference reduction of the formation time delay of the paraphrase signals is ensured basically in the gain circuit of the input matching element by means of branching the signals with the help of the phase separating transistor (Fig. 3.39a) [5]. With the high level of the input signal U_{IH} the emitter current of the phase separating transistor VT1 supports the open state of the output transistor VT2 and forms the inverse one as to the input output signal $U_{OL}^{\overline{Y}}$ of the low level.

The voltage low-level U_C^{VT1} of the collector of the phase separating transistor VT1 closes the diode VD, the output transistor VT3, and forms the high-level output signal U_{OH}^{Y}, similar to the input one. During alteration of the input signal level at U_{IL}, the phase separating transistor VT1 and the output transistor VT2 will close up, and the transistor VT3 will open up. Meanwhile, at the outputs Y, $\overline{Y}$, there will be formed such paraphrase signals from the input signal U_{IL}.

In Fig. 3.39b the electric diagram the matching element with formation of the paraphrase signals by means of application of the transistors of the different type of conductivity: n-p-n VT1 and p-n-p VT4, one of which functions in the mode with the “common emitter” and the second one with the “common collector” [2].

With the low level of the input signal U_{IL}, the transistor VT1 is closed, and at the output Y, the high level signal will be formed (U^{Y}_{OH}), the transistor VT2 is open, and there will be formed the signal $U^{\overline{Y}}_{OL}$ of the low level. With the high level of the input signal U_{IH}, the transistor VT1 is open, and VT4 is closed, and at the outputs Y, $\overline{Y}$, the opposite levels of signals will be formed. The resistor R1 in the circuit performs the function of the current common source for the output transistors VT2 and VT3, which makes it possible to cut down the consumption current. The shortcoming of the electric diagrams in Fig. 3.39 is the low load capacitance and the impossibility of construction on their basis of the power output circuits with the push-pull output.

The simplest method of enhancing the load capacitance of the paraphrase outputs of the matching element is connection of the additional gain circuits to the paraphrase outputs of the low power input matching element. In order to enhance the load capacitance of the input matching elements with the paraphrase outputs, we use the parallel switching two phase separating transistors VT1-1 and VT1-2 (Fig. 3.44a).

The first from these transistors VT1-1 forms the signal to control the base of the transistor VT4 of the circuit for formation of the forward signal Y. The second transistor VT 1-2 controls the output transistors VT2, VT5, and VT6 of the push-pull formation circuit of the inverse signal $\overline{Y}$. In order to reduce the time difference of the formation delay of the paraphrase signals, the circuit had additionally the transistors VT4-1, VT7 diode VDS, and resistor R3, reducing the base current of the output transistor VT3 during its disabling and stepping up during activation. In the electric diagram in Fig. 3.40b [1], the phase separating transistor VT1-1 operates in the activation circuit with the "common collector," and formation of the forward signal Y, plotted from the transistor's collector VT3, is attained by means of the additional transistor VT7. Taking into account that the transistor collector current VT1-2 is limited with the resistor R1, one can achieve by the rated value selection of the resistors R3, R5, and R6 the minimum difference in the formation delay time of the paraphrase signals Y, $\overline{Y}$.

3.3.4 Input ME TTLS Integrated Circuits with Memory

Input ME TTLS Integrated Circuits with Elements of Memory, Clocked by the Signal Level

Matching elements of such type are built on the basis of the memory elements, clocked by the synchrosignal edge and reviewed in Sect. 3.2.2. As for such matching elements, being "transparent" for the data signal the transfer time of the signal t_{PHL}, t_{PLH} should be minimum, and then during their design development, the memory elements are matched with the ME input circuits.

In Fig. 3.41a is an example of the electric diagram of the input matching element with the memory, based on the D flip-flop of the "latch" type, in which the input matching circuit is aligned with the write key of KW D-flip-flop. The schematic

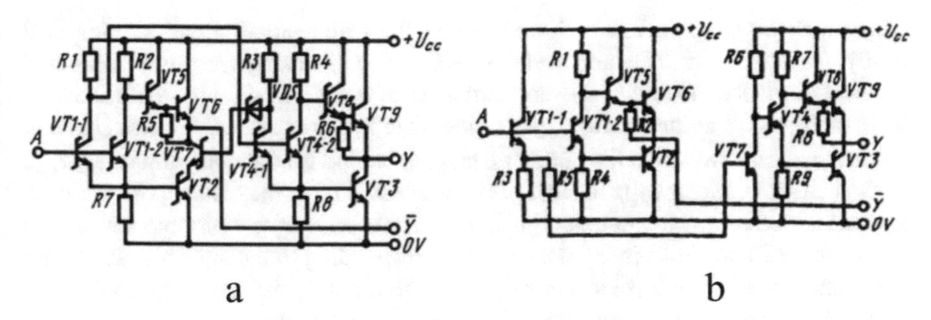

Fig. 3.40 Gain diagrams of the input ME TTLS integrated circuits with the paraphrase outputs and enhanced load capacitance [1]

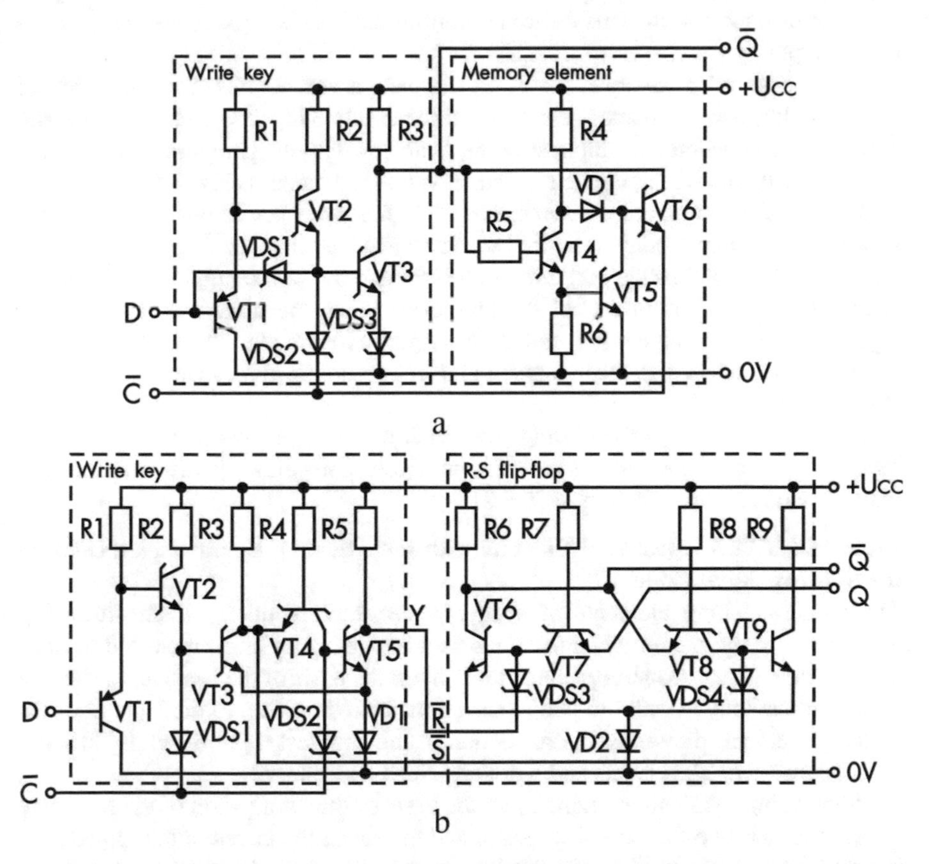

Fig. 3.41 Input ME diagrams with memory: (**a**) with D flip-flop of the "latch" type; (**b**) with R-S flip-flop

diagram differs from the one reviewed in Sect. 3.2 with availability of the additional transistors VT1 and VT2 and resistors R1 and R2, raising the input switching threshold of the write key to the standard level of the TTLS circuits. With the high level of the signal at the synchroinput *C*, the write key is open, and the data signal in the inverse form arrives to the output $\overline{Q}$, lagging behind for the delay time t_{PHL}, t_{PLH} of the write key. Meanwhile the feedback transistor VT6 of the bistable cell is closed and has no effect on the output signal, and the inverter of the bistable cell on the transistors VT4 and VT5 is set to the state, corresponding to the input signal. During application of the low-level synchrosignal to the input *C*, the write key gets closed and locks reception of the informational signal from the inputs.

Meanwhile there happens activation of the feedback transistor VT6. As a result, a bistable cell is included into the memory element in which is stored the last arrived to the inverter information. This information from the transistor collector VT6 arrives to the output $\overline{Q}$.

In the case of necessity of forming the paraphrase outputs in such matching elements, they use the memory element on the basis R-S flip-flop, to the inputs $\overline{R}$ and $\overline{S}$ of which is connected the input strobing write key with the paraphrase outputs. An example of the matching element of such type is indicated in Fig. 3.41b. With the high level of the signal at the synchroinput *C*, the write key is open, and the data signal from the input, converting into the forward *Y* and inverse $\overline{Y}$ signals, arrives to the inputs $\overline{R}$ and $\overline{S}$ flip-flop and through it is transferred to the outputs *Q*, $\overline{Q}$. On the arrival to the synchroinput *C* of the low-level signal, the write key will lock the signal reception from the input, and at the outputs *Y*, $\overline{Y}$ the high level signals are set. These signals, arriving to the inputs $\overline{R}$ and $\overline{S}$ R-S of the flip-flop, switch it over to the storage mode.

In case of necessity of acquiring from such matching elements the logic capacitance as a memory element, there can be used the appropriate circuits with the high load capacitance, described in Sect. 3.2.

Input ME TTLS Integrated Circuits with the Memory Elements, Clocked by the Synchrosignal Edge

The input matching elements of such type are built similarly to the matching elements with the memory elements, clocked by the signal level, reviewed in Sect. 3.2.1. When using the memory element, built on the basis of the basic logic element integrated circuits, in order to obtain such a matching element to the input D of the memory element, they make a connection of the simplest input matching element with the standard TTL levels and a not high load capacitance.

When using the memory element on the basis of the Dt flip-flop of the M-S type in order to improve the rate of response and to reduce the consumption power, the input threshold circuit of the Dt-flip-flop is to be aligned with the input matching circuit. The diagram in Fig. 3.42. differs with availability of the transistors VT15 and VT16 and resistors R15 and R16, enhancing the input switching threshold to the level of the standard TTL circuits. Functioning of such a matching element does not differ from functioning of the Dt-flip-flop, described in Sect. 3.2.1.

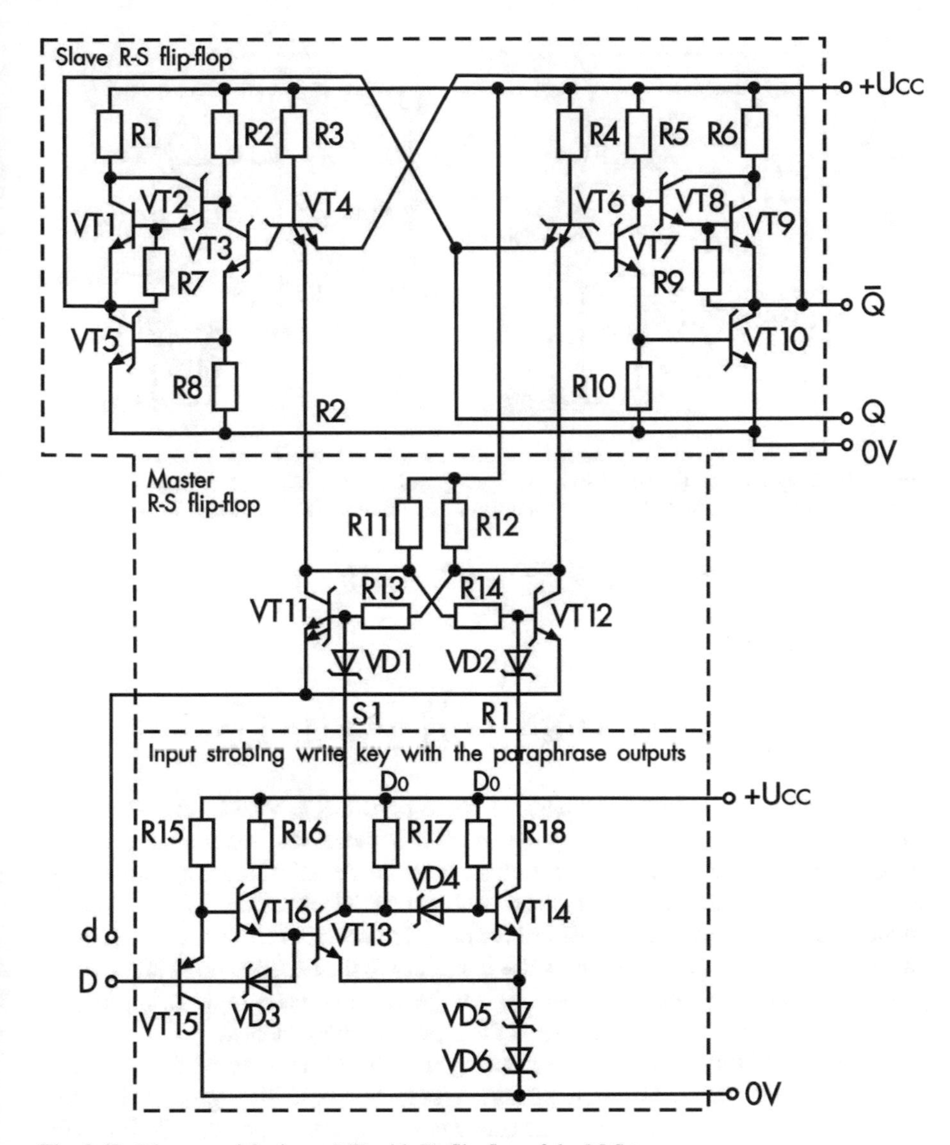

Fig. 3.42 Diagram of the input ME with Dt flip-flop of the M-S type

3.3.5 Input BE TTLS of Integrated Circuits with the Enhanced Noise Immunity

The most simple method of enhancing the noise immunity of TTLS of the input matching elements is raising the input switching threshold U_T.

For this there can be used bias elements—diodes or transistors [1] (Fig. 3.43a), or bias resistors [5] (Fig. 3.43b). In the first case, the value of the threshold voltage is as follows:

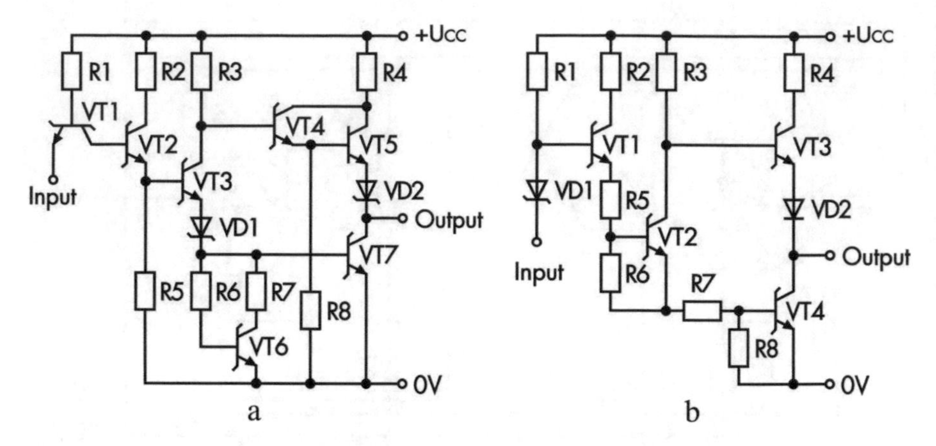

Fig. 3.43 Diagrams of the input BE TTLS integrated circuits with the enhanced switching threshold with application of the diode or transistor bias elements (**a**) or resistive separators (**b**)

$$U_\mathrm{T} = U_{BE}^{VT2} + U_{BE}^{VT3} + U_{P}^{VD1} + U_{BE}^{VT7} - U_{CE}^{VT1};$$

in the second case:

$$U_\mathrm{T} = U_{BE}^{VT2} + U_{BE}^{VT3}\left(1 + \frac{R5}{R6}\right) + U_{BE}^{VT7}\left(1 + \frac{R7}{R6}\right).$$

Application of the resistive-transistor chain of R6, R7, VT6 in the diagram in Fig. 3.43a, apart from smoothening the transfer characteristic, ensures enhancement of resistance of the input matching element to influence of the short-time pulse noises at the expense of holding in the closed state of the transistors VT3 and VT7 up to the input voltages, close to the threshold ones. However, in all cases, enhancement of U_T is related to reduction of the noise immunity to the negative noise ΔU_T, which requires the supply voltage increase U_{CC} and the input voltages logic swing ΔU_T, as well as worsens the rate of response of the input matching element.

The most effective method of enhancing the noise immunity of the input bias elements is application of the internal reverse bonds [1], making it possible to create the "hysteresis" on the transfer characteristic of the input bias element. Owing to the "hysteresis," the input voltage of unlocking U_{TL} exceeds the input voltage of locking U_{TH}, and the static noise immunity relative to the noises of both signs ΔU_T^+, ΔU_T^- is enhanced. In order to obtain the "hysteresis," they use the chains of the positive feedback. Meanwhile, switching a matching element from one state to the other bears a regenerative nature; therefore, the width of the junction area on the transfer characteristic is small.

In order to ensure the maximum noise immunity, the transfer characteristic is selected in such a way that it should be symmetrical relative to the input voltages U_{TL}, U_{TH}. In Fig. 3.44 the diagram is indicated with the internal feedback, in which the feedback chain contains the resistors R4, R5, and R7 and the transistor VT4.

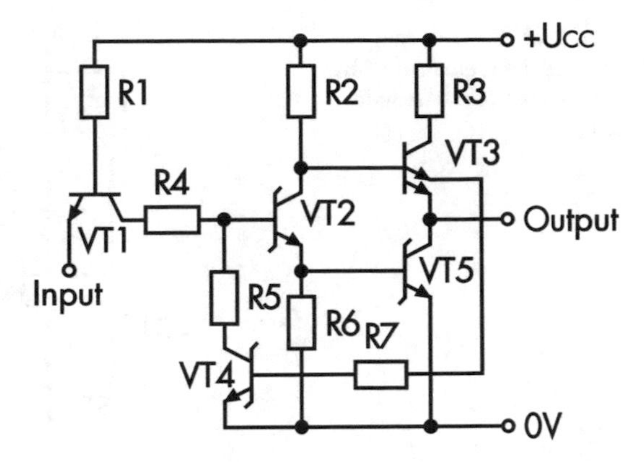

Fig. 3.44 The diagram of the input matching element of the enhanced noise immunity with the internal feedback

With the low voltage level U_{IL} at the input of the circuit, the transistors VT1, VT2, and VT5 are closed, and at the output of the circuit is the high voltage level U_{OH}. This level, while getting through the resistor R7 into the base of the transistor VT4, opens it and activates the resistive divider R4, R5 in the base of the transistor VT2. Therefore, when stepping up the input voltage, the matching element is switched over with the voltage:

$$U_{TH} \approx U_{BE}^{VT2} + U_{BE}^{VT5} \cdot \left(1 + \frac{R4}{R5}\right) - U_{CE}^{VT1}.$$

Sinking of the output voltage U_{OH} will result in locking the transistor VT4 and the regenerative activation of the transistors VT2 and VT5. Meanwhile, the chains of open link are disabled and those of the resistive divider R4, R5, because of which the switch-over of the matching element into the reverse state during reduction of the input voltage will occur at the voltage:

$$U_{TL} \approx U_{BE}^{VT2} + U_{BE}^{VT5} - U_{CE}^{VT1}.$$

Consequently, the hysteresis loop width is as follows:

$$\Delta U_H \approx \left(U_{BE}^{VT2} + U_{BE}^{VT5}\right) \cdot \frac{R4}{R5}.$$

While changing the values of the rated nominal values of the resistors, it is possible to obtain the required width of the hysteresis loop.

It is known [1] that the transfer characteristic of the “hysteresis” type is intrinsic to the Schmitt circuit. The diagram (Fig. 3.45) comprises the input part on the components R2, R4, VT1, VT2 and the output part on the gain inverter D1.

The positive feedback is effected by connection of the emitters of the transistors VT2 and VT3 and the resistor R4. With the low voltage level at the input of the

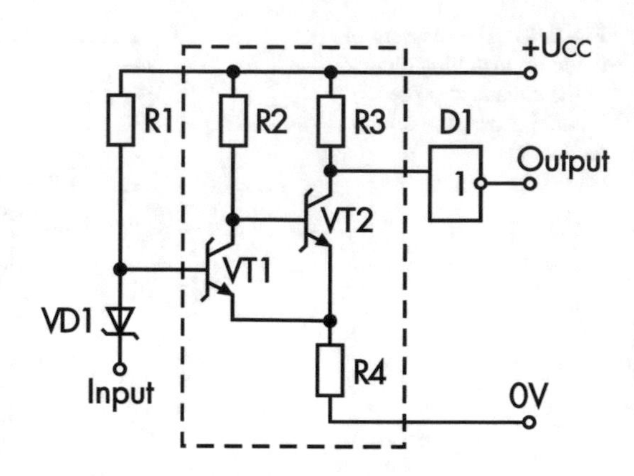

Fig. 3.45 The diagram of the matching element with application of the Schmitt circuit

matching element, the transistor VT1 is closed, VT2 is open, and at the output of the matching element, the high level voltage U_{OH}. will be set. In this state on the feedback resistor R4, the following voltage will be set:

$$U^{/}_{R4} = \frac{U_{CC} - U^{VT3}_{BE}}{R2 + R4} \cdot R4 + \frac{U_{CC} - U^{VT2}_{CE}}{R2 + R4} \cdot R4.$$

While stepping up the input voltage U_{IL} the switch-over of the matching element will occur at the voltage:

$$U_{TH} = \frac{U_{CC} - U^{VT3}_{BE}}{R2 + R4} \cdot R4 + \frac{U_{CC} - U^{VT2}_{CE}}{R2 + R4} \cdot R4 + U^{VT1}_{BE} - U^{VD1}_{P}.$$

During activation of the matching element, the transistor VT1 goes over to the open state and VT2 to the closed one, at the output of the matching element the low level voltage U_{OL} will be set. On the feedback resistor R4, the voltage will be set:

$$U^{//}_{R4} = \frac{U_{CC} - U^{VT1}_{BE}}{R2 + R4} \cdot R4 + \frac{U_{CC} - U^{VT1}_{CE}}{R2 + R4} \cdot R4.$$

During reduction of the input voltage U_{IH}, the switch-over of the matching element into the reverse state will occur at the voltage:

$$U_{TL} = \frac{U_{CC} - U^{VT1}_{BE}}{R1 + R4} \cdot R4 + \frac{U_{CC} - U^{VT1}_{CE}}{R2 + R4} \cdot R4 + U^{VT1}_{BE} - U^{VD1}_{P}.$$

On the assumption, that $U^{VT1}_{BE} = U^{VT2}_{BE} = U_{BE}$; $U^{VT1}_{CE} = U^{VT2}_{CE} = U_{CE}$ we will acquire the width of the "hysteresis loop":

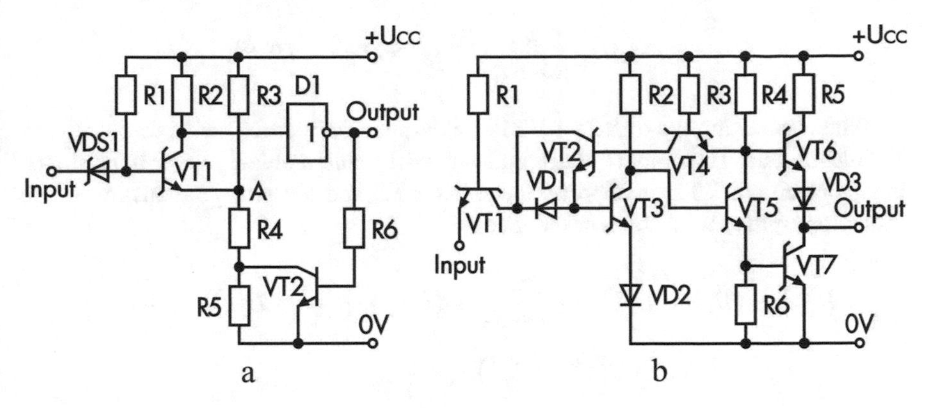

Fig. 3.46 Diagrams of the input matching element of the enhanced noise immunity with application of the shunting reverse links—with shunting the resistive divider (**a**), zener diode (**b**)

$$\Delta U_H \approx R4\left[\frac{1}{R2+R4}-\frac{1}{R1+R4}\right]\cdot(2U_{CC}-U_{BE}-U_{CE});$$

From the last expression, it follows that on the transfer characteristic of the matching element, the hysteresis is observed, and the switching process will bear the regenerative nature on the condition that *R1* > *R2*.

As the expression for U_{TH}, U_{TL} contains the relations of resistances, then ΔU_H weakly depends on the temperature and the technological scatter of the parameters. However, both threshold voltages U_{TH}, U_{TL} and the width of the hysteresis loop ΔU_H depend on the supply voltage U_{CC}.

The known method of formation of the feedback circuits for formation of the transfer characteristic of the "hysteresis" type (Fig. 3.46) is essentially in bypassing the components, stepping up the threshold voltage U_{TH} [5]. As indicated in Fig. 3.46 of the matching element, the bypassing circuit contains the components R3-R6, VT3. With the low level of the input voltage and $U_{IL}=0$ the transistor VT2 is closed, and at the output of the matching element, there will be the low-level voltage U_{OL}. As the base of the feedback transistor VT2 is connected with the output via the resistor R6, then the transistor VT3 is closed and has no significance on distribution of potentials on the matching element circuit. In this state to the emitter of the transistor VT1 via the voltage divider, the following voltage will be applied:

$$U_E^{VT1}=U_{CC}\frac{R4+R5}{R3+R4+R5};$$

With increase of the input voltage U_{IL} the transistor unlocking and the matching element switching will occur on attaining the voltage level:

$$U_{TH} = U_{CC}\frac{R4 + R5}{R3 + R4 + R5} + U_{BE}^{VT2} - U_{P}^{VDS1};$$

When unlocking the transistor VT1, the output amplifier D1 switches over to the high-level state. The output voltage build-up will result in unlocking the transistor of the shunt circuit VT2, in shorting the resistor R5 and the voltage reduction at the transistor emitter VT1 down to the level:

$$U_{E}^{VT1} = \left(U_{CC} - U_{CE}^{VT1}\right) \cdot \frac{R4}{R3 + R4} + \left(U_{CC} - U_{CE}^{VT1} - U_{CE}^{VT2}\right) \cdot \frac{R4}{R2 + R4} + \left(U_{CC} - U_{CE}^{VT2} - U_{BE}^{VT1}\right) \cdot \frac{R4}{R1 + R4}.$$

Therefore, during reduction of the input voltage, a switch-over of the input matching element will occur at the level:

$$U_{TL} = \left(\frac{U_{CC} - U_{CE}^{VT2}}{R3 + R4} + \frac{U_{CC} - U_{CE}^{VT1} - U_{CE}^{VT2}}{R2 + R4} + \frac{U_{CC} - U_{CE}^{VT2} - U_{BE}^{VT1}}{R1 + R4} + \right) \cdot R4 + U_{BE}^{VT11} - U_{P}^{VDS1}$$

By value selection of the resistances R1–R5, it is possible to set the required switching thresholds, U_{TL} U_{TH} and the required width of the "hysteresis loop" and ΔU_{H}. Despite the advantage (an insignificant scatter of the voltage values U_{TL}, U_{TH} and their weak temperature dependence), such a diagram has a substantial drawback—dependence of the threshold voltages U_{TL}, U_{TH} on the supply voltage U_{CC}.

In the diagram in Fig. 3.46b as a shunt element, zener diode VD1 [4] is applied. When stepping up the input voltage, the activation threshold voltage of such a matching element will be as follows:

$$U_{TH} = U_{C}^{VD1} + U_{BE}^{VT3} + U_{P}^{VD2} - U_{CE}^{VT1};$$

where: U_{C}^{VD1}—voltage value of the Zener diode VD1

As during the low-level input voltage on the collector of the transistor VT5 there is available the low-level voltage, then the shunt circuit, containing the transistors VT2 and VT4 and the resistor R3, is disabled and has no effect on the functionality of the circuit. During activation of the matching element at its output, the high-level voltage U_{OH} is set. Meanwhile, the high voltage level on the collector of the transistor VT5 will activate the shunt circuit and will shunt the zener diode VD1 by means of the open transistor VT2. Therefore, during reduction of the input voltage U_{IH}, deactivation of the transistor VT3 will take place not through the zener diode VD1, but through the open feedback transistor VT2.

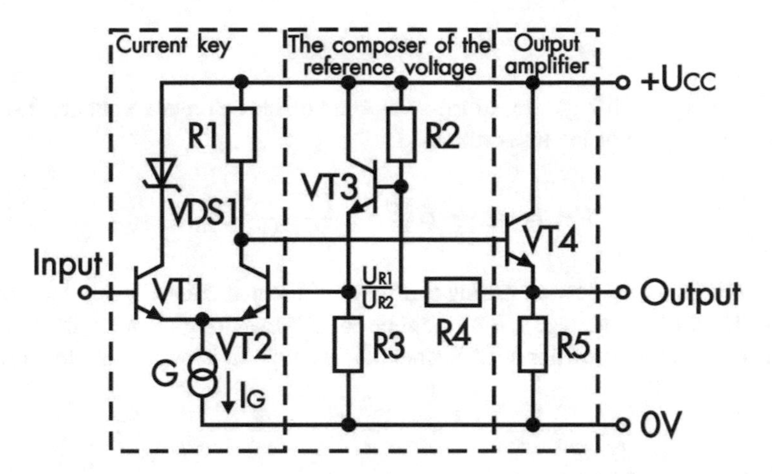

Fig. 3.47 The diagram of the input matching element of the enhanced noise immunity on the basis of the current key

The threshold switch-off voltage:

$$U_{TH} = U_{BE}^{VT3} + U_{P}^{VD2} + U_{CE}^{VT2} - U_{CE}^{VT1};$$

and thc hysteresis loop width is as follows:

$$\Delta U_H = U_C^{VD1} - U_{CE}^{VT2}.$$

As a zener diode one can use the emitter junction of the n-p-n transistor, which in dependence on the design layout has the reverse breakdown voltage $U_Э$=3 ÷ 4 V and weakly depends on the temperature.

The latter ensures the stable switching thresholds U_{TL}, U_{TH} within the range of the operating temperatures and voltages. In all above described circuits, formation of the feedback circuits was ensured by means of the transistors of a single type of conductance (n-p-n).

The "hysteresis" transfer rate can be formed by using as an input matching element as a double input current key, one from whose inputs is used as an input of the matching element, and to the second one are applied the state-dependent standard voltages U_{R1}, U_{R2}, determining the threshold switch-on voltages U_{TH} and the switch-off ones U_{TL}. In Fig. 3.47 the electric diagram of such a matching element is indicated, containing the tone key on the transistors VT1 and VT2 and the current generator G, the output amplifier on the transistor VT4, the resistor R5, and the composer of the reference voltages U_{R1}, U_{R2}.

With the high voltage level U_{IH} at the input of the matching element, the transistor VT1 is open, the VT2 transistor is closed, and at the output the high level voltage is set:

$$U_{OH} = U_{CC} - U_{BE}^{VT4}.$$

This voltage, while getting to the composer of the reference voltage, forms the voltage in the base of the transistor VT2:

$$U_{R1} = U_{CC} - U_{BE}^{VT3} \cdot \frac{R2}{R2 + R4} - U_{BE}^{VT3};$$

which determines the low switching threshold of the matching element U_{TL}. During the low level of the voltage U_{IL} at the input of the matching element, the transistor VT1 is closed, the transistor VT2 is open, the voltage falls on the resistor R1:

$$U_T = I_G \cdot R1;$$

and at the output the low-level output voltage is set:

$$U_{OL} \approx U_{CC} - U_{БЭ}^{VT4} - I_G \cdot R1.$$

This voltage, while getting to the composer of the reference voltage, forms the reference voltage in the base of the transistor VT2:

$$U_{R2} = U_{CC} - U_{BE}^{VT3} \cdot \frac{R2}{R2 + R4} - I_G \cdot \frac{R2}{R2 + R4} - U_{BE}^{VT3};$$

which determines the upper switching threshold of the matching element U_{TH}. The width of the "hysteresis loop":

$$\Delta U_H = I_G \cdot R1 \frac{R2}{R2 + R4};$$

is determined by means of the value of the standard current and the resistance values of the resistors and does not depend on the supply voltage of the matching element. The Schottky diode ensures the input isolation from the power supply bus with the turned off power supply U_{CC}.

3.3.6 Input Matching Element with Conversion of the Signal Levels

Input Matching Elements with Conversion of ECL Signal Levels

Matching complexity of the ECL outputs of the integrated circuits with the TTLS integrated circuit inputs is related to the various polarity of the input (output) signal levels. Thus, the TTL-input matching elements, having the switching threshold of $U_T = 1.4$ V, steadily differentiate the input signals of low $U_{TL} = 0.8$ V and high

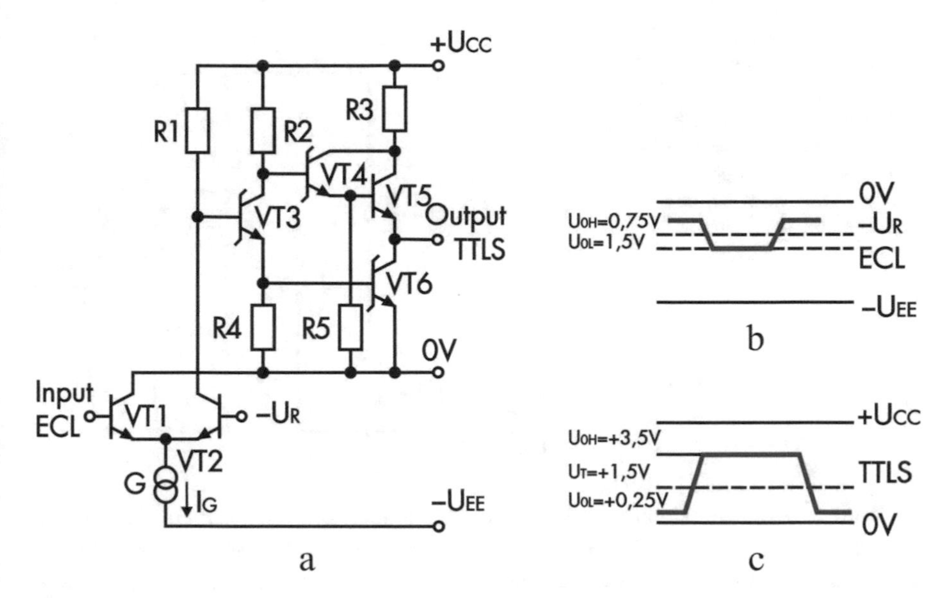

Fig. 3.48 Diagrams of the input matching element (**a**), converting the ECL signals of the different polarity and the time charts of the input (**b**), and output (**c**) signals of the matching element on the transistors VT1 and VT2 and the current generator G and TTLS-inverter

$U_{TH} = 2.0$ V levels, while the ECL output signals have the values of the low $U_{OL} = -1.5$ V and high $U_{OH} = -0.75$ V levels, recognized by the TTL inputs, as a low-level signal.

An option is possible, when both ECL and TTLS integrated circuits have the similar supply voltage: positive, $+U_{CC}$*–0V*, or negative *0V–*U_{EE} ("pseudo-ECL"). However, even in this case, despite the similar polarity of the output signals of ECL and the input TTLS switching threshold, also it is required to apply the special input matching elements, as the output signals of the integrated circuits will be acknowledged by TTLS of the integrated circuits as a high-level signal.

TTLS Matching Elements, Recognizing the ECL Signals of the Different Polarity

An elementary diagram of the input matching element of such type [1] is indicated in Fig. 3.48a, the levels of the converted input ECL signals in Fig. 3.48b, output TTLS – in Fig. 3.48c.

One "branch" of the current switch is connected with the common bus of 0 *V* while the other one with the TTLS input of the inverter. One input of the current switch forms the input of the matching element, and on the second one the standard voltage was applied, whose value in dependence on the input levels is selected within the range of -0.75 V<$-U_R$<1.5 V. The level matching is ensured by selection of the current value I_G of the current generator and the nominal rated value of the resistor R1.

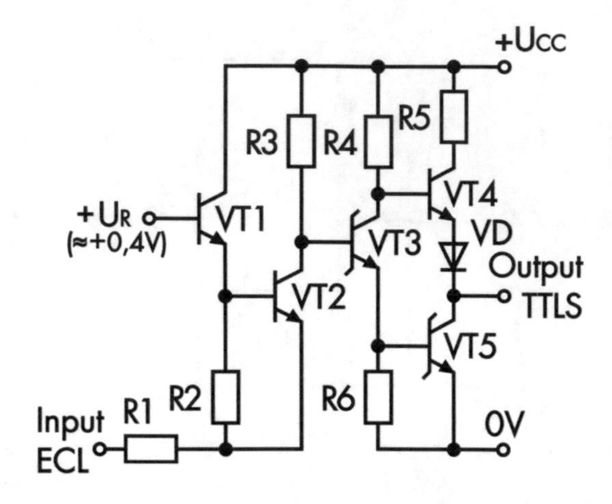

Fig. 3.49 The diagram of the input ECL TTLS integrated circuits, converting the ECL signals of the different polarity and having the power supply source of the single polarity with the integrated circuits.

With the high signal value at the input of the matching element of U_I> 0.75 V, the transistor VT1 is open, and via it from the common bus flows the current I_G. As meanwhile the transistor VT2 is closed, then the source current through the resistor R1 created by the power supply opens the transistors VT3 and VT6 and creates at the TTL output the low-level signal $U_{OL}^{ТТЛ}$ =0.25 V. During application to the matching element input of the low-level signal of U_I<−1.5 V, the transistor VT1 closes, and the transistor VT2 opens, and through it and the resistor R1 from the supply source U_{CC} flows the current I_G of the generator, creating the voltage drop on the resistor R1:

$$U_I = I_G \cdot R1;$$

This voltage drop lowers the potentials of the transistor base VT3 to the level:

$$U_B^{VT3} = U_{CC} - I_G \cdot R1;$$

Selecting the values of the current I_G and the resistor *R1*, it is possible to ensure the closed state of the transistors VT3 and VT6. Meanwhile, at the output of the matching element, the high-level TTL signal of U_{OH}=3.5 V.

As the potential in the base of the transistor VT3 does not lower, below the potential *OV*, the transistor VT2 operates in the active fast response mode, and it can be applied without the Schottky diode. The diagrams of the considered type have received a certain popularity in the integrated circuits because of the necessity of the three power supply pins: the positive one +U_{CC}, the common one *0* V, and the negative one $-U_{EE}$. The most prospective ones are the matching elements, also using the similar power supply voltage, as the internal blocks of the integrated circuits—the positive one +U_{CC}. In Fig. 3.49 the electric diagram is indicated of such matching element [1], consisting of the TTLS inverter on VT3–VT5 transistors and the level offset circuit on the transistors VT1 and VT2. As the offset voltage is U_R = +0.4 V, then the base potential of the transistor VT2 is fixed as $U_B = -0.35$ V.

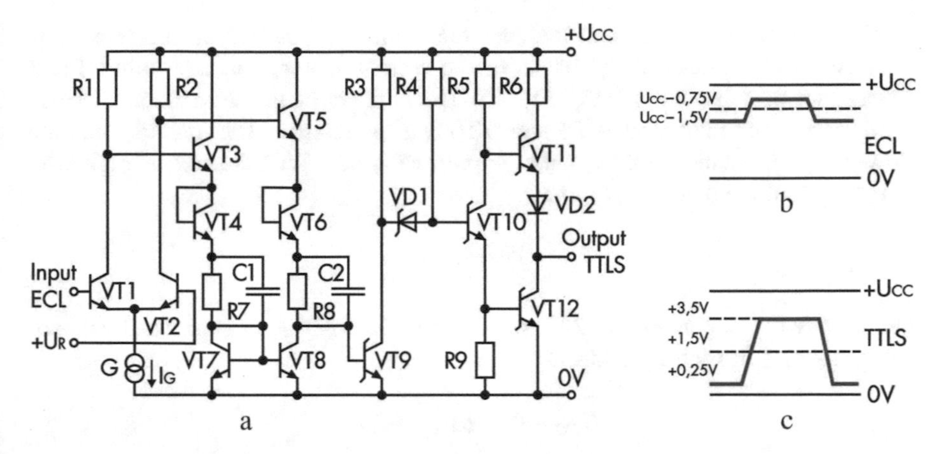

Fig. 3.50 The diagram of the input ME of the TTLS integrated circuits, converting the ECL signals of the similar with the ME polarity (**a**) and the time charts of the input (**b**) and output (**c**) signals

Then during application to the input of the matching element of the high-level voltage $U_{IH}^{ЭСЛ} \leq -0.75$ V, the voltage on the junction of thc basc-cmitter of the transistor VT2 will constitute $U_B^{VT2} = 0.4$ V, which will sustain its closed state.

Because of this, the transistors VT3 and VT5 are open, and at the output of the matching element, the low-level voltage $U_{OL}^{TTL} = 0.25$ V. During application to the input of the matching element of the low-level voltage of $U_{IL}^{ECL} \leq 1.75$ V on the resistors Rl and R2, the voltage drop of $\Delta U = -1.4$ V will be set, and the transistor VT2 will switch over to the open state. Meanwhile in the circuit of its collector current will flow:

$$I_C^{VT2} \approx \frac{0,75B}{R1} - \frac{U_{BE}^{VT2}}{R2};$$

This current, creating the voltage drop on the resistor R3, will result in the potential reduction in the transistor base VT3. One can control the value of this potential by means of the resistor R3. During fulfillment of the condition:

$$U_{CC} - I_K^{VT2} \cdot R3 < 2U_{BE}$$

the output TTLS inverter will switch over to the closed state, and at its output, the high-level voltage $U_{OH}^{TTL}$of =3.5 V will settle.

In this case TTLS of the integrated circuits has the similar with the control matching element of the integrated circuit the supply voltage of +*Ucc* and *0* V, or *0* V and $-U_{EE}$. The diagram of the input matching element is indicated in Fig. 3.50a [3] and the converted signals in Fig. 3.50b and Fig. 3.50c (for the positive power supply case). The matching element consists of the current switch on the transistors VT1 and VT2 and the current generator G. One input of the switch forms the input of the matching element, and to the second one is applied the standard potential U_R.

Besides, the matching element includes the emitter "repeaters" on the transistors VT3 and VT5, controlled by means of the current switch, and the output TTLS inverter on the transistors VT10÷VT12. The emitters of the transistors VT3 and VT5 include the transistors VT4 and VT6 and the resistors R7 and R8, bringing down the levels of the output signal of the current switch. With the voltage high level at the input of the matching element:

$$U_{IH}^{ECL} > R_H;$$

transistor VT1 is open and VT2-closed. When selecting the nominal rated value of the resistor R1 in such a way that:

$$U_{CC} - I_G \cdot R1 > 3U_{BE};$$

the emitter "repeater" VT3 will be closed, because of which the "mirror" of the currents on the transistors VT7 and VT8 will also be closed. Meanwhile the emitter "repeater" VT5 via the transistor VT6 and the resistor R8 opens the transistor VT9. The low voltage level on the collector of the transistor VT9 via the diode VD1 will switch over the output TTLS inverter to the closed state, and at its output the high-level signal U_{OH}=3.5 V will be formed. With the low level of the input signal of U_{IL}^{ECL}, the emitter one will be switched over to the open state: the "repeater" VT3.

On the assumption that:

$$U_{CC} - I_G \cdot R1 > 3U_{BE};$$

the transistor VT5 will be closed, and the open transistor VT3 through the transistor VT4 and the resistor R10 will open the transistor VT8. The potential on its collector will diminish below U_{BE}^{VT9}, the transistor VT9 will get closed, and the potential on its collector will rise, which will result in the unlocking of the transistors VT10, VT12 TTLS of the inverter and formation at its output of the low-level signal:

$$U_{OL}^{TTL} = 0.25B;$$

It is possible to simplify a similar diagram when using as the emitter "repeaters" of the transistors of both types of conductivity, as indicated in Fig. 3.51 [3].

Input ME TTLS of the Integrated Circuits with the CMOS Conversion of the Signal Levels

Matching CMOS of the integrated circuits with the TTLS of the integrated circuits is not a complex problem, as with the similar supply voltages of U_{CC} the CMOS outputs of the integrated circuits permit the direct load by means of the TTL inputs. This is related to the fact that TTLS of the input ME, having the standard switching threshold $U_T^{TTLS} = 1.5$ V (Fig. 3.52), differentiates the output CMOS high-level signals of $U_{OH} = 4.1$ V and of the low level of $U_{OL} = 0.4$ V. Meanwhile, the shortcoming of the direct matching is a certain deterioration of the switch-off time of

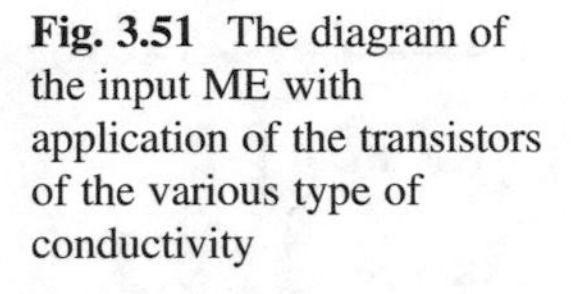

Fig. 3.51 The diagram of the input ME with application of the transistors of the various type of conductivity

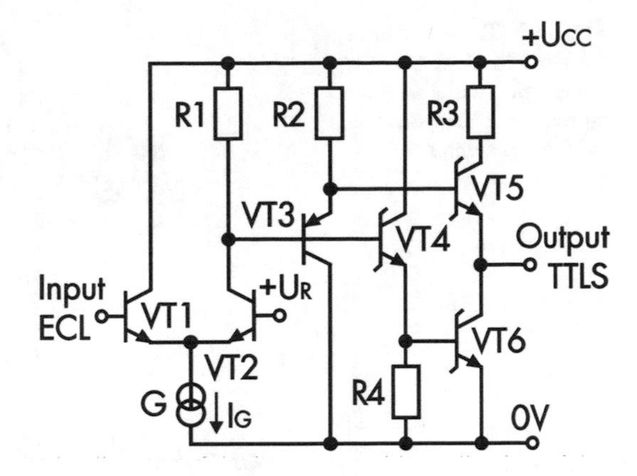

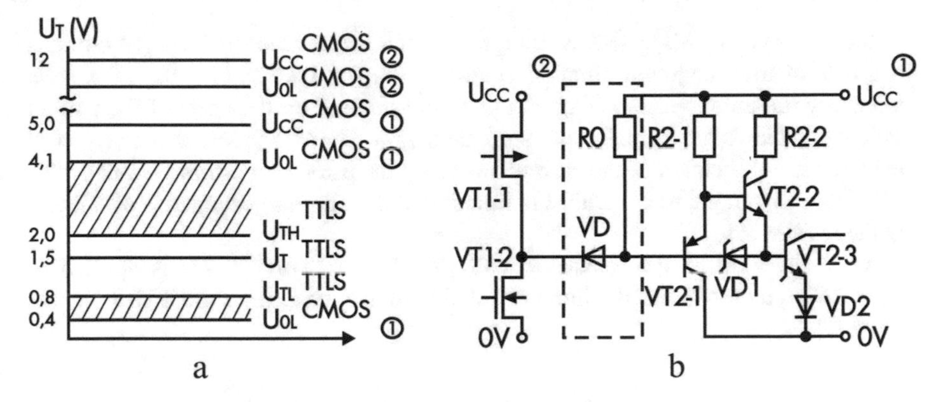

Fig. 3.52 The diagram of the threshold levels and output voltages (**a**) CMOS and TTLS integrated circuit and the conversion circuit of the signals CMOS-TTLS (И)

TTLS of the input ME (matching element) because of the higher value of the high-level output voltage of $U_{OH}^{CMOS} = 1.5$ V. Matching is complex, when the control CMOS circuit has a higher power supply voltage (for instance, $U_{CC} = 12$ V for CMOS of the circuits, series CD4000 A,C), as the output high level voltage of CMOS integrated circuits $U_{OH}^{CMOS} = 9$ V surpasses the limit permissible input voltage of TTLS of the integrated circuits with the power supply of $U_{CC} = 5$ V and $U_{Imax} = 7$ V.

In this case the matching can be ensured as per the diagram, indicated in Fig. 3.52b with application of the external or internal components: the resistor R0 and the diode VD. Meanwhile, it is required that the reverse breakdown voltage of the diode U_{BR}^{VD} substantially exceeded the output voltage U_{OH}^{CMOS}:

$$U_{BR}^{VD} >> U_{OH}^{CMOS};$$

the direct voltage drop:

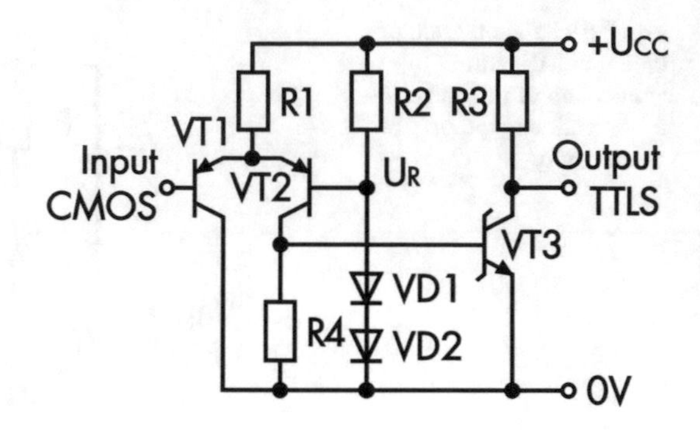

Fig. 3.53 The diagram of the input ME TTLS of the integrated circuits on the current switches, converting the CMOS signals

$$U_{\Pi P}^{VD} < U_{T}^{\text{ТТЛШ}} - U_{OL}^{\text{КМОП}}.$$

As the majority of CMOS (MOS) integrated circuits have the small output currents in TTLS of the integrated circuits, controlled by the CMOS circuits, of a wide popularity are the input matching elements, constructed on the basis of the current switches and having the small input currents I_{IL}, I_{IH}. The ME circuit of such type can be built with application of the current switch on the p-n-p-transistors VT1 and VT2 (Fig. 3.53) [3]. In the ME circuit, the function of the current generator is performed by the resistor R1.

As in the base of the transistor VT2 the standard voltage VD is applied of $U_R \approx 2U_P^{VD}$, then ME has the input threshold voltage, typical for the TTLS circuits $UT \approx 1.5$ V.

3.3.7 Protection Diagrams of the Input ME TTLS Integrated Circuits

TTLS of the integrated circuits in comparison with other types of the integrated circuits are relatively resistant to influence of the static electricity. However, the further advancement of the technology and reduction of the linear sizes of the elements of the integrated circuits resulted in enhancement of their sensitivity to influence of the static electricity. Application of the p-n-p transistors at the ME inputs made it possible to improve the load capacitance; however, it increased the input resistance and enhanced their sensitivity to influence of the static electricity, which required introduction of the additional protection means of the inputs of the integrated circuits.

The most applied protection element of the TTLS inputs of the integrated circuits is the Schottky diode VDS2 (Fig. 3.54a), participating in the breakdown mode at the overvoltage of the input and directing the current of the static charge to the common bus *0* V. The shortcoming of such circuit—the steadily rising IV, characteristic of the diode breakdown (Fig. 3.55a, a curve 1), which with the large power of the static

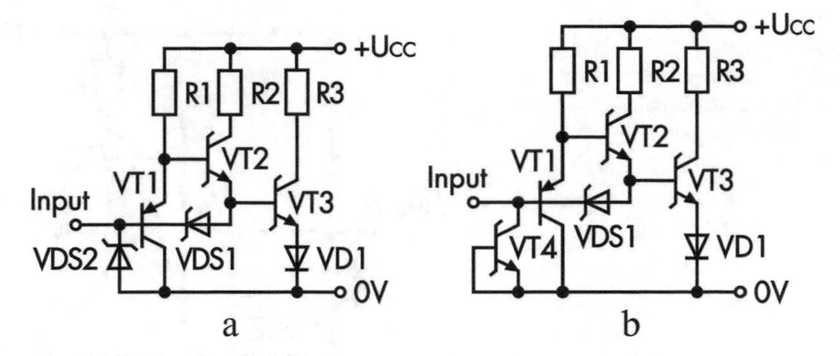

Fig. 3.54 The protection diagrams from the static electricity of the input ME TTLS

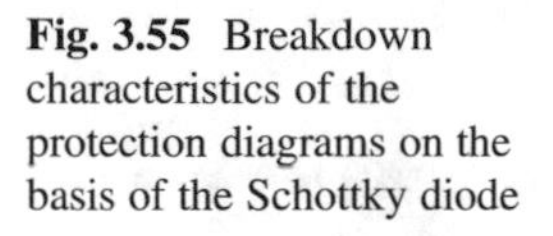
Fig. 3.55 Breakdown characteristics of the protection diagrams on the basis of the Schottky diode

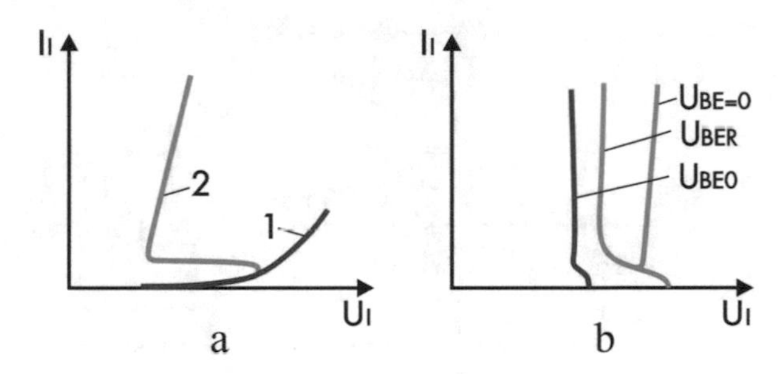

charge and the continuous flowing of the current through the diode VDS2—may cause its disintegration and preclude the reliable protection of the ME internal components.

The most advanced is the protection diagram, indicated in Fig. 3.54b, in which the protective element is the n-p-n-transistor with the "shortened" base VT4. During the voltage step-up at the ME input, the transistor switches over to the mode of the impact ionization, the voltage on its collector abruptly drops, and in its IV characteristic, the section appears with the negative resistance (Fig. 3.55a, the curve 2). Meanwhile, at the input the low-Ohmic discharge circuit is created, for the accumulated at the input of the static discharge, thus preventing the destruction of the ME internal components. During the shorting of the junction of the base-emitter of the transistor VT4 through the resistor R4 (Fig. 3.55b), it is possible to reduce the residual voltage of the transistor VT4 in the breakdown mode (U_{BER}) and with the disruption of its base to attain the minimum value of the residual voltage. However, in this case, the maximum permissible voltage at the ME input lowers.

Besides the n-p-n transistors, for protection of the ME inputs from the static electricity, it is possible to use the p-n-p transistors both with the shortened via the resistor base [36] and with the "disrupted" base (Fig. 3.56) [3]; meanwhile, this functionality mechanism of the protection circuits is similar to the protection circuits on the n-p-n-transistors.

In the above indicated ME circuits, the protection circuits are introduced between the ME input and the common pin *0* V. In order to enhance reliability from the static

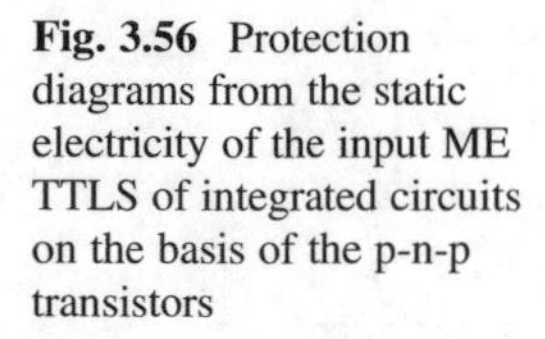

Fig. 3.56 Protection diagrams from the static electricity of the input ME TTLS of integrated circuits on the basis of the p-n-p transistors

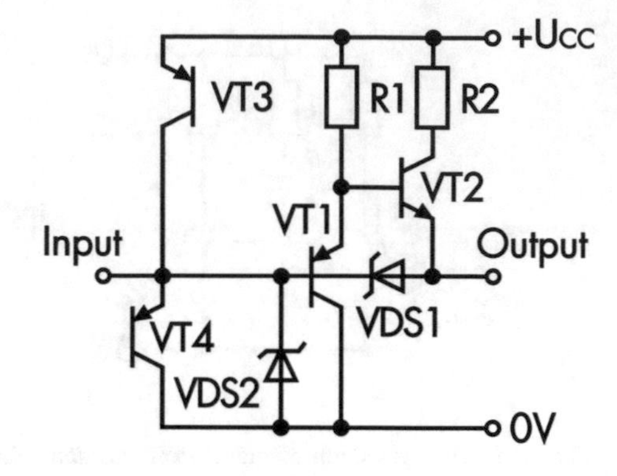

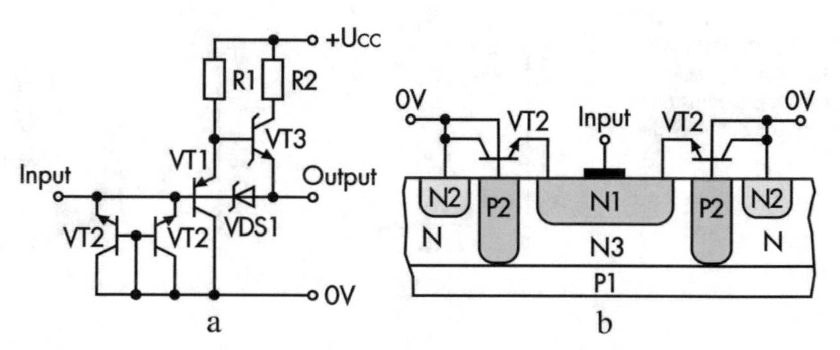

Fig. 3.57 Protection diagrams of the input ME from overloads on the basis of the n-p-n-transistor by means of the isolation with a locked n-p-n-transistor (**a**) and a lateral cross-cut of the active structure (**b**)

electricity protection circuits may be positioned between the supply input and output Ucc , as indicated in the diagram in Fig. 3.56 [3]. Such protection circuit creates the low Ohmic circuit, through which the built-up at the input static charge is tapped into the ME supply circuit.

When getting to the input of the bipolar integrated circuits of the negative input voltages (relative to the pin *0* V) and exceeding the voltage of the forward shift of isolation U_{IS} the opening is possible of the isolating junctions of components of the input ME, inclusion of the parasitic transistor structures, and the performance disruption of the input MEs and the integrated circuits as a whole. Foe exclusion of this parasitic effect the negative input voltages should be limited to the level $U_P^{VDS} < 0,5$ V, at which the isolating junctions in the integrated circuits are closed.

The effective method of protection of the input MEs from influence of the negative input voltages is isolation of the ME input (for instance, of the base of the p-n-p-transistor VT1, (in Fig. 3.57a) from other elements of the circuit by means of the reverse offset horizontal n-p-n-transistor) [1]. In Fig. 3.57b, a lateral cross-cut is shown of the active structure of the ME input. Area №1 forms, for instance, the base of the horizontal p-n-p-transistor VT1, isolated from other elements of the circuit by junctions P2. The isolated areas P1 are surrounded by the circular area N2, connected with the bus *0* V. As a result, the base of the p-n-p-transistor VT1 is isolated from other elements of the circuit by means of the horizontal transistor VT2 of the type n-p-n (N1-P2-N2) with the "earthed" collector. In such a structure even with the forward shift (the negative voltage at the input) of the emitter junction of the protective transistor VT2 its earthed collector prevents the forward shift of the isolating junctions of other components of the ME circuit and disruption of their performance.

3.4 Schematics of the Output Matching Elements of TTLS Integrated Circuits

3.4.1 Output ME TTLS of Integrated Circuits with Standard TTL Output Levels

The basis of the output ME TTLS integrated circuits, ensuring the requirements, is a known push-pull circuit of the signal amplifier (SA) with the TTL levels, indicated in Fig. 3.58a: the principles of functioning and the main parameters of the circuit are considered in detail in [1, 2]. The element Z_O in the diagram is essentially a circuit to discharge the capacitance of base-emitter of the transistor VT3. In order to match the amplifier as per the signal levels with the internal logic elements, they use the additional matching circuits (MC). The basic versions of such diagrams are listed in Fig. 3.58b–d.

The combined matching circuit and the amplifier circuit form the output ME with the active output (AO), as the charge and discharge of the load capacitance is performed by the switchable active ME elements: output transistors VT2 and VT3. In the switched-on state, ME is characterized by the low-level output voltage:

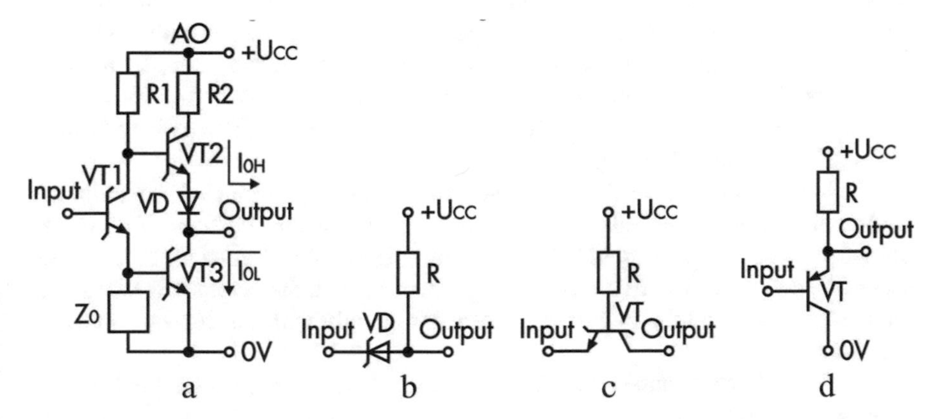

Fig. 3.58 Diagram of the signal amplifier of the ME type of the "active output" of the TTLS integrated circuits (**a**) and options of inputs (**b–d**)

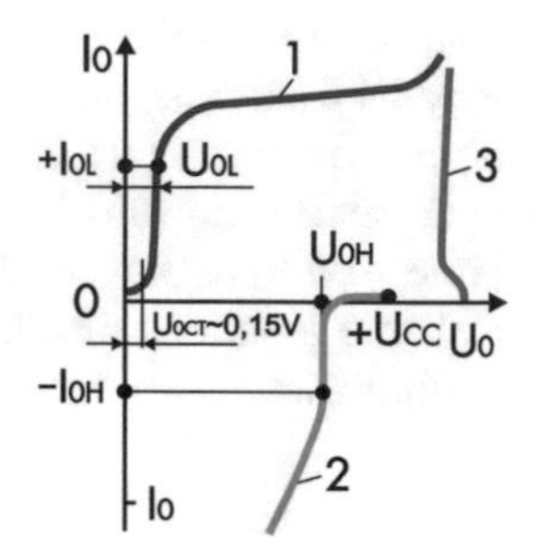

Fig. 3.59 Output characteristics of output MEs of TTLS integrated circuits

$$U_{OL}^{AB} = U_{CE}^{VT3} + I_{OL} \cdot r_K;;$$

where: r_K—collector's resistance of the output transistor VT3

The output ME characteristic in the activated state is reflected by the curve 1 in Fig. 3.59. The output ME of TTLS integrated circuits differ with the increased output voltage U_{OL} =-0.3÷0.35 V, which is related with the stepped-up residual voltage $U_{КЭН0}^{VT3}$ of the output transistor VT3 because of application in it of the "shunting" Schottky diode:

$$U_{CE0}^{VT3} \approx U_{BE}^{VT3} - U_{CB}^{VT3} \approx 0,75B - 0,55B = 0,2B;$$

In the deactivated state ME is characterized with the high-level output voltage:

$$U_{OH}^{AB} = U_{CC} - U_{BE}^{VT2} - U_P^{VD} - \frac{I_{OH}}{\beta_N^{VT2}} R1.$$

The output ME characteristic for the disabled state complies with the curve 3. The type value of the parameter U_{OH} is comparable with the values U_{OH} of other TTL circuits and constitutes $U_{OH}^{AB} = 2.9 \div 3.0$ V. In the TTLS integrated circuits, they widely apply the output MEs, in which apart from two active states with the voltage levels U_{OH}, U_{OL}, there may be formed the "third" state. In this state, both output transistors VT2 and VT3 are in the closed state, owing to which the ME output is disconnected from the load. The version of an elementary ME circuit with triple states (TS) is indicated in Fig. 3.60a.

The ME switch-over to the third state is performed by means of the additional control input EN, connected with the bases of the transistors VT2 and VT3 via the Schottky diodes VDS1 and VDS2 by applying the low-level voltage. In the active states, the output levels of voltage U_{OH},U_{OL} comply with the output ME levels with an active output, and the output characteristics comply with the curves 1 and 2 in Fig. 3.59.

In the "third" state (passive), ME is characterized by the output currents I_{OZL}, I_{OZH} representing the leakage currents of the closed transistors VT2 and VT3, and the output characteristic complies with the curve 3 in Fig. 3.59.

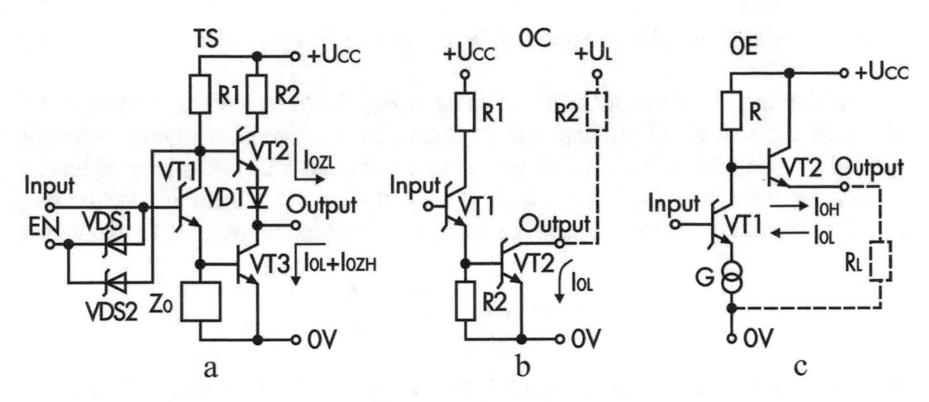

Fig. 3.60 Diagrams of the output TTLS integrated circuits with triple states (**a**); "open collector" (**b**) and "open emitter" (**c**)

Apart from these circuits, there exist ME versions, which have one active component, forming the output levels. These circuits are less popular and are used mainly in TTLS of the integrated circuits, intended for application in the computing systems, where it is permitted to have the "mounting" combination of outputs. The diagram in Fig. 3.60b has only the active element for discharge of the load capacitance-transistor VT2, and the ME output is formed by its collector. The diagrams of such type are called the "open collector" (OC). In the deactivated state, the circuit is characterized by the output voltage U_{OL}^{OK}, similar as per the value of U_{OL}^{AB}. And the output characteristic in the enabled state of such ME complies with the curve 1 in Fig. 3.59. In the switched-off state, the output high-level voltage U_{OH}^{OK} is formed at the expense of the external power supply source U_L and the load resistor R_L:

$$U_{OH}^{OK} = U_L - I_{OH} \cdot R_L.$$

The output ME characteristic in the closed state (without the load resistor R_L) complies with the curve 3 in Fig. 3.59.

In the diagram in Fig. 3.60c, there is only active component of the load capacitance charge—the transistor VT2—and the ME output is formed by its emitter. The circuits of such type are called the "open emitter." In the switched-on state, the circuit is characterized by the output voltage $U_{OHS}^{OЭ}$, analogous as per the value U_{OH}^{AB}:

$$U_{OH}^{OЭ} \approx U_{CC} - U_{BE}^{VT2} - U_P^{VD} - \frac{I_{OH}}{\beta_N^{VT2}} R.$$

The output ME characteristic in this state complies with the curve 2 in Fig. 3.64. In the deactivated state the output voltage of the low level U_{OH}^{OK} is formed at the expense of the external load resistor $\mathrm{R_L} U_{OH}^{OK} = U_L - I_{OH} \cdot R_L$.

The output ME characteristic in the closed state (without the load resistor) complies with the curve 3 in Fig. 3.64.

Modifications of the output MEs of the AO type with the improved characteristics

The diagram of the output ME, indicated in Fig. 3.63a during application as the discharge element Z_0 of an elementary resistor, has the reduced high-level output voltage U_{OH}. As the voltage, while getting to the ME input from the internal blocks of the integrated circuits, has the final value and $U_I > 0$, then the phase dividing transistor is in the active mode, and the voltage on its collector reduces for the value:

$$\Delta U_K = \left(U_I - U_{BE}^{VT1}\right) \cdot \frac{R1}{R_{ZO}}.$$

For the same value, the output high-level voltage U_{OH} is also reduced. In order to eliminate this shortcoming as an element Z_0, the circuits are used, ensuring the closed state of the transistor VT1. The versions of executions of the electric diagrams of the element Z_0 are indicated in Fig. 3.61a–c.

In order to improve the switching delays t_{PLH}, t_{PHL} in the main ME circuit are introduced the additional circuits, increasing during switching the output currents of the low I_{OL} and high I_{OH} levels. Thus, in the ME diagram in Fig. 3.62a for increase of the output high-level current I_{OH}, the transistor Darlington circuit VT2, VT3 is used. This makes it possible with one and the same level of the output voltage U_{OH} to increase by b_N times the output high-level current I_{OH}. In order to increase the output low-level current I_{OL} with the load capacitance discharge, the Schottky diode VDS1 is envisaged in the circuit, forming the feedback between the output and the collector of the phase dividing transistor VT1. When activating the transistor VT1 and reducing the voltage on its collector, the Schottky diode VDS1 opens and from the ME output via the transistor VT1 into the base of the output transistor VT4 flows the additional current I_D. This current increases the output low-level current I_{OL} and boosts the load capacitance discharge.

With the output voltage:

$$U_O < U_{BE}^{VT4} + U_{CE}^{VT1} + U_P^{VDS};$$

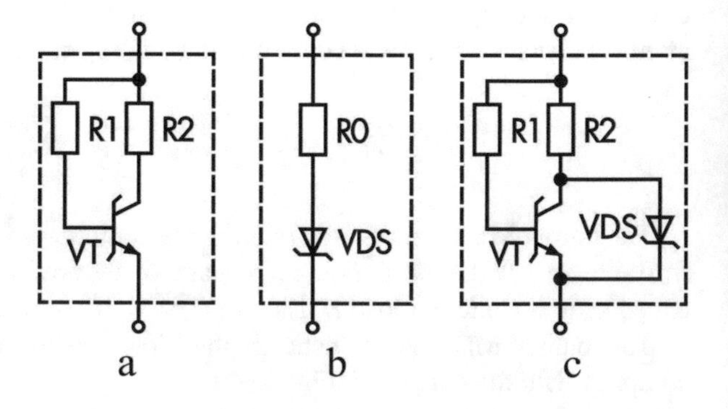

Fig. 3.61 Diagrams of the discharge element Z_0 of the base capacitance of the output transistor: transistor (**a**), diode (**b**), and diode-transistor (**c**)

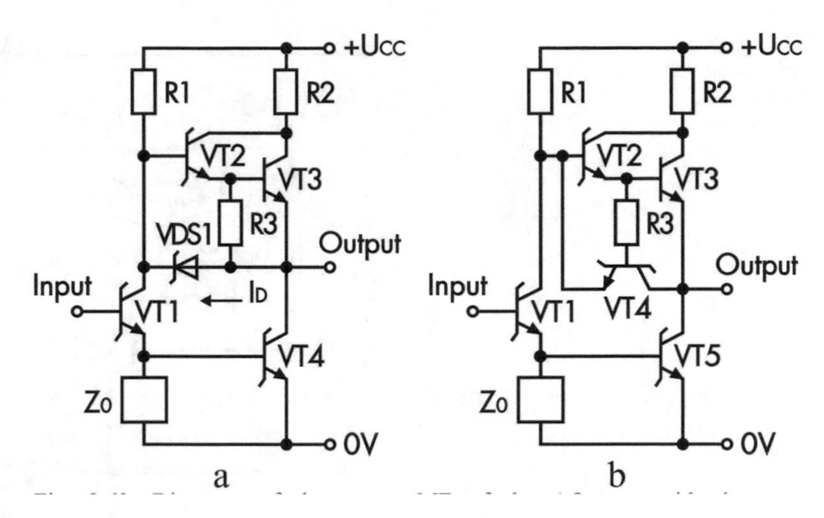

Fig. 3.62 Diagrams of the output ME of the AO type with the accelerating reverse connection by diode (**a**), by transistor (**b**)

The Schottky diode closes and does not have influence of the ME functionality. The similar effect is attained during application instead of the reverse connection diode VDS1 transistor VT4, connected as per the diagram, indicated in Fig. 3.62b [4]. However, application of the transistor VT4 makes it possible during the ME activation to simultaneously accelerate the base capacitance discharge of the output transistor VT3 via its junction of base-emitter. During this duration of the activation edge t_{HL} is improved, and the dynamic consumption current I_{CC}^{G} ME is reduced.

The main method of improving duration of disabling TTLS of the output MEs is acceleration of the base capacitance discharge of the output transistor. This is ensured by creation of the additional control circuits, which during ME activation create the low Ohmic discharge circuits of the base capacitance of the output transistor. During ME activation, these circuits close and do not reduce the base current of the output transistor VT2, thus retaining the ME load capacitance. The example of the electric ME circuit, using this principle of enhancing the rate of response, is indicated in Fig. 3.68. In the circuit, the input transistor VT1, controlling the phase dividing transistor VT2, simultaneously performs during ME deactivation the accelerated base capacitance discharge of the output transistor VT5. During ME activation and locking the transistor VT1, both Schottky diodes VDS1 and VDS2 get closed and have no influence on the ME operation.

Modifications of the Output ME with Triple States of Output (TS).
The diagram of the output ME of TS type, indicated in Fig. 3.60a, has a large input current as per the control input EN, as in order to ensure the high output current I_{OL} the resistor R1 has a small resistance value. This shortcoming can be eliminated by application of the p-n-p transistor VT2 in order to control the "third state" (Fig. 3.60a). Meanwhile, the input current as per the pin EN reduces. [4]. The other substantial shortcoming of the ME circuit, indicated in Fig. 3.60a, is the

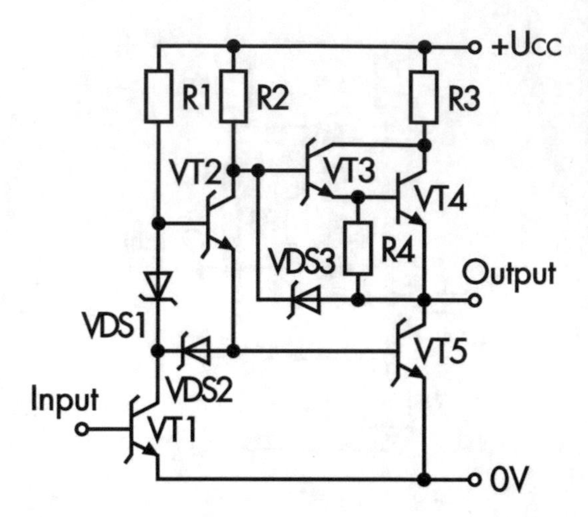

Fig. 3.63 Diagram of the output ME of the AO type with the improved deactivation time

increased consumption power in the passive third state P_{CCZ}, substantially exceeding the power in the active states P_{CCL}, P_{CCH} . Therefore, the majority of the schematic technical methods in MEs of the TS type is aimed at the power consumption reduction P_{CCZ} (Fig. 3.63).

In the ME diagram in Fig. 3.64b [4], the divided load is used of the phase dividing transistors VT1 and VT2, consisting of the resistors R1 and R2, having 2 times greater resistance and controlled by means of the phase dividing transistors VT1 and VT2. Therefore, in the active switched-on state, the total current of their emitters, i.e., the base current of the output transistor VT5, is retained, as that of the type circuit. When switching over to the "third" state via the diode VDS1, the current will flow of only one resistor R1, having twice as much resistance, which makes it possible to reduce the consumption power P_{CCZ}. It is according to the similar principle that the ME circuit functions, indicated in Fig. 3.64c [4]; however, in it for the sake of dividing the load resistors R1 and R2 of the phase dividing transistor VT1, the Schottky diode VDS1 is used. The consumption current of the output ME in the third state can be practically brought to the minimum when applying the p-n-p transistor VT1, as it is indicated in the diagram in Fig. 3.64d [5]. With the low voltage level U_{IL} at the input EN, the transistor VT1is open and sets up the currents of the input transistor VT3 and the output transistor VT5, thus ensuring the active ME output states. When applying the high-level U_{IH} to the input EN, the transistor VT1 closes, which results in locking the output transistors VT5 and VT6 and the "third state" at the ME output. Meanwhile the ME consumption current I_{CCZ} will be practically equal to the leakage current of the control transistor VT1. It is necessary to note that the circuits of the TS type do not permit inclusion of the resistor between the base of the output transistor VT4 and the output, which ensures the capacitance discharge of the transistor base VT5, which is related to the fact that during cutting off supply to ME through this resistor, the junction of base-collector of the transistor VT5, the resistor R6, and the supply bus U_{CC}, the voltage level will lower down on the bus, to which is connected the ME output. In order to eliminate this, this resistor

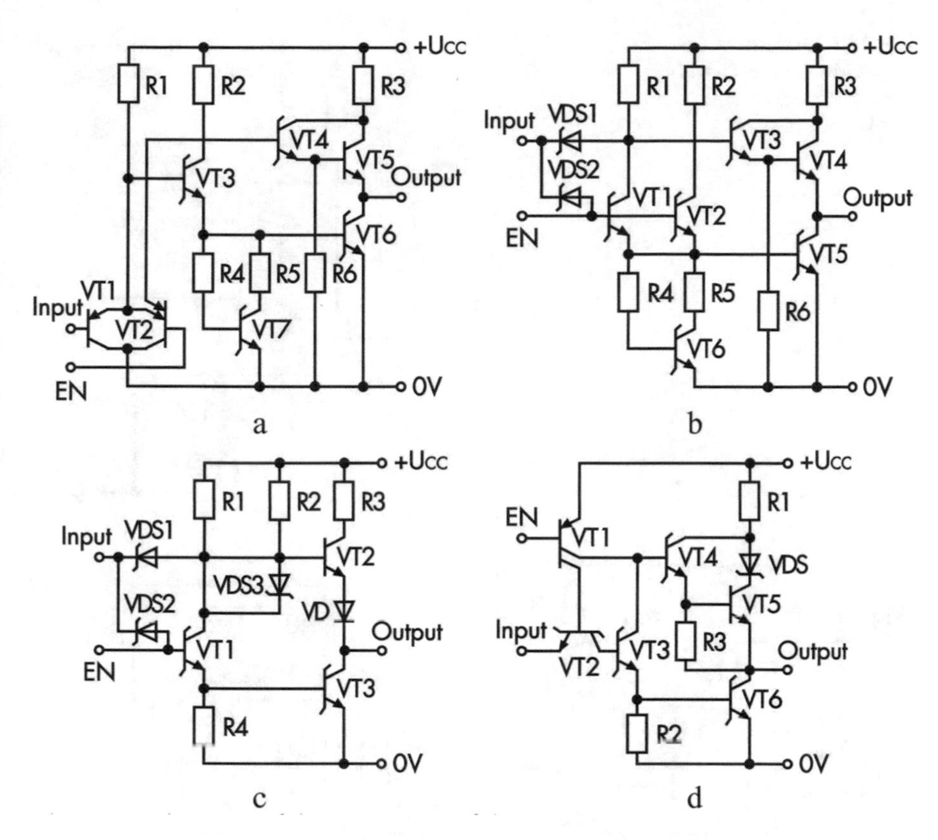

Fig. 3.64 Diagrams of the output ME of the TS type: (**a**) with p-n-p transistor at input; (**b**) with the divided resistive load and two phase dividing transistors; (**c**) with the divided diode-resistive load; (**d**) with the p-n-p-transistor in the ME supply circuit

R4 is connected either to the common bus *0V* (Fig. 3.64b), or the Schottky diode is introduced into the collector's circuit of the output transistor (diodes VDS3, VDS4 in Fig. 3.65).

Availability of the "third" state controlling circuit in the collector of the phase dividing transistor does not permit to improve duration of ME activation by means of the feedback Schottky diode, as indicated in Fig. 3.62a. For elimination of this drawback, it is possible to apply the ME circuit, indicated in Fig. 3.65a [1] and using for creation of the accelerating circuit via the diode VDS6 the separated phase dividing transistors VT1 and VT2 with the resistors R1 and R2. Meanwhile the transistor VT1 ensures control of the output transistors VT3, VT4, and VT2—with the Schottky diodes VDS5, VDS6 accelerating activation of the output transistor VT6. In order to eliminate the parasitic ties of the output with the supply bus $+U_{CC}$, the Schottky diodes VDS3, VDS4 are introduced into the circuit. The circuit in Fig. 3.66a [4] serves as an example of application in the circuit of the reverse bond (feedback), accelerating activation of the output transistor VT5, transistor with the

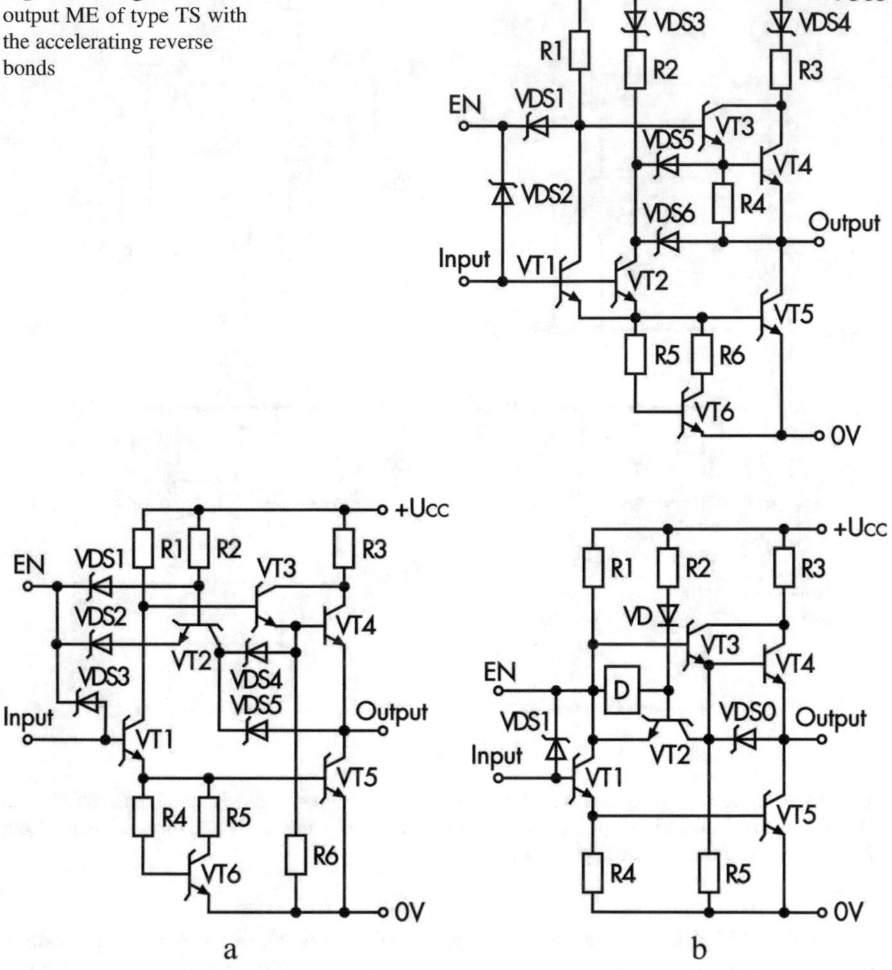

Fig. 3.65 Diagram of the output ME of type TS with the accelerating reverse bonds

Fig. 3.66 Diagrams of the output ME of type TS with the improved transition time into the "third" state

Schottky diode VT2. During transition to the "third" state, deactivation of the accelerating transistor VT2 is ensured by means of the additional Schottky diode VDS1.

The circuit of the similar type makes it possible to improve the parameters of the "third" state, for instance, of the time transition from the high level state into the "third" state. The ME diagram of this type is indicated in Fig. 3.66b [4], where an additional delay element D is included between the base and the emitter of the accelerating transistor VT2. Resultant from this, there emerges in the control circuit the difference control element (DCE), having the conductive state during the short period of time. Therefore, during the ME transition, being in the high-level state,

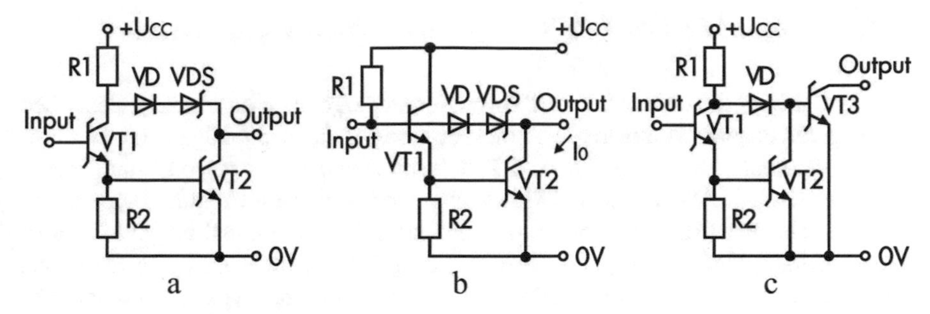

Fig. 3.67 Diagrams of output ME of the OC type with feedback (the reverse connections): (**a**) with the built-in load resistor; (**b**) with the controlled consumption current; (**c**) with the improved deactivation time

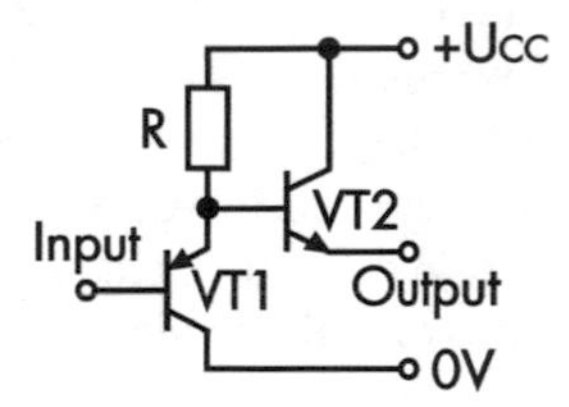

Fig. 3.68 Diagram of the output ME of the OE with the p-n-p-transistor

with application of the low-level signal to the input EN the transistor VT2 during a certain period of time, Dt will be in the conductive state. This formation lowers the voltage level at the ME output and accelerates its transition to the level of the "third" state. In the future, the transistor VT2 closes and has no effect on the functionality of the circuit.

Modifications of the Output MEs of the "Open Collector" Type

The ME circuit in Fig. 3.67a has no the external resistor, setting the high-level output voltage at the expense of using the circuit, consisting of the diodes VD, VDS. The ME circuit in Fig. 3.67b [4] apart from the improved time of activation owing to application of the transistor VT1, operating in the mode OC, possess an additional property: the consumption current of such circuit I_{CCL} in the open state depends on the load current I_{OL}. This is attained by means of the circuit of the reverse connection on the diodes VD, VDS, current controlling the transistor base VT1 in dependence on the load current I_{OL} .

The switch-off time in the "OC" circuits can be improved also as in the circuits of the AO type at the expense of the base capacitance discharge acceleration of the output transistor VT3. Thus, for instance, during deactivation of the output transistor VT3, Fig. 3.67c, the capacitance discharge of its base is ensured by the transistor VT2, possessing the output resistance several times lower than the resistance value of the resistor R2 of the circuit in Fig. 3.67a.

From the elementary circuits of interest is application of the p-n-p-transistor VT1, making it possible to raise the load capacity as per the ME output (Fig. 3.68).

3.4.2 Output ME of TTLS Integrated Circuits with Memory

A schematics of the output ME of TTLS integrated circuits with memory, described in the given chapter, is illustrated on the examples of the electric diagrams, combining an elementary bistable cell or a R-S flipflop and an output push-pull signal amplifier (SA) of AO type with TTLS-levels. Activation of the required type of the memory element (ME) into the output matching element (signal edge clocked or level clocked, etc.) can be performed in compliance with the electric diagrams, described in Sect. 3.2, and formation of the required output type is in compliance with the electric diagrams, described in p. 3.4.1.

The most simple technique of arrangement of the output ME of TTLS integrated circuits with memory is matching the output signal amplifier with one from the logic element of the base cell (BS), as indicated in Fig. 3.69a. In the circuit, the inverter on the transistors VT3–VT8 performs both the function of the output signal amplifier, forming the output signals, and the function of the R-S flipflop "arm." Meanwhile, with the purpose of cutting down a number of elements, the second logic element of the base cell (transistors VT1 and VT2) is made without the output push-pull stage, and in order to exclude influence of interferences from the output to the state of the base cell, the feedback circuits are made from the collectors of the phase dividing

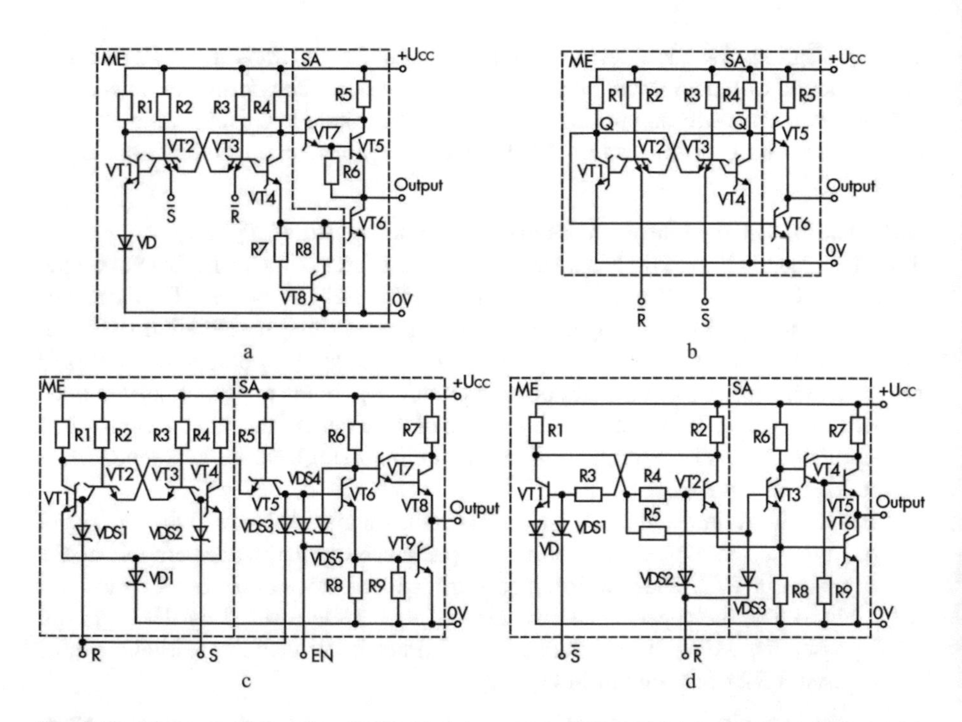

Fig. 3.69 Diagrams of the output ME with the memory of the TTLS integrated circuits: (**a**) with matching the output signal amplifier with the base cell arm; (**b**) with the paraphrase control of the output ME transistors; (**c, d**) with the serial and parallel connected power signal amplifier

transistors VT1 and VT4. The ME circuit of such type has a minimum switch-over delay time t_P^{ME}, equal to the base cell switch-over time t_P^{BS}.

The described ME circuit can be simplified, if to use for the paraphrase control of the output transistors VT5 and VT6 not a phase dividing transistor VT4, but the paraphrase signals of outputs $\overline{Q}$,Q, base cell (BC) (Fig. 3.69b) [5]. Meanwhile, the switch-over delay time of the output ME is as follows:

$$t_P^{ME} \approx t_P^{BS}.$$

The circuits of the output ME with memory, indicated in Fig. 3.69a, b, possess a minimum number of elements, signal switch-over delay time, equal to the delay time of the base cell switch-over and are suitable for application in the integrated circuits. However, their substantial drawback is the impossibility of forming the output signal amplifier of the "triple state" type. In such cases they use the serial (Fig. 3.69c) or parallel (Fig. 3.69d) connections of the base cell and output ME.

With the serial connection, the base cell is made on the low-power TTLS LE, to one of whose outputs (collector of the transistor VT4) is connected the power output signal amplifier of the type "triple state" with TTL-levels [1]. Such connection of the signal amplifier worsens the total ME switch-over time delay; therefore, for acceleration of the deactivation process, an additional Schottky diode VDS was introduced into the circuit. Meanwhile, the ME activation and deactivation delay times are as follows:

$$t_{PHL}^{ME} \approx t_{PLH}^{BS} + t_{PHL}^{SA};$$
$$t_{PLH}^{SA} \approx t_{PLH}^{BS};$$

where: t_{PLH}^{BS}, t_{PHL}^{BS}—delay times for activation, deactivation of the base cell

t_{PLH}^{SA}, t_{PHL}^{SA}—delay times for activation, deactivation of the output signal amplifier

With the high signal level on the control input, EN ME functions as a regular output ME with memory, and with the low level of signal, the output signal amplifier switches over to the "third" state. Meanwhile, the transistor VT5 is closed, and the state of the base cell will not change.

The dynamic parameters of the ME circuit with "triple states" of the output can be improved with the parallel execution of the base cell and output signal amplifier, as it is indicated in Fig. 3.69d [5]. In this circuit the input of the signal amplifier (base of the transistor VT3) is connected parallel to the input of the logic element of the base cell (base of the transistor VT2). ME control from the input R is performed simultaneously both with the arm of the base cell (via the diode VDS2) and with the output signal amplifier (via the diode VDS3). In such circuit the delay times for activation and deactivation will constitute:

$$t_{PHL}^{ME} \approx t_{PLH}^{BS} + t_{PHL}^{SA};$$
$$t_{PLH}^{ME} \approx t_{PLH}^{SA};$$

For formation at the output of the third «third» state into the circuit of the signal amplifier, it is suffice to introduce two Schottky diodes as per the diagram, similar to the diagram, indicated in Fig. 3.69c.

3.4.3 *Output ME of TTLS Integrated Circuits with Conversion of signal levels*

Output MEs of TTLS Integrated Circuits with Formation of the Output ECL-Signal Levels

Output MEs with the output ECL signal levels, similar to the TTLS polarity integrated circuits.

The output MEs with the output ECL signal levels, of the similar polarity with the internal signals of the TTLS integrated circuits of "pseudo-ECL," are constructed basically as per the diagram, indicated in Fig. 3.70a [5]. MEs contain the current switch on the transistors VT1 and VT2, the current generator G, and the emitter "repeater" on the transistor VT3. In order to ensure the input switching threshold of the TTL type of ME to the base of the transistor VT2, the shift voltage of U_R =1.45÷1.5 V was applied. With the input voltage of U_I <1.5 V (Fig. 3.70b), the transistor VT1 is closed, the transistor VT2 is opened, and at the ME output, there will be formed the low-level output voltage:

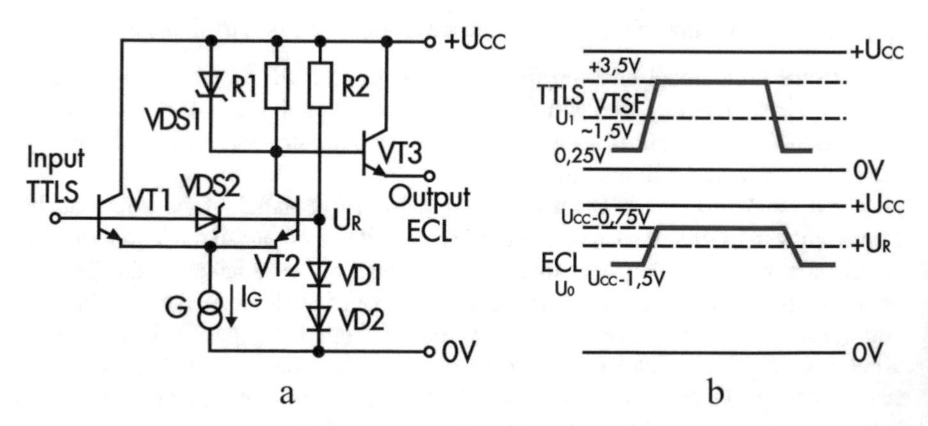

Fig. 3.70 Circuit of the output ME of TTLS integrated circuits, forming the output ECL signal of the similar with TTLS polarity (**a**) and time charts (**b**)

$$U_{OL} = U_{CC} - I_G \cdot R1 - U_{BE}^{VT3}.$$

For formation of the required rate of response and noise immunity, the values of current I_G and resistance of the resistor R1 are selected from the following relation:

$$I_G \cdot R1 - \approx U_{BE}^{VT};$$
$$U_{OL} \approx U_{CC} - 2U_{BE}^{VT3} \approx U_{CC} - 1.5B;$$

With U_I >1.5 V, the transistor VT2 gets closed, and at the ME output, the high voltage level is set:

$$U_{OH}\, U_{CC}\, U_{BE}^{VT3}.$$

Meanwhile, the logic voltage drop at the ME output is as follows:

$$\Delta U_T \approx U_{BE}^{VT3}.$$

It is possible to enhance the rate of response of the circuit of the described type at the expense of lowering the logic voltage drop at the output by means of the parallel connection to the resistor R of the Schottky diode VDS, meanwhile:

$$\Delta U_T^{ME} \approx U_P^{VDS}.$$

Modifications of the output MEs with the "pseudo-ECL" by means of the output signal levels differ basically with the enhanced circuits of the current generators G and with the circuits for formation of the standard levels U_R.

Output MEs of TTLS integrated circuits with the output ECL-signal levels of different polarity.

The more complex schematic technical task is conversion of the signals of the TTL type of positive polarity into the signals of the ECL type of negative polarity. The basic option of the electric diagram of such ME is indicated in Fig. 3.71a [5] and contains two current switches, the first of which is the input one (transistors VTI and VT2, generator G1) and has the positive power supply +U_{CC}, while the second one in output (transistors VT5 and VT6, generator G4) with the negative power supply U_{EE}, and the circuit of the levels offsetting to the negative voltage area (transistors VT3 and VT4; resistors R3 and R4, generators G2 and G3). The TTL switching threshold of the input switch is ensured by application to the transistor base VT2 of the standard offset voltage +U_R. With the input voltage U_I<1.5 V (Fig. 3.71b) transistor VT1 is closed, and transistor VT2 is open; consequently, on the emitters of transistors VT3 and VT4, the output voltage will be as follows:

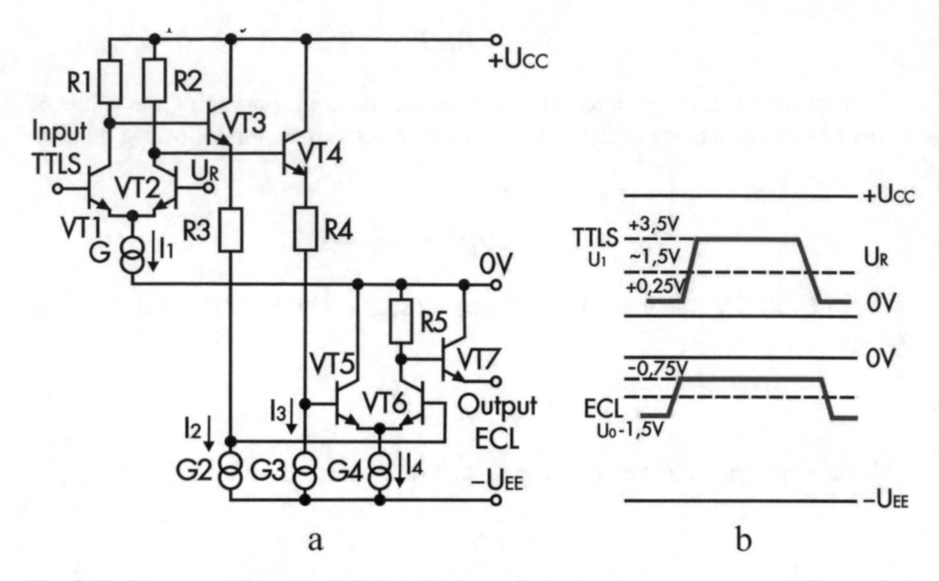

Fig. 3.71 Diagram of the output ME TTLS integrated circuits, forming the output ECL signal of different with the integrated circuits polarity (**a**) and time charts (**b**)

$$U_E^{VT3} \approx U_{CC} - U_{BE}^{VT3};$$
$$U_E^{VT4} \approx U_{CC} - U_{BE}^{VT4} - I_I \cdot R2;$$

As the emitters of transistors VT3 and VT4, operating in the mode of the emitter's "repeater," are connected to the current generators G2, G3 via R3, R4, then at the expense of the voltage drop on the resistors R3 and R4, there occurs the voltage level offset at the transistor bases VT5, VT6 of the output switch:

$$U^{VT5} \approx U_{CC} - U^{VT4} - I_1 \cdot R2 - I_3 \cdot R4;$$
$$U_B^{VT6} \approx U_{CC} - U_{BE}^{VT3} - I_2 \cdot R3;$$
$$U_B^{VT5} < U_B^{VT6};$$

As $U_B^{VT5} < U_B^{VT6}$, then the transistor VT5 is closed, the transistor VT6 is open, and at the output, there will be a low voltage level, equal to the following:

$$U_{OL} = -\left(U_{BE}^{VT7} + I_1 \cdot R5\right).$$

Meanwhile, in order to ensure the "non-saturated" mode of operation of the output switch (key), it is required to meet the following conditions:

$$\left|U_{CC} - U_{BE}^{VT4} - I_1 \cdot R2 - I_3 \cdot R4\right| > \left|2U_{BE}^{VT3}\right|;$$

During alteration of the input voltage for U>1.5 V, the voltage levels change for the opposite ones, and at the output, there will be high-level voltage, as follows:

$$U_{OH} = -U_{BE}^{VT7};$$

The output MEs of the TTLS integrated circuits do not permit the direct control of the CMOS IC inputs basically because of the non-compliance of the high-level output voltage $U_{OH}^{TTLS} \approx 3.0$ V with the high-level input threshold voltage $U_{TH}^{CMOS} \approx 3.7 \div 4.1$ V, i.e., $U_{OH}^{TTCL} < U_{TH}^{CMOS}$, which does not ensure the complete locking of the input ME of the CMOS integrated circuits and causes the massive through consumption current of the CMOS integrated circuits. With the similar supply voltage of $U_{CC}^{TTLS} = U_{CC}^{CMOS} = +5$ V for control of CMOS, one can use the output MEs of the type "open collector" with the resistor, connected to the supply +U_{CC} and stepping up the high level output voltage U_{OH} up to the level, close to +U_{CC}. With the supply voltage of U_{CC2}^{CMOS}, differing from the voltage U_{CC1}^{TTLS}, $U_{CC2}^{CMOS} > U_{CC1}^{TTLS}$, as the output ME, one can also use ME of the type "open collector." The required high-level output voltage U_{OH} for control of the CMOS integrated circuits is possible to ensure connection to the ME output of the external resistor R_O and the supply source U_{CC2}, meanwhile for the reliable operation of the integrated circuits, the following requirements should be met:

$$U_{O\max}^{TTLS} > U_{CC2}^{CMOS};$$

where: $U_{O\max}^{TTLS}$—maximum permissible voltage

3.4.4 Schematics of the Protection Circuits of the Output ME of TTLS Integrated Circuits

Protection of the ME Output Circuits from the "Dynamic" Miller Current

The parasite "Miller effect" [2], to a considerable extent, influences the performance of output MEs of the TTLS integrated circuits in the dynamic mode. In Fig. 3.72a, an electric diagram is depicted of the output ME. Let's assume that the output ME is switched over from the low-level state to the high one. Meanwhile, the transistor VT1 gets closed, and the input capacitance of the transistor VT4 discharges via the resistor R4, which causes the rise of the output voltage U_O.

As the output transistor VT4 has a considerable capacitance of the junction collector-base C_{CB}^{VT4}, then during alteration of the output voltage U_O, the capacitive current passes through it:

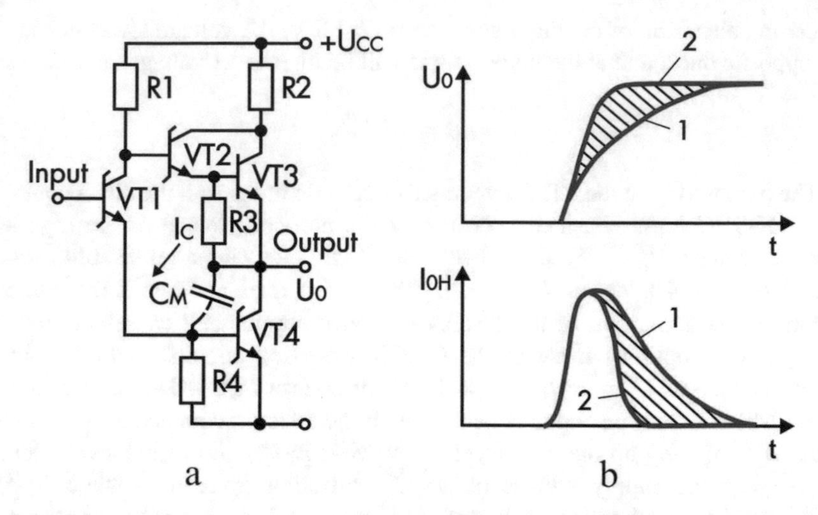

Fig. 3.72 Diagram of the output ME of TTLS integrated circuits, with consideration of the Miller capacitance (**a**) and diagrams of the output voltages U_O (**b**) and current I_{OH} (**c**)

$$I_C = C^{VT4} \cdot \frac{\Delta U_O}{\Delta t},$$

which creates the voltage drop on the resistor R4:

$$U_{R4} = I_C R4.$$

This voltage maintains the output transistor VT4 in the open state and increases its output current, which is equivalent to increase of the output capacitance of the transistor VT4 (Miller capacitance) for β_N^{VT4} times:

$$C_O = C_M = (C_{CB} \cdot \beta_N)^{VT4};$$

where: β_N—current gain ratio of the transistor base VT4

This, in its turn, will increase the transition duration of the output voltage U_O to the high-level state (the curve in Fig. 3.72b).

As the influence extent of the Miller effect is determined by the resistance value of the resistor R4 in the base of the output transistor VT4, the basic method of protection from the dynamic Miller current is creation of the low Ohmic circuits of draining the capacitive current I_C in the base of the output transistor VT4. For elimination of this effect, they use the controlled transistor circuits, switched over during transition of the output ME to the high-level state. Thus, into the circuit in Fig. 3.73a [5], an additional circuit is introduced from the transistor VT6, resistors R2 and R7, and diodes VD1, VD2, and VDS. In the switched-on state of ME, the transistor VT6 is closed and does not influence the performance of the transistor VT4, as the open transistor VT1 via the diode VDS intercepts the current of the

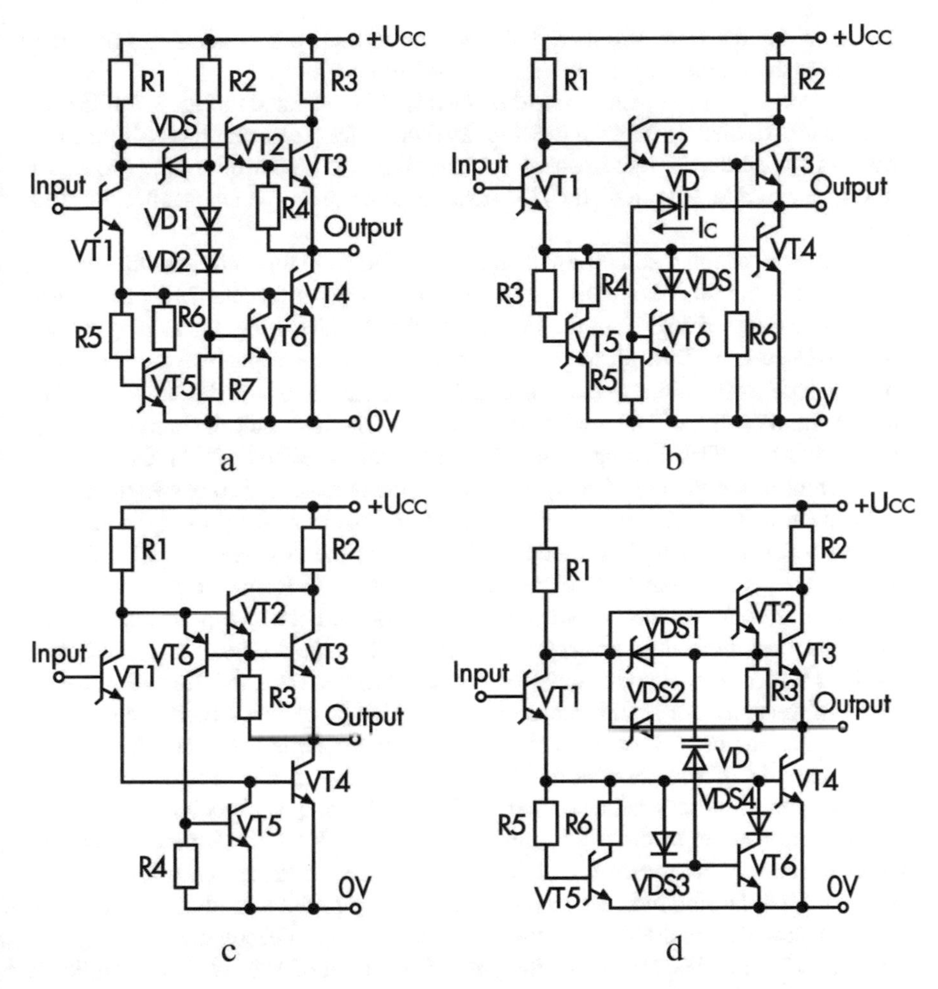

Fig. 3.73 Diagrams of the output ME of the TTLS integrated circuits with protection from the "dynamic" Miller current: (**a**) with the diode-transistor circuit, controlled by the collector of the phase dividing transistor; (**b**) with the dynamic control as per the ME output; (**c**) with the dynamic control of the p-n-p transistor; (**d**) with the dynamic control from the collector of the phase dividing transistor

resistor R1. During disabling ME voltage on the collector of the transistor VT1 increases, which results in locking the Schottky diode VDS and unlocking the transistor VT6, which drains the capacitive Miller current of the transistor VT4 to the common bus *0* V and maintains the small value voltage on the base of the transistor VT4 in the turned-off state of ME. During this, the disabling edge duration of the output signal t_{LH} abruptly falls (the curve 2 in Fig. 3.72b).

The shortcoming of such circuits is that they consume power in the disabled state of the output ME, which increases the total consumption power of ME. In literature a number of circuits were proposed, using the capacitance components and functioning only at the moment of deactivation of the output ME.

The electric diagram with a circuit of such type is indicated in Fig. 3.73b [5] and contains the transistor VT6 with the Shottky diode VDS, representing itself the drain circuit of the Miller current, resistor R5, and diode VD, controlling the protection circuit. Meanwhile, the p-n diode VD is used as capacitance, which is reflected in the figure.

In the static enabled and disabled states, the diode VD does not pass the current, and the drain circuit VT6 VDS is closed and does not affect the ME parameters. During transition of ME from the enabled to the disabled state, the output voltage swing ΔU_0 causes flow through the diode VD of the capacitive current I_{CC}^{VD}, proportionate to the diode capacitance C^{VD}. This current opens the transistor VT6 and the Shottky diode VDS and creates the low Ohmic circuit to drain the Miller output transistor VT4. During transition to the static switched-off state, the capacitive current of the diode I_C^{VD} sharply falls, and the transistor VT6 is switched off.

Application in the circuits of the output ME transistors of the p-n-p type makes it possible to create a simple protection circuit from the "dynamic" Miller current [5]. In the diagram in Fig. 3.73c, the n-p-n transistor VT5 forms the drain circuit of the Miller current of the output transistor VT4, and the p-n-p transistor VT6 represents itself the control protection circuit. In the enabled state of ME, the output transistor VT4 is open, and the output transistor VT3 is closed; therefore, the emitter current of the transistor VT6 is close to zero, and the protection circuit is turned off. During transition of ME from the activated state to the turned off state, through the emitter circuits of the transistors VT2 and VT3 the current flows the current, controlling the load capacitance charge C_L and the output capacitance C_O. This causes the unlocking of the p-n-p transistors VT6 and VT5; the latter drains to the common bus *0* V the emerging at that moment the Miller current of the output transistor VT4. During increase of the output voltage to the settled high-level U_{OH}, the transistors VT2 and VT3 get closed, which results in deactivation of the p-n-p transistor VT6 and deactivation of the protection circuit of ME. With availability at output of the static output current I_{OH}, the deactivated state of the protection circuit is ensured by the appropriate selection of the resistance value of the resistor R4.

Of all types of the protection circuits of the output ME from the dynamic Miller current, the most widely spread is the circuit, indicated in Fig. 3.73d [5]. The protection circuit includes the transistor VT6, Schottky diodes VDS5 and VDS6, and the diode VD, used as a capacitance. The main distinction of the given circuit is in controlling the protection circuit on alteration of the voltage level on the collector of the phase dividing transistor, which makes it possible to match to the optimum time-wise the moments of the maximum unlocking of the transistor of the Miller current draining and the peak value of the Miller current. In the established activated state, the current through the diode is practically equal to zero, the transistor VT6, and the Schottky diode VDS4 are closed and have no effect on the state of the circuit.

During transition of ME from the enabled to the disabled state, the voltage rise on the collector of the phase dividing transistor VT4 causes flow through the diode capacitance C^{VD} of the current, opening the transistor VT6, which drains the Miller current of the output transistor VT4 to the common bus *0* V. This results in the abrupt deterioration of the edge duration of disabling the output signal. During transition of ME to the settled high-level state, the current through the diode sharply falls, which results in deactivation of the protection circuit. It is to be noted that introduction of the protection circuits results in a certain increase of capacitance in the base of the output transistor VT4, which may extend the delay time of activation of the output ME t_{PHL}. However, during this time is substantially gained in the deactivation delay, which improves the overall performance of the output ME.

Protection of the ME Output Circuits from the Miller Current in the Static State

Opposite to the above described Miller effect, reading in the dynamic mode, Miller effect may manifest itself in the static mode of operation of the output ME, influencing the ME performance, received the identification as «dynamic», the Miller effect may manifest itself also in the static mode of operation of the output ME. In Fig. 3.74a the ME diagram is indicated connected to the bus of the signals transfer and being in the "third state."

As to the bus there may be connected other transmitters DN and receivers of the R1–RN, and then during the signal transfer along the bus by means of the transmitter DN in the output circuit of the signal transmitter R1 (transmitter VT4), there may flow the capacitive current *Ic*, proportionate to the capacitance collector-base of the C_{KB} transistor VT4. This current, flowing through the resistor R3, will create on it the voltage drop, which may result in the slight opening of the output transistor of the output transistor VT4 and stepping up the output current of the signal transmitter D1. And this is equivalent to increase of the output capacitance *Co* ME for *C* times, which will result in deterioration of the signal transmission along the bus *C*. For elimination of this effect in the output MEs, they use the control transistor circuits, creating the low Ohmic circuit of draining the Miller current of the output transistor VT4 in the "third" state.

The electric diagram of the output ME of the type "triple states" with protection from the "static" Miller current is indicated in Fig. 3.74b. In the ME circuit [5] during application to the control input EN of the high-level signal, the Schottky diodes VDS1 and VDS2 and the transistor VT7 are closed, and ME functions as a regular ME of the type "Active Output." With application of the low-level signal to the input EN, the Schottky diodes VDS1 and VDS2 get open, and voltage on the base of the transistors VT1–VT3 drops, which results in transition of ME to the "third state." However, meanwhile, the Miller current drain circuit is activated, containing the transistors VT5–VT7 and draining the Miller current of the output transistor VT4 to the common pin *0* V.

Protection of the Output MEs of TTLS Integrated Circuits from Noise

Schematics of the noise suppression in the supply circuits of the output MEs. One of the basic methods of noise suppression, originating during switch-over of the output ME in the supply circuits, is formation of the flat (or linear changing) edges of the output signal by means of controlling the build-up rate of the output current $\frac{dI_O}{dt}$.

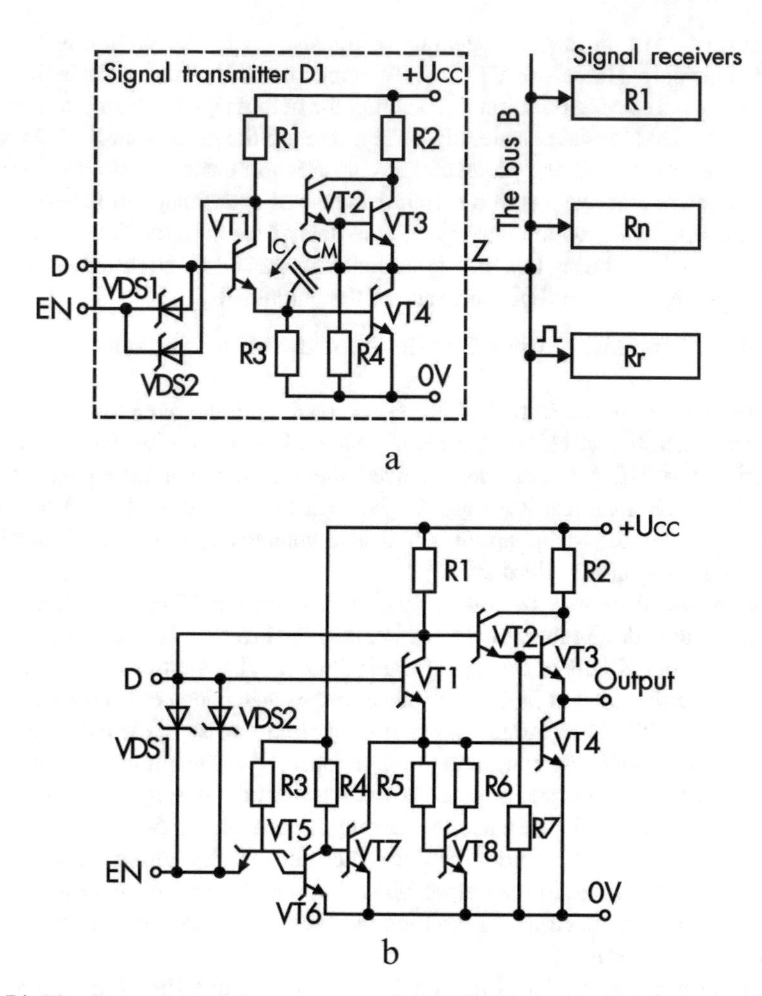

Fig. 3.74 The diagram, explaining emergence of the Miller current in the output ME of the TS type, being in the static state (**a**) and the diagram of the output ME of the type TS TTLS integrated circuits with protection from the "static" Miller current (**b**)

In Fig. 3.75a, the diagram is indicated of the output ME, which serves to ensure the duration control of the output signal activation edge [1]. The element contains the output circuit on the transistors VT3–VT5, forming the output voltage levels, input circuit on the transistors VT1 and VT2 (delay element DLO), and a number of delay element DL1-DLN, built as per one diagram, but differing from the input circuit and from each other with a delay in the increasing order for the delay time of $\Delta\tau$, $2\Delta\tau$, .., $n\Delta\tau$. The specific delay time in the elements is set by an appropriate selection of the resistors R1 and R3.

During application to the ME input of the positive voltage swing ΔU_I because of the time difference of activating the delay elements DL0...DLN in the base of the output transistor VT5, there will be formed a steady, ramp building-up current signal

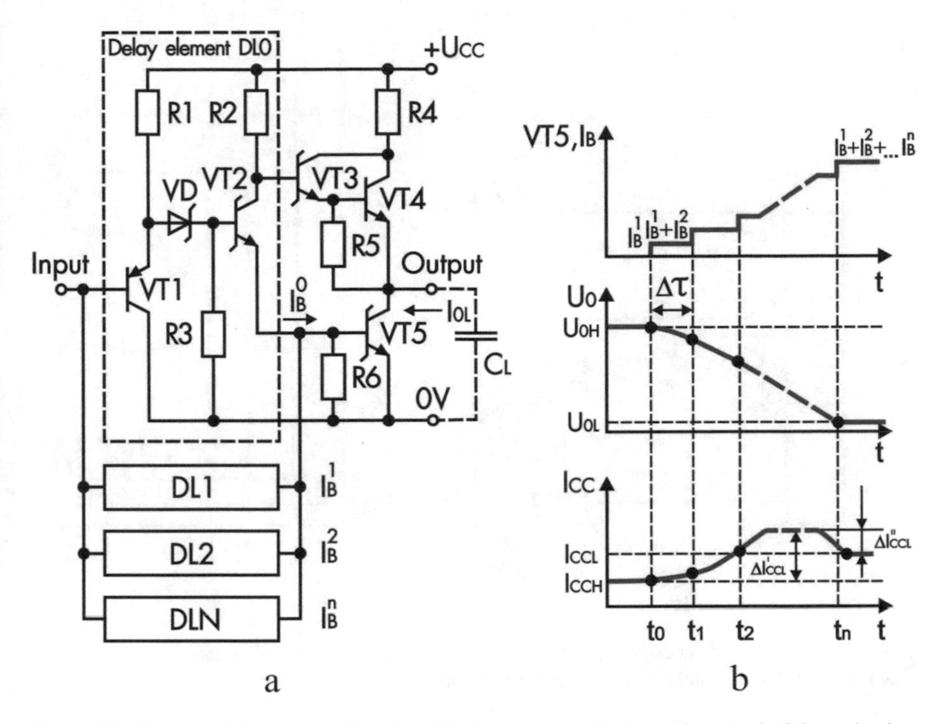

Fig. 3.75 Diagram of the output ME of the TTLS integrated circuits with control of the activation edge duration (**a**) diagrams of signals in the circuit (**b**)

of the base I_B^{VT5} with a period of $\Delta\tau$ (Fig. 3.75b). This will result in the unlocking of the output transistor VT5, i.e., the output current will increase, not jump-wise, but gradually, which will result in the monotonous discharge of the load capacitance C_L, via the output transistor VT5 and in the gradual fall of the output voltage. Absence of the abrupt changes of the consumption current ΔI_{CC} at the expense of the load capacitance discharge C_L will result in reduction of the voltage fall ΔU_G on the parasite inductivity L_G of the common pin of the integrated circuit *0V*, reduction of interferences in the common bus *0V*.

Similarly, one can execute control of the rise rate of the load capacitance charge C_L during the ME deactivation. In Fig. 3.76a, the diagram is indicated of the output ME, in which one performs control of the rise edge duration of the output signal t_{LH}. In this circuit, ME during activation of the control unit (CU) performs the serial activation (with a delay, Dt multiple) of the charge transistors VT2-VTN (Fig. 3.76b) while gradually increasing the output current *Io*, charging the load capacitance C_L. This results in the monotonous rise of the output voltage *Uo* and of the consumption current *Ice* during switch-over. Absence of the abrupt changes of the consumption current I_{CC} during the switch-over procedure because of the currents of the load capacitance charge C_L will result in reduction of the voltage drop ΔUs on the parasite inductivity *Ls* of the supply pin of integrated circuits and in the noise level reduction in the supply bus *+Ucc*.

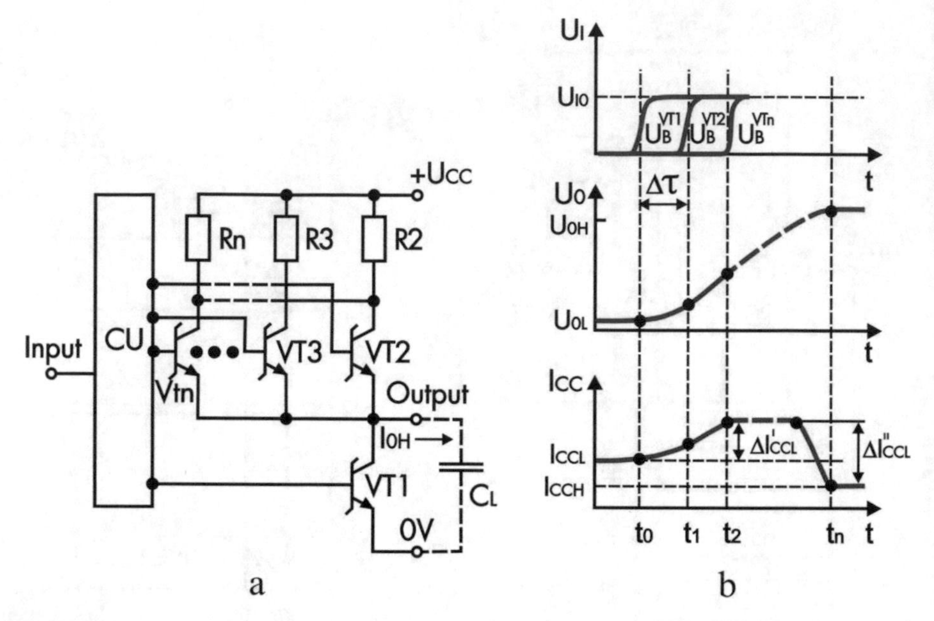

Fig. 3.76 Diagram of the output ME of the TTLS integrated circuits with control of the deactivation edge duration (**a**) and signal diagrams in the circuit (**b**)

Control of the rise rate of the output current $\frac{dI_O}{dt}$ can be performed by means of the nonlinear load into the collectors of the phase dividing transistor [1]. The ME diagram of such type is indicated in Fig. 3.77. The collector of the phase dividing transistor VT1 includes the nonlinear load, containing the low Ohmic R1 and the high Ohmic R2 resistors and a set of the diodes VD1–VD3. During deactivation of ME and the voltage drop on the collector of the transistor VT1 down to the level of $Ucc\text{-}3U_P$, the alteration rate of the output transistor base VT4 (and, appropriately) will be determined by the high Ohmic resistor R2; therefore, at the initial section of activation, the edge of the output signal is flat.

During the voltage change on the transistor collector VT1 below the level of *Ucc-3Unp*, there will happen activation of the low Ohmic circuit with the resistor R1, and the rise rate of the output current increases. Owing to such alteration of the output current, the noise level in the common bus *0V* will be reduced. Altering a number of diodes in the transistor's collector circuit VT1, it is possible to alter the activation threshold of the low Ohmic circuit and, appropriately, the shape of the output signal. The analogous process occurs also at activation of the transistor VT1, which also reduces the level of noises in the supply bus *+Ucc*.

The efficient method of controlling the rise rate of the output current $\frac{dI_O}{dt}$ is application of the reverse relations (feedbacks) [2]·

In Fig. 3.78a, the diagram of the output ME is indicated, using this principle of control during activation. Opposite to the type diagram of the output ME TTLS, the given circuit contains the feedback circuit on the transistor VT5 and the resistor R6.

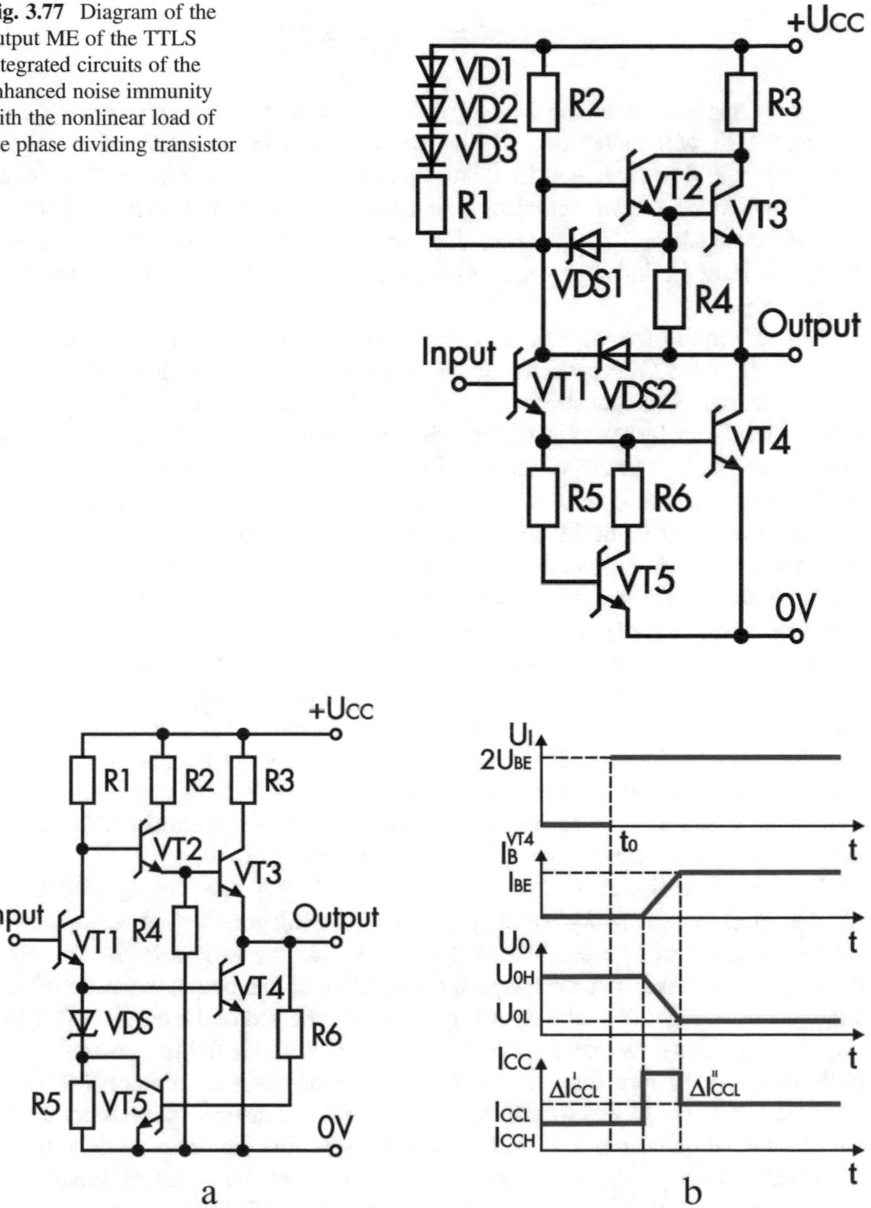

Fig. 3.77 Diagram of the output ME of the TTLS integrated circuits of the enhanced noise immunity with the nonlinear load of the phase dividing transistor

Fig. 3.78 Diagram of the output ME with the controlled rise rate of the output current $\frac{dI_O}{dt}$ (**a**) and the diagram of signals in the circuit (**b**)

In the disabled state, the transistors VT1 and VT4 are closed, and the high voltage level at the output U_{OH} through the resistor R6 sets up the open state of the feedback transistor VT5, which can drain current:

$$I_c^{VT5} \approx \frac{U_{OH} - U_{BE}^{VT5}}{R5} \cdot \beta_N^{VT5};$$

During application to the ME input of the voltage swing $\Delta U_I = 2U_{BE}^{VT}$, the transistor VT1 will switch over to the open state, and in the circuit of its emitter, the current will flow I_E^{VT1}, branching into the collector of the transistor VT6. After the time Dt, when the emitter's current I_E^{VT1} exceeds the current, drained by the collector I_C^{VT5} of the transistor VT5, a part of this current gets into the base of the transistor VT4, resulting in its slight opening and reduction of the output voltage Uo (Fig. 3.78b).

This reduction of voltage through the resistor R6 will result in reduction of the current, drained by the collector of the transistor VT5 and, appropriately, in the current increase of the transistor base VT4, and the current rise process is repeatable. It is possible to set by selection of the resistor values R1, R5, and R6 the required rise rate of the output current discharge $\frac{dI_O}{dt}$ and to avoid the sharp changes in the consumption current Ice during activation and thus to cut down the noise level in the common bus *0* V. In the activated state, the low voltage level disconnects the feedback circuit. After deactivation of ME, activation of feedback occurs in the reverse order and ensures $\frac{dI_O}{dt}$ control of the rise rate of the charge output current and reduction of the noise level in the supply bus U_{CC}.

The other known cause of the noise generation in the TTLS supply buses of the output ME is availability of the "through" consumption currents during switch-over. This is related to the fact that with the equality (or a small difference) in the delays of deactivation $t_{PLH}^{VT4,VT5}$ and activation $t_{PHL}^{VT4,VT5}$ of the control circuits of the output transistors (Fig. 3.79a), an intermediate state emerges in ME, at which both output transistors VT4 and VT5 are open, and in the circuit of the emitter VT4 and the collector VT5, the current flows (Fig. 3.79b) (the curve 1).

This causes an abrupt consumption current build-up I_{CC}^F generating noise on the parasite inductances L_S, L_G, of supply pins of the integrated circuits. In order to reduce the level of interferences in the supply circuits, related to the "through" consumption currents, use the advance deactivation of one output transistor relative to the other. In Fig. 3.79a, the electric diagram is indicated of the output TTLS ME, implementing this principle [3]. The diagram contains two individual control circuits of the output transistors: for control of the output transistor VT5, resistors R2 and R3 and the transistor VT1 and for control of the output transistor VT4, resistors Rl, R4, and R5 and the transistors VT2 and VT3. While selecting the values of the resistors, the switch-off delay t_{PLH1}^{VT4} of the transistor control circuit VT4 is made smaller than the activation delay t_{PHL0}^{VT5} of the transistor control circuit VT5 and the shutdown delay t_{PLH1}^{VT5} of the transistor control circuit VT5 to be made smaller than activation delay t_{PHL0}^{VT4} of the output transistor control circuit VT4. In such a situation the unlocking of the output transistor VT5 will happen with the closed output transistor VT4 and the unlocking of the output transistor VT4 with the closed output transistor VT5, which eliminates in the ME the intermediate "through" state and reduces the "surge" of the consumption current Ice during the switch-over ME (the curve 2 in Fig. 3.79b).

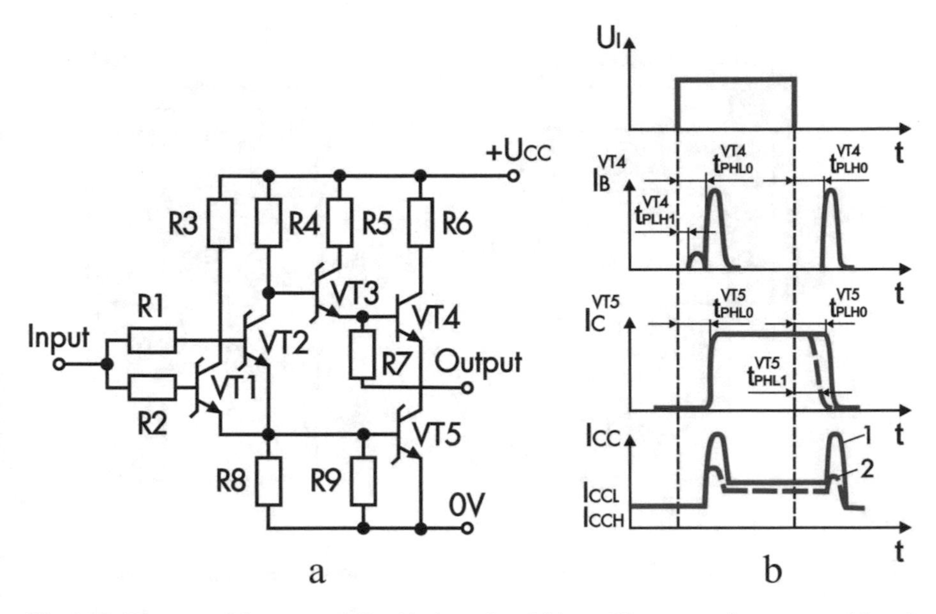

Fig. 3.79 Diagram of the output ME with the reduced "through" consumption currents (**a**) and diagrams of voltages and currents in the circuit (**b**)

One can attain the similar effect of lowering the "through" consumption currents and the related noise level at the expense of using the circuits, accelerating deactivation of the ME output transistors. The accelerated discharge is ensured by application of the additional transistors VT3, including in the diagram in Fig. 3.80a, in-parallel to the phase-dividing transistor VT1 and, in the diagram in Fig. 3.80b, in-parallel to the output VT5. During activation of ME, these transistors create the low Ohmic discharge circuit of the base capacitance of the output transistors VT4.

Schematics of the Noise Suppression Circuits of the Cross Type in the Output ME

Availability in the integrated circuits of the cross-capacitive distributed connection among the ME outputs can result in appearance of the negative-type noises, affecting the functionality of the system, using the integrated circuits. In Fig. 3.81a, a fragment is indicated of the integrated circuits with two output ME: D1 and D2 between whose outputs was formed the cross bond of the capacitive type (capacitances C_X^1, C_X^n). Element D1 is in the low-level state U_{OL}, while element D2 is in the high-level state U_{OH}. During switch-over of the D1 element into the high-level state, and after the time ΔT of the D2 element into the low-level state, at the output D1 due to the cross link the interference ΔU is formed rather long (Fig. 3.81b), extending to the negative area of voltages. When the noise ΔU attains the value, surpassing the isolation voltage of the integrated circuit elements, there will occur the forward offset of the isolating junctions of the ME elements and disruption of its operation. One of the simplest methods of reducing the level of the cross noises of such type is connection to the output of the damper Schottky diode VDS1 (Fig. 3.81c), limiting the value of

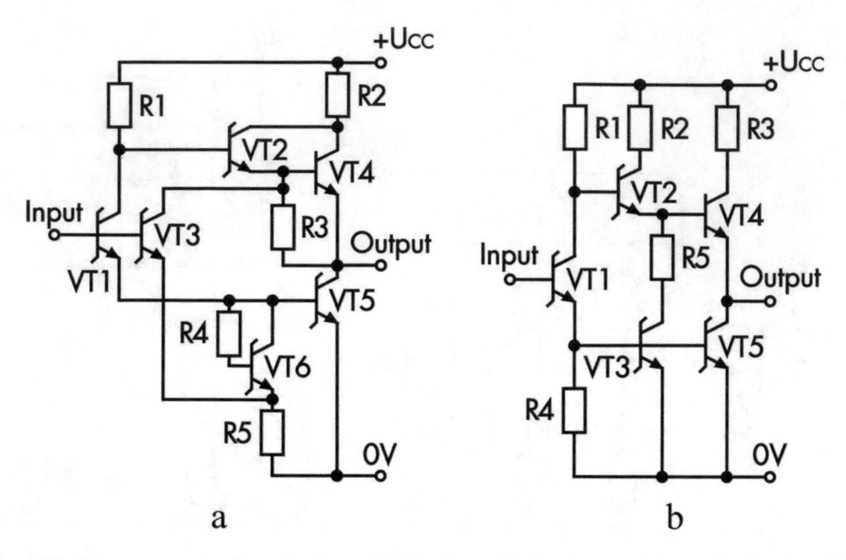

Fig. 3.80 Diagrams, using for reduction of the devices with the shortened time of activation of the output transistors

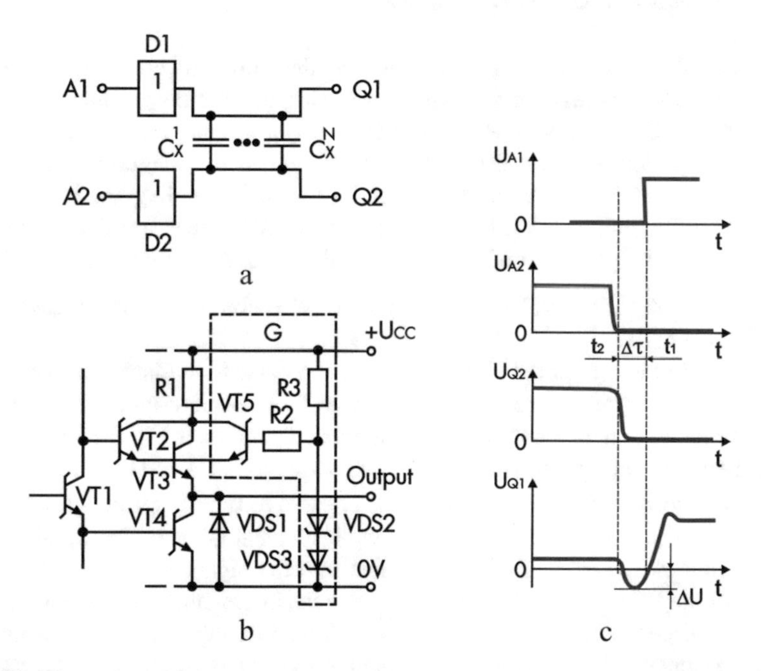

Fig. 3.81 Diagram explaining emergence of the cross-capacitive connections among the ME outputs, resulting in appearance of the negative interferences (**a**), diagrams of signals (**b**), and the electric diagram of their suppression (**c**)

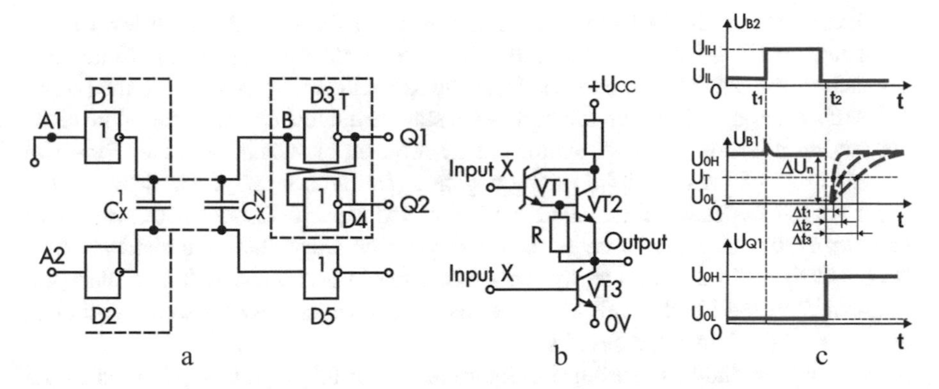

Fig. 3.82 Diagram explaining emergence of the cross-capacitive links among MEs, resulting in appearance of noises at the high level of the signal (**a**), diagrams of signals (**b**), and the electric circuits of their suppression (**c**)

the negative interference ΔU. The most effective method of suppression is application of the special purpose built-in noise suppression circuits (Fig. 3.81c). The suppression circuit G contains the transistor VT5, the resistors R2 and R3, and the Schottky diodes VDS2 and VDS3. In the high-level state, the voltage on the base of the output transistor VT3 closes the transistor VT5, and the suppression circuit has no effect on the ME functionality. With the low level of the output voltage, both closed transistors VT2 and VT3 maintain the closed state of the transistor VT5, and the suppression circuit also does not influence the ME functionality. With emergence at the ME output of the negative-type noise and at its attainment of the following level:

$$\Delta U \approx 2U_{BE} - 2U_P^{VDS};$$

the transistor VT5 gets open, and through it flows the large current, suppressing the negative "surge" of the output voltage. The described circuit ensures suppression of the negative interferences to the level:

$$\Delta U = 2\left(U_{BE} - U_P^{VDS}\right) \approx 300 \text{ mV};$$

The cross-type noises may occur not only at the low level of the output voltage U_{OL}, as indicated above, but also at the high-level U_{OH} (Fig. 3.82b), and after attainment of the threshold voltages of switching the load circuits, they may result in disruptions of their operation.

Figure 3.82a indicates a fragment of the integrated circuits, containing two output MEs: D1 loaded with the bistable cell T and D2 loaded with the logic element D5. Among the ME outputs, the cross-capacitive link is formed, indicated as $C_X^1 \ldots C_X^n$. The element D1 is in the high-level state U_{OH}, maintaining the low-level state of the bistable cell T. The element D2 is in the high-level state. Alteration of the ME state D2 from the high level to the low one does not influence the state of the bistable cell and results in the insignificant voltage rise at the output *Q1*. ME switch-over D1 from the high-level state to the low one because of the cross connection it results in the appearance at the moment of switching at the high-level output signal of the output *Q2* of the negative interference ΔU_{Π}. The value of the noise voltage depends on the value of the coupled capacity C_x and with the greater capacity C_x the noise level may exceed the switching threshold of the bistable cell T (Fig. 3.82b). With the sufficiently large duration of noise, there will take place the "flip-over" of the bistable cell to the opposite state which may cause disruption of operation of the integrated circuits.

The main method of tackling the interferences of this type is application in the output MEs of the low Ohmic current sources, forming the high-level output voltage U_{OH} ME. Such current source can be by the Darlington transistor circuit, or the transistor circuit on the emitter "repeaters" with feedback (Fig. 3.82c). Such circuits, having the large output current, make it possible to efficiently suppress the negative interferences at the high level of the output signal.

Suppression Circuits of Mismatch Interferences

This type of interferences to a considerable extent influences the reliability of functioning owing to the shape distortion of the output signals because of reflections in the transmission lines of the output signals of the integrated circuits. The problems of reducing the level of the mismatch interferences are basically solved by the methods of design development of the matched loads of the integrated circuits. However, in certain cases, the means of the mismatch noise suppression can be introduced into the output MEs of the integrated circuits.

The simplest method of reducing the mismatch noise level, effectively used in TTLS of the integrated circuits, is the serial connection to the output of the integrated circuits of the low Ohmic resistor $R_O = 25–30$ Ohm, as indicated in Fig. 3.83a). Such resistor makes it possible to reduce for 2 times the level of the mismatch noise at the output signal (Fig. 3.83c). Other effective means, applied both in the output ME of integrated circuits and in the external devices relative to the integrated circuits, are the special purpose circuits for signal level limitation. The simplest circuit of such type is connection to the ME output of the Schottky diodes VDS1, VDS2 (Fig. 3.83b). Such a circuit ensures limitation of the noises down to the level: the positive noises $U_{CC} + U_P^{VDS} = 5,5$ V and the negative noises $U_P^{VDS} = 0,5$ V.

3.5 Digital Integrated Circuits on the Basis of the Integrated Injection Logic

Appearance of the elements of the integrated injection logic (I^2L) created the element base for realization of the complex-functional bipolar integrated circuits by the degree of integration of the closely resembling MOS integrated circuits. The first

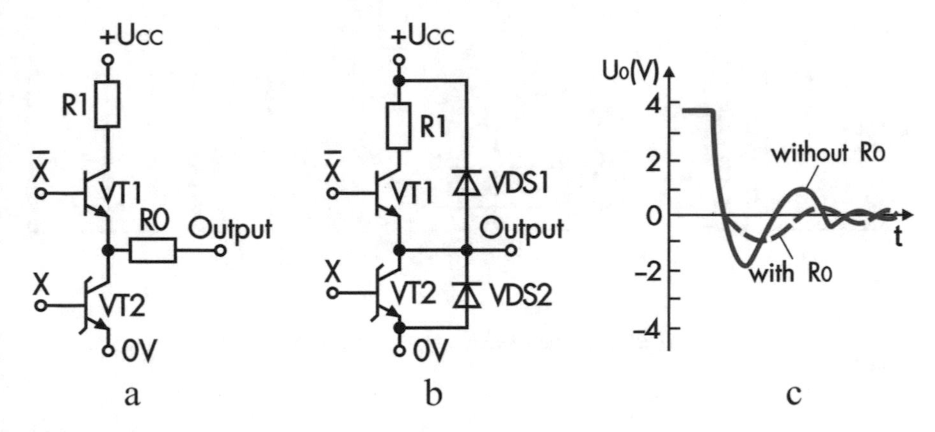

Fig. 3.83 Diagram of suppression of mismatch noises (**a**), limitation of the signal level (**b**), diagram of signals (**c**)

injection microprocessors (1972–1975), alongside with a simple fabrication technology, differed with a high degree of integration (up to 40,000 elements), a low value of the dissipated power (up to 0.1 mW/el), and a wide range of the supply voltages (from 1.0 V to 15.0 V) with the substantially high clock frequency of operation of these integrated circuits (up to 20 MHz). In 1976–1978, there appeared the first I^2L integrated circuits of the second generation of the series [5], differing with the higher technical characteristics, and, primarily, with the high performance. This task was tacked by two parallel methods—stepping up the high performance of the basic logic elements by means of realization of the new technological solutions [1] and application of a whole number of new schematic technical solutions [4].

Standard I^2L LE

In Fig. 3.84a is shown the type diagram of the element I^2L LE, comprising the current setting p-n-p transistor VT1 and the switchable n-p-n transistor VT2.

In the static mode, the serviceability of the elements I^2L LE is assessed by means of the families of the transfer parameters by voltage and current. By them it is easy to discern the levels of voltages of the logic zero U_{OL} and the unit U_{OH}, the switching thresholds, and noise immunity to the unlocking and locking interferences. Expression for the transfer characteristic by voltage (Fig. 3.84b) in the common case has the expression [4]:

$$\exp\left(\frac{U_I}{\phi_T}\right) = \frac{[\exp(U_{IE}/\phi_T) - \exp(U_O/\phi_T)]/r - [\exp(U_O/\phi_T)]/\beta_N}{1 - \exp(U_O/\phi_T)};$$

where: U_I, U_O—input and output voltages

$r = \frac{I_S^n}{I_S^P}$—constant value

$I_S^n; I_S^P$—cutoff currents of the transistors VT1 and VT2 correspondingly

U_{IE}—voltage between the emitters of the p-n-p and n-p-n transistors

β_N—gain ratio of the base current of the n-p-n transistor VT2

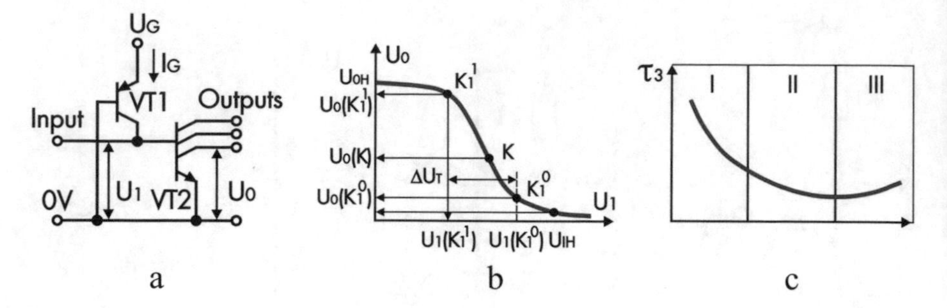

Fig. 3.84 Type diagram (**a**), transfer characteristic (**b**), type dependence of the switching delay (**c**) I2L LE

From this expression it is easy to obtain the values of the output levels of the low (U_{OL}) and high (U_{OH}) levels and the value of the logic swing (ΔU.):

$$U_{OH} = U_{IE} - \phi_T \cdot \ln \cdot (r/\beta_N);$$
$$U_{OL} = U_{CES}^{VT2} \approx \phi_T/\beta_N;$$
$$\Delta U_o = U_{OH} - U_{OL} == 0,6 \div 0,7\text{B};$$

With analysis of the static noise immunity of the elements I^{2}L as the boundaries of their closed and open states are assumed by the points of single amplification K_1^0, K_1^1, in which:

$$\left| {dU_0}/{dU_1} \right| = 1$$

These same points (Fig. 3.84b) determine the input threshold voltages and the margin values of the output logic levels of signals, and the values as such, in their turn, are determined by the design-technological and electrical physical parameters:

$$U_I\left(K_1^0\right) = U_{IE} - \phi_T \ln(r/2);$$

$$U_0\left(K_1^0\right) = \phi_T \ln 2;$$

$$U_I\left(K_1^1\right) = U_{IE} - \phi_T \ln(2r);$$

$$U_I\left(K_1^1\right) = U_{IE} - \phi_T \ln(2r/\beta_N);$$

Serviceability of I^{2}L LE is ensured until the transfer characteristic accommodates the area, in which: $\left| {dU_0}/{dU_1} \right| \rangle 1$.

The width of this area is $\Delta U_I = \phi_T \ln 4$, and the central point K of the area is characterized by the coordinates:

$$U_I(K) = U_{IE} - \phi_T \ln r;$$
$$U_O(K) = 0,5[U_{IE} - \phi_T \ln (r/\beta_N)];$$

The switching of the element load in the serial circuit of I^2L can be ensured on condition that the output voltage is as follows:

$$U_O > U_I(K);$$

Noise immunity of the element of I^2L relative to the locking noises can be determined from the expression:

$$\Delta U_{\overline{T}} = U_{OH} - U_I(K) \approx \phi_T \ln \beta_N$$

For the type value β_N=5, the value is $\Delta U_{\overline{T}} = 40$ mV. Consequently, the elements of I^2L possess the low margin of the noise immunity relative to the locking interference, which is to be taken into consideration during the design development of the I^2L integrated circuits.

The margin of the noise immunity relative to the unlocking elements of I^2L is far greater:

$$\Delta U_T^+ = U_I(K_1^1) - U_{OL} \approx U_{OH};$$

The noise immunity of the element by the current relative to the locking interferences can be determined from the expression:

$$\Delta I_{\overline{T}} = \alpha_N^P \cdot I_G(1 - 1/\beta_N)$$

where: α_N^P—normal gain ratio of the p-n-p transistor VT1

I_G—current of the injector

Thus, the noise immunity of the basic elements of I^2L during the schematic technical computations should be characterized by the parameters ΔU_T^+, $\Delta U_{\overline{T}}$, $\Delta I_{\overline{T}}$, which are the increasing functions of the current of the injector I_G.

The elements of I^2L possess the unique capability—a potentiality of controlling the element performance by altering the value of the current value of the injector, meanwhile the range of alteration of performance of the logic element can attain the magnitude of several orders. The performance of the element is determined by the processes of altering the charges, accumulated and dissipated in the active areas of the key n-p-n transistors. In dependence on the dynamics of the processes of building up and dissipation of charges, they differentiate three modes of operation of the injection element: the modes of small, medium, and large currents of the injector. In the mode of the small currents, the charges are substantial, localized on the barrier capacitances of the transistor junctions (influence of the charge of the suspended carriers can be ignored). The basic dynamic parameter of the element I^2L is the average delay time of switching:

$$t_P = 0,5(t_{PLH} + t_{PHL});$$

For the mode of the small currents of the injector (area 1) (I_G<50 uA), we get the expression: [2]

$$\tau_p^M = \frac{C_\Sigma \Delta U_T}{(2I_G \alpha_N^P)};$$

where C_Σ is the total of the mean barrier capacitances of the emitter and collector junctions of the n-p-n transistor VT2 and also of the parasite elements, connected to the base VT2.

From this expression it follows that in the mode of the small currents, the delay time is reversely proportionate to the current value I_G. In Fig. 3.84c, the common nature of dependence t_p is indicated on the injector current. Indeed. The total charge $C_\Sigma \Delta U_T$, accumulated on the barrier capacitances of the transistors VT1 and VT2, practically does not depend on the currents, passing through their p-n junctions. In the mode of the large currents of the injector (the area III) (I_G>150 uA), the charge build-up of the mobile carriers (holes) in the active areas of the п-p-п transistor occurs primarily in the area of the emitter. In this mode t_P reaches its minimum value t_P, not dependent on the current I_G, which is determined by the design technological factors and can be assessed from the expression:

$$t_p^з \approx \sqrt{\beta_N(N_K)}[2\pi f_{TM}(N_K)];$$

where: $f_{TM}(N_K)$—dependent on the number of collectors N_K margin frequency of single amplification of the n-p-n transistor

$\beta_N(N_K)$—dependent on N_K gain ratio value of the n-p-n transistor

In the area of the average values of the injector currents (50 uA < I_G < 150 uA), the charges, accumulated on the barrier capacitances and in the active areas of the transistors, become co-measurable; therefore, with the acceptable for the practical evaluations accuracy, the value t_p^{cp}can be determined by using the semi-total $(t_p^M + t_p^B)$.

3.5.1 *Varieties of the Basic Elements of the I^2L Integrated Circuits*

Modifications of the Basic Element I^2L

Modifications of the schematic technical solutions are aimed at improvement of the technical characteristics of the elements I^2L–enhancement of performance, reduction of the consumed power, increase of noise immunity, expansion of the functional potentialities, etc.

Increase of performance of the elements I^2L is attained as per three basic directions: reduction of the logic swing of the output signals, lowering (or exclusion) the saturation degrees of the n-p-n and p-n-p transistors, and introduction of the additional components.

In Fig. 3.84a, the diagram is shown of the base element I^2L with the saturation limitation of the key n-p-n transistor VT1.

The design of this element is realized in such a way, so that the low Ohmic n^+ areas of the collectors VT1 directly blend with the low Ohmic area of the base of p^+-type, forming the equivalent p^+-n^+ diode, bypassing the more high Ohmic circuit of the collector junction. The logic signal swing of this element reduces as compared with the element in Fig. 3.84, which ensures the switching delay reduction by 20–25%.

In order to limit saturation of the current setting p-n-p transistors, the following two main methods are used. The first technique presupposes the structural (physical) separation of the p-n-p transistors in the volume of the semiconductor structure; meanwhile, the base of the p-n-p transistor is connected with the emitter of the n-p-n transistor via the resistor R1 (see Fig. 3.85b), setting the offsets of their emitter junctions. The second method (Fig. 3.85b) is based on introduction of the individual high Ohmic resistors R2 in the supply circuit of each injector I^2L of the integrated circuits, in parallel with the junctions of emitter-collector of the p-n-p transistors.

In Fig. 3.84b, the element I^2L is represented with the simultaneous saturation limitation of the current setting and switchable transistors at the expense of activation of the diode VD in parallel to the junction of emitter-collector of each p-n-p transistor I^2L of the integrated circuits. As the diode VD, there can be used both the Schottky diode and the p-n diode.

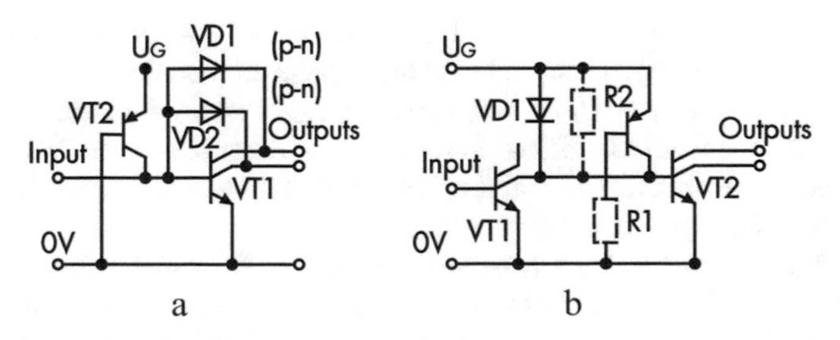

Fig. 3.85 Varieties of the basic elements I^2L with p^+-n^+ diodes (**a**) and resistors and a diode (**b**)

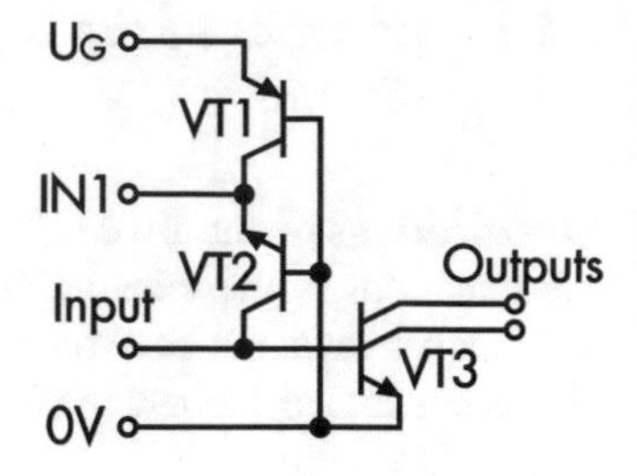

Fig. 3.86 Diagram of the multi-pin injection logic MI²L

I²L LE with Current Hogging One of the I²L modifications is the injection logic with current hogging (current hogging injection logic, CHIL), which is based on the effect of intercepting a portion of the injected by the emitter of the p-n-p transistor carriers by the additional p-area of semiconductor, located between the injector and the base of the p-n-p and n-p-n transistors [4]. Of practical interest is only one from the varieties of CHIL–the multi-input/output injection logic (multi-input/ multi-output integrated injection) or MI²L, represented in Fig. 3.86. The element consists of the current setting p-n-p transistor VT1, the n-p-n transistor of "interception" VT2, and the p-n-p transistor VT3. The injector current VT1 can be "intercepted" from the emitter circuit VT2 by means of applying the appropriate signal to the additional input IN1. Introduction of the input IN1 makes it possible to realize the function of "conjunction," and availability of the multi-collector transistor VT3 makes it possible to electrically uncouple the output signals during construction of more complex combinatorial circuits. The given technical solution represents the theoretical interest, as the practical application in the integrated circuits is limited with the technological complexities of realization of the p-n-p Schottky transistor VT3.

I²L LE with the Schottky Diodes

Introduction of the Schottky diodes in the circuit of the I²L elements ensures attainment of the following goals: a step-up of performance (at the expense of reduction of the logic swing and the saturation limitation of the switchable transistors) and ensuring the electric uncoupling of the logic circuits.

In Fig. 3.87 are depicted the I²L elements with the "uncoupling" Schottky diodes, created in the base vicinity (Fig. 3.87a) or the collector (Fig. 3.87b) of the n-p-n transistor.

A step-up of performance is attained by reduction of the logic swing as compared with the basic element I²L for the value of the voltage drop on the Schottky U_{DS}. The logic swing $\Delta U_T = U_{BE} - U_{CES} - U_{DS}$, where U_{CES} is the saturation voltage VT1.

As it is obvious from Fig. 3.86a, introduction of the input diodes VD1 and VD2 makes it possible on the base of the transistor VT1 to realize the functions of conjunction of the input signals.

In Fig. 3.86c is represented a version of the element I²L with the Schottky diodes, limiting the saturation degree of the n-p-n transistor at the expense of the "classic" solution–by connection in parallel with each collector junction of the transistor VT1 of the "shunt" Schottky diodes VD1 and VD2.

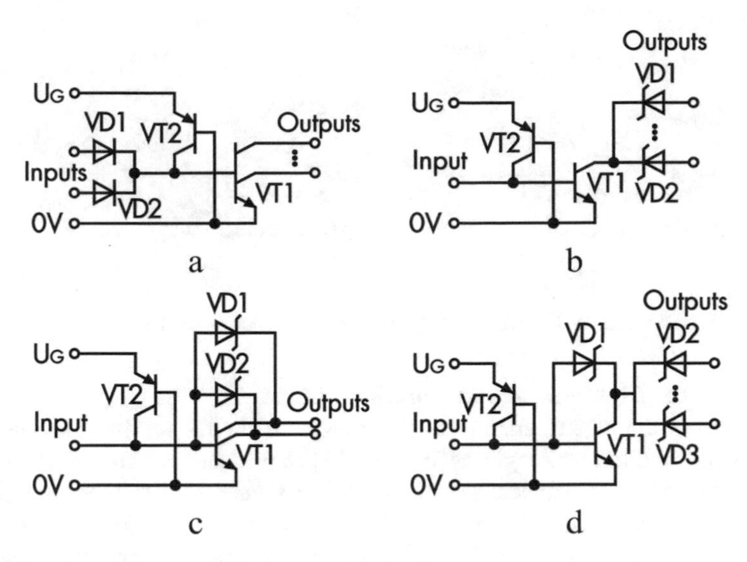

Fig. 3.87 Versions of the base I^2L elements with the Schottky diodes

The further increase of performance of I^2L element, shown in Fig. 3.87b, is possible at the expense of introduction into the circuit of the additional Schottky diode VD1 between the collector and the base of the transistor VT1 (Fig. 3.86d).

Such inclusion reduces the charge build-up in its base, as well as reduces the logic swing at the expense of the voltage increase on the collector VT1 (it exits from saturation to the boundary with the active mode). In such a circuit, the value of the logic swing is $\Delta U_T \approx U_{DS1} - U_{DS2}$, where U_{DS1} is the voltage on the "shunt" diode VD1. It is natural that the diode VD1 should have the value of the potential barrier height by 200–250 mV higher than VD2. The technological fabrication process of I^2L integrated circuits with two types of the Schottky diodes is complex, which is a deficiency of this solution.

The elements I^2L with the base Schottky diodes (Fig. 3.87a) make it possible to successfully solve such schematic technical problems, unsolvable in the standard I^2L, as connection of several outputs of the preceding elements to one input, connection of one output to several inputs with Schottky diodes of the subsequent elements.

The elements I^2L with the "uncoupling" Schottky diodes in the collector purposes are possible in the versions both with a single (Fig. 3.88a) and several (Fig. 3.88b) isolated collectors; meanwhile, in both cases, a possibility can be ensured of application of the electric connection between the elements without the Schottky diodes.

The issues of analysis and the schematic technical specific features of the above perused elements of I^2L with the Schottky diodes are reviewed in quite a detail in [1].

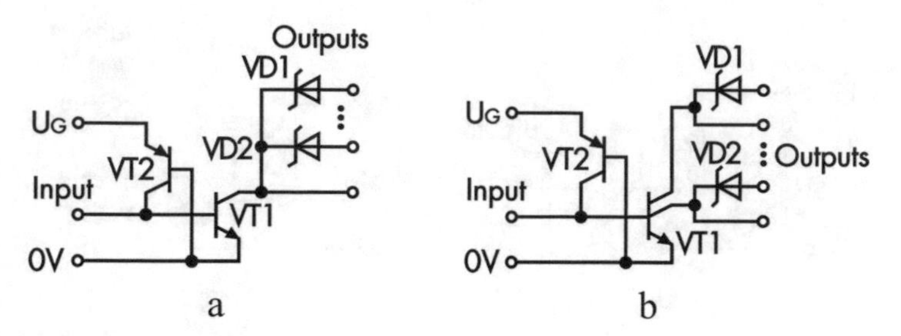

Fig. 3.88 Modifications of I²L LE with the "uncoupling" Schottky diodes

I²L LE for the Multivalued and Threshold Logic

Realization of the blocks, fragments and integrated circuits on the elements of the threshold and multivalued logic, as indicated in [5], provides definite advantages: the consumption power is reduced, as well as the chip's area (at the expense of reduction of a number of elements), and functional potentialities are augmented of the basic elements. Meanwhile, with increase of values of the logic elements, the advantages become more substantial. I²L LEs are more prospective for realization of the threshold and multivalued logic elements of the integrated circuits.

For representation of the multivalued variables in the multivalued injection integrated logic (multivalued injection integrated logic (MVI²L), they use the preset "weight" value of the current. The common principle of functionality of an element of the multivalued logic is explained in Fig. 3.89a, in which is represented a multi-output current generator on the basis of the current "mirror." Current I from the external circuit via the resistor R and the transistor VT1 in the diode switching presets the fixed voltage values of the emitter-base junctions VT2-VT1N. Thus, the *I* multiple currents (*2I*, *3I...NI*) can easily be obtained by means of the parallel connection of the appropriate number of collectors of the I²L transistors.

The logic functions in the MVI²L logic are performed by means of the "mirror reflection," adding and presetting the threshold values of currents. These operations are performed, in their turn, by means of the current "mirrors" and the threshold elements. The current "mirror" (Fig. 3.89b) has the transfer function:

$$I_{OUT}/I_{IN} = \beta/(1 + \beta_F);$$

where: β–the gain ratio by current of each from the two collectors

β_F–the gain ratio by current of the feedback collector

The ideal current mirror should have the single transfer function for each collector; however, in practice, the transfer ratio of the threshold I²L element (Fig. 3.89d) is different from the ideal one, because the current setting p-п-p transistor VT1 (Fig. 3.89c) in dependence on the collector voltage can operate in two modes, active one or passive, which determines the alteration (by 20% and over) of the p-n-p source current.

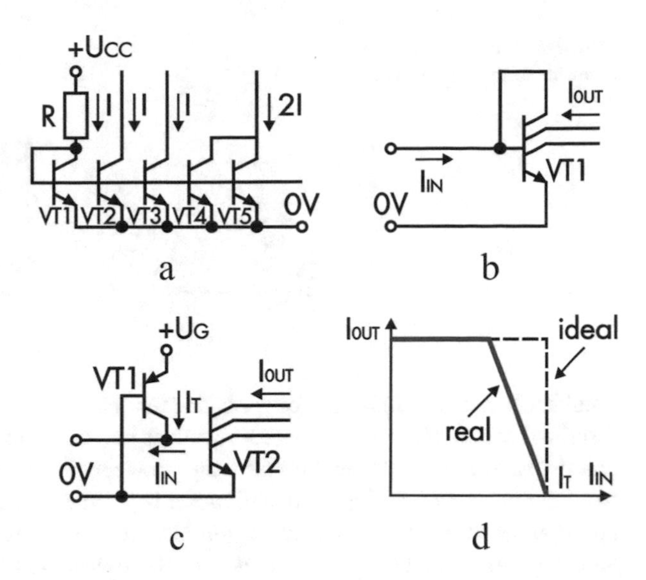

Fig. 3.89 Standard elements of the injection multivalued logic

The threshold I^{2}L element (Fig. 3.89c) has the current source VT1, offsetting the transistor VT2 for the required threshold value: if the current I_{IN} is smaller than the threshold value I_T, then the transistor VT2 is activated; if it exceeds its value or is equal to the threshold value, then VT2 is deactivated. The current source should be permanent for exclusion of the fluctuations of the transfer characteristic, and the gain ratio β should be maximum high, so that to preclude the source's exit from the active mode, and for securing the transfer characteristic to the ideal view.

The standard design-schematic solutions of the I^{2}L elements do not make it possible to attain the quite high values of the gain ratios of the inversely switched n-p-n transistors. In Fig. 3.90a is depicted the threshold element MVI^{2}L on the basis of the Darlington circuit [5] with power supply from the isolated n-p-n transistor VT1. Meanwhile, the p-n-p transistor of the current source can be offset to preclude its transition to the saturation mode. The transfer function of the current mirror has the following view:

$$I_{OUT}/I_{IN} = \beta \cdot (\beta_I + 1)/\beta \cdot (\beta_I + 1) + 1;$$

where: β_I, β –gain ratios of the transistor currents VT1 and VT2

The threshold element MVI^{2}L, made as per the Darlington circuit, proposed by Dao [1], is represented in Fig. 3.90b. This element ensures the higher noise immunity as compared with other known solutions. It is attained by the unsaturated mode of operation of the current setting p-n-p transistor VT1, whose base is offset relative to the common bus at the expense of the voltage drop on two series connected into the base diodes VD1 and VD2. This offsetting is of a sufficient value, so as to make VT1 operate as a DC current source in all modes of operation of the element.

Transistor VT1 may have many collectors, each from them will form half a value of the total source current; for presetting a definite threshold value, one needs to

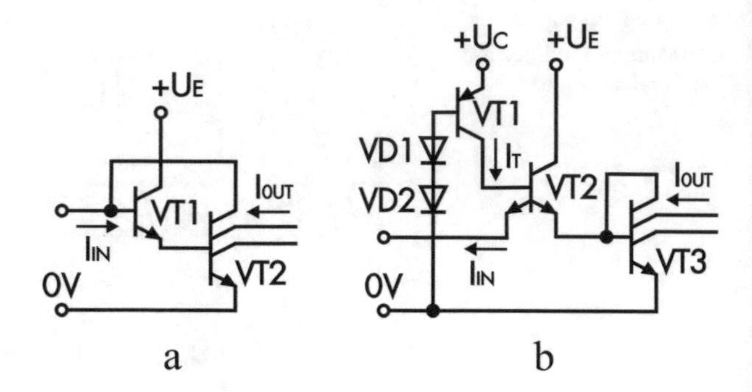

Fig. 3.90 Modernized logic elements of the multivalued logic

combine the required number of such "half-value" currents, whose total ensures the threshold value. However, this base element has one intrinsic shortcoming–a comparatively long time of deactivation–the process of dispersal of the minor carriers from the base VT3 during its deactivation is hampered due to unavailability of the appropriate low Ohmic discharge circuit. Indeed, at the time of arrival at the element's input at the time moment t_0 of the low logic level signal, the current I_T. (Fig. 3.90b) starts switching from thc base circuit VT2 to the input external circuit. After some time t_1 the emitter current VT2 will reduce to zero, and VT2 will enter the cutoff mode; during the time t_1 at the outputs VT3, there will be reduction of the collector currents (a build-up of the output voltage). Duration of this process is determined by the time of dispersal of the charge of the minor carriers, accumulated in the base VT3. As the low Ohmic discharge circuit of this charge is not available, the time of dispersal is determined by the recombination processes and can result in reduction of the element's performance.

Injection-Coupled Synchronous Logic

LE of thc injcction-coupled synchronous logic IC^{2}L (injection-coupled synchronous logic–ICSL) [1] represented in Fig. 3.91a differs from the logic with the current hogging injector (CHIL, current hogging injector logic) by the following:

- The injecting p-n-p transistor never switches the current to the common bus in process of operation.
- The input current, controlling the additional injector, can be used for arrangement of the dynamic effects of remembering the information at the expense of controlling the time of the charge build-up in the p-n-p transistor between the injector control input and the base of the inverting transistor.
- The injector control input increases the functional potentialities of LE.

In the circuit of IC^{2}L LE, the input C, connected with the base of the n-p-n transistor VT1, is considered as a synchronization input, and the one connected with an injector (VT2) input D is considered as an informational one.

LE can operate in two modes–the main and in the memory mode (C). The first mode is used in the circuits of the current comparators and the second one during arrangement of the synchronous logic circuits. With $D = 1$ and $C = 1$ on the base, VT1: $I_B = I_I + I_2$.

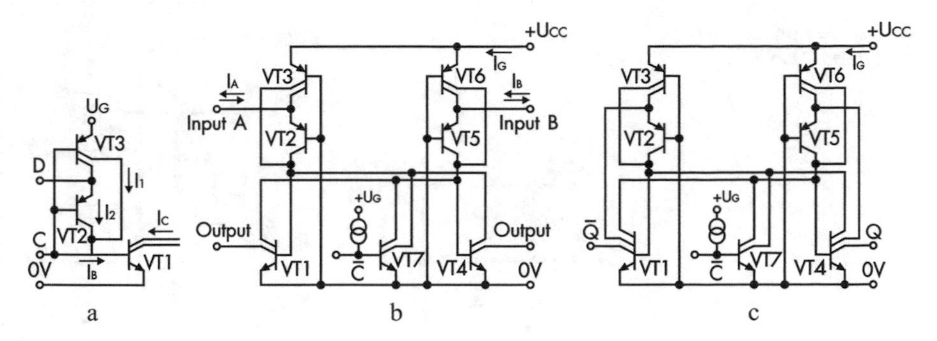

Fig. 3.91 Element of IC^{2}L (**a**) and examples of its realization on its basis of the current comparator (**b**) and T flipflop (**c**)

If the control element of the input C is activated, the base potential VT1 is converging to the potential $0V$, the base current I_B flows from the element, on the collector VT1 current $I_0=0$. If the preceding transistor at the input C is switched off with transition of the synchrosignal from the level $C = 0$ to $C = 1$, the voltage on the basc VT1 rises from zero up to $U_B = (I_1 + I_2)/C_B$, where C_B is the total capacitance by the input C. Meanwhile, on the collector VT4, the current rises from zero to I_B. If now $D = 0$ and $C = 1$, on the base VT1, the current is equal to I_1; meanwhile with transition of the synchrosignal from $C - 1$ to $C = 0$, the transistor VT1 switches off. When the synchrosignal C goes retreats from the level "0" to "1," the voltage on the base VT1 rises from zero to $U_B = I_1/C_B$; meanwhile, the collector current VT4 rises from 0 to I_G.

Operation of the element in the second mode is based on the fact that the deactivation time of the p-n-p transistor VT2 with the lateral structure is substantially longer than the time periods of activation and deactivation of VT1. In Fig. 3.91b, c are shown the typical examples of application of IC^{2}L LE in the schematics of the integrated circuits. Thus, in Fig. 3.91b is shown the diagram of the current comparator, realized on IC^{2}L LE, and in Fig. 3.91c is represented the diagram of the T flip-flop.

With the entire originality of the schematic solution of IC^{2}L LE, it has a significant deficiency, limiting its possibility of application in the integrated circuits–the slow performance, determined by duration of the processes in progress of building up, and dissipation of charges in the active areas of the p-n-p transistors VT2 and VT3.

Functionally Integrated I^{2}L LE

The concept of the structural-functional integration is natural for the injection logic, as it ensues from the fundamental principle of the injection logic–matching the various functions in the volume of semiconductor. In Fig. 3.92a is represented the equivalent diagram of the functionally integrated structure of I^{2}L LE of the enhanced performance with the internal feedback circuit. LE contains the additional n-p-n transistor VT3, which is structurally located directly near the base of the basic

Fig. 3.92 Functionally integrated I^2L LE

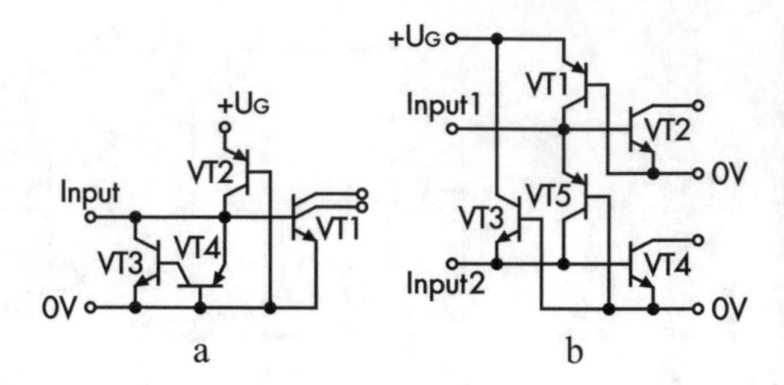

switchable n-p-n transistor VT1 and is connected with it by means of the internal structural link in the form of the equivalent p-n-p transistor VT4; collector VT3 is electrically connected with the base of the element VT1. During application of the supply voltage U_O, the n-p-n transistor VT1 is activated and comes into the mode of saturation, recharging the load capacitances with the collector current:

$$I_C = I_G \cdot \alpha_\beta \cdot \beta_I;$$

where: I_G–injector current

α_β–current transfer ratio of the p-n-p transistor VT2
β_I–current gain ratio of the base VT1

The open base-emitter junction of the transistor VT1, in its turn, injects the carriers, which offset in the forward direction the emitter junction of the additional transistor VT3, which is activated and reduces potential of the base of the main n-p-n transistor VT1, thus, bringing it out of the saturation mode. Owing to the existent feedback, the value of the base potential VT1 will always be sufficient to its maintenance in the open state.

The condition of serviceability of such element can be expressed in the form:

$$(1 - \alpha_{PI} \cdot \beta_{IS}) \cdot \beta_I \geq 1;$$

where: α_{PI}–inverse current transfer ratio of the current VT2

β_{IS}–current gain ratio of the base VT3
β_I–current gain ratio of the base VT1

The feedback depth depends on the current transfer ratio of the additional p-n-p transistor VT3 and the base current gain ratio of the additional n-p-n transistor VT4.

On this condition, the n-p-n transistor VT1 operates in the active mode, which makes it possible to increase the performance of the element without increasing consumption power.

The similar method of enhancing performance at the expense of the saturation limitation of the n-p-n transistors in I^2L LE is attained structurally: in the isolation layer of the integrated circuit elements, in between the elements, the "windows" are made, which is equivalent to inclusion between the bases of the n-p-n transistors of I^2L VT2, VT4 elements of the bidirectional p-n-p transistors VT5 (Fig. 3.92b).

If one of the "adjacent" IL elements is activated, and the other is turned off, then the equivalent p-n-p transistor VT5 draws from the base of the activated element a share of the current, reducing the degree of its saturation.

3.5.2 Memory Elements of I^2L Integrated Circuits

The basis for construction of the memory circuits I^2L of the integrated circuits is represented by the flip-flops. The information writing and reading process can be performed in series or in parallel, in the forward, reverse, or paraphrase codes, and introduction of the additional control circuits makes it possible to shift the contents of the memory circuits.

The logic potentialities of the memory circuits are determined by selection of the appropriate types of flip-flops. From the entire multitude of the existent varieties of the flip-flops, the most popular with the IC designers are two basic classes of the synchronous flip-flops, differing with the principle of the synchronization circuit arrangement. The first class is formed by the flip-flops with synchronization by the level of synchrosignal and the second one flip-flops with synchronization by the synchrosignal edge (swing).

Commonly accepted is classification of flip-flops by the functional distinction as per the form of the logic expression, characterizing the state of the circuit inputs and outputs at the moments of time before and after actuation. Of the most simple and widely popular implementation on I^2L LE is with asynchronous RS and synchronous R-S and also with level synchronized D flip-flops. R-S flip-flop (Fig. 3.93a) contains four I^2L LE, and D-flip-flop (Fig. 3.93b) contains eight I^2L LE and is built on the basis of the R-S flip-flops.

Information storage into the flip-flop as per the input D is performed with $C = 0$. The switching delay is by the control inputs (R, S, D): $t_n = 3t_3$, the maximum synchronization frequency:

$$f_{\max} = 1/3t_3.$$

The distinctive feature of the flip-flops I^2L (Fig. 3.93b) is application of one inverter of the synchrosignal to control operation of two flip-flops.

The highest performance is ensured by the improved R-S and D-flip-flops (Fig. 3.93c) whose maximum frequency is $f_{\max} = 1/2t_3$; the flip-flops are synchronized by the level of the synchrosignal $C = 0$ and in total contain five I^2L LEs.

The flip-flops, synchronized by the synchropulse edge, are constructed on the basis of the asynchronous R-S flip-flop, whose switching is performed either by means of the control R-S flip-flop (CF) or by two commutating R-S flip-flops (CF). Synchronous flip-flops on the basis of CF are built on two level synchronized R-S or D-flip-flops. Two versions are possible of the flip-flop circuits on this type of

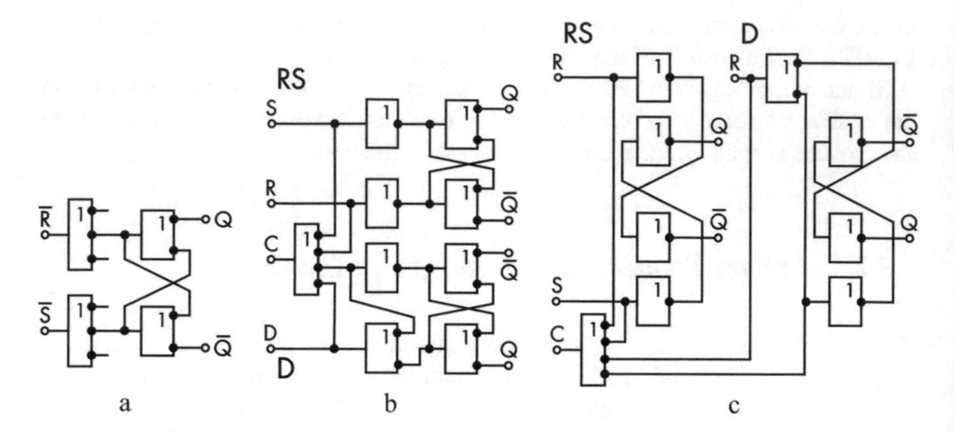

Fig. 3.93 Standard asynchronous RS flip-flop (**a**), synchronized by the level D flip-flop (**b**), the improved RS (**c**), and D flip-flop

element I²L: with the paraphrase synchronization (the circuit with the inverter) and with the denying links.

In Fig. 3.94a–d, there are diagrams of the synchronous flip-flops of the types R-S, J-K, and D-V with the paraphrase synchronization (FPS), in which are used the level synchronized basic R-S flip-flops. The flip-flops are synchronized by the positive edge of the synchrosignal. By means of altering the activation circuit of the synchrosignal inverters, as it is indicated in Fig. 3.94c, it is possible to easily realize synchronization by the negative edge. The asynchronous inputs of setting and resetting S_d, R_d perform the setting of the output stages in the state of $Q = 1$, $Q = 0$ with $S_d = l$; $R_d = 0$ or reset to the state of $Q = 1$, $Q = 0$ with $S_d = 0$, $R_d = 1$.

Combination of $S_d = R_d = 1$ (deactivation of the inputs R_d, S_d) does not render influence on the operation of the flip-flops, combinations of $S_d = R_d = 0$ being denied. In the synchronous flip-flops with the disabling links, one inverter of the synchropulses is excluded. The diagrams R-S(J-K), D(D-V) of these flip-flops of this type are represented in Fig. 3.95a, b. The flip-flops are synchronized by the positive edge and peculiar with the enhanced reliability of storage.

With application of the level synchronized R-S and D flip-flops, it is also possible to construct the synchronous flip-flops with the paraphrase control. In Fig. 3.95c, d, the diagrams are indicated with the R-S and D flip-flops of this type, synchronized by the positive signal edge at the input C. The negative edge synchronization is realized with connection of the inverters of the synchropulses similar to the diagram in Fig. 3.94f. In Fig. 3.95, *a* and *b*, the dotted lines indicate realization of the new J-K and D-V-flip-flops, having the maximum switching frequency $f_{\max}=1/4\tau_3$, i.e., 1.5 times more than the flip-flops, considered above.

In all considered diagrams, the output and control flip-flops, being level synchronized, can be both the single type (both R-S or both D) and also of different type (one R-S, the other D). T flip-flop, rarely applied in the integrated circuits, can be realized on the basis of the J-K flip-flop, if to leave the inputs *J* and *K* vacant (it complies with

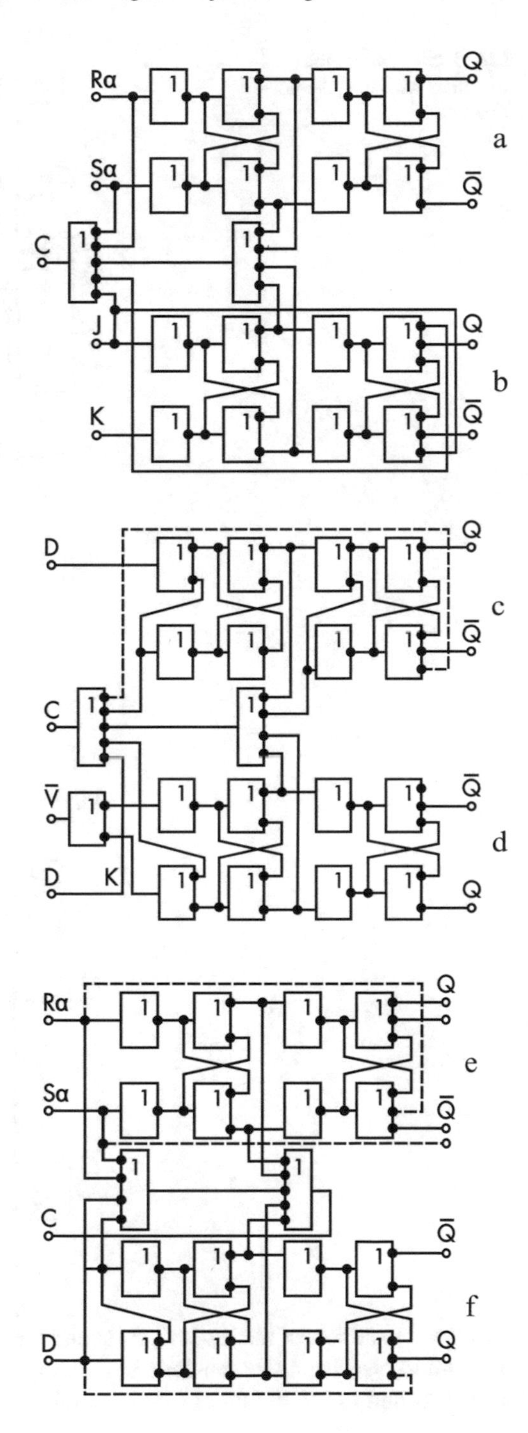

Fig. 3.94 Synchronous I^2L flip-flop of types RS (**a**), JK (**b**), D (**c**), DV (**d**), (**e**), (**f**)

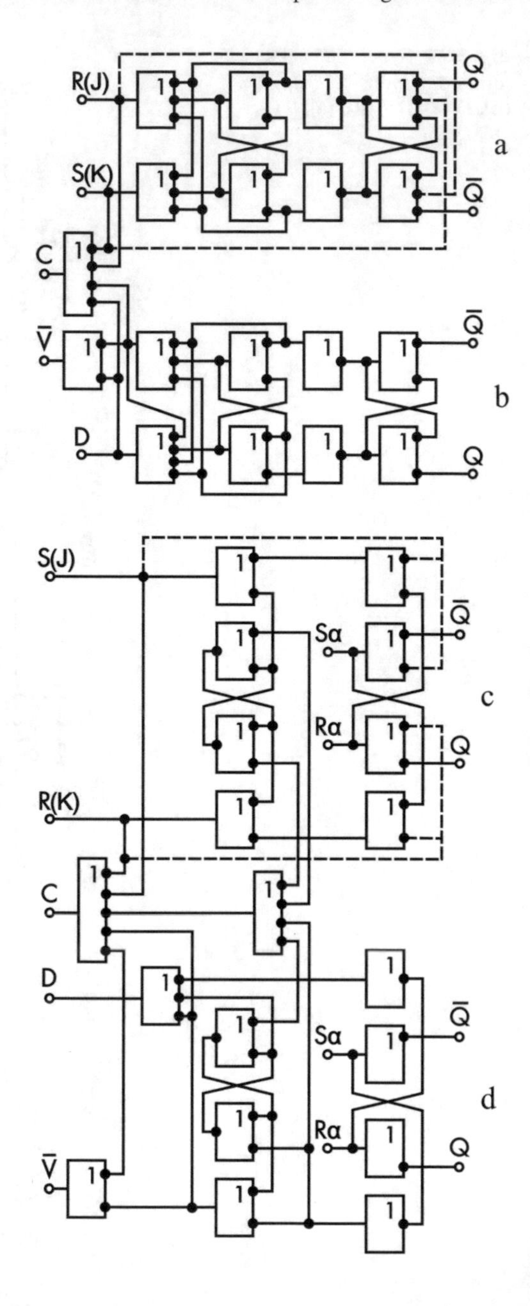

Fig. 3.95 Synchronous flip-flops with the "denial" links

$J = K = 1$) and to use the input C, as countable one ($C = T$). It is possible to get the T flip-flop also from the standard D flip-flop, for which it is necessary to connect D with Q and to use the input C as countable.

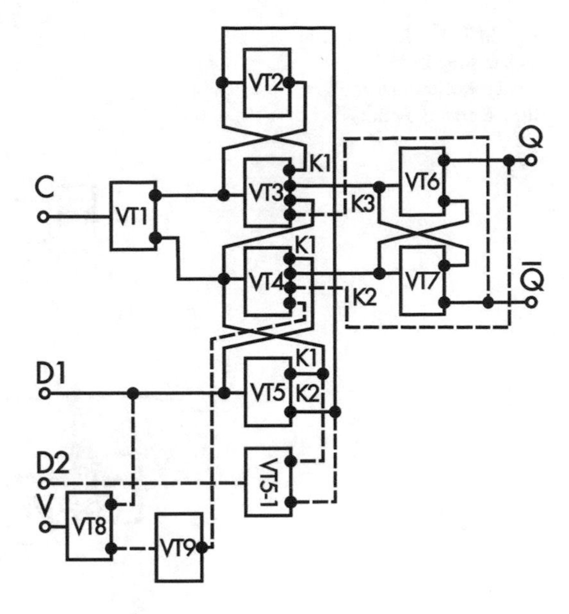

Fig. 3.96 Injection D-V flip-flop with the internal delay with expansion of runs by OR

Figure 3.96 represents the D-V flip-flop with the internal delay *b* with the expansion of inputs by "OR." The feature of the internal delay, as a rule, makes it possible to perform in one clock cycle of the write operation of the new information into the flip-flop and to read from it the earlier written information.

A substantial problem with the schematic design development of the I^{2}L integrated circuits is arrangement of the reliable functioning of a large number of flip-flops, located at various distances from the power supply pads and the synchronization pads on the die. It is impossible to connect to the common synchronization bus directly the majority of the considered flip-flops–it is necessary for the sake of avoiding the "races" to use the individual synchronization circuits to apply the clock signals to the inputs from each of the flip-flops. Figure 3.97 represents the injection D-V flip-flop with the internal delay [6], making it possible to simplify the system of the internal synchronization of the I^{2}L integrated circuits. The principal difference of this flip-flop from those, considered above, is the control arrangement of its operation by the information input instead of the classic operational control by the synchronization input.

This D-V flip-flop operates in the following way. With the low signal level at the input *V*, the outputs *K1* and *K2* of the element VT3 are in the high state while the output of the element VT10 in the low state.

The high level at the output *K1* of the element VT8, assembled by OR with the signal at the input of the D flip-flop, passes this signal at the element input VT5, while the low level at the element output VD9, assembled by OR with the level at the output *K2* of the element VT6, yields at the element input VT10 the low level irrespective of the state of VT6. Thus, with arrival of the synchrosignal *C* in the flip-flop, the information will be stored, arriving at the input of D flip-flop, as the outputs

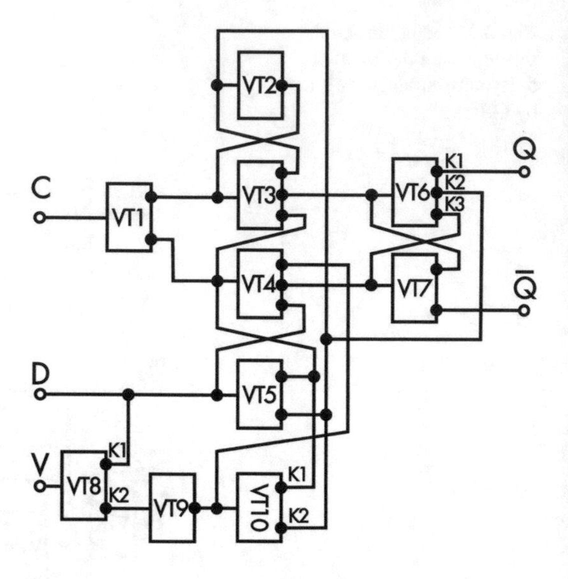

Fig. 3.97 I^{2}L D-V flip-flop for the simplified synchronization systems of the integrated circuits

K1, K2 of the element VT10 will be in the high state. At the high level of the signal at the input *Y*, the outputs *K1* and *K2* of the element VT8 are in the low state and the element output VT9 in the high state. The low level at the output K1 of the element VT8 ensures the low level at the input VT5 irrespective of the signal level at the input of the D flip-flop, i.e., actually, it disconnects the information input of the D flip-flop. The high level at the flip-flop output VT9 permits the passage to the element input VT 10 of the level from the output *K2* of the element VT6. In this case, with arrival of the synchrosignal to the input C, the flip-flop will write from the element input VT10 the state, which was written in the previous cycle, i.e., it will confirm its state and will keep on confirming this state until at the input of the *V* flip-flop, the low level appears of the control signal *V*=0.

Thus, the diagram of D-V flip-flop arranges its operation control at the cost of commutating the data, arriving to its information input, in which the flip-flop synchronization input remains "clean," and this input can be directly connected to the common synchronization circuit. In particular, with availability of the power shaper of the synchrosignal, application of this flip-flop makes it possible to connect the synchronization inputs of entirely all flip-flops in the integrated circuit to the unified common synchronization bus, i.e., to simplify to maximum the synchronization circuit of the integrated circuit by elimination from this circuit of the additional uncoupling elements, via which the synchrosignal arrives to the synchronization inputs of the D-V flip-flops. Elimination of these elements guarantees also absence of "races" along the synchronization circuits. The pulses and synchronizations will be arriving to all flip-flops simultaneously, irrespective of the place of their location on the die of the integrated circuit and the possible technological scatter of the dynamic parameters of the individual elements of the synchronization circuit, as the entire synchronization circuit consists of one common bus and the shaper, loaded on this bus.

It is necessary to note also that in the complex synchronization circuits, consisting of the various elements, the scatter of the switching delays of the synchronization circuit elements depends on the ambient temperature alteration or the integrated circuit supply voltage. As a result of this, even if the "races" in the integrated circuit synchronization in its operation under the normal conditions do not exceed the permissible limits, then, during the ambient temperature or supply voltage variation, the switching delays scatter may exceed the permissible limit, which will result in the failure of the integrated circuit. Such recurrences can be excluded during application of the reviewed D-V flip-flop, as alteration of the dynamic parameters of the common shaper, which can be used in that case, turns out similar for all flip-flops of the circuit, connected to this shaper via the unified common bus of synchronization.

All above considered flip-flops [6] possess, in their turn, one common deficiency– a relatively low load capacity, determined by the inverse connection of the n-p-n transistors of the I^2L elements. In Fig. 3.98 a diagram is presented of the injection D-V flip-flop with the enhanced load capacity. A distinctive feature of this flip-flop being that the performance of this flip-flop meanwhile does not deteriorate owing to absence of the additional signal delay, when information is sent to the output. I^2L elements VT1–VT7 form the classic reconfiguration of the D flip-flop, and the circuits on the elements VT8–VT11 make it possible to ensure a step-up in the load capacity when keeping the rate of performance.

As the circuits have no additional logic switching delay of the device, then presentation of the result occurs during one and the same time, as in the case of the "classic" D flip-flop.

3.5.3 *Schematics of the Input Matching Elements of the I^2L Integrated Circuits*

Input MEs with Conversion of the TTL Signal Levels

The I^2L integrated circuits have a distinctive feature of the substantial dependence of the ME dependence on the accepted design-technological basis.

One differentiates the schematics of MEs of the I^2L integrated circuits on the basis of the "non-isolated" I^2L and I^2L with isolation.

It should be noted that the input MEs with the "clean" I^2L signal levels ($U_{OL} < 0.4$ V; $U_{OH} > 0.7$ V) in the integrated circuits are not used, as a rule, due to their low load capacity, high sensitivity to the input interferences, and dependence of the input switching thresholds on the technological factors and the temperature conditions.

In Figs. 3.99 and 3.100 are represented the ME circuits with conversion of the TTL signal levels on the basis of the "non-isolated" I^2L.

The most elementary input ME with conversion of the TTL levels to the I^2L levels is realized (Fig. 3.99a) and contains the resistive divider R1, R2 and the inversely

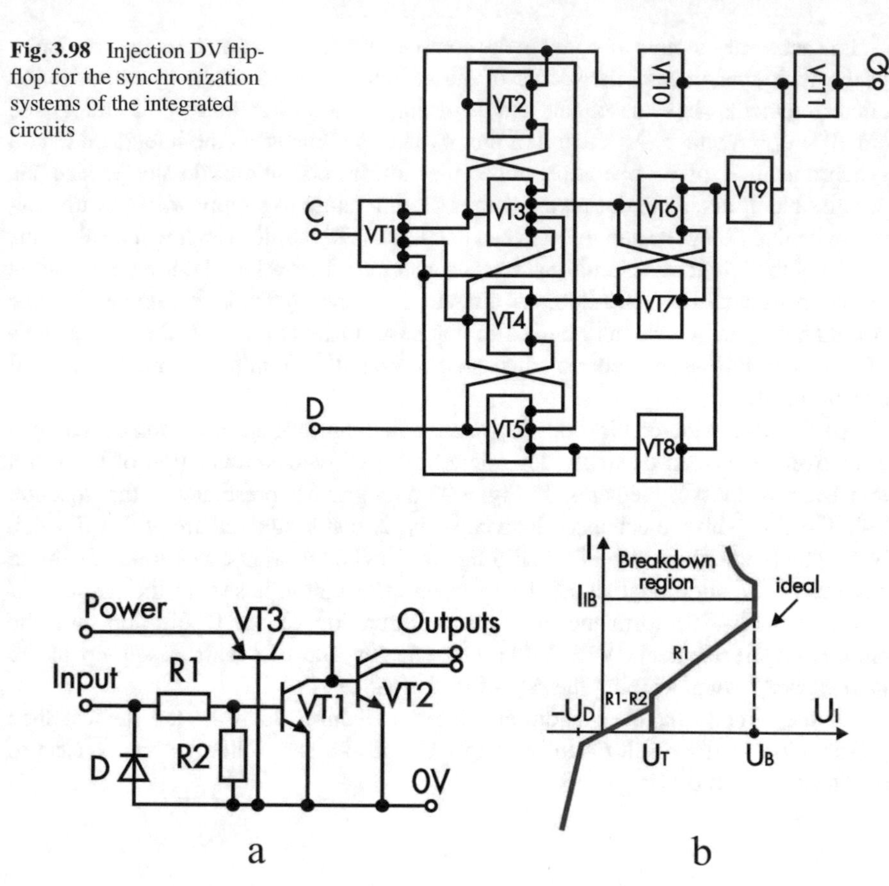

Fig. 3.98 Injection DV flip-flop for the synchronization systems of the integrated circuits

Fig. 3.99 Diagram of the input ME on the basis of the non-isolated injection logic: the most elementary ME (**a**) and its input characteristic (**b**)

connected n-p-n transistor VT1, whose load is the p-n-p transistor VT3 of the internal multi-collector injection transistor VT2. ME ensures the direct electric matching of the integrated circuits as per the input characteristics of both I^2L of both integrated circuits and TTL integrated circuits. The distinction of the given ME is that it consumes power only from the external power supply source. In Fig. 3.98b an input characteristic of ME is presented:

$$I_I = f(U_I).$$

If $U_I < U_T$, where U_T is the switching threshold voltage, the input current of the integrated circuits is determined by the expression:

$$I_{IL} = U_I/(R_1 - R_2).$$

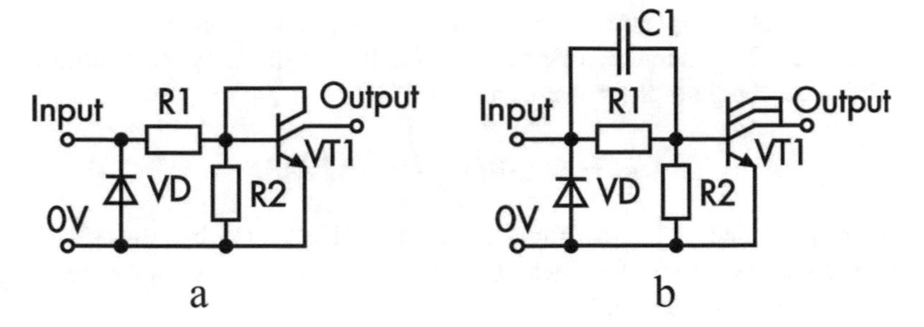

Fig. 3.100 Input ME with the opposite saturating collector and the accelerating capacitance

With the input signal $U_I > U_T$, the input current of the integrated circuits is as follows:

$$I_{IH} = \frac{(U_1 - U_E)}{R_1}.$$

Inversely connected transistor VT1 comes into the saturation mode, characterized by the parameter "saturation degree":

$$S = \frac{\beta_N}{I_{CP}} \cdot \frac{U_T - U_{IHB}}{R_1} + 1;$$

where: β_N–gain ratio of the base current of the inversely connected n-p-n transistor VT1

I_{CP}–current collector of the p-n-p transistor VT3

Nominal rated values of the input divider resistors are determined from the expressions:

$$R_1 \geq \frac{U_I - U_{IHB}}{I_{I\max}} + 1;$$

$$R_2 = R_1 \cdot \frac{U_E}{U_{IHB} - U_E - (I_{CP}/\beta_N)R_1};$$

where: $I_{\max}$–maximum input current of ME

Availability of the saturation mode determines a low performance of ME. During arrival to the signal input of the high logic level U_{IH}, the element switching delay:

$$t_{PHL} = [R_1 \cdot R_2 \cdot (R_1 + R_2)] \cdot (U_E/U_{IH}) \cdot (C_E + nC_C + C_P);$$

where: C_E, C_C, C_P–capacitances of the emitter and collector junctions, as well as the total metallization capacitance of the mount pad pins of the integrated circuits output

Switching delay t_{PLH} during arrival to the input of the low-level signal U_{IL} is determined by the saturation degree, specifically by the dispersion time of the excessive base charge of the n-p-n transistor VT1:

$$t_{PLH} = \tau_B \cdot ln\,(1 + I_{IH} \cdot R_1/U_E);$$

For the type values of the resistors $R1 = R2$ from 10 to 20 kOhm, the delay time is t_{PHL}=15–25 ns, t_{PLH}=20–30 ns, which substantially limits the velocity characteristics of the I^2L integrated circuits.

Improvement of the dynamic characteristics of the input ME of the I^2L integrated circuits on the basis of the "non-isolated I^2L" can be attained by means of application into the quality of the VT1 transistor with the additional "counter-saturation" collector, as it is indicated in Fig. 3.100a [6].

The saturation degree S and the rated nominal values R_1, R_2 can be computed as per the same expressions, as for ME, Fig. 3.99a, where instead of the ratio β_N, it is necessary to use the effective gain ratio:

$$\beta_{ef} = \frac{\beta_N}{(1 + \beta_S)};$$

where: β_S–gain ratio as per the feedback collector

Application of the counter-saturation collector makes it possible to reduce by 4–6 times the saturation degree VT1 and to accelerate the dispersion process of the accumulated carriers by means of formation additionally to the resistor circuits of the surging circuit of the feedback collector. For this case the switching delay is as follows:

$$t_{PHL} = \frac{\tau_H}{1 + \beta_S} \ln \frac{I_{B1} + I_{B2}}{I_C[1 + \beta_S/\beta_N] + I_{B2}};$$

where I_{B1}, I_{B2} are the incoming and outgoing currents of the base VT1.

However, the given schematic and technical solution of the input ME has a substantial deficiency–sensitivity to the technological scatter of the numerical values of the gain ratios β_N, β_S, expressed in the appropriate non-controlled variation of t_{PHL}, t_{PLH}. Therefore, such ME is advisable to apply in the input circuits, not critical to alteration of the dynamic parameters.

In order to enhance performance of the input circuits of the integrated circuits on the basis of the "non-isolated" injection logic, widely applied is the diagram, indicated in Fig. 3.100b [6], with the "accelerating" capacitance C1, included in parallel to the resistor R1. The positive effect of the given schematic technical solution is in the fact that at the time in progress, at the input of signal "1" the capacitance C1 is under charging, and after completion of the pulse (which corresponds to arrival of the signal "0"), this capacitance by its negative pin turns out to be connected directly to the base with transistors, accelerating the dispersion process of the accumulated charge of the minor carriers in the saturation mode. Application of the “accelerating” capacitance with the value of 2–3 pF makes it possible to reduce the ME switching delay almost by an order, up to t_{PLH}=3÷6 ns. However, activation of the capacitance C1 can reduce the noise immunity of the input ME at the cost of reducing the dynamic value of its threshold activation voltage–the false actuation of the integrated circuits is possible from the “prompt” dynamic interferences at the input, when the noise pulse duration constitutes 1÷2 ns. With selection of the optimum values of the capacitance C1 with the preset resistance values of the resistors R1 and R2, permissible levels of the dynamic noises, and the preset durations of the input signal edges, it is possible to use the following differential expression [69], describing the transitional processes in the given input ME:

$$\frac{dU_E}{dU} = \frac{U_{IO}}{R_1C}\left[\left(\frac{R_1C_1}{\tau_u} - 1\right)e^{-\frac{t}{\tau_n}+1}\right] - \frac{U_E}{R_EC_E} - \frac{I_O}{C_E}\left(e^{\frac{U_E}{m}-1}\right);$$

where: $C_Э = C_1 + C_2$–capacitance of the emitter junction VT1

$R_Э = \frac{R_1 \cdot R_2}{(R_1+R_2)}$–equivalent resistance

τ_u–exponent parameter, representing the input pulse in the form

$$U_I = U_{IO}\left(1 - e^{-\frac{t}{\tau_u}}\right);$$

I_O–the reverse current of the equivalent diode of the emitter junction VT1

In the practical computations of the input ME the given expression is solved by the Euler method; meanwhile, for each preset value of the pulse edge τ_u and the amplitude edge U_{IO} by means of the circuit response computation by the method of the subsequent approximations, the capacitance C1 value is determined, at which the pulse signal at the base VT1 does not exceed the preset threshold unlocking voltage VT1. The obtained dependences of the minimum value of the capacitance $C_{l\,\min}$, at which the noise signal still does not directed to output, make it possible to select the optimum value of the capacitance C1 from the noise amplitude and duration of the input edges.

It should be noted that in dependence on the requirement of the specific conditions of application, it is possible to control the value of the switching threshold U_T by means of the design alteration of the ratio R1/R2 (asymmetrical input ME).

Inclusion into the design of the I^2L integrated circuits of the complete isolation made it possible to realize on one die of the integrated circuits both the elements I^2L, TTL, and ECL and their different combinations. The necessity of such design solution is determined by the requirements of enhancing performance, load capacity and noise immunity of the input MEs, reduction of the input currents of the I^2L integrated circuits, and expansion of their functional features. The resistive-transistor input ME on the transistor VT1 Fig. 3.101a [6] differs with the sheer simplicity, area on the die, and the enhanced (as compared with ME in Fig. 3.99a) performance. The circuit operates in the following way: with increase of the input signal voltage up to

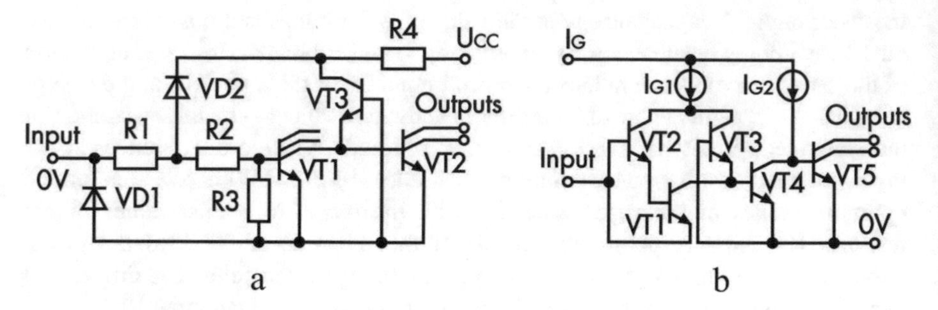

Fig. 3.101 The most elementary input MEs of the I^{2}L integrated circuits with the complete isolation: (**a**) with the resistive divider; (**b**) non-resistive

the value of U_T, corresponding to the activation threshold voltage, the transistor base current VT1 will be as follows:

$$I_B = \frac{U_{D2} + U_B - U_E}{R_2} - \frac{U_E}{R_3};$$

where: U_{D2}–voltage on the limit diode VD2

U_B–voltage on the injectors supply bus of the I^{2}L elements (correspond to the voltage on the junction of emitter-base of the open transistor of the p-n-p type VT3)

U_E–voltage on the junction of emitter-base of the open n-p-n transistor VT1

The saturation degree of VT1 is limited in this case and does not depend on the further increase of the input voltage, which ensures a small deactivation delay of ME. Meanwhile, the input current, releasing by the external control circuit to the input of the stage, is not wasted, but is directed to the supply bus of the I^{2}L integrated circuits, also promoting the performance enhancement of the I^{2}L elements.

In Fig. 3.101b [6] is shown the original diagram of the input ME of the enhanced noise immunity, differing by absence of the resistors in its composition. The ME input is connected simultaneously with the base of the p-n-p transistor VT2 and the emitter of the n-p-n transistor VT1. When arriving to input, ME of the low-level signal VT2 is activated with current of the generator $0.5I_{G1}$, ensuring current of the base VT1, required for its activation. As the base current of the I^{2}L transistor VT4, formed by VT3, is drawn by the activated transistor VTI, the multi-collector I^{2}L transistor is activated by current of the generator I_{G2}; at the element output, the input signal logic level is repeated–the level “0”.

When arriving to the ME input of the level “1”, the transistors VT2 and VT1 are locked, and the controlled by the full-value current I_{01} transistor VT3 promptly activates the transistor VT4, draining the base current I_{G2} and I^{2}L transistor VT5–the level “1” appears at the ME output. Activation of the circuit VT1–VT2 will have place already in the case, when the voltage at the ME input exceeds the potential of the common bus by 60÷80 mV. The deficiencies of this ME are the high value of the input (outgoing) current and dependence of the switching threshold on the gain ratio of the base current VT2 and the current I_{G2}.

On the other hand, while controlling these values, it is possible to control also the value of the noise immunity of the input ME. Figure 3.101 shows a group of the input MEs, combined by their common peculiarity: application of the multi-emitter n-p-n transistors.

In Fig. 3.102a is shown a "classic" diagram of ME [6], specific for the first types of the I^2L integrated circuits. The circuit is constructed on the basis of the multi-emitter transistor VT1, transistors VT2 and VT3 playing the role of the bias diodes, and the output transistor VT4, whose collector is connected to the input of the injection element. For the case when it is required to have a small load capacity, one can use as VT4 the injection multi-collector n-p-n transistor.

In order to reduce the input incoming current I_{IH}, determined by the intrinsic I^2l high values of the inverse gain ratios β_I, the transistor base VT1 is connected with its collector. For compensation and reduction of the circuit switching threshold to the value of the voltage drop on the collector junction VT1 into the circuit was introduced the bias transistor VT2 with the short-circuited collector junction. When arriving to any from the ME inputs of the signal U_{IL}, the appropriate emitter junction VT1 gets open, forming the low Ohmic circuit for the source U_{CE}, and the current through R2 reduces to zero, deactivating the transistors VT3 and VT4, with the high signal level appearing at the output.

In Fig. 3.102b [1] ME is constructed on the minimum quantity of elements: two multi emitter transistors VT1 and VT2 and two resistors; meanwhile, the additional emitter of the output transistor VT2 forms the feedback circuit as per the base VT1.

The common ME deficiency in Fig. 3.102a, b is their comparatively slow performance (long time t_{PLH}). In fact, during the ME deactivation (Fig. 3.102a), there happens a current switch-over from the circuit of the second emitter VT1, connected with the base VT2, to the first emitter VT1, connected with the ME input; after which VT2 and VT3 are subsequently deactivated. Meanwhile, the major contribution to the delay of t_{PLH} is made by the dispersion time of the minor carriers, accumulated in the bases VT3 and VT2, as the low Ohmic circuits are not available, which could have accelerated the processes of the charge dispersion and the discharge of the barrier capacitances VT2.

This deficiency is not evident in the ME diagram in Fig. 3.102c [1], where the design of VT2 has the second emitter introduced, connected with the ME input. On arrival to the ME input of the high potential in the forward direction, the emitter junctions of VT1 are activated. The current, flowing via the second emitter to the base of VT2, results in the unlocking of VT2 and VT3 and appearance of the low-level voltage at the output. With the subsequent application to the ME input of the low potential, there occurs a current switch-over of VT1 from the emitter circuit, connected with the base VT2 to the circuit.

Meanwhile, the switching time is determined by duration of the recharging process of the barrier capacitance of the emitter junction VT2. The ME deactivation time is practically determined by the deactivation time VT3 and is reduced approximately twice. A similar method was placed into the basis of the ME operation with

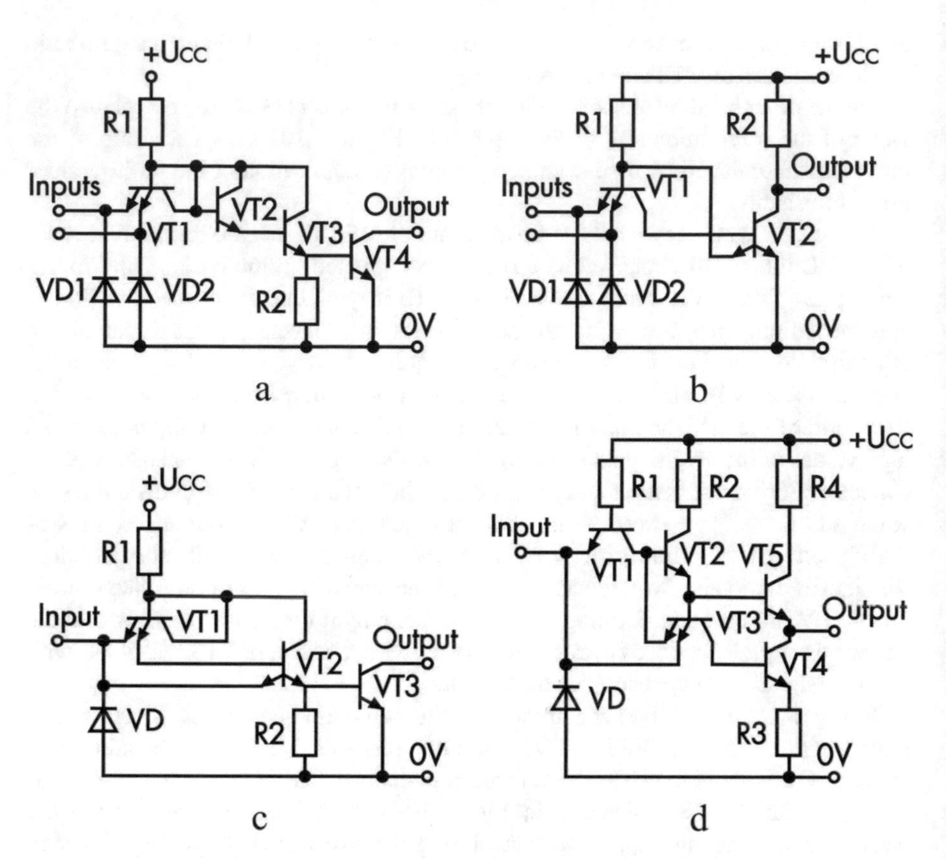

Fig. 3.102 Input ME TTL on the basis of the multi-emitter n-p-n transistors

the enhanced load capacity (Fig. 3.102d) [6]. Introduction into the design layout VT3 of the second emitter, connected directly with the ME input, makes it possible to substantially cut down the time t_{PLH} and somewhat shorten up the time t_{PHL}.

A substantial deficiency of the represented in Fig. 3.102 ME group, making use of the input multi-emitter transistors, is presence of large values (up to hundreds of microamperes) of input incoming and outgoing currents. The effective method of reducing the values of the input currents is introduction into the composition of the input ME stage of the key p-n-p transistors, base controlled.

In Fig. 3.103a is shown the base diagram of the input ME, in which they use activation of the p-n-p transistor as per the diagram with the common collector. The additional emitters of the transistors VT2 and VT4, connected with the base, ensure reduction of the saturation degree of the n-p-n transistors.

When the low-level signal U_{IL} arrives at the ME input, the p-n-p transistor gets open, with the outgoing current appearing at the external pin (input of the integrated circuits):

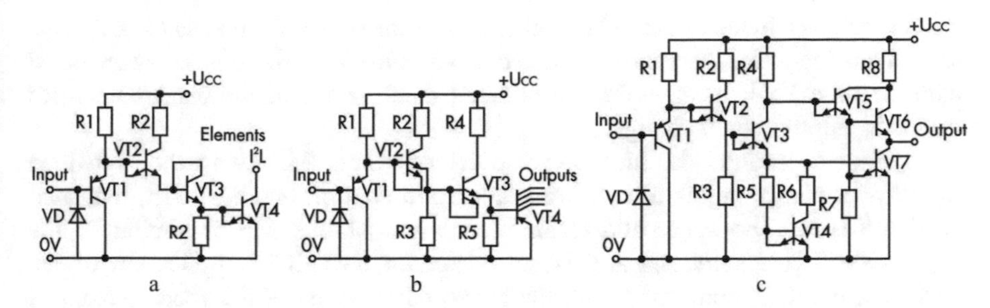

Fig. 3.103 Input ME TTL on the basis of the p-n-p transistors with control as per the emitter base, connected with the VT2 base to the emitter circuit, connected with the ME input

$$I_{IL} = (U_{CC}-U_{IL}-U_{EP})/(\beta_P+1)R1;$$

where: U_{EP}–voltage of the forward bias emitter junction of the p-n-p transistor

β_P–gain ratio of the base current of the transistor VT1, connected as per the diagram with the common collector

Due to a number of the design-technological limitations in the I^2L integrated circuits, one usually selects not a vertical but a lateral structure of the p-n-p transistor, ensuring somewhat smaller values $\beta_p = 10$–20 within the range of the operating temperatures.

With the activated transistor VT1, the base current VT2 is equal to zero. Transistor VTI is off, which corresponds to the high signal level at the output; the potential of its collector is determined by the load characteristic.

During arrival at the input of the high-level voltage signal U_{IH}, the input p-n-p transistor VT1 gets locked, the transistors VT2–VT4 get open, and the normally activated transistor VT4 comes into the saturation mode, ensuring the output voltage $U_{OL} = U_{CES}$ (the voltage collector-emitter in the saturation mode). The input current I_{IH} is determined by the leakage currents of the back-biased junctions and constitutes 10–20 uA. The input ME inverts as to the input signals.

The input ME in Fig. 3.103a ensures the pyramiding factor at the output $N < 10$, i.e., to the collector VT4, one can connect not fewer than ten bases of the injection transistors, connected in parallel.

In Fig. 3.103b is presented the input that has instead of the normally activated output n-p-n transistor with the short-circuited emitter a multi-collector injection n-p-n transistor VT4. Each collector VT4 controls the base of the appropriate I^2L element, ensuring the required electric uncoupling among them. When designing the logic blocks of the integrated circuits, the transistor VT4 is used in composition of these blocks as the input logic inverter, making it possible to cut down the logic depth of the integrated circuits, to raise its performance.

The ME pyramiding factor is determined by a number of the inversely connected n-p-n transistor VT4 (usually $n = 4$); to its output one can connect four logically isolated bases of the injection transistors of the logic blocks. In order to enhance the load capacitance twice as much in parallel to the transistor VT4, one can connect the second quadcollector inversely connected n-p-n transistor, whose base is connected with the base VT4; meanwhile, the nominal rated values of the resistors R1–R5 should be appropriately changed.

In order to arrange the high-power efficiency input MEs with the high load capacity, primarily inputs of synchronization and control, one uses ME, indicated in Fig. 3.103c. The schematic technical solution of this ME is similar to the appropriate TTLS solutions and is considered in Sect. 3.1. Application of the additional emitters, connected with the bases, takes notice of the specific features of the forward connected n-p-n transistors in the I^2L integrated circuits–the high inverse values of their gain ratios. The stage output ensures the load current up to 20 mA and the control possibility simultaneously of up to 100 injection elements, which is widely resorted to in arrangement of the internal synchronization circuits of the I^2L integrated circuit dice.

With the purpose of reduction of the consumed power in the multi-pin injection integrated circuits, one widely uses the input MEs with the reduced supply voltage (U_{CC}=1.5÷3 V). In Fig. 3.104a [6] is presented the most elementary inverting ME on the basis of the resistive divider R1 and R2 and n-p-n transistors VT1 and VT2 in their forward connection, emitter controlled. As the ME output transistor, they apply the multi-collector injection transistor VT3, which usually is essentially the input element-signal multiplier of the logic block of the integrated circuits. The shortcoming of this ME–the low load capacity–is excluded from the inverting element in Fig. 3.104b, where the normally activated output n-p-n transistor VT4 can control a large number of loads, bases of the injection transistors. The number of the I^2L elements–the load–is determined by selection of the nominal rated value of the resistor R5 from the formula:

$$N = {}^{(U_{CC}-U_{EB})}\!/_{R_5 \cdot I_B};$$

where: I_C–the base current of the n-p-n transistor of the I^2l element

The maximum number N of the combined bases is limited by the effect of interception (redistribution) of the common current.

The input MEs (Fig. 3.104a, b) have the supply voltage of U_{CC}=+3 V. The problem of the power reduction, consumed by the input MEs of the integrated circuits, is exclusively actual for the multi-pin ones with a number of pins of the package of over a hundred (for instance, 128, 144, 256).

Meanwhile, reduction of the consumed power should not result in reduction of performance, which is a complex technical task. In Fig. 3.104c is presented the high-efficiency input ME of the type TTL-I^2L with the minimum supply voltage, equal to the supply voltage of the "pure" I^2L element (U_{CC}=1.0÷1.5 V). The input transistor VT1 is made dual-emitter; meanwhile, the first emitter is connected directly with the ME input and the second one with the base of VT4. Let's consider the ME operation.

Suppose the voltage at the input is U_{IH} = 2.4 V. While passing via R1, the input current opens the transistor VT1, whose emitter current ensures activation of VT2, at

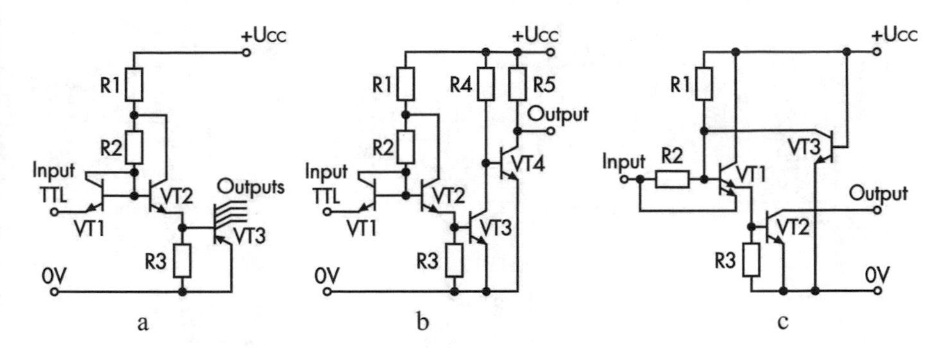

Fig. 3.104 Input MEs of the I²L integrated circuits with the reduced supply voltage

the output is set the low-level voltage $U_{OL} = 0.4$ V. A share of the input current is drawn by the collector of the activated regulating transistor VT3, because of which the mode of operation of VT1 is changed–from the saturation mode, it switches over to the active area on the boundary with the cutoff mode. The transistor VT2 also changes the mode of operation; on its collector–ME output–the following voltage is set:

$$U_{OUT} = U_G - U_{BE}^{VT3};$$

For supply voltage of injectors of the I²L elements $U_G - 1.2$ V with the standard value of the bias voltage of the emitter junction VT3 $U_{BE} = 0.8$ V, the value U_{OL}=0.4 V.

In this settled state, not a single one from the transistors VT1–VT3 is in the saturation mode, which ensures the high rate of the ME shutting down. With the arrival to the input of the low-level voltage of $U_{IL} = 0.4$ V, the emitter current of the transistor VT1 switches over from the base circuit VT2 to the input circuit, deactivating VT2 and VT3, and the following voltage is set at the output $U_{OH} = 0.8$ V, determined by the load inputs of the I²L base elements.

The considered input ME is used in the I²L of the integrated circuits with one source of the lowered supply voltage U_C, common for the input, output, and base I²L logic elements.

Input MEs with Conversion of the ECL Levels to the I²L Levels

The most elementary version of the diagram of the inverting input ME with conversion of the ECL levels to the I²L levels [5] can be realized just on two transistors VT1 and VT2 and the resistors with the power supply from the sources $U_{CI} = -5.0$ V; $U_{C2} = 1.5$ V (Fig. 3.105a). This ME has some intrinsic deficiencies–the narrow area of the stable operation and the high sensitivity to the tolerances of the components. In fact, the resistance value of the resistor R3 is related to the above indicated ratios with the resistance values of the resistors R1 and R2; therefore, the nominal rated value R3 should not be increased for reduction of the current,

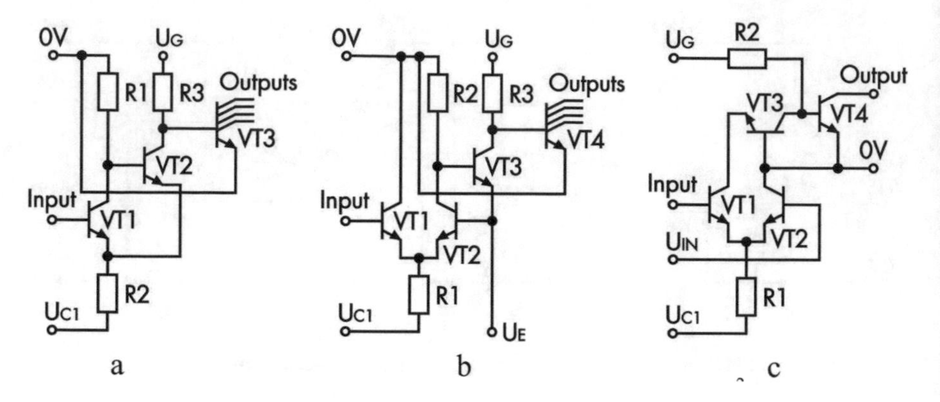

Fig. 3.105 Input MEs with conversion of the ECL levels to the I²L levels

inflowing into the base VT3. Transistor VT3 operates in the mode of deep saturation with the saturation degree:

$$S_H = \frac{\beta(U_G - U_{EB})}{I_C};$$

where: β, I_C–gain ratio of the base current and the total current of the collectors VT3

The high value of the saturation degree S determines the low performance (long time of the VT3 deactivation). This deficiency is partially removed in the diagram in Fig. 3.105b, which may be realized in the form of the non-inverting input stage of the integrated circuits with the supply voltage pins $U_{C2} = +1.5$ V; $U_{CI} = -5$ V.

One can use instead of the circuit U_G-$R3$ the generator of the stabilized current of the I²L of the integrated circuits (or the injection p-n-p transistor of the I²L element in the transistor base VT3). By changing the current value, it is possible, if necessary, to control the ECL input threshold. The circuit operates in a simple way. With the input voltage $U_{IL} = -1.6$ V, the transistor VT1 is closed, and VT2 is open and passes the current:

$$I_I = (U_E - U_{EB} - U_{C1})/R_1;$$

where: $U_E = 1.2$ V–the reference voltage source

If the condition is met $I_I\, R_2 < U_E$, then the transistor VT3 is securely locked, and VT4 is open and ensures at the output as per any collector $U_O = U_{OL} = 0.4$ V. With $U_I = -0.8$ V, the transistor VT1 gets open, and VT2 gets closed, which ensures $U_O = U_{OH}$ (it is determined by the nature of load, in case of I²L, the load $U_{OH} = 0.8$ V).

However, and in this ME circuit, the probability of the deep saturation of the transistor VT3 is also not excluded as compared with the solution, provided in Fig. 3.105a.

Arrival of the transistor VT3 into the saturation mode corresponds to appearance at the base VT4 of the negative voltage U_E, which disrupts the operational mode of

VT4 and also worsens the Me performance. The second deficiency of the given ME is the necessity of application of the additional reference voltage source.

The common ME deficiency, indicated in Fig. 3.105a, b, is the single phase input, not permitting to obtain the output signal of the random polarity. In Fig. 3.105c is represented ME with the paraphrase inputs. The paraphrase outputs of the integrated circuits are widely used during arrangement of the dual communication lines among the circuit boards. If the preceding ECL integrated circuit has a paraphrase output, then one from two ME inputs is connected to the forward output and the other to the inverse one. If the preceding ECL integrated circuit has only one (forward or inverse) output, then one from the ME inputs is connected the ECL output of the integrated circuit and the second one to the reference voltage source. In dependence on the method of connection of inputs, ME will invert or translate (repeat) the input logic signal. The transistors VT1 and VT2 of the differential couple determine the state of the output. If VT1 is closed (VT2 is open), then the transistor VT3 is also closed, and VT4 is open and ensures at the ME output the level $U_{OL} = 0.4$ V. For this, the following condition should be met:

$$(U_G - U_{EB})/R_2 = S_H(I_{CH}/\beta_4);$$

where: U_G=+1.5 V–the supply voltage of the injection blocks of the I^2L integrated circuits

β_4, I_{CH}–the gain ratio of the base current and the value of the collector current VT4

For the case, if VT1 is open, then the transistor VT3 is also open, draining the base current VT4; at the ME output, the voltage U_{OH2} is ensured at the I^2L levels. Instead of the circuit U_G-R7, one can use the current setting p-n-p transistor of the I^2L element.

Input MEs with Conversion of the Input CMOS Levels to the I^2L Levels
The schematic technical ME solutions with the signal conversion of the CMOS levels to the I^2L levels to a great extent are similar with the solutions of the elements, performing conversion of the TTLS levels to the I^2L levels. However, the most effective are the input MEs of the resistive type, indicated in Fig. 3.98 and 3.99. Meanwhile, in order to ensure reliability of the ME operation, the following conditions should be met:

$$U_{OH}^{CMOS} < U_{\mathrm{I\,max}}^{I^2L};$$

Output MEs with Conversion of the I2L Levels to the TTL Levels
The absolute majority of the schematic and technical solutions of the output ME I^2L integrated circuits of the non-isolated injection logic are related to the OC type of circuits. Absence of the electric isolation between the active components of the I^2L integrated circuits of this type substantially limits the possible schematic and technical options, and practically the only schematic and technical solution of the output ME is represented in Fig. 3.106a. The element performs the function of the current amplifier from the working levels of 10–100 mA of the internal I^2L elements up to

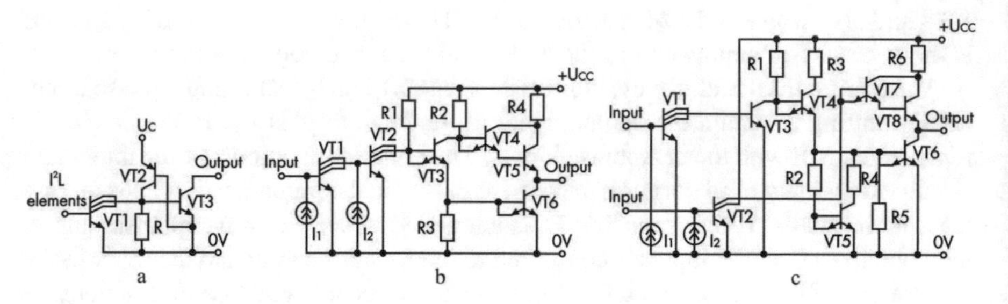

Fig. 3.106 Standard output MEs with conversion of the I^2L levels to the TTL levels

the required output level of 4–10 mA and over. The injection transistor VT1 with the extended area of the collector performs the function of the intermediate current gain stage while the inversely connected n-p-n transistor VT3 the function of the final gain stage. The resistor R is intended for stepping up the output breakdown voltage.

Application of the full-scale isolation makes it possible to substantially expand the solutions of the input MEs.

The basic diagram of the output ME (Fig. 3.106c) contains the preliminary current gain stage on the injection transistors VT1 and VT2 with the stepped up collector current from 10–50 to 100–500 uA, required for operation of the output TTL stage on the normally activated n-p-n transistors VT3–VT6. The design structure of the n-p-n transistors VT3, VT4, and VT6 of the output TTL stage of ME has the introduced additional emitters. Connection of the additional emitter of the n-p-n transistor with the base makes it possible to reduce duration of the processes of dissipation of the excessive charges of the minor carriers, accumulated in the base. The effectiveness of such a solution is all the more higher with the higher values of the inverse gain ratio of the n-p-n transistor. In order to enhance the performance of ME during formation at the high-level output, one uses the "accelerating" link: the additional collector of the injection transistor VT2, the additional emitter of the output power n-p-n transistor VT6 (indicated with a dotted line). On arrival to the input VT2 with the high-level voltage, the power basic and the additional collectors VT2 switch over to the saturation mode, their potential coming close to the potential *0V*. Thus the preconditions are created for acceleration of the process of activation of VT6. The enhanced performance of ME is determined by the fact that during activation of VT2, the emitter current of the phase dividing transistor VT3 switches over from the basic emitter circuit, connected with the base VT6, to the low Ohmic circuit of the "additional emitter" VT3–collector VT2. These schematic and technical solutions ensure the time shortening, required for deactivation of ME.

The diagram in Fig. 3.106 of the output MEs with the TTL levels is characterized with a high load capacity and high performance rate; however, they consume rather large vale of power and occupy a vast area on the die. Therefore, they are applied in the integrated circuits with a limited number of digits and in those cases when the requirements for performance and the high load capacity are decisive.

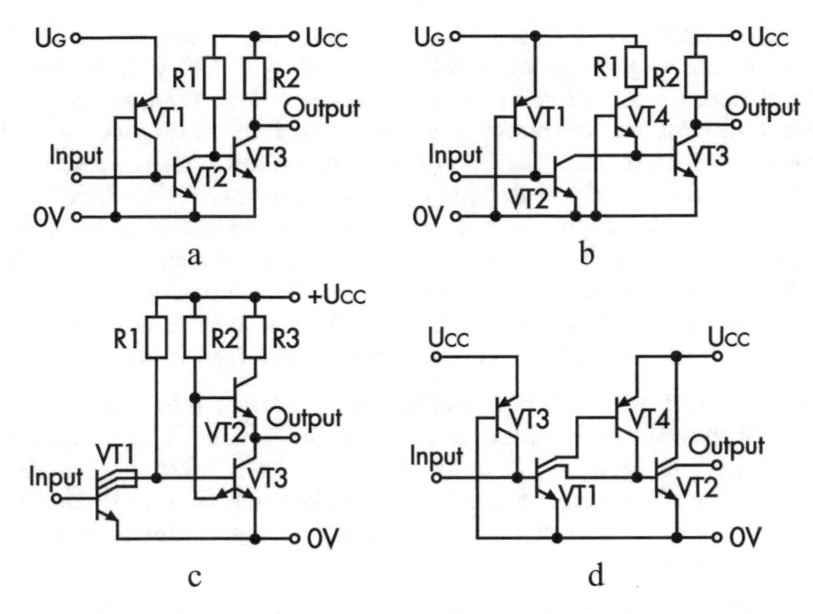

Fig. 3.107 The improved output MEs with conversion of the I²L levels to the TTL levels

Let's consider the improved schematic and technical solutions of the output MEs, implemented with a fewer number of components.

In Fig. 3.107a is presented an elementary ME on the basis of the injection n-p-n транзистора VT 1 и нормально сключенноdo n-p-n транзистора VT3 transistor with the resistors R1 and R2, where R2 performs the function of the load resistor of the OC circuit. As R2 is connected to the power supply source $U_{CC} = 5$ C, then the amplitude of the output signal is sufficient for control of the standard TTLs.

The deficiency of the given ME is in the large values of the consumed power and the area, occupied on the die by the ME high value (in Ohm) resistors.

As a matter of a fact, on the basis of the transistor VT1 for the typical $I_B^{VT1} =$ 0.1 mA, *bVT1*=10, and *UCC*=5 V' the nominal rated value being *R1*=5 kOhm, which requires quite an expansive area and precludes enhancement of the integration degree of the integrated circuits.

In ME in Fig. 3.107b the resistor R1 is replaced with the current setting p-n-p transistor VT4, which makes it possible to somewhat reduce the consumed by ME of the OC-type power and the required for it area. However, this circuit possesses the other deficiency–dependence of the operational mode of the p-n-p transistor VT4 on the conditions VT1 and VT3. When VT1 is closed, VT4 comes into the mode of deep saturation, the effective current of the base VT3 reduces, and the signal transfer to the output becomes unstable.

In order to do away with this deficiency and to enhance reliability of the signal transferring vie ME, the circuit was proposed, indicated in Fig. 3.107c of the AO type with the reduced supply voltage [1] on the basis of the multi-emitter normally activated n-p-n transistor VT3. If VT1 is open, then VT3 is closed, and the transistor VT2 sustains the high-level output voltage: U_{OH}–U_{BE} = 2.25 V. If the transistor VT1 gets closed, then VT3 gets open; on its second emitter owing to a large value inverse gain ratio, the low potential will be maintained, which locks VT2. The diagram of the output ME of the AO type, but not containing resistors, is shown in Fig. 3.107d and is actually an analogue of the above reviewed ME, whose resistors Rl and R2 are replaced with the working in the active mode current setting p-n-p transistors VT4 and VT5. Be changing the ME supply current, it is possible to consider in wide scope the consumption power and the performance of the circuit, as well as to choose the values of the output current VT3, required to ensure the electric matching with the TTL inputs of the integrated circuits of the different series.

Output MEs with Conversion of the I²L Levels to the ECL Levels
The task of conversion of the I²L output levels to the ECL levels in practice is seen exceptionally rare. However, in those cases, when it is necessary, the electric matching of the I²L outputs of the integrated circuits should be made in compliance with the solutions, used when matching TTLSs of the output MEs of the OC type with the ECL input.

Output MEs with Conversion of the I²L Levels of Signals to the CMOS Levels
For formation in the I²L integrated circuits of the output levels of the signals, capable of controlling the CMOS integrated circuits, it is most effective to use the elementary circuit of the output ME of the OC type. In this case the output voltage of the low-level U_{OL}=U_{CES} ensures locking the input of the CMOS integrated circuits. Formation of the high-level output voltage is performed by connection to the ME output of the supply voltage source U_L via the load resistor R_L. Meanwhile, the voltage value U_L is selected on the supposition that:

$U_L > U_{TH}$,which ensures formation of the required output level and, consequently, the reliable locking of the input of the CMOS integrated circuits.

3.5.4 Protection of the I²L Pins of the Integrated Circuits from Overvoltage and Static Electricity

Protection of the pins from the static electricity is necessary for prevention of destruction of the I²L integrated circuits both at any stage of the technological operations of their production and in the process of their mounting on the printed circuit boards or in the process of debugging the systems. Also in the design layout of the integrated circuits, care should be taken with regard to introduction of the elements of protection of the pins from destruction with emergence of the unsanctioned pulses of voltage (currents) of the positive or negative polarity. As influence of the static electricity or overvoltages renders effect mainly on the input or output MEs of the integrated circuits, and their schematics is similar with the schematics of the TTLS integrated circuits, then the methods of protection of the

I^2L integrated circuits from the indicated influences are similar with the methods of protection of the TTLS integrated circuits. Meanwhile, the schematic and technical solutions of the protection circuits, reviewed in Sect. 3.1, are applicable for protection of the I^2L integrated circuits.

References

1. Belous, A. I., Emelyanov, V. A., & Turtsevich, A. S. (2012). *Fundamentals of schematics of microelectronic devices* (472 pp). Technosphere (in Russian).
2. Alekseenko, A. G., & Shagurin, I. I. (1990). *Microschematic engineering: Educational manual for higher institutions* (2nd Ed., revised and supplemented). Radio and Communications (in Russian).
3. Belous, A. I., Blinkov, O. E., & Silin, A. V. (1990). *Bipolar integrated circuits for interfaces of automated control systems.* Machine Building, Leningrad Affiliate (in Russian).
4. Belous, A. I., Ponomar, V. N., & Silin, A. V. (1998). *Schematic of the bipolar integrated circuits for the high efficiency information processing systems* (162 pp). Polyfact (in Russian).
5. Belous, A. I., & Yarzhembitsky, V. B. (2001). *Schematics engineering of digital integrated circuits for the information processing and transmission systems* (116 pp). UE Tekhnoprint. ISBN 985-464-064-7 (in Russian).
6. Belous, A. I., Podrubny, O. V., & Zhurba, V. M. (1993). *Microprocessor kit of digital signal processing K1815: Reference book* (45pp). Rotoprint, BSPA (in Russian).

Chapter 4
Circuit Engineering of Bi-CMOS IC

This chapter delves into circuitry engineering of Bi-CMOS microcircuits. It is a common knowledge that CMOS circuitry engineering is the most convenient way to design high-speed low power consumption microcircuits with large scale integration. However, while chip complexity enhances, there emerges the problem of controlling comparatively large aggregated capacities formed by capacities of multiple lines of interconnections wiring and loads control capacitors on chips and at microcircuit outputs. It is shown that Bi-CMOS circuitry engineering normally provides win-win trade-off solution.

There are available well-arranged results of reviewing schematic solutions of base standard logic cells, memory cells, and input and output interfacing cells.

Considerable portion of the chapter explores non-standard schematic solutions, which are efficient, however, and well proven in mass production: input matching components with signal conversion, level translators with enhanced loading capacity, enhanced interference protection, built-in memory, etc.

Similarly, there are depicted output modified cells: with generation of output CMOS levels, TTL and ECL levels, output Bi-CMOS cells with memory, and plenty of other smart schematic solutions.

4.1 Basic Logic Elements of Bi-CMOS IC

CMOS circuit engineering is most convenient for creating modern digital ICs due to the low power consumption and high packaging density. However, with the increasing complexity of digital ICs, there is a problem of managing large capacitances formed by the capacitances of long-distance interconnects and the capacitances of controlled loads on the chip. Since, as was shown in Chap. 1, the maximum output current of MOSFETs is significantly limited by the geometric dimensions of the components W, L, and the supply voltage U_{CC}, the load capacity charge rate C_L in CMOS ICs and, consequently, the performance are limited by equation [1]:

A. Belous, V. Saladukha, *The Art and Science of Microelectronic Circuit Design*,
https://doi.org/10.1007/978-3-030-89854-0_4

$$dt_p/dC_L = 1/C_{IX}W, \quad (4.1)$$

where C_L is the MOSFET load capacity and W is the MOSFET width.

It also follows from this formula that MOSFETs, with the appropriate selection of the width of the MOSFET W, are in principle capable of switching large currents. However, this increases the input capacitance C, which again limits the performance of MOSFETs. In addition, MOSFETs due to the low transconductance of the gain and low output conductivity are ineffective in implementing a number of typical digital ICs blocks, for example, CMOS-ECL matching component (MC), etc. In contrast, bipolar transistors, due to the exponential dependence of the collector current on the input voltage and high gain factors, in principle make it possible to implement faster digital circuits than MOSFETs, even in the presence of large capacitive loads:

$$dt_p/dC_L = \phi_T/I_K,$$

where φ_T is the temperature potential and I_C is the collector current.

It follows from this formula that thc chargc rate of the load capacity C_L can be controlled by changing the collector current I_C. At the same time, the input capacitance C of a bipolar transistor depends on its collector current, while that of a MOSFET is constant.

The widespread use of bipolar transistors in VLSI is constrained by the high level of power consumption of circuits based on them. As a result, in digital ICs, the direction of the element base, which uses the advantages of both bipolar and complementary MOSFETs (Bi-CMOS), has become difficult.

Since the LE output transistors should operate in antiphase when controlling any load, the simplest implementation of Bi-CMOS LE is a CMOS LE with paraphase outputs that control a push-pull bipolar circuit. The MC circuit of this type is shown in Fig. 4.1 a [2] and contains two inverters (the first on transistors VT1 and VT2, and the second on transistors VT3 and VT4). The first one controls the down-level n-p-n transistor VT5, and the second one controls the up-level n-p-n transistor VT6.

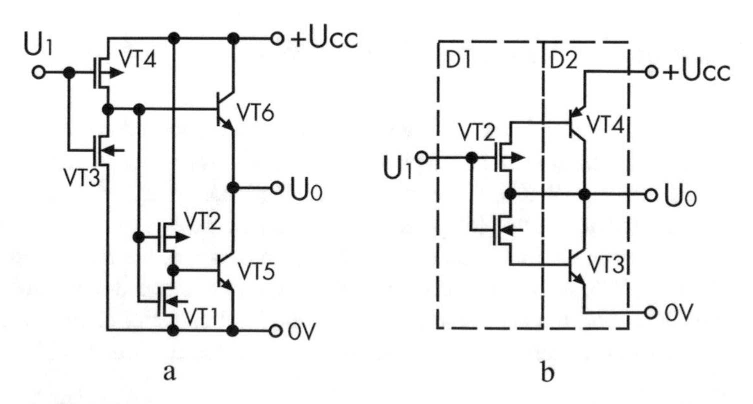

Fig. 4.1 Bi-CMOS LE circuits based on CMOS circuit with paraphase outputs based on bipolar transistors of the n-p-n (**a**) and p-n-p (**b**) types

However, this circuit is not widely used due to the presence of a static current consumption I_{CCL} in the low-level state, flowing through the open p-MOSFET VT2 and the base-emitter junction of the n-p-n transistor VT5. The CMOS LE circuit, free from this disadvantage, is shown in (Fig. 4.1 b) [3] and is constructed according to a two-stage structure: the first stage D1 is formed by the CMOS LE (transistors VT1 and VT2), and the second stage D2 is formed by bipolar p-n-p and n-p-n transistors VT3 and VT4. The logic functions of the Bi-CMOS LE are usually implemented in the CMOS stage, and the methods of their circuit engineering implementation are described in detail in [1]. Therefore, the architecture of Bi-CMOS LE on the example of an inverter is discussed below.

In this circuit, the p-MOSFET VT2 and the p-n-p-transistor VT4 provide the charge of the load capacity, and the n-MOSFET VT1 and the n-p-n-transistor VT3 provide the discharge of the load capacity. However, such a circuit is also not widely used due to the large difference in the characteristics (performance, gain) of the p-n-p and n-p-n transistors and, consequently, the large difference in the delays of turning on the t_{pHL} and turning off the t_{pLH} of the Bi-CMOS LE. In addition, this circuit is technologically complex due to the presence of four types of transistors and the need to control a large complex of technological parameters.

The circuit using one type of n-p-n transistors is more efficient and has received practical use (Fig. 4.2a). The first n-p-n transistor VT4 is designed to charge the load capacity and form the high-level output voltage U_{OH}, and the second n-p-n transistor VT3 is designed to discharge the load capacity and form the low-level output voltage U_{OL}.

To discharge the capacitance of the base C_{BE}^{VT3} of the n-p-n transistor VT3, when the LE is turned off, a discharge element Z1 is introduced into the circuit, and to discharge the capacitance of the base C_{BE}^{VT3} of the n-p-n transistor VT4, when the LE is turned on, a discharge element Z2 is introduced. Rl and R2 resistors or permanently switched-on n-MOSFET or p-MOSFET transistors replacing them can be used as elements of the discharge Z1, Z2 [1] (Fig. 4.2b), or their replacement permanently switched-on n-MOSFET or p-MOSFET transistors. The structure of such a circuit received the designation R+R type. In another variant, Fig. 4.2c [4], controlled n-MOSFET transistors VT1 and VT3 are used as elements of the Z1 and Z2 discharge. The structure of such a circuit is called an N+N type circuit. In addition, a mixed version of the use of discharge elements is possible [4]: resistor R and n-MOSFET VT1 (Fig. 4.2d). The structure of such a circuit is designated as R +N type. In ICs, simplified upgrades of this Bi-CMOS LE circuit can be applied, depending on the application conditions and requirements for the signal edge durations.

The first of these circuits in Fig. 4.3a [1] uses a single *n-p-n* transistor VT4, which accelerates the charge of the capacitance and improves the duration of the X_{LH} cut-off front. The discharge of the load capacity is provided by the n-MOSFET VT3, which is similar to the CMOS LE. In the second circuit in Fig. 4.3b [4], the n-p-n transistor VT3 is used to discharge the load capacitance $C_{£}$ and improves the duration of the turn-on edge THL. The charge of the load capacity is provided by the p-MOSFET VT3 and stored in a similar CMOS LE. Of the considered types of circuits in the literature, the most detailed circuit is the LE of N+N type (Fig. 4.2c), despite the fact that all three types of LE are used in ICs.

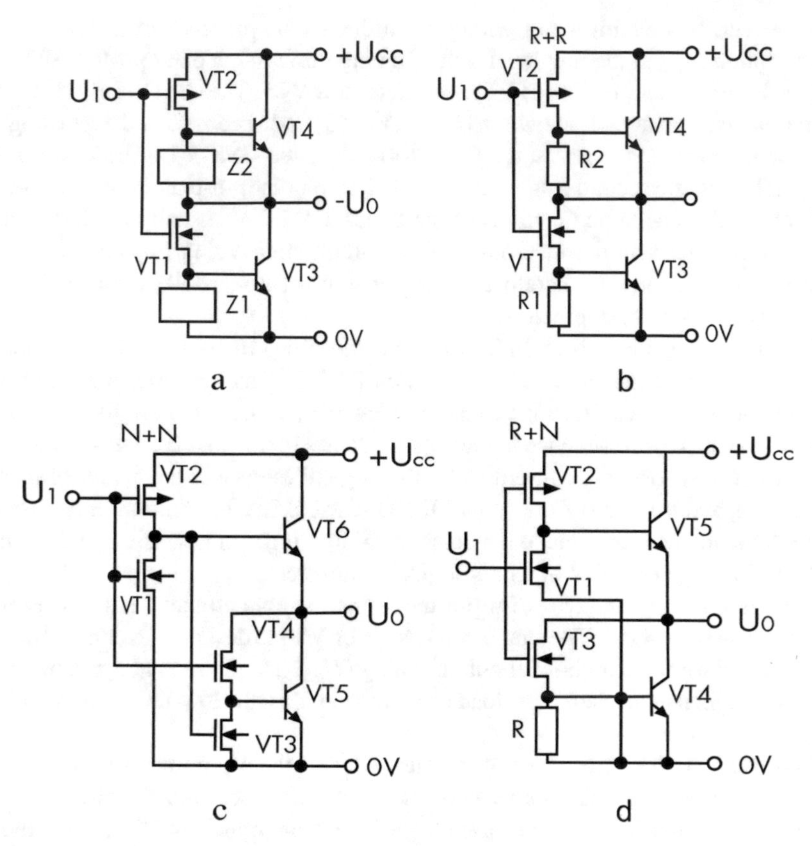

Fig. 4.2 Electrical circuits of Bi-CMOS LE using a pair of *n-p-n* transistors

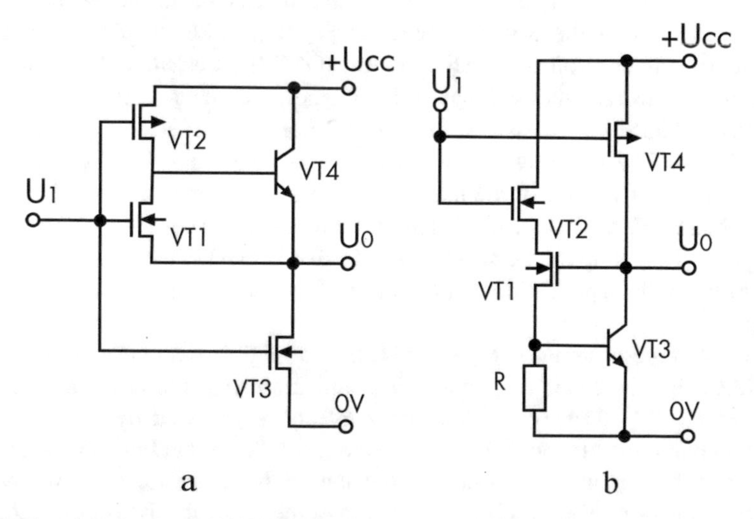

Fig. 4.3 Electrical circuits of Bi-CMOS LE with one step-up (**a**) and step-down (**b**) n-p-n transistor

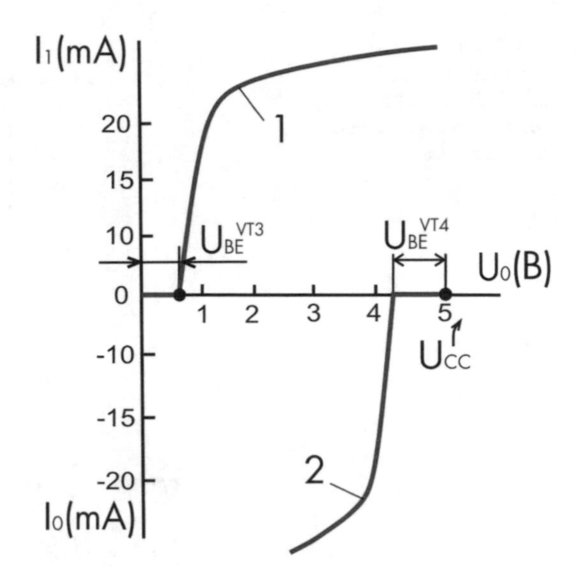

Fig. 4.4 Output characteristics of the N+N type Bi-CMOS LE

Let us analyze the basic principles of the functioning of this type of circuit. At a low voltage level $U_{IL} < U_T^n$ at the input of the circuit, the n-MOSFET transistors VT1 and VT4 are closed, the n-MOSFET transistor VT3 is open and holds the output p-p-p transistor VT5 in the closed state. In this case, the open p-MOSFET VT2 keeps the output n-p-n transistor VT6 open, which generates a high voltage level at the output:

$$U_{OH} \approx U_{CC} - U_{BE}^{VT6}. \tag{4.2}$$

The output characteristic of LE in this state is shown by curve *1* in Fig. 4.4.

At a high voltage level at the input of the circuit: $U_{IH} \approx U_{CC}\text{-}U_T^p$ MOSFETs VT2 and VT3 are closed, so that the output n-p-n transistor VT6 is closed. The open n-MOSFET VT1 keeps the level close to zero in the base of the output n-p-n transistor VT6, and the open MOSFET VT4 provides the open state of the output n-p-n transistor VT5. In this case, the LE output will be set to a low voltage level:

$$U_{OL} \approx U_{BE}^{VT5}. \tag{4.3}$$

The output characteristic of LE in the open state is shown by curve *2* in Fig. 4.4.

Comparing the output voltage levels U_{OH}, U_{OL} of CMOS and Bi-CMOS LE, we can conclude that Bi-CMOS LE has the worst values of the output levels, so that the noise immunity ΔU_T^+, ΔU_T^- of such a circuit is lower. At the same time, in order to exclude "through" consumption currents I_{CC}^c in the Bi-CMOS LE and to ensure the closed state of the MOSFETs in the LE circuit, the following requirements must be met:

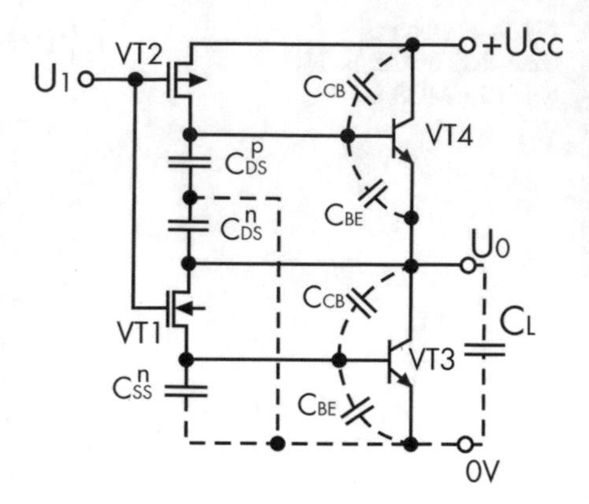

Fig. 4.5 Simplified electrical circuit of the Bi-CMOS LE of N+N type for calculating dynamic characteristics

$$U_T^{VT1,VT3} > U_{BE}^{VT5}; \tag{4.4}$$

$$U_T^{VT2} > U_{BE}^{VT6}. \tag{4.5}$$

The shutdown delay times of the Bi-CMOS LE t_{pLH}, t_{pHL} can be estimated using the circuit in Fig. 4.5. The circuit includes the following designations: C_{CB}—collector-base junction capacitance of n-p-n transistors; C_{BE}—base-emitter junction capacitance of n-p-n transistors; C_{CS}—isolation capacitance (collector-substrate junction) of n-p-n transistors; C_{DS}^n—source-substrate capacitance of n-MOSFET; C_{DS}^n—drain-substrate capacitance of n-MOSFET; C_{DS}^P —drain-substrate capacitance of n-MOSFET. Shutdown delay time:

$$\begin{aligned} t_{pLH}^{BiCMOS} &= U_{BE}(C_{CB} + C_{BE} + C_{DS}^p)/I_C^p + U_{OH}(C_{CB} + C_{DS}^p) + I_C^p + \\ &+ U_{OH}(C_L + C_{CB} + C_{DS}^n + \beta C_{CB})/(\beta \bullet I_C^p). \end{aligned} \tag{4.6}$$

The first component of the delay time t_{pLH} is the time required for the potential in the base of the VT4 bipolar transistor to increase to the U_{BE} level. The second component is the time during which the potential in the base of the n-p-n transistor VT4 is set from the U_{BE} level to the U_{BE}+ U_{OH}. The third component is the time during which the U_{OH}. level is set on the output pin. Power-on delay time:

$$\begin{aligned} t_{pHL}^{BiCMOS} &= U_{BE}(C_{CB} + C_{BE} + C_{DS}^n)/I_C^n + U_{OH}(C_{CB} + C_{DS}^n) + I_C^n + \\ &+ U_{OH}(C_L + C_{CB} + C_{DS}^n + \beta(C_{CB} + C_{DS}^p))/I_C^n. \end{aligned} \tag{4.7}$$

The first component t_{pHL} of Bi-CMOS is the time required to establish the potential in the base of the output n-p-n transistor VT3 at the U_{BE} level. The second component is the turn-on time of the output n-p-n transistor VT3, and the third component is the time during which the voltage at the LE output will decrease to the level U_{OL}. Summing up the delay times t_{pHL} and t_{pLH} to determine the average delay t_p=0.5(t_{pLH}+t_{pHL}) we get for CMOS LE:

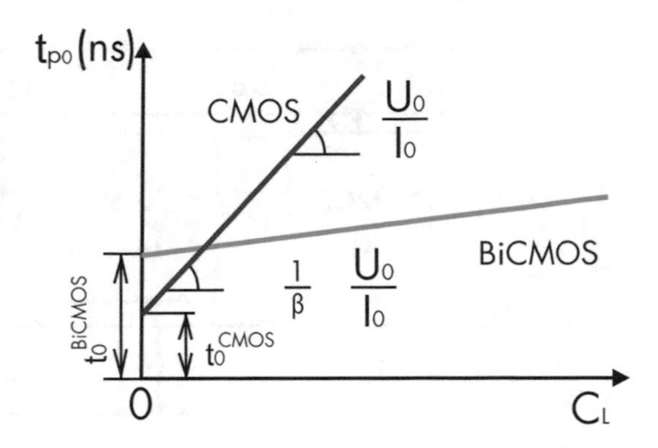

Fig. 4.6 Comparative dependences of the performance of Bi-CMOS LE and CMOS on the load capacity

– for CMOS LE:

$$t_p^{CMOS} = t_0^{CMOS} + U_{OH} C_L / I_C^{n(p)}; \tag{4.8}$$

– for Bi-CMOS LE:

$$t_p^{BiCMOS} = t_0^{BiCMOS} + U_{OH} C_L / \left(\beta \bullet I_C^{n(p)}\right); \tag{4.9}$$

and presenting them in graphical form (Fig. 4.6), we can make two main conclusions:

1. The dependence of the delay time t_p on the load capacity for the Bi-CMOS LE

$$dt_p^{BiCMOS} / dC_L \approx U_{OH} / \beta I_C^{n(p)}; \tag{4.10}$$

is more flat than for CMOS LE:

$$dt_p^{CMOS} / dC_L \tilde{U}_{OH} / I_C^{n(p)}.. \tag{4.11}$$

As a result, at large load capacities, the C_L of Bi-CMOS LE has a higher performance than the CMOS.

2. For small load capacities C_L, the performance of the CMOS LE is higher than that of the Bi-CMOS LE.

Due to such dependencies, to ensure maximum performance of the digital Bi-CMOS IC, the selective use of internal LE is necessary. For small load capacities C_L, the use of the CMOS LE is more efficient, for large capacitances C_L, the CMOS LE, the circuit of which is shown in Fig. 4.2c, as well as the CMOS LE, practically

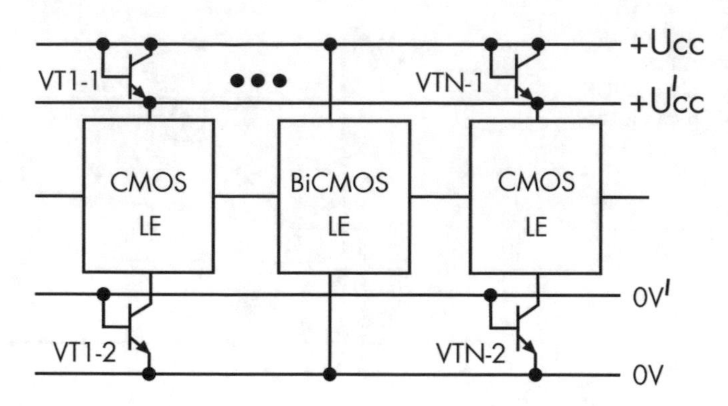

Fig. 4.7 Power supply diagram for CMOS LE and Bi-CMOS LE with a limited output voltage drop

does not consume current in the static state, which is associated with the antiphase operation of the transistors in the LE.

So, at a low voltage level U_{IL} at the input, the MOSFETs VT1, VT4, and n-p-n transistor VT5 are closed, the MOSFET VT3 and n-p-n transistor VT6 are open, and there is no current flow between the pins $+U_{CC}$–$0V$. Similarly, at a high voltage level U_{IH} at the input, the LE MOSFETs VT2 and VT3 are closed, VT1 and VT4 are open, the n-p-n transistor VT5 is open, VT6 is closed, and the current flow circuit between the pins *+Ucc-0V* is also absent. In dynamic mode, the Bi-CMOS LE consumes the power:

$$P_{CCF} \approx S * C_L * U_{CC}^2 * F; \tag{4.12}$$

where S is the pulse duty cycle and F is the switching frequency.

Since in the considered LE circuit the low-level voltage U_{OL} is increased by a value U_{BE}^{VT3} in comparison with the CMOS LE, and the high-level voltage value is lowered by the value U_{BE}^{VT4}, such a circuit is called "LE with a limited voltage drop." To ensure reliable joint operation of CMOS and Bi-CMOS of this type, it is possible to build LE circuits according to the diagram shown in Fig. 4.7 [1].

So, the CMOS LE are connected directly to the power pins $+U_{CC}$, $0V$, and the CMOS LE are connected to the intermediate buses $+U_{CC}$, $0V$, connected to the power pins $+U_{CC}$, $0V$ through the offset elements-transistors VT1-1-VTN-1 and VT1-2-VTN-2. In this case, the output voltage drops and the thresholds for unlocking (locking) of the CMOS LE are limited and the reliability of the circuit functioning is increased. The output levels U_{OL}, U_{OH} can be improved by LE circuits using the resistors of R+R type (Fig. 4.2b) and R+N type (Fig. 4.2d). The R+N type circuit reduces the low-level output voltage to $U_{OL} \approx 0\ V$, and the R+R type circuit allows you to get the output levels $U_{OL} \approx U_{CC}$; $U_{OL} \approx 0$. However, these types of circuits differ in the performance characteristics t_p (Fig. 4.8a) and the power consumption P_{CC} (Fig. 4.8b).

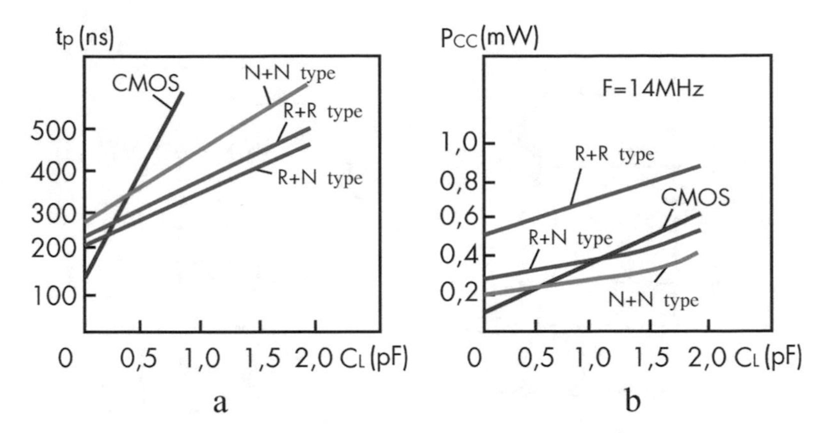

Fig. 4.8 Comparative characteristics of the performance (**a**) and dynamic power consumption (**b**) of different types of Bi-CMOS LE depending on the load capacity

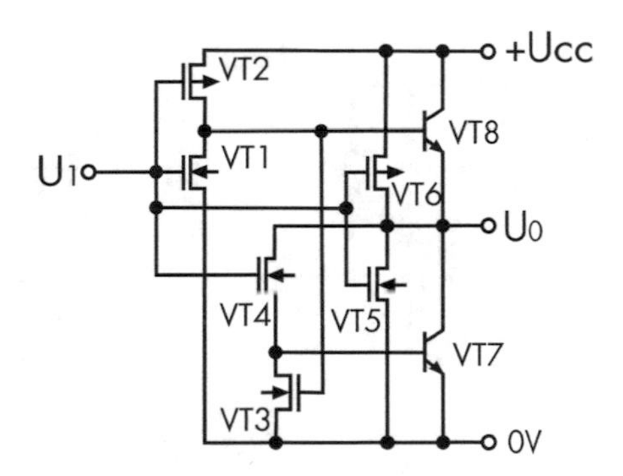

Fig. 4.9 Bi-CMOS LE circuit with improved output voltage levels

It is also possible to improve the output levels of the Bi-CMOS LE by connecting a CMOS LE operating in the same phase as the Bi-CMOS LE to its output in parallel. The wiring diagram of such a LE is shown in Fig. 4.9 [1] and is distinguished by the presence of a p-MOSFET VT6, which increases the high-level output voltage to U_{OH} to $+U_{CC}$ and n-MOSFET VT5, which lowers the output voltage of the low level U_{OL} to a value close to $0V$.

Circuit engineering methods for increasing the speed of Bi-CMOS are mainly aimed at accelerating the charge (discharge) of the capacitances of the LE components and increasing the output currents of the LE. One of the high-speed Bi-CMOS circuits is shown in Fig. 4.10 a [2] and is distinguished by the presence of an additional n-MOSFET VT3.

When the input voltage U_{IL} is low, a high voltage level appears on the drain of the n-MOSFET VT1, which opens the additional transistor VT3. This creates an additional low-resistance discharge circuit of the input capacitance n-p-n of the transistor VT5, which improves the duration of its shutdown. When the input voltage goes to a high level, the transistor VT3 closes and does not affect the operation of the circuit.

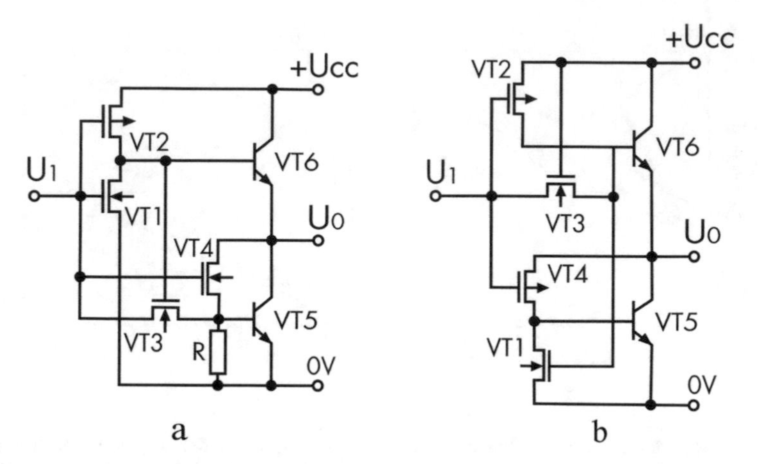

Fig. 4.10 Bi-CMOS LE circuits with improved shutdown time

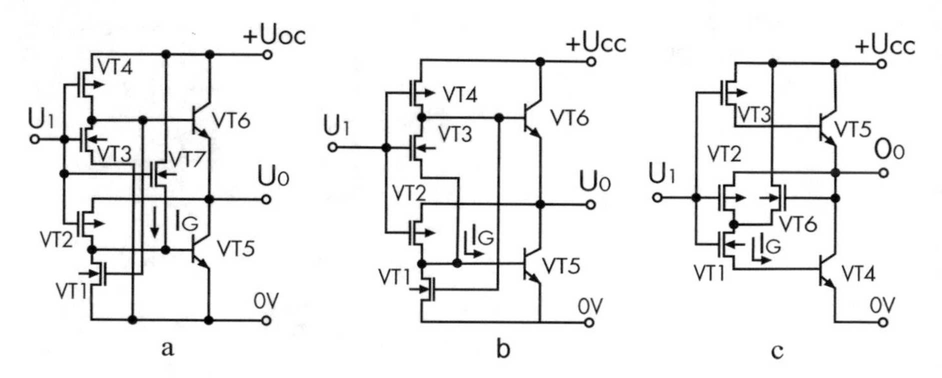

Fig. 4.11 Bi-CMOS circuits with improved turn-on time

According to a similar circuit, a circuit for accelerating the discharge of the input capacitance of the boost n-p-n transistor VT6 can be constructed (Fig. 4.10b).

To accelerate the charge of the capacity of the n-p-n transistor VT5 lowering the output level, an additional current source should be used, the function of which is performed by the n-MOSFET VT7, which is included in the circuit shown in Fig. 4.11a [1]. This transistor turns on when a high-level input voltage U_{IH} is applied, generates an additional current *Ia*, accelerates the charge of the input capacitance n-p-n of the VT5 transistor, and improves its turn-on time. However, when switched on, this circuit has a large static current consumption I_{CC} flowing through the switched on n-MOSFET transistor VT7 and the base-emitter junction of the output transistor VT5. To eliminate this drawback, another connection of the n-MOSFET is possible to speed up the switching-on of the n-p-n-transistor VT5. In this circuit, Fig. 4.11b, the n-MOSFET VT3, which accelerates the process of switching on the n-p-n transistor VT5, is a part of the input CMOS inverter and generates an additional current I_G, the charge only when it is turned on. In the static

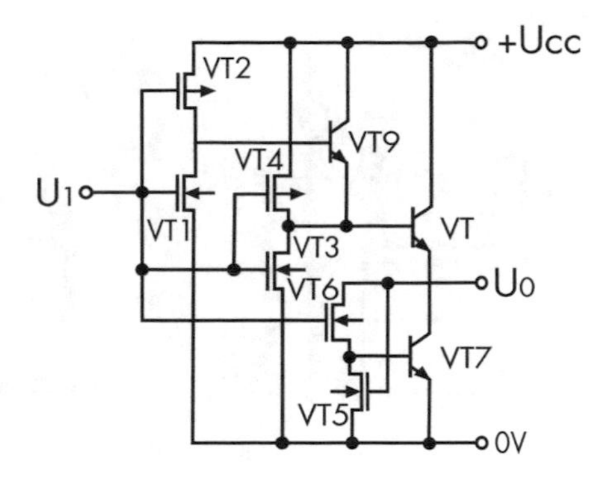

Fig. 4.12 Bi-CMOS LE circuit based on the Darlington circuit

on state, the p-MOSFET VT4 breaks the current flow circuit and the current consumption I_{CCL} is low. In another version, shown in Fig. 4.11c [1], an additional source of accelerating current is formed by the n-MOSFET transistors VT1 and VT6 and functions when the LE is switched on. When the input voltage U_{IL} is low, a high voltage level is set at the LE output, opening the transistor VT6. Since the voltage level at the LE input is low, the transistor VT1 is closed and there is no additional current. When a high voltage level is applied to the input, the transistor VT1 opens and when the transistor VT6 is open, an additional current I_G will be formed in the base of the output n-p-n transistor VT4, accelerating its activation. When the output voltage U_O is lowered, the n-MOSFET VT6 is closed and when the LE goes into a static state, the current I_G is reduced to an insignificant level that does not affect the operation of the LE.

Of the circuit engineering methods for improving the turn-on time of the output step-up n-p-n transistor VT8 (Fig. 4.12), the most effective is the inclusion of an additional n-p-n transistor VT9 according to the Darlington circuit [1]. In this case, the discharge of the input capacitance of the n-p-n transistor VT8 is provided by an additional n-MOSFET transistor VT3, and the increase of the high output voltage U_{OH} to the value $Ucc\text{-}U_{BE}^{VT8}$ is provided by an additional p-MOSFET transistor VT4, controlled from the LE input.

Effective methods of increasing the performance and reducing the dynamic current consumption of Bi-CMOS LE are the use of feedbacks that control the processes of recharging the capacitances of the output n-p-n transistors. One of the circuits of this type is shown in Fig. 4.13a [1] and contains a feedback circuit on the LE D1, D2 of n-MOSFET VT3, which controls the base circuit of the output n-p-n transistor VT6. When the input voltage U_{IH} is high, the low level of the output voltage U_{OL} through the LE D1, D2 turns off the n-MOSFET VT3 and disables the feedback circuit. Therefore, when a low voltage level U_{IL} is applied to the LE input, the n-MOSFET VT2 is closed, and a high voltage level U_{OH} is set at the LE output, which turns on the feedback circuit. Since the switching on of the n-MOSFET VT3 will occur with a delay in relation to the switching off of the n-MOSFET VT2, the feedback circuit practically does not affect the switching off process.

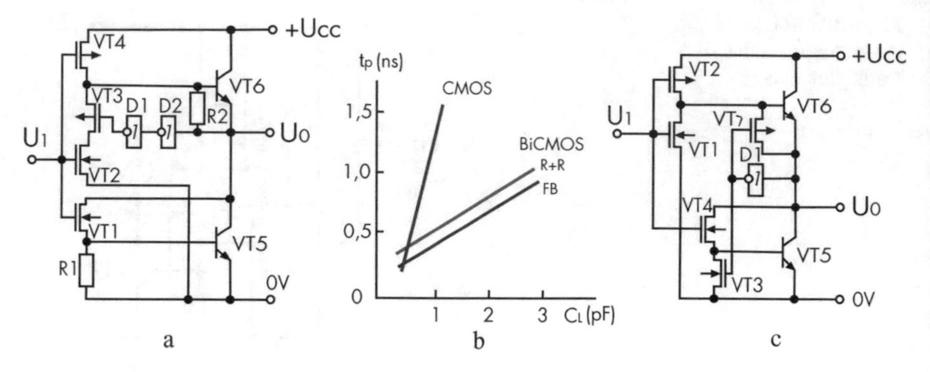

Fig. 4.13 Bi-CMOS LE circuits with feedbacks (**a** and **b**) and comparative performance characteristics of Bi-CMOS LE (**b**)

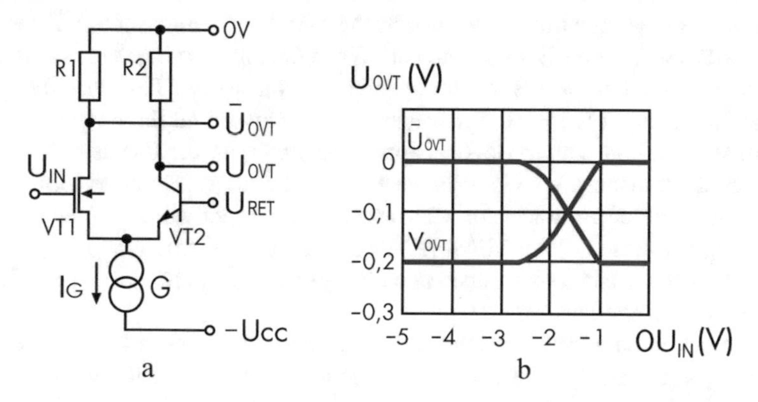

Fig. 4.14 Bi-CMOS LE of the current logic (**a**) is its transfer characteristic (**b**)

When a low voltage level U_{IL} is applied to the LE input, the n-MOSFET VT2 is turned on, and when the transistor VT3 is turned on, an additional low-resistance discharge circuit of the input capacitance of n-p-n transistor VT6 is formed. When the output level of LE D1, D2 is lowered, the n-MOSFET VT3 is switched off, as a result of which the additional discharge circuit is switched off. The comparative dependences of the delay time t_P on the load capacity C_L for CMOS LE, Bi-CMOS LE of R+R type and Bi-CMOS LE with feedback (FB) are shown in Fig. 4.13b. A similar feedback loop can be formed for the step-down output n-p-n transistor VT5. The electrical diagram of the Bi-CMOS LE of this type, described in [1], is shown in Fig. 4.13c.

One of the directions of Bi-CMOS LE is the application of bipolar circuit engineering, in which the disadvantages of bipolar transistors are eliminated by replacing bipolar transistors with MOSFETs. One of the types of such circuits is the Bi-CMOS current key logic, which is the Bi-CMOS analog of LE ECL [1]. The wiring diagram of LE (Fig. 4.14a) contains an n-MOSFET VT1, which determines the circuit through which the current of the generator I_G flows. Small voltage drops on the resistors R1 and R2 allow you to connect the LE outputs to the LE ECL inputs. The input LE is controlled directly by the voltage drop of the CMOS LE, and the reference voltage U_R is selected so that the supply voltage is distributed equally between the transistors and the current source G. The transmission characteristic of the LE is shown in Fig. 4.14b.

4.2 Bi-CMOS IC Memory Elements

As memory elements in Bi-CMOS ICs, well-known configurations of CMOS triggers are used, in which outputs are formed using bipolar transistors. The most common options for forming a bipolar push-pull output for a bistable cell are shown in Fig. 4.15a, b.

Among the features of the Bi-CMOS ME, it is necessary to note the introduction of bipolar n-p-n transistors in the synchronization and write/read lines of the ME, which allows to increase their performance. In the diagram in Fig. 4.16, n-p-n transistors VT4 and VT6 are introduced into the synchronization circuit of the slave R-S-trigger, which reduces the time for rewriting information from the master R-S-trigger to the slave one [1]. In the option of the ME circuit for RAM-type circuits (Fig. 4.17), a bipolar transistor VT6 is introduced in the ME read lines, which allows you to speed up the charge (discharge) of the read line and increase the performance of the circuit.

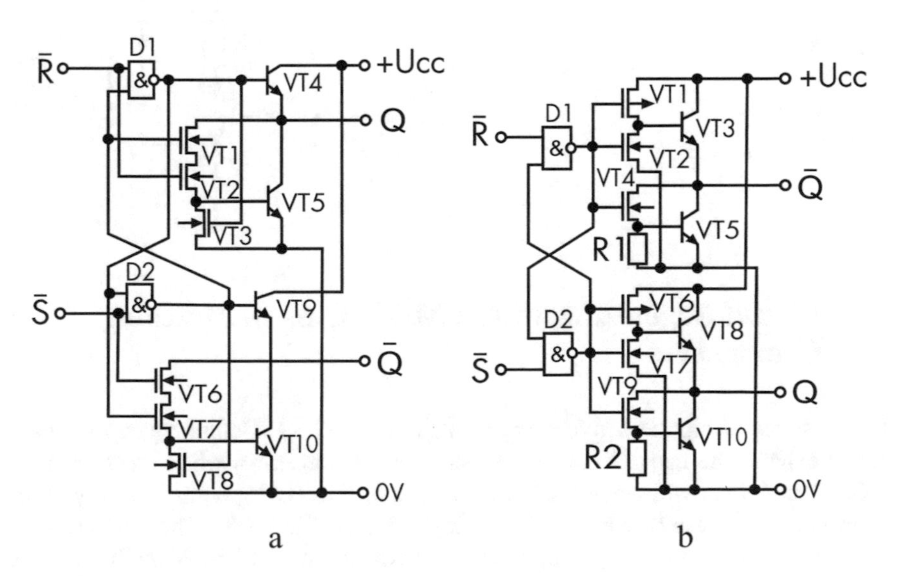

Fig. 4.15 Bi-CMOS ME circuits with bipolar outputs

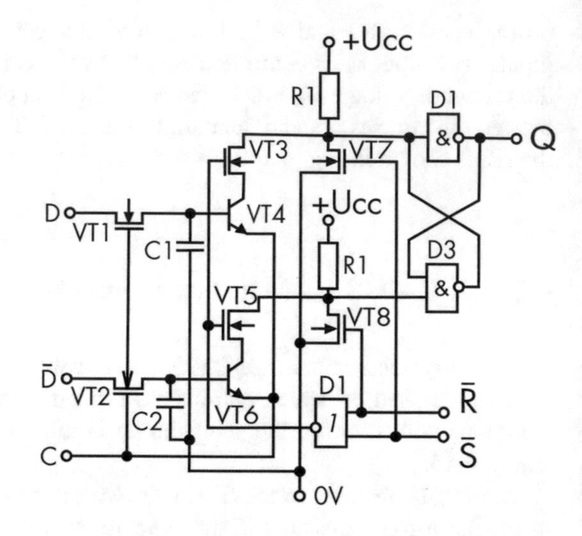

Fig. 4.16 Bi-CMOS ME with bipolar n-p-n transistors in the synchronization circuit

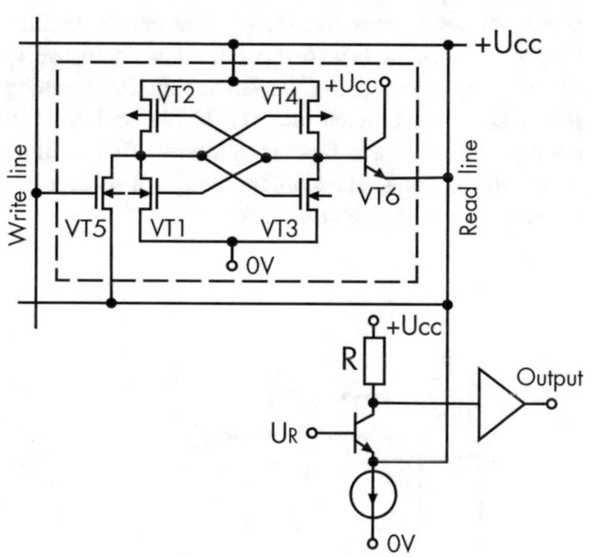

Fig. 4.17 ME circuit for RAM blocks, general-purpose registers

4.3 Circuit Engineering of Bi-CMOS IC Input Matching Components

The presence of bipolar active components in the CMOS IC allows you to significantly expand the capabilities of CMOS circuit engineering when creating input MCs for matching with any of the known types of ICs. At the same time, any of the known input MC can be combined on a single chip Bi-CMOS IC. Therefore, it is not possible to single out a single input MC as a typical one for Bi-CMOS IC, and the main one for Bi-CMOS IC will be a specialized input MC introduced on the chip, depending on the conditions of IC application (with TTL levels, CMOS, or ECL).

4.3.1 Input MC of Bi-CMOS IC with Signal Level Conversion

Input MC with Signal Levels CMOS Conversion

Since the input switching thresholds U_{TH}, U_{TL} of the Bi-CMOS LE are close to the threshold voltages U_{TH}, U_{TL} of the CMOS ICs (with the same supply voltage $+U_{CC}$), no special MC circuits are required in the Bi-CMOS ICs. As such an MC that perceives CMOS signal levels, you can use a simple CMOS inverter on transistors VT1 and VT2 (Fig. 4.18). Sometimes, at the MC input, an additional resistor R is introduced into the Bi-CMOS IC, connected either to the U_{CC} power line or to the common *0V* line, fixing the potential to the MC input.

Input MC with TTL Conversion of Signal Levels

Since the input MC on bipolar transistors have a high level of static power consumption for Bi-CMOS ICs, it is more efficient to use MC-translators of TTL levels based on the CMOS element base. However, the disadvantage of this type of circuit is the asymmetry in the switching delays t_{pHL}, t_{pLH}, which can be eliminated by using MC, Fig. 4.19 a [1]. The circuit is distinguished by the presence of an n-p-n transistor VT4, switched on according to the circuit with a "common emitter," which, when the MC is turned on, increases its output current I_O, reduces the shutdown duration τ_{LH}, and reduces the asymmetry in the delay. The asymmetry in the switching delays t_{pLH}, t_{pHL} can be eliminated using the MC electrical circuit, shown in Fig. 4.19b [3], where the offset of the switching threshold U_T of the Bi-CMOS inverter is achieved by introducing an offset element (transistor VT1).

Circuit switching threshold $U_T \approx U_{\mathrm{T}}^{CMOS}$-$U_{BE}^{VT1}$. The input characteristic has the form shown in Fig. 4.19b, curve 1 and differs from the typical TTL IC characteristic

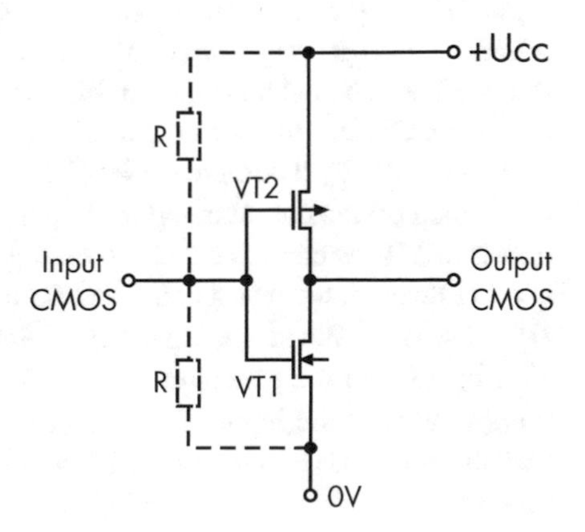

Fig. 4.18 Diagram of the input MC with CMOS switching threshold

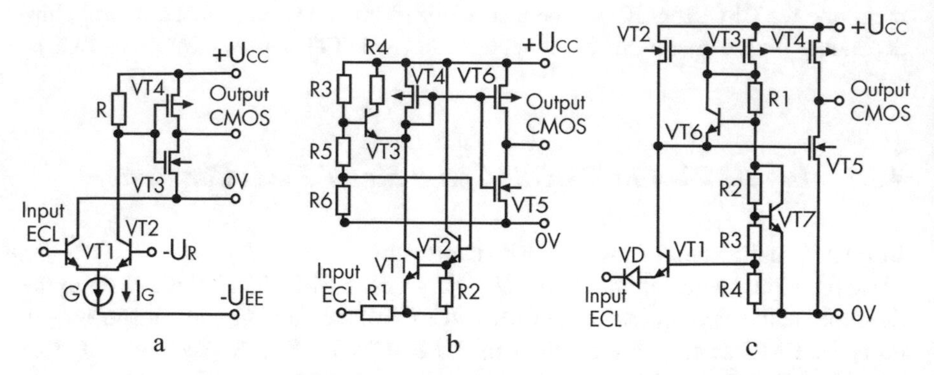

Fig. 4.19 Bi-CMOS circuits of input MC with TTL switching threshold (**a** and **b**) and input characteristic (**b**)

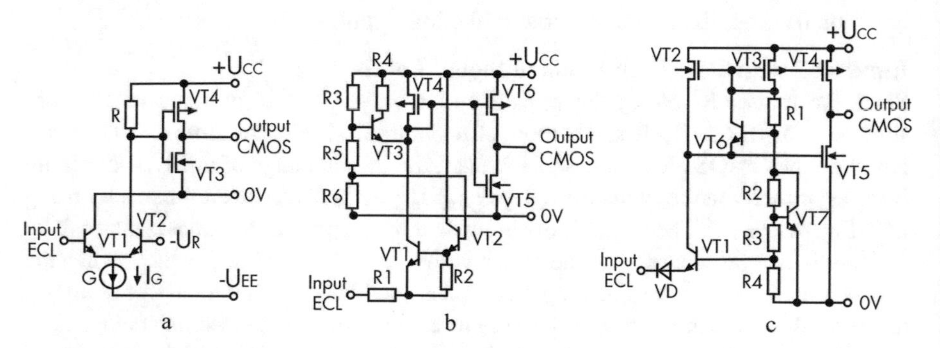

Fig. 4.20 Bi-CMOS circuits of the input MC receiving ECL levels of negative polarity (**a**) and receiving ECL levels of negative polarity and a single power supply $+U_{CC}$ (**b**, **c**)

(curve 2). The closed state of the p-MOSFET VT4 transistor is provided by the resistor R1 (Fig. 4.19b).

Input MC with ECL Conversion of Signal Levels

Input MC with conversion of ECL of negative polarity levels to CMOS levels of positive polarity. An example of the simplest MC circuit of this type is shown in Fig. 4.20, and it contains an input bipolar switch, a level offset circuit (resistor), and an output CMOS inverter on MOSFETs. The required threshold for switching the ECL key is set by the reference level U_R, and the output CMOS LE levels U_{OH}, U_{OL} are formed by the output inverter. At a high input voltage level $U_{IH}^{ECL} >| -U_R |$ n-p-n transistor VT2 is closed and a high voltage level $-U_{CC}$ is set at the input of the output inverter through the resistor R. The MC output will set to a low-level output voltage $U_{OL}^{CMOS} \approx 0$ V. When the input MC voltage is $U_{IL}^{ECL} >| -U_R |$, n-p-n the transistor VT2 opens and the generator current I_G flows through the resistor R, shifting the voltage level at the input of the inverter. When the condition $U_{CC}\text{-}I_G R < U_T^{VT3}$ is met, the output CMOS inverter will switch to the closed state and the MC output will be set to a high-level output voltage $U_{\mathrm{OH}}^{CMOS} \approx +U_{CC}$.

However, this type of circuit is rarely used in digital ICS due to the need for two power sources: positive and negative $-U_{EE}$. The diagram in Fig. 4.20b [4] illustrates an MC of this type, for which only one power source is needed: the positive $+U_{CC}$. The circuit contains an output CMOS inverter on MOSFETs VT6, VT7, a level shift circuit on transistors VT1-VT5, and a reference voltage source $U_R \approx +0.4$ V on resistors R3, R5, R6, and a diode VD.

When the MC input voltage is high $U_{IH} \geq$-0.75 V, the voltage is $U_B^{VT1} \approx 2U_{BG} - \mid U_{IH} \mid + U_R \approx 0.35$ V. In the base of the n-p-n transistor, VT1 keeps it in the closed state, as a result of which a high voltage level $U_B^{VT4} \approx U_{CC} - U_T^p$ is set at the input of the output inverter. A low voltage level of $U_{OL}^{CMOS} \approx 0$ V will be generated at the MC output. When applying a low voltage level $U_{IL} \leq -1.5$ V at the MC input, the transistor VT1 will open and current will flow through it:

$$U_K = 2U_{BE} + U_R \text{-} |U_{IL}|/Rl \approx 0,4 \ \ B/R1. \tag{4.13}$$

This current, creating a voltage drop on the transistor VT4 and resistor R4, will shift the input voltage on the CMOS inverter below the level U_T^{VT6} and a high voltage level $U_{OH}^{CMOS} \approx +U_{CC}$ will be formed at the MC output.

The MC circuit shown in Fig. 4.20 b is an advanced modification of the MC with a single supply voltage $+U_{CC}$. The principles of operation of this circuit are similar to the circuit in Fig. 4.20b.

A common disadvantage of this type of MC is the long level conversion time associated with a significant voltage drop and their shift from the negative stress region to the positive one.

Input MC with ECL conversion of the same polarity as CMOS For this type of MC, the input (output) levels are in the same polarity region, in most cases - in ECL, negative 0V $- -U_{EE}$. In this case, the MC is based on a bipolar ECL key (Fig. 4.21) on n-p-n transistors VT1 and VT2, a level shift circuit (resistor R, n-p-n transistor VT3, and diode VD2) and an output driver (MOSFETs VT4, VT5, and diode VD3).

When the input voltage is low $U_{IL}^{ECL} <$ -U_R, n-p-n transistor VT2 is open and the generator current I_G creates a voltage drop on the resistor that shifts the voltage at the input of the output driver. At the level of:

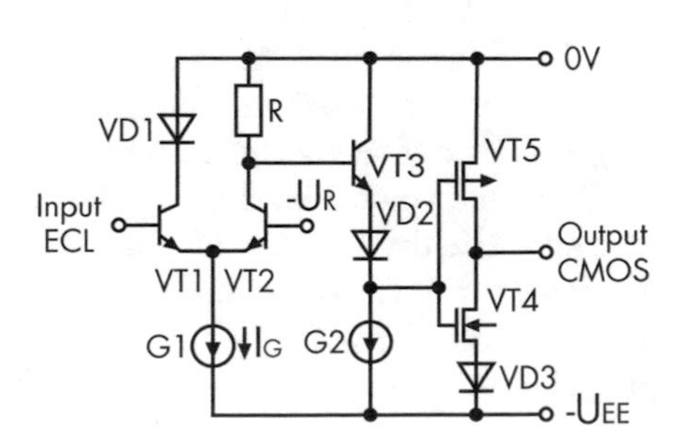

Fig. 4.21 The circuit of the Bi-CMOS of the input MC of the ECL-CMOS type with input (output) levels of the same polarity

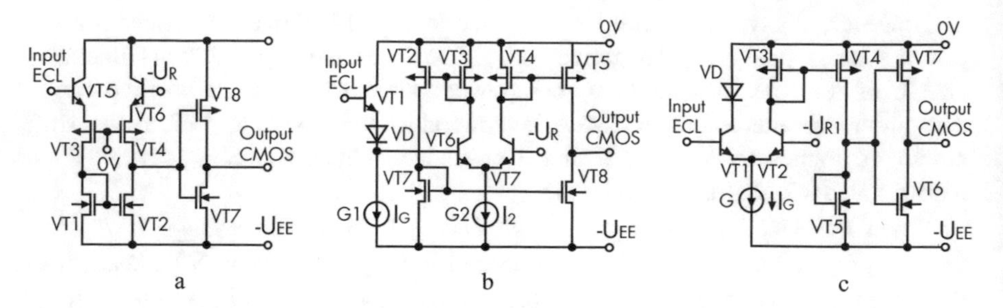

Fig. 4.22 Diagrams of input MC receiving ECL levels with a single power supply

$$U_{CC} - I_G R - U_{BE}^{VT3} - U_P^{VD2} < U_T^{VT4} + U_P^{VD3}. \tag{4.14}$$

The output driver is closed and a high voltage level of $U_{OH}^{CMOS} \approx 0$ will be generated at the output.

When a high voltage level $U_{IH}>|-U_R|$ is applied, n-p-n transistor VT2 will switch to the closed state and the generator G2 will set a high voltage level at the input of the output driver:

$$U_I \approx U_{CC} - U_{BE}^{VT3} - U_P^{VD2} > U_T; \tag{4.15}$$

where
U_T is the switching threshold of the output CMOS driver.

If this condition is met, the MC output will be set to a low voltage level U_{OL}^{CMOS}. The disadvantage of the circuit is the increased level of the low-level output voltage $U_{OL} \approx U_{np}^{VD3}$. The need to introduce a VD3 diode is associated with the exclusion of the saturation mode of the bipolar transistors VT1 and VT2 of the ECL key. In addition, due to the reduced voltage level at the input of the output driver in the on state, the driver has an increased current consumption. These disadvantages are eliminated in the MC circuit shown in Fig. 4.22 a [5], where the level shift circuit is made on the VT3-VT5 MOSFETs. The ECL key load function is performed by the p-MOSFET VT3, which is connected to the shift transistor VT4 according to the "current mirror" circuit. A current source on the n-MOSFET VT5 is connected to the drain of the transistor VT4.

At a high voltage level at the MC input $U_{IH}>|-U_R|$ the transistor VT2 is closed and the drain current of the transistor VT4 is low. Therefore, a low voltage level will be set at the input of the output CMOS driver (MOSFETs VT6 and VT7). At $U_{IL} < U_T^{VT6}$ a high voltage level will be generated at the MC output.

When the voltage level $U_{IL}>|-U_R|$ is low, the transistor VT2 is open, a current will appear in the drain circuit of the MOSFET VT4, which creates a voltage drop on the transistor VT5. When the voltage on the transistor VT5 falls on the input of the output driver $U_{IH} > U_{CC} - U_T^{VT7}$, a low voltage level $U_{OL}{\approx}0$ will be formed at the output of the MC.

To increase stability under conditions of changes in the supply voltage U_{CC}, temperature T_A° it is possible to use more complex modifications of the considered circuit, for example, the circuit in Fig. 4.22 b. In the MC circuit shown in Fig. 4.22 c [1], a differential amplifier on transistors VT5 and VT6 with a load that shifts the output levels in the circuit of their emitters is used at the input. When the voltage level at the input $U_{IL}>|-U_R|$ is low, the transistor VT5 is closed and the n-p-n transistor VT6 sets the level $U_{IH} \approx U_R - U_{BE}^{VT6}$ at the input of the output CMOS driver (transistors VT7 and VT8). Provided that this voltage is above the switching threshold of the UT driver, the MC output will be set to a low voltage level U_{0L} of the CMOS. A high voltage level at the MC input $U_{IH}^{ECL} \approx -U_R$ opens the n-p-n transistor VT5, which opens the MOSFET VT2. In this case, the voltage at the input of the output driver drops and at a level below U_T^{VT7}, and a high voltage level U_{OH} will be formed at the output of the MC.

4.3.2 *Input MC of Bi-CMOS IC with Increased Load Capacity*

Input MC with increased load capacity in Bi-CMOS ICs can be formed by connecting in series to an input MC with the required type of input levels of the output buffer stage based on bipolar n-p-n transistors. Figure 4.23 shows a diagram of the input MC with increased load capacity and CMOS input levels.

4.3.3 *Input MC of Bi-CMOS IC with Paraphase Outputs*

In Bi-CMOS ICs, as well as in other types, one of the ways to expand the functionality is the formation of *paraphase output signals*. Circuit engineering techniques for constructing such Bi-CMOS circuits are distinguished by the use of bipolar n-p-n transistors as output circuits. Any of the circuits with the required input levels

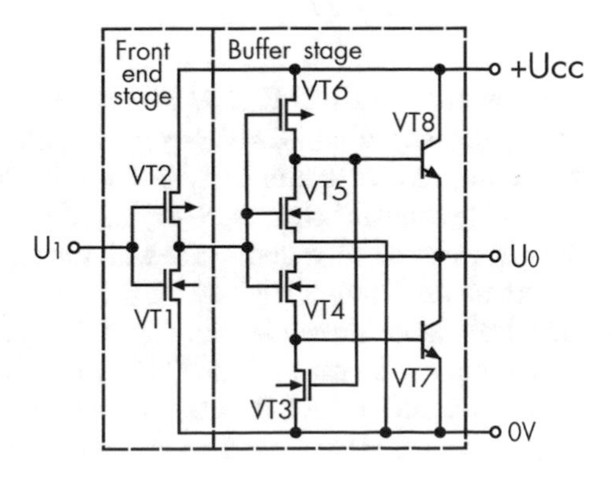

Fig. 4.23 Bi-CMOS circuit of the input MC with increased load capacity

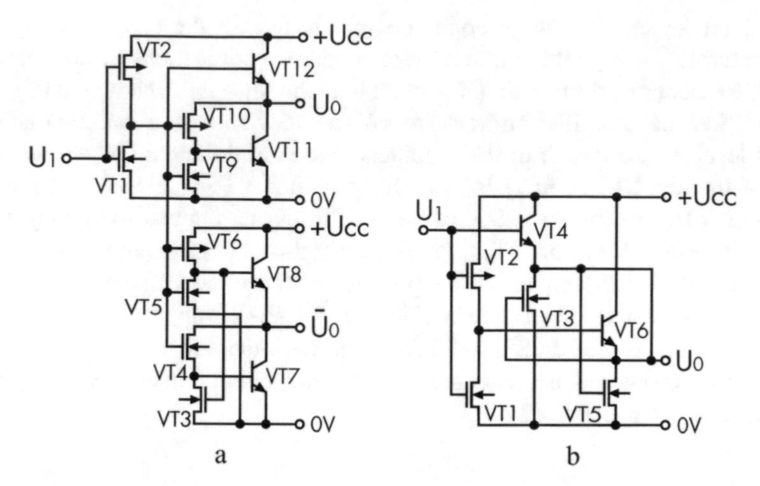

Fig. 4.24 Bi-CMOS circuit of the input MC with paraphase outputs

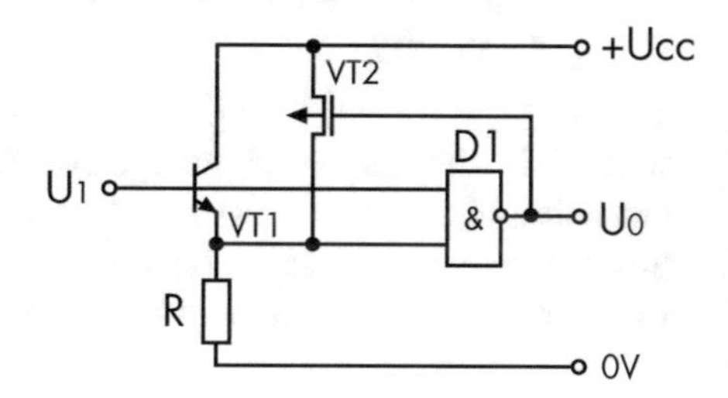

Fig. 4.25 Schmitt's trigger Bi-CMOS diagram

(CMOS, TTL, and ECL) can be used as input. Examples of Bi-CMOS circuits with paraphase outputs and CMOS input switching threshold are shown in Fig. 4.24 a, b [1].

4.3.4 Input MC of Bi-CMOS IC with Increased Noise Immunity

The input MC of Bi-CMOS IC of increased noise immunity can be constructed using both bipolar-type and CMOS-type circuits. However, the most effective means of increasing the noise immunity of CMOS ICs is the use of Schmitt triggers, which have a "hysteresis" characteristic, based on CMOS circuit engineering. To increase the performance, an output stage with bipolar n-p-n transistors can be installed at the output of the Schmitt trigger. Of the other variants of Schmitt trigger circuits, the electrical circuit shown in Fig. 4.25 [1] is practically interesting.

The circuit is based on the CMOS LE D1 and contains an input n-p-n transistor VT1 and a p-MOSFET feedback transistor VT2. At a low voltage level at the input $U_{IL} < U_T^{D1}$ LE D1 is closed and at its output a high voltage level p-MOSFET VT2.

Therefore, when the switching voltage increases, the MC will occur at $U_{TH} \approx U_T^{D1} + U_{BE}^{VT1}$.

After switching LE D1, a low voltage level at its input will turn on the p-MOSFET VT2, which, due to the voltage drop at the resistor R, will close the input n-p-n transistor VT1. As a result, reverse switching of the MC will occur at the voltage level $U_{TL} \approx U_T^{D1}$.

Width of the "hysteresis" loop:

$$\Delta U_H \approx U_{BE}^{VT1} \tag{4.16}$$

does not depend on the MC supply voltage.

4.3.5 Input MC of Bi-CMOS Memory IC

The main method of constructing such MCs is the sequential cascading connection of the simplest input MCs based on the Bi-CMOS discussed in this chapter with CMOS memory elements. If necessary, outputs based on bipolar transistors can be formed at the output of the memory elements.

4.3.6 Circuit Engineering of Input MC of Bi-CMOS IC Protection

Protection of the Input Bi-CMOS MC from Static Electricity

Bi-CMOS ICs are distinguished by a variety of input MCs that are combined on a single chip, both bipolar and CMOS. As is known, bipolar (in particular, TTL) input MCs have a sufficiently high resistance to the effects of static electricity and for their protection in Bi-CMOS ICs, you can use the simplest means of protection, such as "anti-ring diodes." Input CMOS MC in Bi-CMOS IC, due to their increased sensitivity to static electricity, should be equipped with protection circuits. However, the presence of bipolar components in Bi-CMOS ICs, due to their higher performance, small size, and ability to reliably dissipate large power levels, allows you to create more effective means of protection against static electricity. The circuit in Fig. 4.26 a contains two n-p-n transistors VT1 and VT2 in the off-state.

Transistor VT1 provides protection against discharge relative to the power supply pin $+U_{CC}$, transistor VT2-relative to the common pin *0V*. To limit the discharge currents, resistors R1 and R2 are introduced into the circuit.

The circuit shown in Fig. 4.26b is built on the thyristor principle and contains a pair of transistors VT1 and VT2 with p-n-p and n-p-n conductivity. To limit the discharge currents, resistors R1-R4 are introduced into the circuit. When an electrostatic discharge occurs and the n-p-n transistor enters the avalanche breakdown

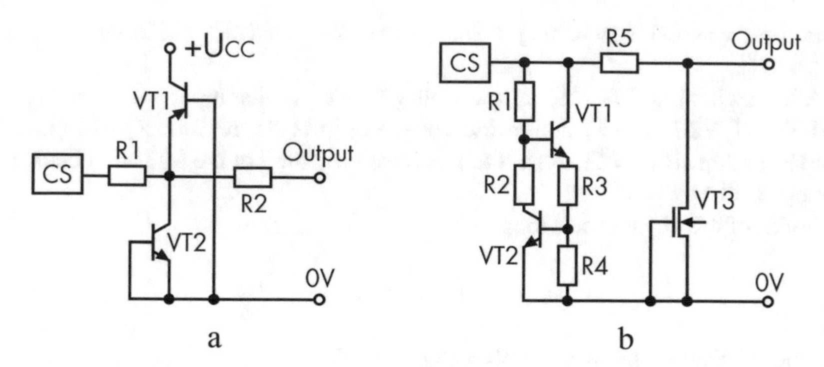

Fig. 4.26 Circuits of input MC Bi-CMOS protection from static electricity

mode, the p-n-p transistor VT1 and the n-p-n transistor VT1 open, through which the electrostatic discharge is diverted to the common pin *0V*. In addition, other protection components, such as the VT3 n-MOSFET, can be added to the circuit.

Protection of the Input MC from Negative Input Levels

Bi-CMOS ICs, as well as bipolar ICs, are sensitive to negative input levels, which can lead to their failures. Therefore, if such a situation is likely to occur, the input MC Bi-CMOS ICS should be equipped with special protection devices. In the case of input MC Bi-CMOS of bipolar type, the TTL and IC protection circuits can be used as a protection circuit.

4.4 Circuit Engineering of Bi-CMOS IC Matching Output Components

4.4.1 Output MC of Bi-CMOS IC with the Formation of CMOS Output Levels

If it is necessary to form CMOS signal levels at the outputs of the Bi-CMOS IC, the circuit of the simplest CMOS output MC can be used as the output MC. However, if high performance and load capacity are required, such a circuit does not provide good characteristics and requires the use of bipolar transistors at the output. Direct use of the Bi-CMOS LE circuit as an output MC, for example, is inefficient due to the degraded output levels $U_{OH}^{ECL} \approx U_{CC} + U_{BE}$.

Therefore, MC circuits with an output on n-p-n transistors are supplemented with MOS-components that increase the output levels to the values of CMOS levels. The diagram of Fig. 4.27 is distinguished by two MOSFETs: p-MOSFET VT4 and n-MOSFET VT5, connected to the MC output in parallel to the output n-p-n transistors VT10 and VT9. Components VT1-VT3 and VT6 provide the necessary on/off phases of MOSFET transistors VT4 and VT5 that are the same for n-p-n

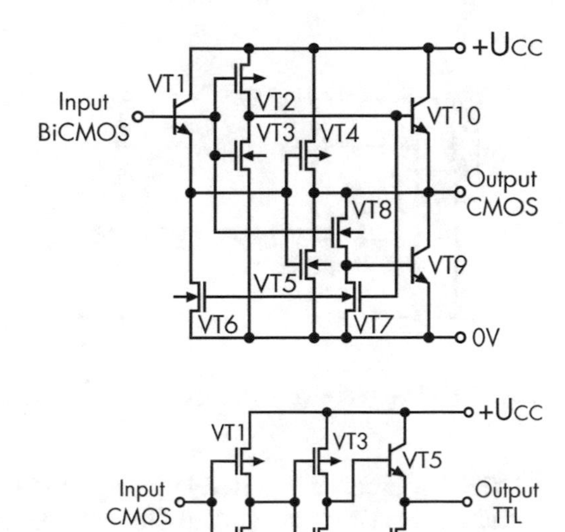

Fig. 4.27 Diagram of output MC CMOS with CMOS output levels

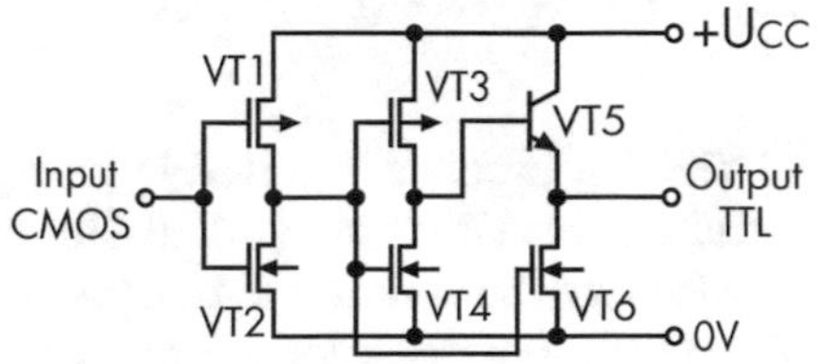

Fig. 4.28 The output MC Bi-CMOS circuit with TTL levels and an n-MOSFET at the output

transistors VT10 and VT9, respectively. The output levels of such an MC are similar to the MC CMOS: $U_{OH}^{ECL} \approx U_{CC}$; $U_{OL}^{ECL} \approx 0V$.

4.4.2 Output MC of Bi-CMOS ICs with the Formation of TTL Output Levels

Output MC CMOS with CMOS levels allow control of IC inputs that have TTL switching thresholds. However, in this case, the output levels are asymmetric with respect to the switching threshold =1.5 V TTL IC, which leads to a loss of MC performance. The simplest circuit of the output MC with TTL output levels is shown in Fig. 4.28 and contains a single n-p-n transistor VT5. Output levels of such MC:

$$U_{OH}^{ECL} \approx U_{CC} - U_{BE}^{VT5} \approx 4,2\ B; \qquad U_{OL}^{ECL} \approx 0. \tag{4.17}$$

In static states with a capacitive load, this circuit does not consume power, but its disadvantage is the use of an n-MOSFET as the output transistor VT6, which reduces the performance of the MC. Using the VT6 transistor as the output n-p-n has two problems:

(a) the need to exclude the saturated mode of operation of the transistor to obtain high performance;
(b) in the static open state, due to the need to set the current to the base of the output transistor, such an MC will consume a current, the value of which should be minimized.

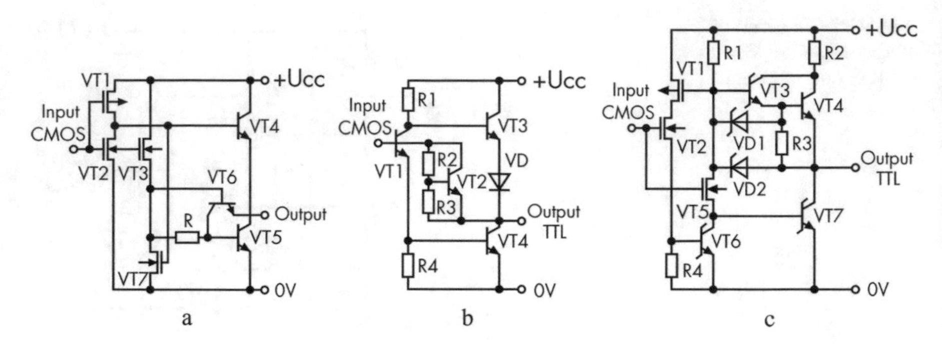

Fig. 4.29 Circuits of output MC Bi-CMOS with TTL levels

The first problem is solved by using various kinds of limiting components and circuits. In the circuit in Fig. 4.29a, the saturation limit of the output transistor VT5 is provided by the collector-base transistor VT6 connected in parallel to its junction. When the n-p-n transistor is turned on, the VT5 drops and opens it. As a result, the voltage at the collector-base junction of the n-p-n transistor VT5 is limited at $U_{CE}^{VT6} \approx 0.1 \div 0.2$ V, which eliminates the forward bias of the junction.

In the diagram in Fig. 4.29b [1], the forward bias of the collector junction of the output transistor VT4 is eliminated by an additional offset element on the transistor VT2, resistors R2 and R3, connected between the collector of the output transistor VT4 and the base of the input transistor VT1. When the output n-p-n transistor VT4 is turned on, lowering the voltage on its collector causes the interception of the base current of the input transistor VT1 and the base current of the output transistor until the voltage on the collector of the output transistor VT4 is stabilized at the following level:

$$U_{CE}^{VT4} \approx U_{BE}^{VT1} + U_{BE}^{VT4} - U_{CM} \approx 0,3 \ \div 0,4 \ B; \tag{4.18}$$

where U_{CM} is the voltage drop on the displacement element.

At the same time, the voltage in the collector-base junction is $U_{CB}^{VT4} \approx 0.35 \div 0.45$ V, which eliminates its forward bias and the saturated mode of operation of the output n-p-n transistor VT4. The values of the resistors R1 and R2 can be calculated using the formula:

$$R2/R1 \approx U_{CM}/U_{BE} - 1; \tag{4.19}$$

where $U_{CM} \approx 2U_{BE}$-$U_{CE} \approx 1.2$ V.

The circuit in Fig. 4.29c [1] is a more complex technological modification of the MC Bi-CMOS IC using n-p-n transistors with Schottky diodes, in which the saturation mode is eliminated structurally by shunting the collector-base transitions of n-p-n transistors with Schottky diodes with a direct voltage of $U_P^{VD} \approx 0,5 \div 0.6$ V

.

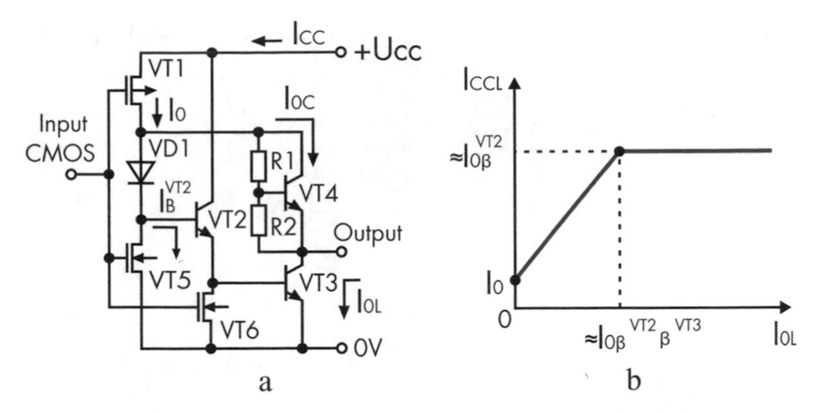

Fig. 4.30 The Bi-CMOS circuit of the output MC with the consumption current control I_{CC} (**a**) and the current control characteristic I_{CC} (**b**)

The second problem can be solved by forming feedback loops in the MC that specify the current consumption of the MC I_{CCL} in the on state, depending on the load current I_{OL}.

The example of this type of MC circuit is shown in Fig. 4.30, and it is distinguished by the n-p-n transistor VT2, which is connected according to the circuit with a common emitter. From the MC output through the offset elements on the n-p-n transistor VT1, resistors R1 and R2 and the diode VD1, a feedback circuit is formed to the base of the n-p-n transistor VT2, which controls its base current. Current consumption in the on state

$$I_{CCL} = U_B^{VT2} + I_{OC} + \beta^{VT2} \bullet I_B^{VT2}; \tag{4.20}$$

where I_{OC} is the current flowing through the feedback circuit and depends on the load current $I_B^{VT2} = (I_{OC} + I_{OL})/\left(\beta^{VT2} \cdot \beta^{VT3}\right)$.

The sum of the currents $I_B^{VT2} + I_{OC}$ is a constant value, depending on the parameters of the MC components:

$$I_B^{VT2} + I_{OC} = I_0 = \left(U_{CC} - U_P^{VD1} - U_{BE}^{VT2} - U_{BE}^{VT3}\right)/R_0^{VT1}; \tag{4.21}$$

where R_0^{VT1} is the output resistance of the open transistor VT1. Assuming I_{OL}=0, $I_{CCLmin} \approx I_O$ (Fig. 4.30b), we get:

$$I_B^{VT2} = \left(I_0 - I_B^{VT2} + I_{OL}\right)/\left(\beta^{VT2} \bullet \beta^{VT3}\right) \approx (I_0 + I_{OL})/\left(\beta^{VT2}\beta^{VT3}\right); \tag{4.22}$$

$$I_{CCL} = I_0 + (I_0 + I_{OL})/\beta^{VT3} \approx I_0 + I_{OL}/\beta^{VT3}. \tag{4.23}$$

When the output current increases to the maximum value, at which the feedback current I_{OC}=0, we write $I_{OLmax} \approx I_O {*} \beta^{VT2} {*} \beta^{VT3}$. When this current is exceeded, the output transistor VT3 goes into active mode and the voltage at its output increases,

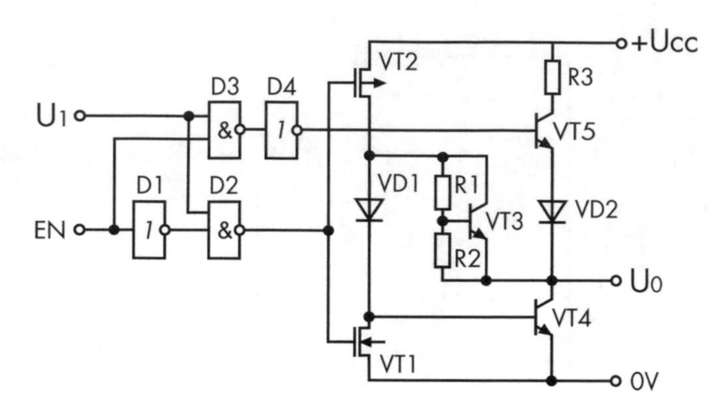

Fig. 4.31 Circuit of the output MC Bi-CMOS of the "three states" type

which is a violation of the MC, and the current consumption of the MC does not change. Then the maximum current consumption $I_{OLmax} \approx I_0 {}^* \beta^{VT2}$ (Fig. 4.30 b).

Thus, the I_{CCL} consumption current is controlled in the MC circuit.

In Bi-CMOS ICs, all the above types of outputs are used. The circuit of the MC with the "three-state" output is shown in Fig. 4.31a, and the MC with the "open collector" output is shown in Fig. 4.30a.

4.4.3 *Output MC of Bi-CMOS IC with the Formation of ECL Output Levels*

The presence of bipolar components in Bi-CMOS ICs allows you to use the advantages of CMOS (Bi-CMOS LE) when they are combined with ultra-fast ECL ICs. This is due to the fact that bipolar n-p-n transistors, due to their high gain, output conductivity, and performance, allow you to form high-speed, output MC-drivers of ECL signal levels on Bi-CMOS chips without significant loss in performance.

Output MC That Converts Bi-CMOS Levels of Positive Polarity to ECL Levels of Negative Polarity

The simplest circuit of such an MC is shown in Fig. 4.32 a [1] and contains an output ECL key on n-p-n transistors VT5-VT7 and a level shift circuit on CMOS transistors VT1-VT4. A reference voltage $U_R \approx 2.5$ V is applied to one of the circuit inputs. At a low voltage level at the MC input, the Bi-CMOS $U_{IL}^{BiCMOS} < 2.5$ V, the MOSFET VT1 is closed, and the MOSFET VT2 is open. In this case, the circuit of the current source on the transistors VT3 and VT4 is closed and the current of the current generator G, creating a voltage drop on the output resistance of the MOSFET VT2, shifts the base n-p-n of the transistor VT5 and opens it. The MC output will generate the ESL low-level output voltage $U_{OL}^{ESL} \approx -U_L^{VT7} - I_G R$. When the input voltage of a high level >2.5 V is applied, the MOSFET VT1 opens and turns on the current source on the transistors VT3 and VT4. The MOSFET VT4 intercepts the current

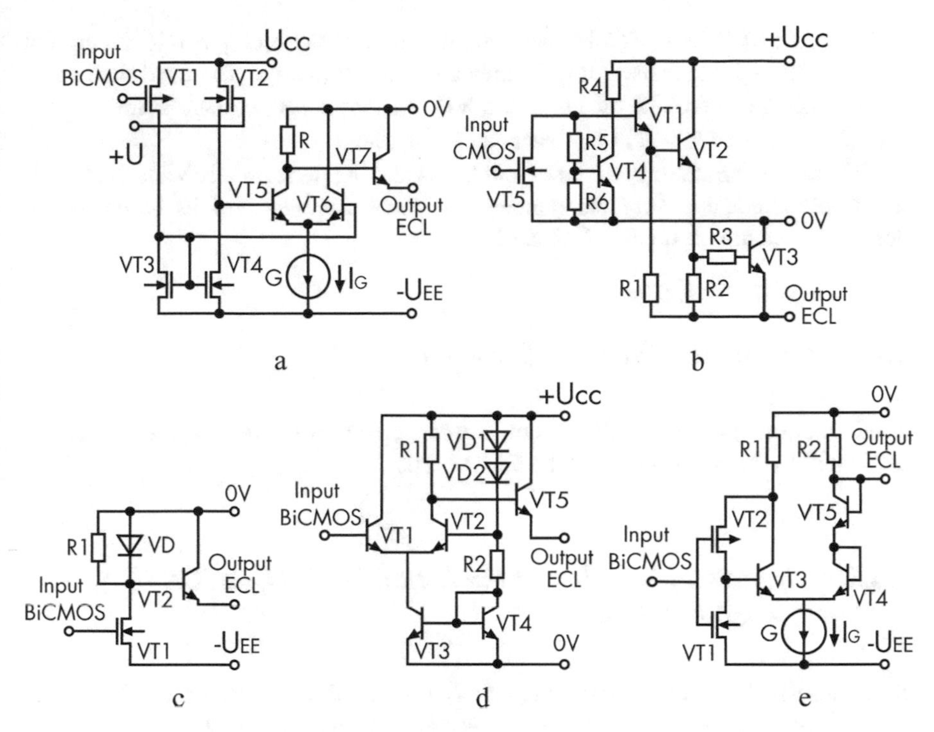

Fig. 4.32 Output MC circuits with the formation of ECL levels of opposite (**a**, **b**) and the same (**c**, **d**, **e**) polarity with the input level

generated by the MOSFET VT2, as a result of which the voltage at its drain drops and the n-p-n transistor VT5 of the current key closes. A high-level voltage $U_{OL}^{ECL} \approx -U_{B}^{VT7}$ will be generated at the MC output.

The disadvantage of this type of circuit is the need for two power sources $+U_{CC}$, $-U_{EE}$ and reduced performance, which is associated with a large voltage drop in the bases of n-p-n transistors. Therefore, in ICS, MC is widely used, having a single supply voltage $+U_{CC}$. A circuit of this type is shown in Fig. 4.32 b. The CMOS shift of the positive levels to the negative region is achieved by the level shift circuit on the n-p-n transistors VT1 and VT2, resistors R1-R3, and the formation of ECL output levels is provided by the output n-p-n transistor VT3.

Output MC That Converts Bi-CMOS Signal Levels to ECL Signal Levels of the Same Polarity

ESC circuits of this type are faster because of the lower level voltage drop when they are converted. The MC circuit in Fig. 4.32 b contains the output n-p-n transistor VT2, which forms the ECL levels and the control circuit based on the n-MOSFET transistor VT1. When the output voltage $U_{IL} < U_{T}^{VT1}$ is low, the MOSFET VT1 is closed and the output n-p-n transistor VT2 will generate a high voltage level $U_{OH}^{ECL} \approx -U_{BE}^{VT2}$ at the output.

When the MOSFET VT1 is open with the input voltage $U_{IH} > U_T^{VT1}$, a current will flow through the resistor R1, creating a voltage drop on it. The maximum value of this voltage is limited by the diode VD, so that a low voltage level $U_{OL}^{ECL} \approx -U_{BE}^{VT7} - U_P^{VD} = 1.5$ V will be formed at the output.

To increase the stability of the output levels and the performance with changes in the supply voltage – U_{EE} and temperature, more complex circuits based on the current switch are used (Fig. 4.32 d, e).

4.4.4 Output MC BI-CMOS Memory ICs

The formation of this type of MC is mainly carried out by connecting the output MC Bi-CMOS to the ME outputs of the CMOS type.

4.4.5 Circuit Engineering of the Output MC BI-CMOS IC Protection Circuits

Since the Bi-CMOS element base is mainly used in ultra-fast ICs, they are characterized by an increased level of interference and the presence of spurious effects of both CMOS and bipolar ICs, so it is advisable to use the methods and circuits considered for CMOS ICs to eliminate and weaken the mentioned effects.

References

1. Belous A. I. Yemelyanov V. A., & Turtsevich A. S. (2012). *Fundamentals of circuit engineering of microelectronic devices* (p. 472). Moscow: Tekhnosfera (in Russian).
2. Alekseenko A. G, & Shagurin I. I. (1990). *Micro-circuit engineering: Manual for graduate students* (2nd ed., reprint). Radio and Communication.
3. Belous A. I., Blinkov O. E., & Silin A.V. (1990). *Bipolar ICs for interfaces of automatic control systems*. Mashinostroenie. Leningrad division (in Russian).
4. Belous A. I., Ponomar V. N., & Silin A.V. (1998). *Circuit engineering of bipolar ICS for high-performance information processing systems* (p. 162). Polifact (in Russian).
5. Belous A. I., & Yarzhembitsky V. B. (2001). *Circuit engineering of digital ICs for information processing and transmission systems* (116 p). UE Technoprint. ISBN 985-464-064-7 (in Russian).

Chapter 5
Structure and Specific Features of Design Libraries for Submicron Microcircuits

This chapter explores the composition, fundamental design rules, and ways of using microcircuit design libraries which among designers are called design kits or PDK (process design kits). For the recent 10 years around the world, there have been being implemented approximately 10,000 IC design projects where every year hundreds and thousands of design engineers are involved. So numerous IC design teams have always used to face the issue of unification and standardization of approaches to creation of such libraries.

Recently in view of emerging autonomous (i.e., independent of IC designers) semiconductor factories engaged in custom IC large-scale production, i.e., IC Foundry Production (Integrated Circuits Foundry—ICF), and also due to obvious apparent need in combined simultaneous use of standard design tools and IP blocks purchased from other companies, design libraries (PDK) particularly turned out to be an interface between IC designers and manufacturers. This chapter briefly reviews standard design flow and standard PDK structures, minimum library composition and minimum list of standard cells, as well as the standard data files of design libraries are reviewed.

The chapter is concluded with the section dedicated to the detailed depiction of the specific training (learning) PDK by *Synopsys* company.

5.1 Development Process Flow and Standard Structure of a Process Design Kit (PDK)

The standard procedure for developing process design kit, PDK, currently includes the following key phases (Fig. 5.1) [1, 2]:

1. Selecting the process.
2. Obtaining the basic process data.
3. Identifying the basic elements for their incorporation into the PDK.

A. Belous, V. Saladukha, *The Art and Science of Microelectronic Circuit Design*,
https://doi.org/10.1007/978-3-030-89854-0_5

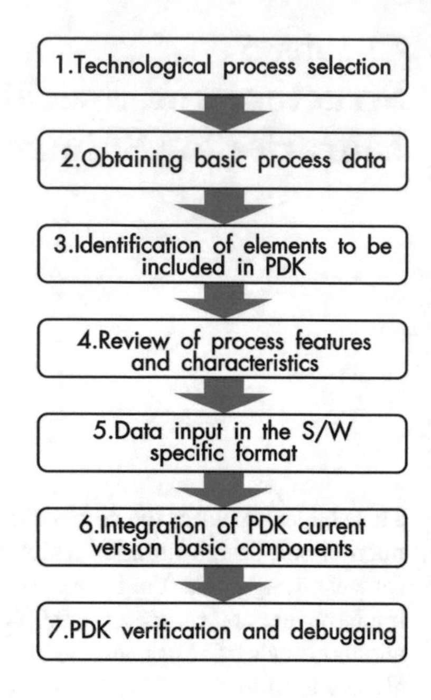

Fig. 5.1 Simplified diagram of the PDK development process

4. Reviewing the process features and characteristics for more accurate identification of the selected elements.
5. Obtaining and incorporating the data into PDK in a format specified by the software selected as a tool for technology/device/circuit/system design.
6. Integrating the basic components of PDK current version.
7. PDK verification and debugging.

Of course, if you create a PDK based only on these standard unified rules, the information obtained only by following the steps shown in Fig. 5.1 will not be enough. Figure 5.2 shows a slightly expanded flowchart of the design library development process, taking into account the accepted rules and standards. It should be noted that the flowchart areas marked with dotted lines are the steps that allow us to standardize the PDK being made.

Thus, section "A" contains procedures and standard rules for describing individual elements to be subsequently integrated in the PDK. Using standard descriptions of their contents, values, and unified translation mechanisms, the designer has the ability to create any other identical components of the design library.

Section "B" contains definitions required for a standardized description of PDK components in terms of ensuring their quality.

If it is necessary to ensure, upon customer's request, that PDK components fully comply with the selected standards, the phases presented in sections "A" and "B" must be performed without fail.

Another well-known methodological approach to the process of creating a design library, based on the design phases, is shown in Fig. 5.3.

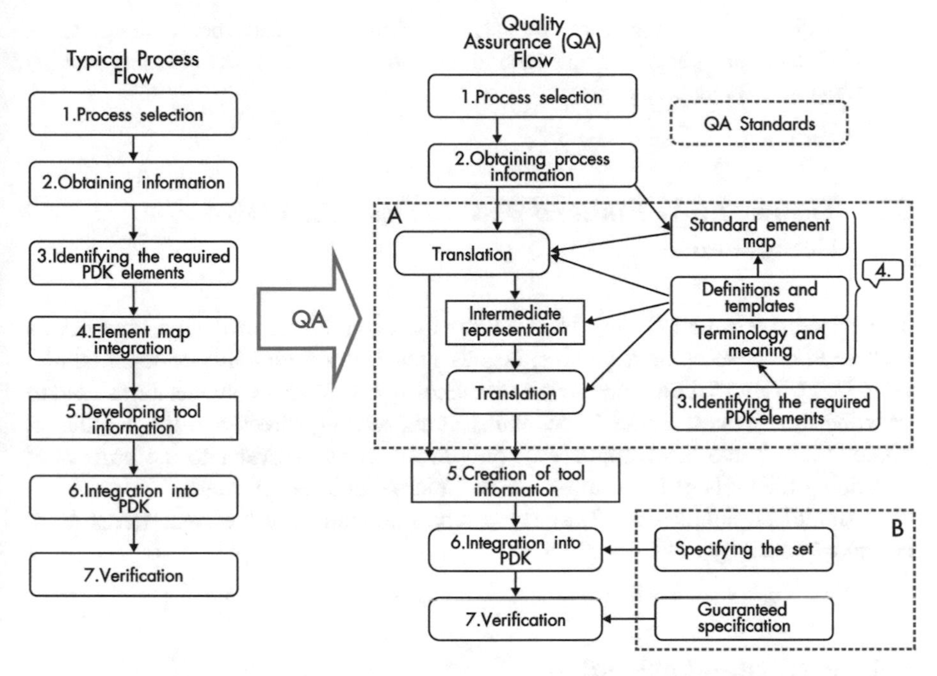

Fig. 5.2 PDK development process flowcharts based on generally accepted rules and standards

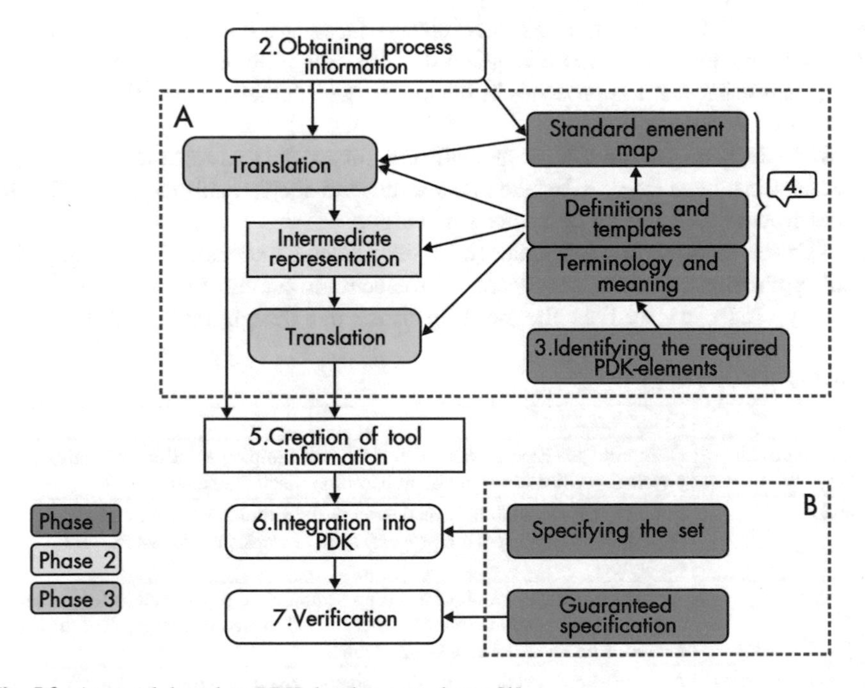

Fig. 5.3 Approach based on PDK development phases [1]

The PDK components to be standardized can also be defined in a multidimensional space. Within each dimension, the PDK components can be described in discrete categories.

5.2 Terms and Definitions Used to Describe PDK Components

The standard terminology used to describe the components of design libraries, as well as PDK development process, is presented below in an easy-to-learn tabular form [2, 3]. Table 5.1 contains general terminology; Table 5.2 contains terms used to describe the devices; Table 5.3 contains terms relating directly to the simulation process; and Table 5.4 comprises of only those terms that refer to the process of describing the tools of the computer-aided microcircuit design stage.

It should be emphasized That These tables contain only the basic terms Most commonly used by PDK designers

5.3 PDK Standardization

Numerous IC design companies have always faced the problem of unifying and standardizing approaches to the creation of PDK design libraries, as well as coordinating and using these approaches in interaction with IC fabrication plants different in terms of technology.

When designing modern ICs, especially that for special and space applications, an important feature is the need to take into account and deeply analyze the peculiarities of the process used during computer simulation.

With the emergence of separate (designer-independent) enterprises engaged in mass production of customer-ordered microelectronic products (Integrated Circuits Foundry, ICF), as well as the need to share the design tools and IP blocks

Table 5.1 General terms

Term	Definition/meaning
Custom circuit design	Designing electrical characteristics, physical implementation, simulation, and physical verification of a circuit based on discrete elements
Device	Main element available for inclusion in the simulated circuit and implemented for a specific process, such as transistors, resistors, and capacitors
Structure	A unit formed as a result of a certain sequence of process steps (diffusion, implantation, metallization, etc.) that is used to create a device. The structure may also be regarded as a single device
Supported device list	A set of supported devices, which characteristics are described for the selected process

Table 5.2 Terms used to describe the devices

Term	Definition/meaning
Intentional device	Devices used for preparation of a device electrical circuit, which are displayed as its contents on the screen of a personal computer
Extracted device	Devices that are not displayed on the screen of a PC, but are included in the electrical circuit (netlist) during its simulation. These devices are typically included in the circuitry when analyzing the results of physical implementation of the designed device
Main device element	Element of the device that performs the main functional purpose or the most significant behavioral role
Parasitic device element	Element of the device resulting from physical implementation data analysis of a device designed on the main components basis
Principal device(s)	One or two devices, on which the architecture is built and whose characteristics are fundamental for optimizing the process parameters, such as n-MOSFET and p-MOSFETs in digital CMOS circuits. It is important to note that the performance of principal devices is affected by spread-specific requirements of the specifications
Primary device(s)	Elements that are included in the supported device list of the process, which are formed on the basis of principal devices, supplemented by some additional structures. Usually, one or more specific well-controlled process steps are included. An example is the formation of a polysilicon resistor based on a polysilicon gate with a specially implanted and masked silicide region
Secondary device	A device from the list of supported devices, which is formed from a structure created during manufacture of principal and/or primary devices. Often such devices are called "free" or "intentionally parasitic," e.g., the use of metal interconnections to form inductors or capacitors. Secondary devices normally have a greater tolerance to the parameter scattering described in the specification
Device class (category)	Basic intended function of a device, such as MOSFET, bipolar transistor, resistor, capacitor, etc.
Device type (vertical)	A unique combination of vertical structures that make up a device of a certain class (category), e.g., poly1 resistor and poly2 resistor
Device style (lateral)	Changes in the horizontal dimensions and shape of a certain type of device designed to control its characteristics, for example, changes in the horizontal (lateral) position of the emitter, collector, and base layers of a bipolar transistor in order to obtain the optimal configuration of current-voltage characteristics
Device size	A special case of the device style, when its functional characteristics are considered as scalable, for example, changing the gate width of a MOSFET

(Intellectual Properties) from various companies, the design libraries have become a major link between the manufacturer and IC design teams [2].

Unfortunately, to date, the unification of requirements and standards has hardly affected the area of PDK creation, where it is necessary to standardize such requirements as nomenclature, models used, interfaces (connection/integration rules), quality characteristics (Q-factor), and, ultimately, the approaches to PDK presentation to end users. The solution to this problem will help to get rid of the confusion and waste of time and money for the adaptation of existing PDKs, which unfortunately occur

Table 5.3 Terms relating to the devices and circuit simulation process

Term	Definition/meaning
Model	Presentation (description as a function) of functional parameters and operating characteristics of a device used in simulation process; models are formed on the basis of behavioral characteristics and corresponding parameters
Model general behavior	Part of model that describes the main functionality for the selected device class and type. It can be implemented on the basis of compact model equations, behavioral models, and/or subcircuits
Model process parameters	Part of model that describes the functionality of a device, based on the relationship between performance and process-specific parameters
Instance parameters (allowed design variables)	Properties related to the instance of a device, which are transferred to a simulator software and a model via the netlist and which directly describe the device parameters, such as length and width of MOSFET channel
DRC (design rule check)	Verification of compliance of a developed topological layout with the rules laid down in the specification
LVS (layout (netlist) versus schematic)	Validation and verification of compliance of the designed circuit layout against the results obtained during the schematic modeling
LPE (layout (netlist) parasitic extraction)	Extraction of a netlist, as well as of capacitance and resistance nominal values in order to check them for compliance with the requirements set out in the terms of reference (ToR)
RCX (RC extraction)	Extraction (search) of new nodes and parasitic elements based on the layout analysis of the designed IC

Table 5.4 Terms used to describe computer-aided design (EDA) tools

Term	Definition/meaning
Primitive	Minimum element used to build an electrical circuit in the environment of EDA software systems for designing IC that is used to create a graphic image of the circuit
Instance	Device placed in the position selected in accordance with the circuit design
Property	Information that is associated with a particular primitive or instance, and is intended to control their characteristics
Instance property (direct, indirect)	User-managed (directly or indirectly) parameters associated with the instance of a device, set by the IC designer. Indirect properties can be calculated or derived from the direct properties

very often in today's electronic industry. Manufacturers will get an opportunity, in a shorter period of time, to provide the developers with the design libraries that are adapted as much as possible to the software packages currently used. As a result, the cost of cell libraries and IP blocks, supplied by companies specializing in this area, will be reduced and their quality will be improved. For companies that develop software packages for electronic design automation, EDA, the process of creating tools for description of components and methods of their integration will be simplified, thus eliminating the need to create "low-level" adjustment tools of libraries.

The greatest advantage will be gained by IC design engineers due to simplification (unification) of the IC design process, the possibility of fast and "transparent" transition to the application of new processes, as well as the reuse of IP blocks [2, 4].

The process of IC design has long had a "bad" reputation as "inefficient," primarily due to the large "bottom-up" dependence of data representation formats in various software packages.

The design data traditionally depend on the tool used because the software developers are actively trying to introduce exclusively their own products and storage formats and make them a "de facto" standard (especially this is typical for domestic design centers), with absolutely no regard for the wishes and concerns of independent developers and designers.

The evolution of IC design technologies has led to the fact that fabless companies (design centers), when purchasing EDA packages, have to acquire and install special data sets on available elements (individual devices or circuits) that are specific to a particular IC manufacturing process. After some time, these data have been called process design kits (IC design libraries) in order to be able to distinguish them from the tools used to describe the characteristics of the process. Initially, these libraries were used by so-called integrated device manufacturers, IDMs, in order to provide the possibility of IC "mixed" design [4].

In recent years, due to clear division of electronic industry companies into manufacturers (foundries) and designers (design centers) of ICs, the development and debugging of PDK is carried out almost independently. A similar situation is present among the developers of EDA and IP blocks. These libraries differ in a fairly wide range of full and abbreviated names (TDK, PDK, etc.); the traditional full name, however, is process design kit (abbreviated as PDK). Nevertheless, due to lack of unified standards, they all differ, albeit slightly, in composition and approaches to describing their contents.

Below we will consider the most well-established description of the standard PDK general structure and the process of its creation and generation of proposals for unification of contents and structure for the design of ICs with design rules of the order of 130 nm.

The advantage of PDK standardization is the obvious fact that the adoption by major companies of a set of standardized requirements to the contents and the rules for PDK interfaces description will reduce the costs of their development and increase their ability to combine basic functionality and performance in new products, and most importantly, will improve their predictability and success in the market. IC manufacturers will be able to offer developers the design libraries, describing the latest technological processes, which are more "flexible" in terms of their modification capabilities and more "filled" in terms of their contents. Library suppliers will have to lower their prices and to simplify licensing terms.

In turn, the companies involved in IC design will gain the following three advantages by increasing the level of PDK standardization [4].

Firstly, a complete, consistent, and logical description of the process makes it possible to eliminate "minor" errors that occur when implementing a product "in silicon." Standardization implies a fuller familiarity with the design process, as well as strict control and versioning of design results.

Secondly, engineers of the departments responsible for the support and use of IC CAD tools will face fewer problems and will have more "flexibility" at their disposal during adaptation of existing software to the requirements and capabilities provided by new design libraries (PDKs), and most importantly, the time frame for implementing advanced technological solutions in the design process will be significantly reduced.

And thirdly, heads of engineering departments will be able to make more cost-effective decisions in the process of planning works on a new product, because they will be better protected against unexpected surprises associated with the use of new processes by having a better understanding of the strengths and weaknesses of implemented solutions.

IC manufacturers (foundries, e.g., X-FAB, TSMC, "INTEGRAL" JSC, "Micron" JSC) and integrated solutions designers (IDM, e.g., Intel), will certainly also benefit from the increased level of PDK standardization.

There are other obvious advantages of this approach. The first advantage is the transparency and clarity of technical interaction, based on the use of standard terminology and data presentation, will improve the effectiveness of cooperation between the teams of designers on both sides and will reduce the time for execution and transfer of new designs from designers to manufacturers.

The second advantage is the use of a standard representation and description of processes significantly reduces the cost of creating and maintaining a wide range of specialized software tools for technology/device/circuit/system design (electronic design automation, EDA) by supplier companies and designers. This is achieved by simplifying the representation process itself through the use of standardized data structures, parameter sets, and their assumed default values—adopting a common *baseline* level for data representation will simplify access to the required data, ultimately meeting the requirements of both the designer and the customer.

The third advantage of standardizing the PDK representation methods is simplification of the version control procedure of the developed products due to unification and documentation of the corresponding CAD modules from the initial version of the process to its modified and upgraded versions, which will have a positive impact on the "life cycle" of the designed solutions.

And the last and fourth advantage, which major IC manufacturers and designers will get, is the ability to directly exchange information about the opportunities and difficulties associated with the implementation of new products based on advanced technological processes. This will reduce financial costs both for the support of the designer by the manufacturer and for the debugging of the process in case it reveals some faults and inaccuracies.

As for the developers of design libraries, the standardization of PDK requirements ultimately helps to minimize the number of differences in the description of components for different technological processes and manufacturers, as well as to reduce the time for creating libraries by quickly understanding the features and differences of the processes [2].

General requirements for PDKs will also greatly simplify the process of developing and testing the library components during their migration from one PDK to another.

On the other hand, for IC design software developers, the standardized PDK requirements provide at least three advantages.

Firstly, standardization will speed up the process of adoption of new software products and their distribution while reducing costs of adapting software to new technological processes. In addition, it will offer excellent opportunities to make significant changes and innovations in mixed and analog IC design tools.

Secondly, there will be less need for in-depth and comprehensive testing of design tools when the process is changed, even for the cases where there is insufficient information about it.

And thirdly, the EDA companies will be able to increase the number of solutions offered, in terms of completeness of the supplied PDKs and the contents of design tools, which will ultimately increase the number of users of software products and create prerequisites for developers of design complexes adapted to the end manufacturer (customer owned tooling, COT).

Thus, the following tasks can be highlighted, which require an urgent solution in order to implement a standardized approach to be used when creating new PDKs and modifying existing ones by all the interested market participants:

1. Simplifying the process of PDK creation/generation/testing.
2. Contribution to PDK support/maintenance.
3. Inclusion of the modules and units from multiple sources.
4. Transferring the proven designs and libraries, including layout of the basic cells, during transition from one design rule to another.
5. Simplifying the replacement of software tools used in design process with products of similar functionality.
6. Introducing the novel software tools with new functionality in the development process.
7. Comparison of the advantages, disadvantages, and capabilities of different technological processes during their selection by the designer.

5.4 Mixed Analog/Digital Microcircuit Design Flow

The goal of PDK standardization is to achieve maximum compatibility between the CAD systems used by IC designers in microelectronics and data included in PDK, so that software modules from different vendors can be standardized.

A generalized design flow for mixed analog/digital circuits is shown below (Fig. 5.4). It describes the basic components (shaded blocks) to be included in the PDK. It is important to remember that the PDKs are not self-contained software, but

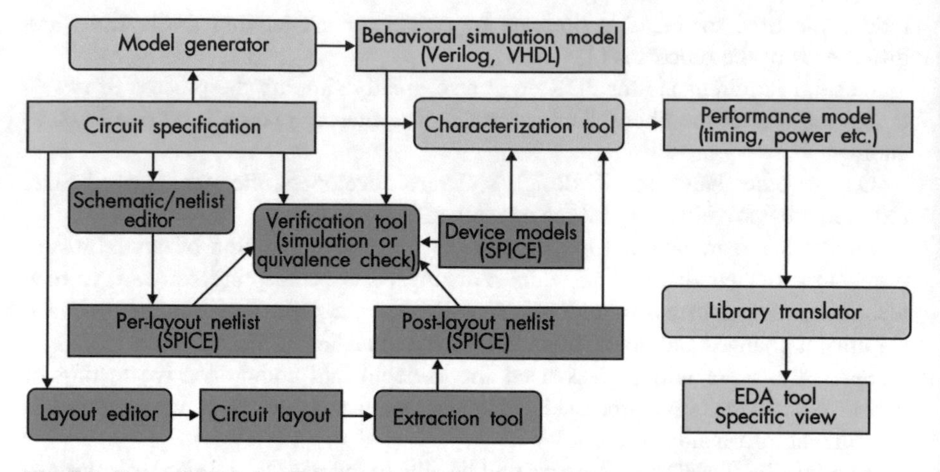

Fig. 5.4 Generalized design flow for mixed analog/digital ICs

are used only in connection with specific CAD software. The suggested design flow explains how the basic PDK components interact with applications that use them.

The design process usually starts with designer's general idea of the functional purpose and characteristics of the IC to be created. Until this moment, there are no formal design descriptions relevant to the design library being used. The IC designer then creates a schematic representation of the electrical circuit, a corresponding netlist (list of interconnections of its elements), and its layout, from which an additional netlist is extracted, in its turn. In addition, the designer develops a behavioral model (either manually or by specialized software modules used to generate such models). A verification tool or a toolkit is used to verify compliance with the requirements of the terms of reference and the correct implementation of a netlist, layout, and, when found appropriate, behavioral model describing the target characteristics of the product as a whole [1].

It should be noted that in practical applications, the same simulation software is quite often used to verify the netlist and to determine the characteristics or parameters (characterization) of the IC. The purpose of characterization is to provide an effective model and data describing all timing and electrical characteristics of a circuit at some abstract level. In addition, in some cases, it is possible to translate (convert) the results of characterization into data formats used in subsequent design steps.

Typically, the basic input data for the circuit simulation software tools are.

physico-mathematical and instrument-technological models of semiconductor devices, in which content is the main object of standardization and unification for different CAD complexes in microelectronics.

In the block diagram shown in Fig. 5.4, some data are already described using the industry standards, such as the language of circuit simulation and description of electronic devices in modern EDA packages (SPICE) and high-level language intended for functional-logical design of digital circuits (Verilog, etc.). The choice

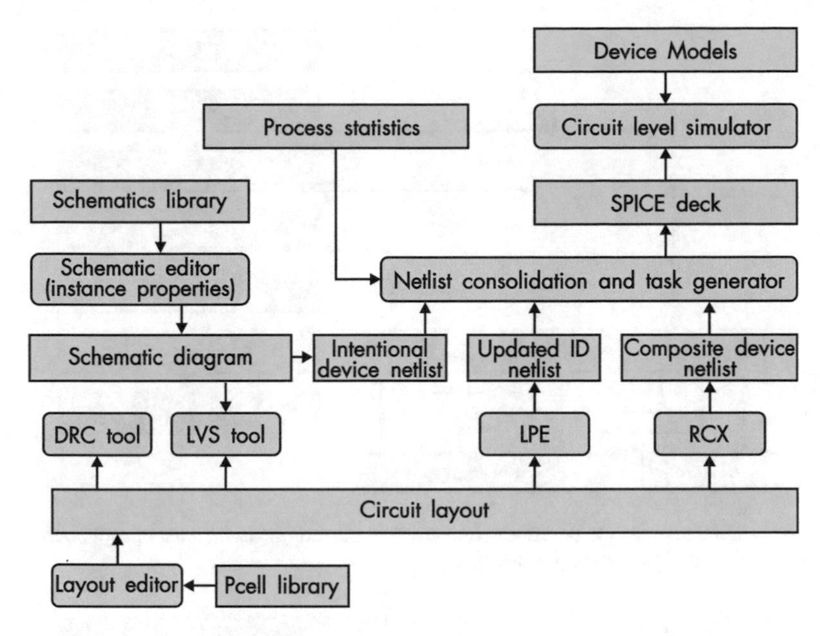

Fig. 5.5 Extended description of mixed analog/digital IC design flow

of identical data representation formats alone, however, does not guarantee compatibility and interoperability of the PDKs in multiple design packages, which is a major challenge.

Figure 5.5 provides more complete information describing the flow of mixed analog/digital IC design at the transistor level. In particular, the "shaded" blocks of Fig. 5.4 are presented here in more detail.

At this level of detailed consideration (Fig. 5.5), the functions, which do not comply with the selected standards, become more obvious and require an appropriate response of the designer.

5.5 Summarized Information Model of Mixed Analog-Digital IC Design

Figure 5.6 presents a general approach to the device description and list of supported devices in the context of a standard design process implementation [2].

Figure 5.7 shows general approach to describing designed IC.

Fig. 5.8 demonstrates a well-known interrelationship between the schematic, layout, and netlist of real and spurious elements in the context of the design process implementation.

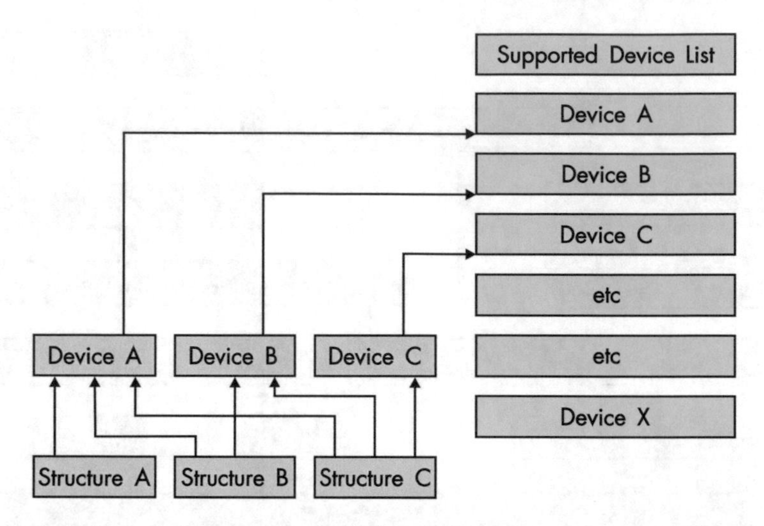

Fig. 5.6 General approach to describing the designed IC and preparing a list of supported devices in the context of design flow implementation

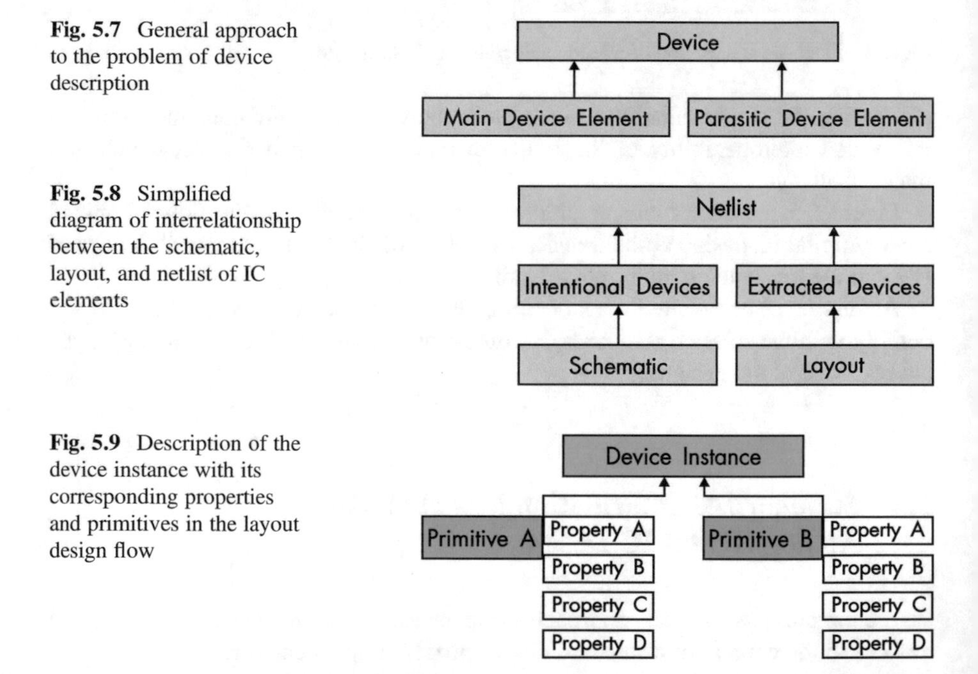

Fig. 5.7 General approach to the problem of device description

Fig. 5.8 Simplified diagram of interrelationship between the schematic, layout, and netlist of IC elements

Fig. 5.9 Description of the device instance with its corresponding properties and primitives in the layout design flow

Figure 5.9 presents an ideology of a general approach to the description of a particular instance of the designed device, as well as its corresponding properties and primitives required for IC layout design (primitives A and B, properties A–D).

Figure 5.10 shows a general approach to the abstraction or creation of a generalized semiconductor device model.

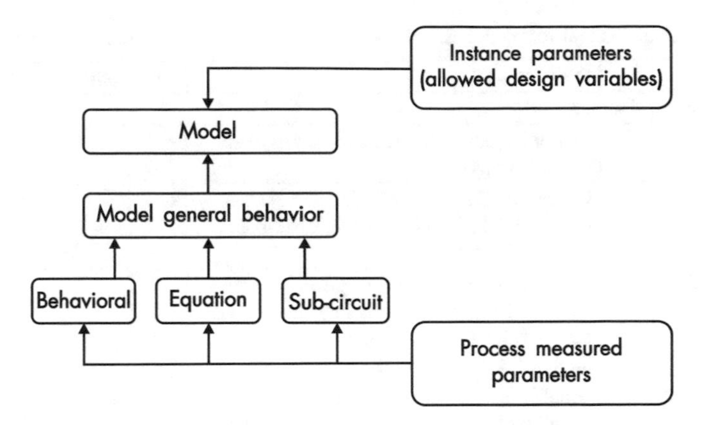

Fig. 5.10 General approach to abstraction or creation of semiconductor device model

5.6 Specifying Basic PDK Components and Standard Elements List

The design library (process design kit) is a representation of a certain process of IC manufacturing in the appropriate software format.

The list of elements presented in Table 5.5 forms the basis of any modern PDK. The designer should have full access to all the components of such a design library, from descriptions of process-related parameters to procedures for PDK implementation in the environment of a selected software package for IC design.

Table 5.6 lists the components of the process-specific design library (PDK) which parameters should be standardized.

5.7 Development Features of Digital Libraries for Designing ASICs with Submicron Design Rules

The components of any application-specific integrated circuit (ASIC) can be divided into three groups. The first group includes IP blocks. These elements are predesigned and are often complex blocks, which in most cases are purchased from third-party suppliers of IP blocks. Examples of such blocks are analog blocks (PLL, DAC), interface blocks (USB, I2C), processors (ARM, PowerPc), memory compilers (RAM, ROM), etc.

The second group consists of standard cells, which are still the basic "building blocks" in system-on-chip (SoC). They are used to serve as a linking logic between several IPs on the same IC and also to create final complex systems.

The last group of building blocks includes I/O elements that form the interface between the IC and the package it is enclosed in.

Table 5.5 Elements that form the basis of the PDK and their source of retrieval (search)

Element	Description	Fab	Software	Documents
Technological documentation				
Device specifications	List of supported devices, including their specifications	+		
Layout rules	Coding standards, layout design rules, peculiarities of device design, examples of designs (layouts)	+		+
Design guidelines	Guidelines describing the optimum procedure for designing ICs, taking into account the specifics of technological process implemented in the PDK	+		+
Device library symbols	Graphic (schematic) representation, including the list of properties of a device library cell		+	+
Examples of designs (parameterized cells)	Examples of "right" projects of layout design and automatic generation of parameterized cells (p-cells)	+		+
Placement calculations	Indirect calculations for dependent parameters (callbacks)	+		+
Technology file	Defining the types and placement order of topological layers, as well as rules for description of the layout	+		
SPICE models	Device models, subcircuits, and behavioral models required to simulate the characteristics of PDK devices	+	+	+
Physical verification				
Rule set for DRC	Verification of compliance of a designed layout with the rules set out in the terms of reference (ToR)	+		
Rule set for LVS	Validation and verification of compliance of the designed circuit layout against the results obtained during the schematic modeling	+		
Rule set for LPE	Extraction of a netlist, as well as of capacitance/resistance nominal values in order to check them for compliance with the requirements set out in the statement of works (SOW)	+		
Rule set for RCX	Extraction (search) of new nodes and parasitic elements based on the results of a designed IC layout analysis	+		

In the recent past, choice of the library was, in fact, the choice of the technology based on the required speed, circuit area, and cost (e.g., 0.35 μm or 0.25 μm).

The technology chosen usually had only one library of logic elements and possibly two libraries of I/O elements. The choice of I/O elements was made by the designer himself based on a compromise between I/O requirements and the

Table 5.6 Standard components of a basic PDK

Standard	Description
List of devices/process types	A set of settings covering the scope of standards to define the class (transistor, resistor, capacitor, etc.) and type (MOSFET, poly-resistor) of components that can be represented within the PDK standard. For example, for digital design, there is no need to include an LDMOS transistor in PDK
Symbol/schematic representation and acceptable dialects	A schematic symbol is used to represent the device graphically. It is typically a symbol (letter) indicating the device, contact configuration and relative positioning of pins for a particular class and type of device, as well as a simplified graphical representation and its acceptable variations
Device instance properties and their names	These are properties and parameters relating to the instance of the device included in the project and are the circuitry and layout properties. A simple example includes L and W for MOS devices. The standard can be extended due to inclusion of additional necessary calculations for the parameter values associated with the device location on a diagram (layout)
Layout viewing (technology file/layer names and numbers with their functions)	It should include a unified methodology, approved by multiple manufacturers (foundries), for coding the layers and structures that make up the device
Standard device encoding	See above
Presentation of the type of data used in simulation and implementation process, including subcircuits	The most complete representation of the device netlist, depending on the selected simulation software. It is also possible to include internal (additional) device parameters, including parasitic elements and representation of the circuit as instances of devices or a set of subcircuits
Organization and methods of design rules structure development	Involvement of as many fabrication plants as possible in the process of creating a standard and unified set of parameters, as well as requirements for the description of device models, layout design rules, etc.
Tools for automatic layout binding (nodes, boundaries)	Development of unified areas of device binding location used in the physical representation of the device, standardized for all the EDA tools
Basic cell layout (templates for p-cell and parameter set)	Standardization of a generalized set of parameterized cell (p-cells) layout for the main types and classes of devices
PDK directory and file naming convention	Setting the list and rules for specifying file names and the list of basic and additional directories used for PDK data storage
PDK quality control methodology	Suggesting the nomenclature and test methods used to determine the degree of compliance (quality) of tested PDK with the requirements of the standard for its development

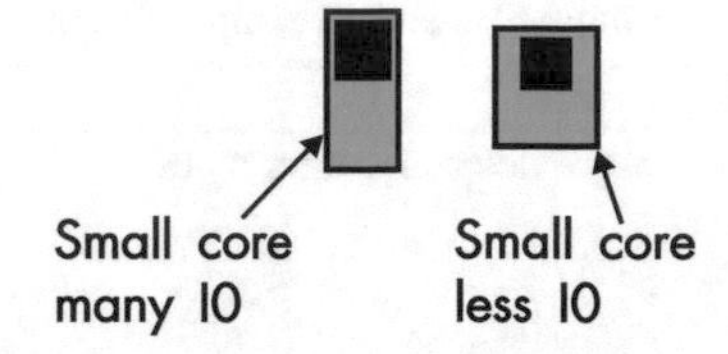

Fig. 5.11 Principle of I/O cell selection based on "core" dimensions

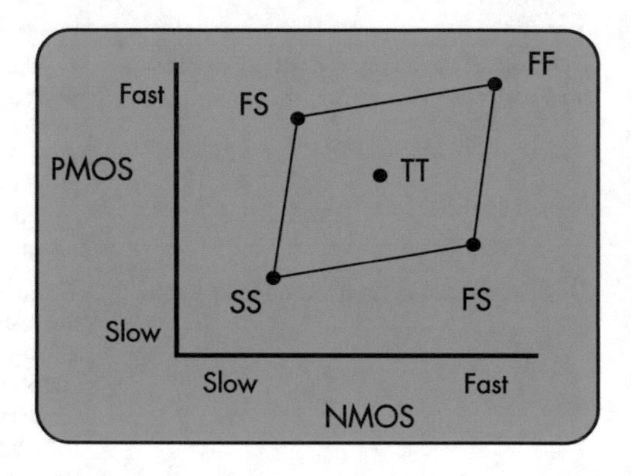

Fig. 5.12 Boundary conditions of standard libraries in "old" technologies

Table 5.7 Minimal parameter set of boundary conditions

Technological process	TT	SS	FF
Supply voltage, V	3.3	3.0	3.6
Temperature, °C	25	125	−40

circuit design solution, taking into account limitations of the logic elements: "small" core, many I/O elements, and "large" core, less I/O elements (Fig. 5.11).

Historically, standard cells have been characterized by an extremely limited number of processes, voltages, and temperatures. Only a few types of timing models were supplied: e.g., a worst-case model (low-speed SS, low voltage, high temperature), a best-case model (high-speed FF, high voltage, low temperature), and standard type, TT. Timing parameters in the worst-case model were used to check the setup time, and in the best-case model to check the hold time.

Figure 5.12 shows five most commonly used typical boundary conditions: TT (nmos-standard, pmos-standard), SS (nmos-slow, pmos-slow), FF (nmos-fast, pmos-fast), FS (nmos-fast, pmos-slow), and SF (nmos-slow, pmos-fast).

Table 5.7 shows basic characteristics of the considered processes.

Moving towards 90 nm technologies, and less, with application of improved supply voltage control techniques, the additional libraries, available to any designer, have been developed, which provided the opportunity to make the right choice depending on criteria such as speed, dynamic power, leakage currents, occupied area, and cost.

Moreover, process-specific libraries have emerged, tailored to particular process parameters (e.g., gate dielectric thickness T_{ox}, threshold voltage V_t, etc.)

Figure 5.13 shows exactly how the application area of the device actually determines the choice of technology. A number of libraries have become available

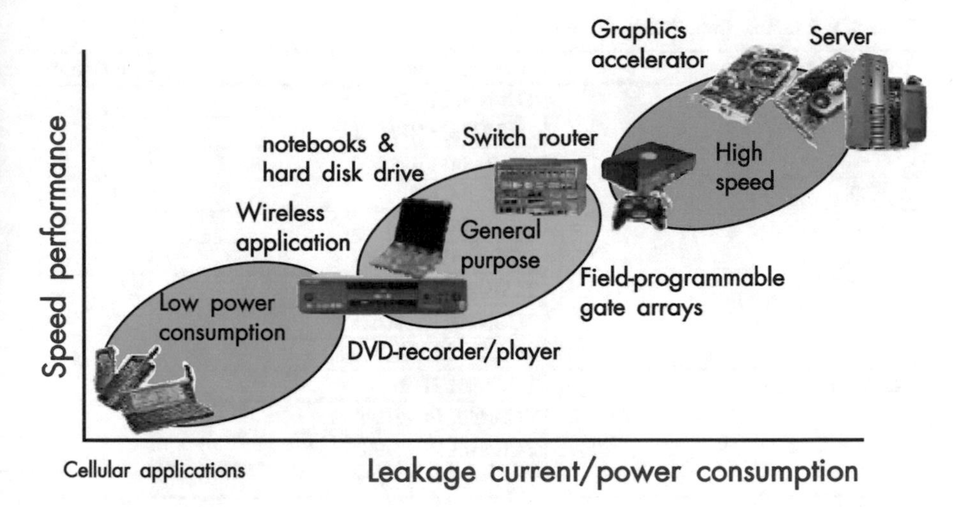

Fig. 5.13 Selection of specific library depending on the designed product application area

Table 5.8 Some criteria of cell library selection for p-channel MOSFETs

	Cell type	Units of measurement	Low power	General cells	High speed
VDD		B	1.2 I 0.84	1.0	1.2
Vt	HVt	B	0.6	0.45	0.4
	SVt	B	0.5	0.35	0.35
	LVt	M	0.4	0.30	0.35
Idsat	HVt	μA/μm	400	500	850
	SVt	μA/μm	500	650	950
	LVt	μA/μm	600	750	1000
Ioff	HVt	nA/μm	0.01	1	10
	SVt	nA/μm	0.2	10	40
	LVt	nA/μm	0.4	80	90

for battery-powered products, which require power-optimized libraries. At the other end of the spectrum, you can see a number of libraries optimized for speed, such as graphics accelerators, but such products consume maximum amount of power. In the middle of the spectrum, there are libraries representing a compromise between speed and power consumption. Each set includes cell libraries with different Vt values. This allows some components of the microcircuit to be optimized in terms of power consumption, and other components to be optimized in terms of speed, all within the same design. The simultaneous use of general-purpose and low-power elements is unacceptable, since in this case, other technological parameters and element levels will be different.

Table 5.8 shows that there is always a trade-off between speed and leakages within single library.

Table 5.9 Example from TSMC 90 nm libraries

Technology (process)	Name of library	Voltage, V
CLN90GT	TCBN90GTHP	1.2
	TCBN90GTHPHVT	
	TCBN90GTHPLVT	
CLN90G	TCBN90GHP	1.0
	TCBN90GHPHVT	
	TCBN90GHPLVT	
	TCBN90GHPOD	1.2
	TCBN90GHPODHVT	
	TCBN90GHPODLVT	
CLN90LP	TCBN90LPHP	
	TCBN90LPHPHVT	
	TCBN90LPHPLVT	
	TCBN90LPHPUHYT	

For example, the highest saturation current (identifying speed of IC) is typical for LVt cells in high-speed cell libraries. The lowest leakage currents can be obtained by selecting an HVt element from a low-power library. There is a certain "overlap" between these two: the speed of the SVt cell in the general library corresponds to the speed of the LVt cell from the low-power library. Furthermore, a certain overlap can be seen from the leakage data between cells from the general library and those from the high-speed library.

It should be noted that cell libraries for low-power consumption are usually characterized by lower voltage values in order to further reduce power consumption [1].

Table 5.9 also provides a list of some 90 nm cell libraries from TSMC, a well-known Taiwan Semiconductor Manufacturing Company, very commonly used by domestic fabless companies to place their foundry orders (see Sect. 5.6).

A number of important points should be noted here:

- General library is also characterized by voltage of 1.2 V, providing the possibility of increased performance,
- Another library with ultrahigh V_t was added to the low-power cell library in order to reduce leakage even further.

Let us briefly review the evolution of the number of boundary conditions required by the designer during transition to 90 nm technology.

Usually, additional boundary conditions, which are described for modern libraries, are presented not only due to a variety of voltages in the design project, but also with due consideration of elements operating at lower voltages.

Thus, it can be seen from the UMC company product catalog extract with 90 nm design rules [] (Table 5.10) that there are different parameters available within the same library. Low-leakage cells have been added to it.

Table 5.10 Minimum set of boundary conditions for 9 nm UMC libraries

Boundary Conditions Library Name	Process (pMOS—pMOS)	Temperature (^{0}C)	Supply voltage (V)	Notes
TTNT1p20v	Typical—Typical	25	1.2	Standard "corner"
SSHT1p08v	Slow—Slow SS	125	1.08	Slow "corner"
FFLT1p32v	Fast—Fast FF	−40	1.32	Fast "corner"
FFHT1p32v	Fast—Fast FF	125	1.32	High-loss "corners"
SSLT1p32v	Slow—Slow SS	−40	1.32	Low-temperature "corners"
SSLT1p08v	Slow—Slow SS	−40	1.08	
Low voltage operating conditions: Same library for low voltages				
TTNT0p80v	Typical—Typical	25	0.80	Standard "corner"
SSHT0p70v	Slow—Slow SS	125	0.70	Slow "corner"
FFLT0p90v	Fast—Fast FF	−40	0.90	Fast "corner"
FFHT0p90v	Fast—Fast FF	125	0.90	High-loss "corners"
SSLT0p90v	Slow—Slow SS	−40	0.90	Low-temperature "corners"
SSLT0p70v	Slow—Slow SS	−40	0.70	

Since for the technologies with the reduced supply voltage requirements and values it is currently difficult to make a clear statement regarding which temperatures produce the slowest and which temperatures produce the fastest cells, the two "corners" (variants of boundary conditions) with ultralow temperatures have been added to the library.

Thus, as a result of microcircuit design, more and more libraries emerge with an increasing number of such basic elements, often referred to as "corners" in the slang of designers (Engl.: "core").

In the latest technologies, the designer finds out an increasing number of cells at his/her disposal, allowing the design tools to choose the cell with the most correct drive signal, which makes some sense in terms of power consumption and performance.

The technologies that have become widespread in recent years (e.g., those providing 65 and 45 nm design rules) offer access not only to basic logic cells and built-in memory but also to radio frequency modules, nonvolatile memory cells, etc.

Let's take a look at the elements that make up the basic digital libraries of the modern ICs. First of all, these are:

- Classic logic gates, like AND, OR, triggers, drivers with different power, etc.
- Special low-power logic elements:
  - Clock signal controls.
  - Multi-Vt elements.
  - Level shift circuits.

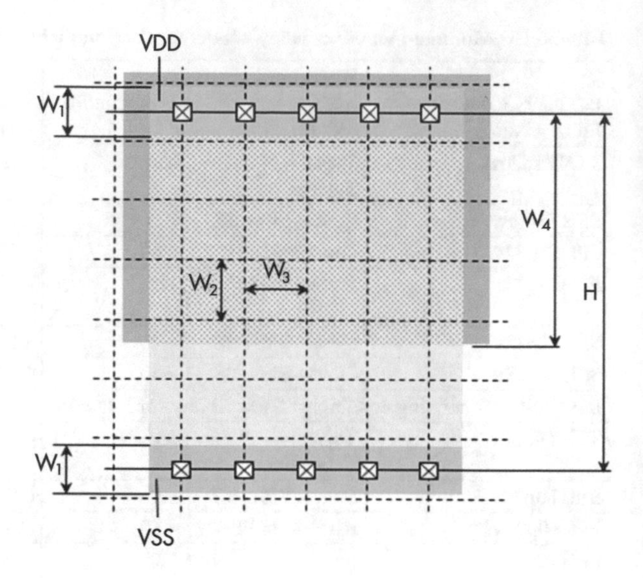

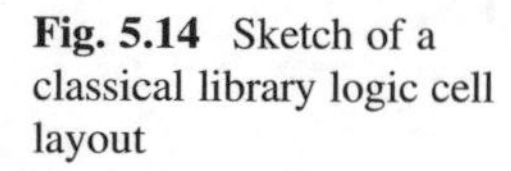
Fig. 5.14 Sketch of a classical library logic cell layout

- Isolators.
- Combined level shifters/isolators.
- Hold registers.
- Keys.
- Power controllers, etc.

First of all, it should be noted that modern libraries contain classic logic gates (AND, OR, flip-flops) necessary to implement the functionality of any designed product.

In addition, there are elements required to support low-power circuitry solutions:

- Clock signal controls are used to reduce the dynamic power in clock circuit.
- Cells functioning at multiple threshold voltages (multi-Vt) are used to meet the trade-offs between increased performance and reduced leakages.
- Level shifters are used in projects with multiple supply voltages.

The remaining elements are required to safely shut down separate power supply domains.

The last on the list presented here are elements with design and process variations (ECO cells), i.e., cells without rigidly defined functions that are sometimes added to a design project. This usually allows making circuitry solution cheaper, in case if functional errors are detected after the pilot samples have been manufactured.

As you know, the layout of a digital circuit starts with routing structure. The routing grid is specified according to the "metal-interlayer contact" principle. The width of power rails and the height of elements are determined based on performance requirements (Fig. 5.14).

The cell height is measured in "tracks" (levels), the formation of which is the first metallization layer (M1). An element of eight tracks is high enough to allow eight

Table 5.11 Typical parameters of library logic cell

Parameter	Symbol
Element height (number of tracks)	H
Power rail width	W_1
Vertical grid	W_2
Horizontal grid	W_3
N-well height	W_4

horizontal M1 conductors passing through it. Cell libraries are designed to a certain number of "tracks" in height, which affects timing and routing characteristics of the library to be taken into account in the design project.

Libraries with high tracks support more complex routing and transistors with higher buffer power and are normally tuned for high performance. However, they can have higher leakage values. A library of 11 or 12 tracks is already considered a high-track library.

Low-track libraries are optimized for space efficiency and are typically designed using transistors with less powerful buffers, so their use is less suited to "high-speed" designs.

Libraries with "standard" track heights are designed to provide a reasonable trade-off between space efficiency and speed. Such libraries are used in most designs. A library of nine or ten levels is considered a "standard track height library."

In turn, complex library cells may be of double or triple height (Table 5.11).

Clock gating controls (Fig. 5.14) are widely used to reduce the dynamic power in the clock circuit. Synthesis tools can automatically replace feedback multiplexer circuits with integrated clock gating (ICG) circuits. Typically, multi-Vt cells are used to give a choice between increasing performance and reducing leakage. However, it must be kept in mind that the use of multi-Vt cells requires additional masks during fabrication. This means that increasing the number of possible threshold voltages (i.e., using multi-Vt cells) leads to an increase in the final chip cost, which cannot be ignored. In practice, more than two values of Vt are very seldom used, since benefit in gain factor due to leakage and speed performance becomes less with increasing number of different threshold voltages [2] (Fig. 5.15).

Figure 5.16 shows the use of semiconductor areas of an IC chip with different supply voltages (multi-VDD). This technique is based on a deviation from the traditional method of applying a single, fixed power rail to all nodes of the device.

The key feature of this approach is the division of the in-chip logic into several areas with the same voltage or power rating. In this case, each of these areas has its own power supply. For example, the speed requirements of a processor may be as high as the semiconductor technology allows. This requires a relatively high supply voltage. On the other hand, a USB unit may operate at a fixed, rather low frequency which is determined by the protocol used, rather than the technology. In such a case, using a rail with lower supply voltage may be sufficient to ensure that USB unit meets the requirements to speed performance. Using a rail with a lower supply voltage means that its dynamic and static power will be lower.

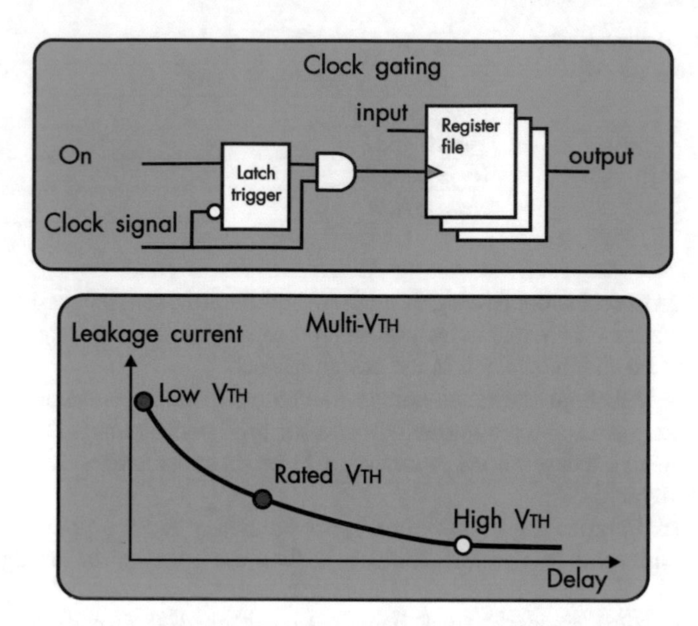

Fig. 5.15 Typical example of using IC clock controls

Fig. 5.16 Application of areas with different supply voltages

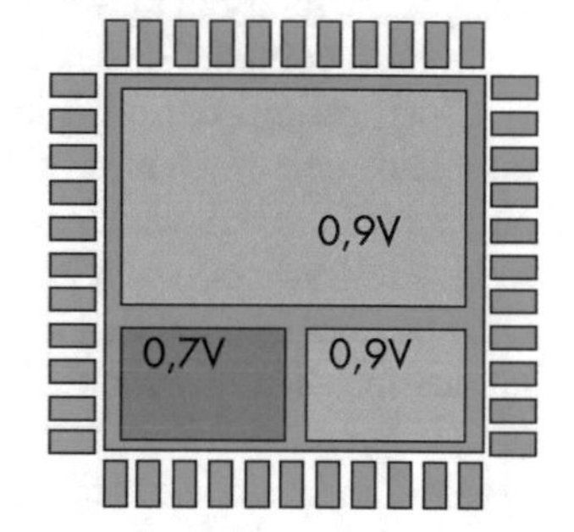

5.8 Structural and Circuit-Level Features of Designing Basic Cells for Submicron Microcircuits Library

Synthesis of even the simplest circuitry using two different voltage values presents certain difficulties to the designers, the key points being as follows [1]:

- Synthesizing voltage level shifter. In order to pass signals between blocks, in which rails have different power values, a built-in level shift circuit is often required, i.e., buffers transferring the signal from one device to another, with different levels of applied voltage.
- Statistical timing analysis (STA). With a single power supply for the entire chip, it is possible to analyze timing parameters in a single functional point. Libraries are characterized particularly for this point, and standard design tools perform

analysis in a normal mode. If some blocks work with different voltages with the use of libraries that cannot be characterized at the accurate values of the applied voltages, then, obviously, the analysis of timing characteristics becomes much more complicated.

- Designing the general layout of a chip and routing the power grid. Application of the methodology of multiple power areas requires a more detailed development of the chip general layout. In this case, the discretization power grid becomes more complex (Fig. 5.16).

5.8.1 Voltage Level Shifters

When the problem should be solved of signal passing between regions with different supply voltages, it is necessary to use voltage level shifters. It can be particularly difficult to pass a signal from 1 V to 5 V region, because there is a high probability that the threshold for 5 V region will not be reached with the difference in 1 V. However, in today's circuits, the internal voltage is firmly tied to 1 V value. Why are voltage level shifters required when transmitting signals from a 0.9 V domain to a 1.2 V domain? The main reason is that a 0.9 V signal supplied to the 1.2 V gate will simultaneously enable the n-channel and p-channel transistors, causing the unnecessary shunt currents.

The best solution to this problem is to have acceptable voltage ranges (as well as rise and fall times) supplied to each domain. This is usually done by setting up special level shifters between any domains using different voltages. This approach limits the problems of voltage swing and voltage domain timing delimitation and leaves the internal clocking of each individual domain unaffected.

At first glance, simply switching from an output buffer to a higher voltage rail will cause no problems. In doing so, there are no short circuits or breakdown problems, and no fast rise times compared to the top-high (header) or top-low (footer) switching levels of CMOS logic. However, timing closure requires "step-down" components specifically designed for this purpose.

HL shifters [1, 3] can be quite simple. Essentially, they are two inverters connected in series. A step-down level shifter introduces only a buffer delay, so its impact on clocking is negligible.

The key problem here is transmission of signals from cells connected to the low voltage power rail to the cell with high supply voltage. There are several known design methods where the direct method uses the buffered and inverted form of a lower voltage signal to drive a cross-coupled transistor structure operating at a higher voltage level.

The simplest HL shifter is shown in Fig. 5.17. It requires only one VDDH voltage.

The step-up level shift circuit (LH shifter) is shown in Fig. 5.18 and requires 2 voltage values and, most often, a two-level system for construction of basic cells.

Additional library cells required for low-power design:

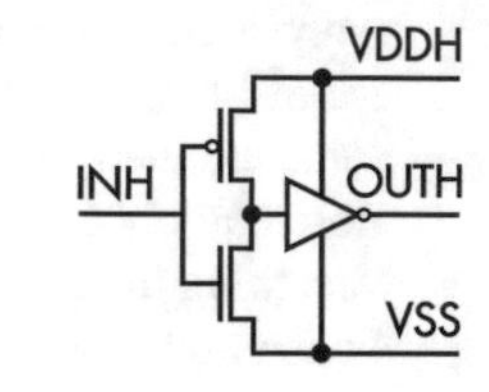

Fig. 5.17 Step-down level shifter (HL shifter)

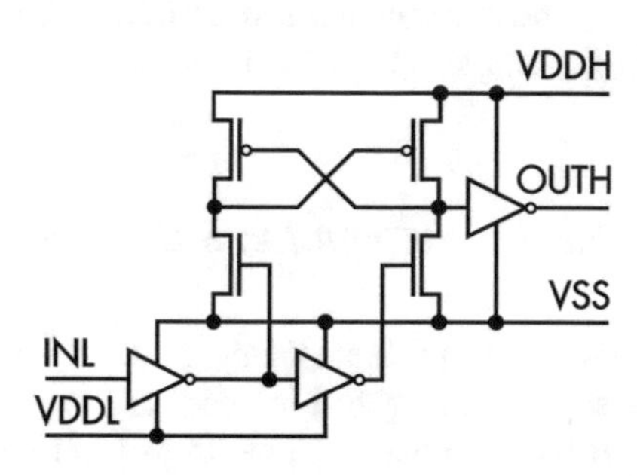

Fig. 5.18 Step-up level shifter (LH shifter)

- Isolation cells.
- Power control circuits.
- Retention flip-flops.
- Always-on buffers.
- Special pad elements.

5.8.2 *Power Gating Circuits*

To reduce the level of on-chip overall leakage current, it is highly advisable to introduce mechanisms for powering down the unused blocks. This technique is known as power gating. Its function is to provide two power modes: power-down and active mode. The purpose is to switch between these modes in appropriate time and manner to maximize energy savings with minimal impact on performance.

The challenge in designing such power gating elements is that the voltage at the outputs of a power-gated block can fall very slowly. The result of this "slow" process can be that the voltage at these outputs will be threshold (unstable) for a significant amount of time, causing large shunt currents in the always-on block. To prevent these shunt currents from occurring, special isolators are placed between the outputs of the power-gated block and the inputs of all the circuits in the block. These isolators must be designed so that they can "quench" the shunt currents when the voltage exceeds its threshold value on one of the inputs, while the control input is powered down. The power gating controller provides this isolating control signal.

For some of these power-gated blocks, it is highly desirable to save the internal state of the block during power-down and to restore this state during power-up. This method can save a considerable amount of time and power during power-up. One way to implement the retention method is to use saving ("lookaside") registers instead of conventional triggers. Saving registers normally have an auxiliary or shadow register as well, which is slower than the main register, but has significantly lower leakage currents. This shadow register is always on and stores the contents of the main register during power gating. Saving registers have to be signaled when to store the current contents of the main register to the shadow register and when to return the data back to the main register. This process is driven by the power-gated controller.

The always-on buffers in the switchable units are used to route the signal from the active block through the powered-down unit to another active block. Special filler cells are used to connect the n-wells and p-wells with the global power and ground rails.

5.8.3 Isolation Library Cells for Submicron Microcircuits

These elements are used to isolate switchable power pads with identical voltage levels.

Each zone interface with power gating requires its own control. It must be ensured that powering-down the zone will not cause shunt currents to flow on any of the inputs of the powered-up blocks. Also, none of the passive outputs of the powered-down blocks must cause spurious behavior of the enabled units.

Outputs of the power gating blocks are a major problem as they can cause electrical or functional problems in other units. Inputs of the power gating blocks are usually not problematic; they can be controlled by the active logic levels supplied by the enabled units without causing electrical problems in the disabled unit.

A simple solution to control the outputs of powered-down blocks is to use isolators to lock the output to a specific, acceptable voltage value. There are three basic types of isolators: latching the signal either to "0," or to "1," or to some required new value. In most cases, locking the output to an inactive state is sufficient. When a high logical level is used, the main problem is to latch the value to "0." The "AND" function on the output solves this problem. When a low logical level is active, the OR function sets the output to a logic "1." The latching elements are designed to eliminate shunt currents and leakages at the passive input of a signal, while the control input is in the corresponding ("isolating") state. In addition, modern models for synthesis usually have additional attributes to ensure that these elements may never remain unoptimized, buffered incorrectly, or inverted as part of logical optimization.

Figure 5.18 shows a low-level isolation option by the "AND" method. When the active gating signal of low isolation level "ISOLN" is in the logic "1" state, the transmitted signal passes to the output; otherwise (the signal is in the logic "0" state) the output is fixed low.

A variant of the OR type high-level logic latching (locking) circuit is also shown in Fig. 5.19. When the isolation control signal "ISOL," active at high logic level, is high, the output is latched at a high level, and when it is low, the signal passes

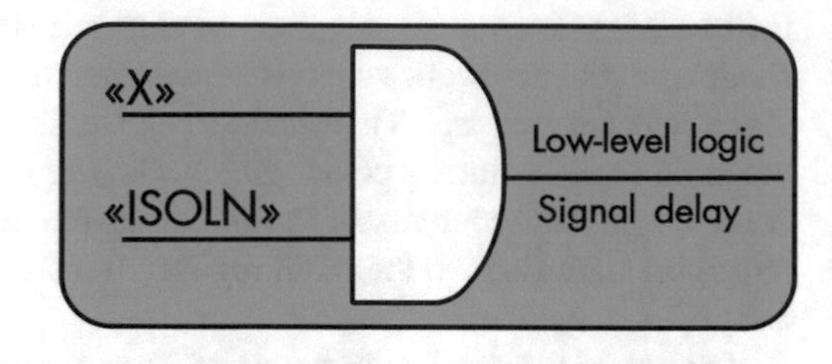

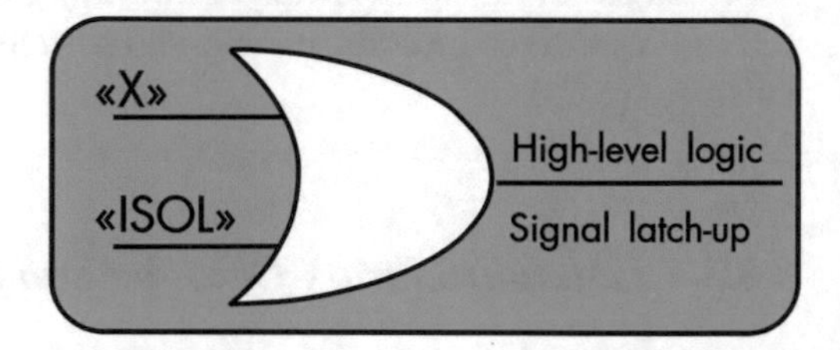

Fig. 5.19 Using isolation elements to control outputs of powered-down blocks

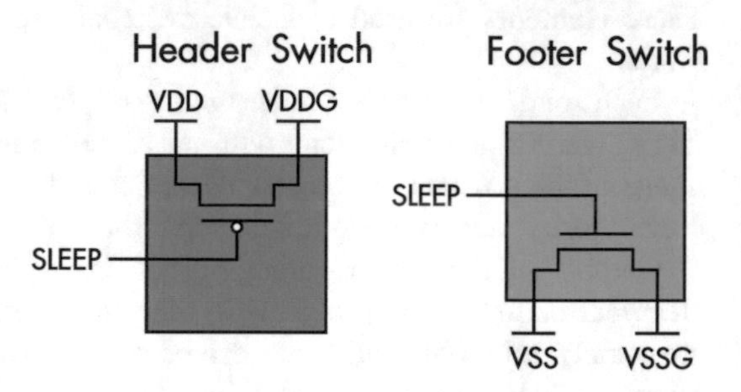

Fig. 5.20 Header switch and footer switch

through the output. These logic locking circuits add a delay to the signals they isolate. For some critical channels, as cache memory interfaces, such added delay might be unacceptable.

Power gating library cells ("sleep transistors") are used as switches to power-down the components of the circuit that are in standby mode. A "sleep" transistor is a p-channel or n-channel transistor with a high Vth (threshold voltage) that connects a constant power supply to a circuit power source, commonly referred to as a "virtual power supply." The p-channel sleep transistor is used to switch the VDD power supply and is referred to as a top-level switch (header switch). The n-channel sleep transistor controls the VSS ground rail and is called a bottom-level switch (footer switch) (see Figs. 5.19 and 5.20) [2].

An example of a "footer" switch layout (height as for standard cells or double height) is shown in Fig. 5.21.

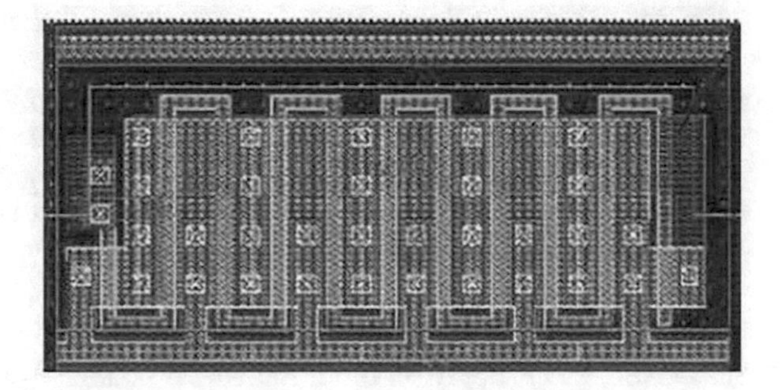

Fig. 5.21 Footer switch layout

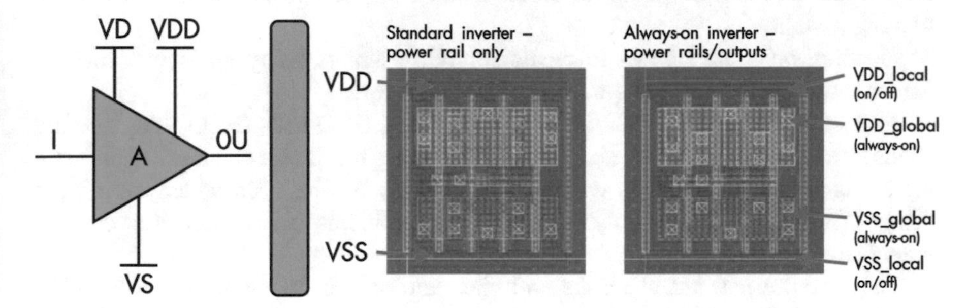

Fig. 5.22 "Always-on" buffer implementation

5.8.4 "Always-on" Buffers

In some cases, there is a need to buffer some signals in powered-down areas of the circuit. "Always-on" buffers are used for this purpose. In such continuously on elements, the switched VDD and/or VSS rails can have a variable value [2].

The always-on VDD/VSS rails in such elements can be represented as additional inputs. These inputs are connected to a nonswitchable supply/ground during the digital circuit routing.

Signal buffering in powered-down blocks is used for:

- Signals passing between active blocks, which require buffering in the powered-down block.
- Power gating signals.

The always-on VDD or VSS outputs are usually characterized by the following differences:

- Not connected directly to the power rails.
- Connected to a nonswitchable supply/ground during routing.

Figure 5.22 shows the layout of such a conventional inverter (buffer) and an always-on buffer. As can be seen, there is no connection to VDD_local/VSS_local in the always-on inverter.

In special filler cells, compared to standard filler elements, the n-well and/or p-well pins are not connected to the VDD/VSS power rails. The voltage on these well pins determines when the cells are in forward or reverse-bias mode. This offset voltage is usually routed as a signal pin or as a special power grid.

Special filler cells play an important role in power gating (Fig. 5.22). If a header filler is used, the VNW output connects to the global VDD rail; if a footer switch is used, the VPW output connects to the global VSS rail. This circuitry keeps the wells "powered up" when the region is powered down. When a header switch is used as a "sleep" transistor in the power gating system, the VNW output is connected to the always-on power supply, in order to avoid a floating n-well. Conversely, if a footer cell is used, the VPW output is connected to the always-on ground in order to avoid a floating p-well.

Floating wells are known to create for IC designers many problems, such as parasitic transistors, leakage currents, or even latches.

As a component of PDK libraries, the so-called ECO cells (ECO—engineering change order) are often used, i.e., elements allowing for design changes. Their main feature is that these are cells without functionality having been added during the design (fillers), and are only used if any problems arise after the chip is manufactured.

They require new metal masks and vias, and only in this case will these cells acquire the desired functionality.

ECO cells (or sets of these cells) can implement more complex functions such as AND, AND-NOT, OR-NOT, XOR, trigger, multiplexer, and inverter; the principle of their use is identical to that of spare cells.

It is known from the authors' experience that ECO cells can fulfill another function related to specific activities of intelligence agencies. They are sometimes used to perform "trojan" functions, when it is required to supply a chip with a "backdoor" to a customer. But that is a topic for other special research outside the scope of this book.

In order to be able to "cheaply" modify the design project by changing the metal layer alone, dummy cells are added to free space on the chip. If a functional problem is detected immediately after the chip is made, these dummy elements are converted into functional cells by changing the metal layers. Of course, the performance of these elements will be lower than that of conventional cells.

In addition, during the layout design process, other dummy cells are sometimes added before the first manufacture. These elements are called spare. They are added at the routing stage so that designers can make changes to the project at a later stage. This provides an opportunity to retain the previous "place-and-route" more accurately by using spare cells, which can significantly reduce the time it takes to send the design to production.

This same class of library cells also includes I/O libraries.

Most low-power I/O devices can be incorporated into standard I/O libraries (Fig. 5.23). Standard I/O devices are the devices that don't have special requirements to the package, connections, or form of signals as special I/O devices do. Standard

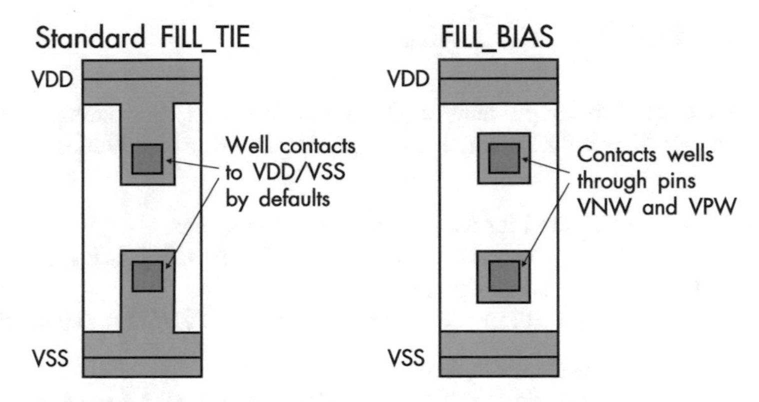

Fig. 5.23 Implementation of special tie elements

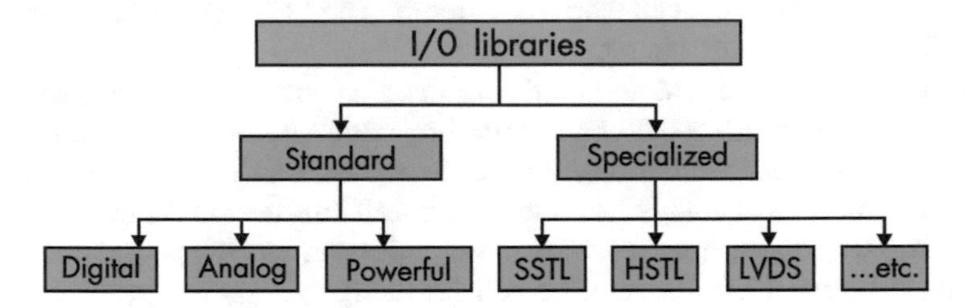

Fig. 5.24 Classification of I/O libraries

I/O devices fall into three main groups: digital, analog, and high-power. These I/O devices can come in many different types, as shown in Fig. 5.23.

The main features of these I/O devices are:

- The ability to drive heavy loads—the contact pads must be able to drive a few pF as opposed to fF loads inside the IC.
- Interface designed to operate at a different voltage due to the use of different supply voltages, on-board and in-chip.
- Low switching noise—due to the inductance of the package and the tracks on the circuit board which can cause excessive voltage deviation from the nominal value at a certain current.
- Electrostatic discharge protection—a person or machine carrying out the assembly can build up a charge of up to 2 kV and 500 V, respectively. Such voltages will damage the IC without proper ESD protection (Fig. 5.24).

5.9 Standard PDK Data Files

The standard cell library contains a set of multiple files. For example, TSMC low-power cell library for 90 nm technology contains over 50,000 files. The main ones are as follows:

- Physical data (LEF) used for the place-and-route purposes.
- Information on timing, power consumption, and functionality: LIB files, used by synthesis tools and layout design tools.
- Register-transfer level (RTL) cell descriptions: Verilog/VHDL for simulation.

Let's consider a specific example of a LEF library file.

The Library Exchange Format (LEF) includes basically all information about a standard cell, as of a "black box," including topological layers, interlayer contacts, placement, node type, and cell macro definitions. The LEF file is actually an ASCII code of the library representation.

It is possible to retrieve almost all the information about the library from a single LEF file. However, this would be a large file, complex, and cumbersome to use. Instead, you can split the information into two files: a technology LEF file and a library LEF file. A technology LEF file contains all information about LEF technology for the design, such as rules for place-and-route, as well as process data on topological layers: LEF file contains cell geometry, pin geometry, gaps, antenna effect data, etc.

A fragment of such a file looks in the following way:

```
CLASS BLOCK ;
FOREIGN single_port_bbb ;
ORIGIN 0 0 ;
SIZE 774 BY 547 ;
SYMMET MACRO single_port_bbb
RY X Y R90 ;
PIN OUT
DIRECTION INPUT ;
USE SIGNAL ;
PORT
LAYER M3 ;
RECT 420.180 625.650 420.960 625.810 ;
END
ANTENNAPARTIALMETALAREA 1.929 LAYER M1 ;
ANTENNAGATEAREA 0.377 LAYER M1 ;
END OUT
OBS
LAYER M1 ;
RECT 0.000 0.000 774.000 547.000 ;
END
END single_port_bbb
```

LIB files are the second group of files where synthesis and layout design tools are used. Let's take a closer look at a specific example of a library file with .lib extension.

This format is designed for simulation, synthesis, and testing. It is generated during description of the library's parameters and contains data on all timing parameters and power consumption of the elements. In addition, this file includes information about the logic functions of the cells, signal propagation delays, duration of their rising and falling edges, times of setting, holding, deleting, recovering, minimum pulse duration values, leakage power, switching power, cell area, pin directions, pin capacitances, and much more. For an example printout of a file with .lib extension, see below.

```
library (Digital_Std_Lib) {
 technology (cmos);
 delay_model : table_lookup;
 capacitive_load_unit (1,pf);
 lu_table_template(cap_tr_table) {
 variable_1 : input_net_transition ;
 variable_2 : total_output_net_capacitance ;
 index_1 ("0.12, 0.24") ;
 index_2 ("0.01, 0.04") ; }
cell (inv) {
 area : 3 ;
 cell_leakage_power : 0.0013 ;
 pin(OUT) {
 direction : output ;
 function : "!IN" ;
 timing() {
 related_pin : "IN" ;
 timing_type : "combinational" ;
 timing_sense : "positive_unate" ;
 cell_rise(cap_tr_table) { values("1.0020, 1.1280", \
 "1.0570, 1.1660"); }
 rise_transition(cap_tr_table) { values("0.2069, 0.3315", \
 "0.1682, 0.3062"); }
 cell_fall(cap_tr_table) { values("1.0720, 1.2060", \
 "1.3230, 1.4420"); }
 fall_transition(cap_tr_table) { values("0.2187, 0.3333", \
 "0.1870, 0.3117"); } } }
```

Unfortunately, for the designers of microcircuits, an urge to create more and more libraries has become the current trend, due to the increasing number of boundary conditions and voltages, by which the libraries need to be characterized. In addition, as more different threshold voltages, Vt, have appeared, the number of available libraries has increased. As a result, more and more files are stored in these libraries.

Another important aspect of libraries is the fact that the timing, noise, and power data of the microcircuits under development must be as accurate as possible, which is especially important for the technologies with submicron-sized elements, where it is no longer sufficient just to have a similar .lib format with non-linear delay models and power consumption schemes. More accurate models are required and this can only be achieved, for example, by using current source models instead of nonlinear models. Since there are many software designers in the market, only two models can be recommended to the designers of contemporary microcircuits: ECSM by Cadence and CCS by Synopsys.

5.10 Standard PDK Current Source Models (CCS)

The top-level metallization has a higher impedance as the metal width decreases, resulting in the impedance of interconnections to be significantly higher than the impedance of an element. For 90 nm dimensions and less, the element capacitance varies considerably between the linear and nonlinear signal regions. In addition, the input capacitance already becomes some function of the steepness of the transmitted signal edge. Because of this problem, the NLDM (nonlinear) method widely used earlier by designers is no longer suitable for simulating input port capacitance. The use of the current source model improves the simulation of the output driver and receiver through greater timing accuracy (Fig. 5.25).

It is worth noting that there are two similar formats of current source models: Effective Current Source Model (ECSM) and Composite Current Source (CCS) model. They have the following capabilities [2]:

- Fully describe the output signals instead of the previously used only rise/delay rate values.
- Provide more advanced receiver models indicating pins capacitances.
- Are supplied by CAD vendors as extensions to existing .lib models.

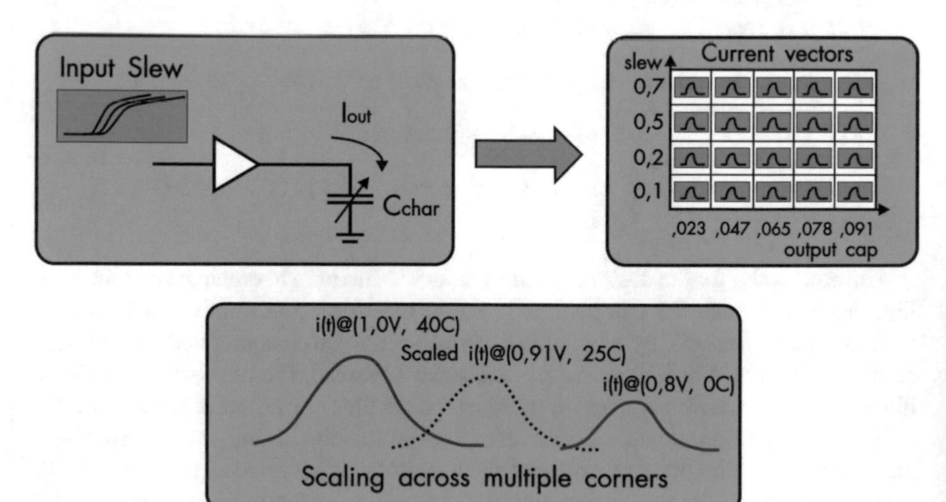

Fig. 5.25 CCS model

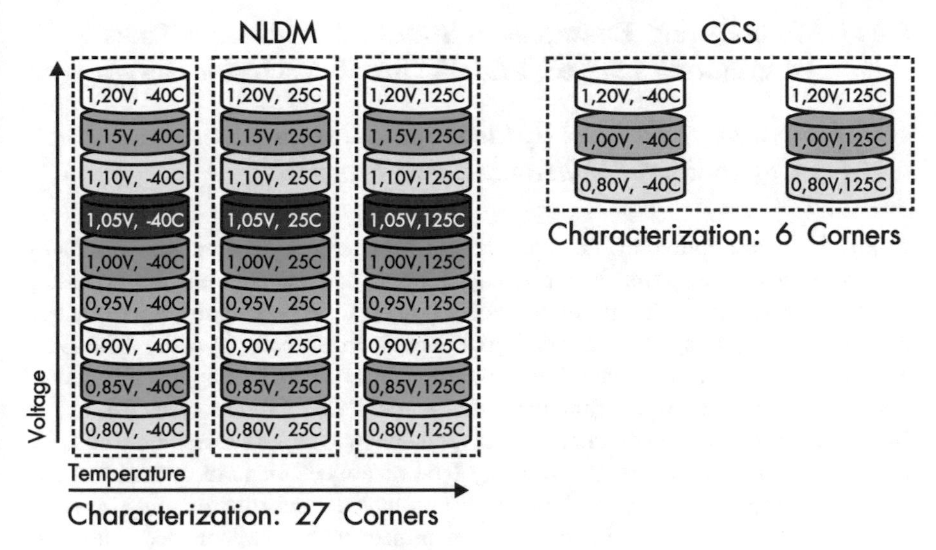

Fig. 5.26 Example of CCS model applications

The first group, CCS model, is constructed as follows. At the time of writing specification for a digital element by varying parameters (edge duration, output capacitance, etc.), the output current and waveform are measured (Fig. 5.25) and automatically stored in a library. Then, knowing the specific modes of cell operation (load, rate of rise, process, supply voltage, temperature, etc.), the CCS models, based on scaling, determine the required signals calculated from adjacent signals characterized for different conditions. Importantly for IC designers, the CCS models are scalable, i.e., as the boundary conditions increase, calculation accuracy goes up and the number of set boundary conditions considerably decreases.

Figure 5.26 shows an example of a PDK library with the variables taken into account for nominal supply voltage and $\pm$ 20% deviation, as well as for low and high temperatures.

CCS scaling allows circuitry to be analyzed for a certain voltage and temperature range with fewer libraries. Only six libraries are required to analyze circuit design in the range from 1.20 V to 0.80 V in 5 mV increments at −40 °C, 25 °C, and 125°C, compared to 27 libraries required to analyze the nonlinear delay model (NLDM) without scaling. In fact, 6VT (voltage-temperature) combinations would be required to fully represent a library with multiple supply voltages.

Of course, these boundary conditions will increase to a minimum of 18 when taking into account variations of the manufacturing process used by a fab. In this case, the number of conditions can be reduced down to six by using statistical models for different process variations.

5.11 Methods and Examples of Standard IC Design Tools Adaptation to 90, 65, and 45 Nm Microcircuit Design

5.11.1 Synopsys Tutorial (Educational) Design Kit: Capabilities, Applications, and Prospects

Below is a description of the open-access Educational Design Kit (EDK) by Synopsys, which supports 90 nm design process and includes all the necessary components, including design rules, models, process files, verification and extraction commands, scripts, symbol libraries, and parameterized cells (p-cells). The EDK also includes Digital Standard Cell Library (DSCL), which supports all modern low-power device design techniques; I/O Standard Cell Library (IOSCL), Set of Memories (SOM) with different capacities and data widths, and Phase-Locked Loop (PLL). The EDK components cover any type of project for both educational and research purposes. Although EDK does not include information on a specific semiconductor factory, it allows a 90 nm project to be implemented with high accuracy and efficiency.

As we know from periodicals, in the era of nanometer technology, universities tend to carry out state-of-the-art research of the highest quality in the field of IC design. In addition to electronic design automation (EDA) tools from leading developers of such software tools, tutorial (educational) design kits (EDKs) for various IC manufacturing technologies are also in demand. The creation of such EDKs, however, is associated with numerous challenges, including the time-consuming design itself and high complexity of validating the design results. However, the most important challenge of these is the intellectual property (IP) restrictions imposed by semiconductor fabrication plants, which do not allow universities to copy their technology into EDKs. That is why, in fact, in order to realize its marketing policy, Synopsys had to create a single open EDK that, on the one hand, would not contain confidential information of the fabrication plants, and, on the other hand, would have characteristics close enough to the real design kits that are provided by IC fabrication plants [4].

5.11.2 Synopsys EDK Overview

Synopsys has developed an open Educational Design Kit (EDK) that is free of intellectual property restrictions and is intended for use in research and education. This EDK focuses on programs designed to train highly skilled microelectronics professionals at various universities, training facilities, and research centers. The EDK is also meant to support trainees in such a way that they can better master today's advanced design methodologies and the capabilities of modern IC design tools from this particular company (Synopsys). The Synopsys design kit makes it possible even for students to design a variety of ICs using 90 nm technology and

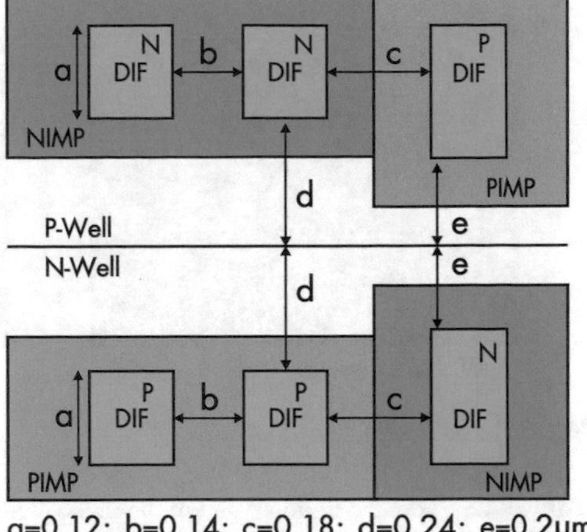

Fig. 5.27 Example of the design rule in EDK

Synopsys' design tools. This design kit also provides the ability to leverage existing approaches for designing low-power devices, which is especially important for space applications.

The Synopsys EDK consists of Technology Kit (TK), Digital Standard Cell Library (DSCL), I/O Standard Cell Library (IOSCL), Set of Memories (SOM), and Phase-Locked Loop (PLL).

For the development of the EDK, a 90 nm abstraction technology was used. The EDK described here contains no actual confidential information from any semiconductor factory. Nevertheless, it is quite close to the "genuine" 90 nm technology in its characteristics. The use of this "abstract" 90 nm technology made it possible for Synopsys to create this open EDK, providing a good opportunity to use it for studying and investigating the actual characteristics of microelectronic devices with 90 nm design rules.

Let us briefly review the *description of the EDK components*.

The backbone of the EDK is the Technology Kit (TK), a set of technology files required to implement physical representations of a design project (in particular, the layout). Standard design Technology Kit consists of:

1. ***Design rules***. These EDK components have been created through the use of design rules of scalable CMOS process by MOSIS. They provide greater "portability" of design projects compared to newly developed rules for 90 nm technology, as the dimensions in the 90 nm rules can be 5-20% larger than the actual rules of the fab. Figure 5.27 illustrates an example of the basic design rules used for a microcircuit

2. ***IC topological layout.*** This part of the TK contains a description of the available basic cells and their layout design rules. This kit contains all the elements offered by a standard 90 nm technology of any fab with 1.2V/2.5V parameters. Figure 5.26 shows examples of such semiconductor structures formation (Fig. 5.28)

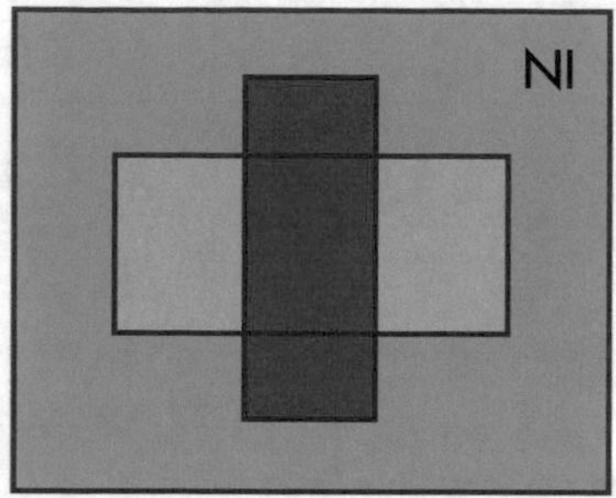

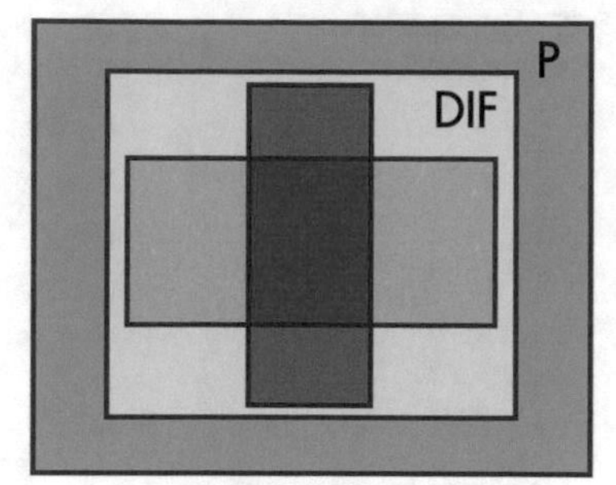

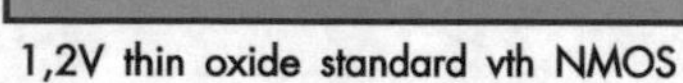

Fig. 5.28 Examples of cell layout

Table 5.12 Example of initial part of the microcircuit topological map of layers

Layer #	Data type	Tape Out Layer	Drawing or Composite Layer	Layer name in TechMap File	Layer Name in DRC	Layer Name in LVS	Layer usage description
1	0	YES	Drawing	NWELL	NWELLi	NWELLi	NWELL
2	0	YES	Drawing	DNW	DNWi	DNWi	Deep NWELL

3. ***Topological Map of GDSII layers.*** This part of the TK contains the names of layers and their numbers in GDSII format used in 90 nm process. Some layers such as 'dummy' layers, markers and text are added to the MOSIS layer map. Any numbers of layers can be selected to form a universal process. Table 5.12 shows an example of a layer map

1. *Description of the technology.* This part of TK contains approximate values for dielectric and metal layer thicknesses.
2. *Universal library of SPICE models.* These models are based on the so-called Predictive Technology Model [2]. The SPICE model library contains the following transistors and diodes:

- *Transistors.*

  (a) Devices with 2.5 V supply voltage: MOSFETs with thick oxide layer.
  (b) Devices with 1.2 V supply voltage: MOSFETs with thin oxide layer and typical (high and low) threshold voltages. Five models of boundary conditions (corners) are defined for each of these devices: TT, both devices are typical; FF, both are fast; SS, both are slow; SF, slow nMOS/fast pMOS; and FS, fast nMOS/slow pMOS.

- *Diode* (P + polysilicon resistor without silicides).

To evaluate the accuracy of SPICE models, the parameters of the models were scaled to 0.25 μm technology node in order to compare them with the characteristics

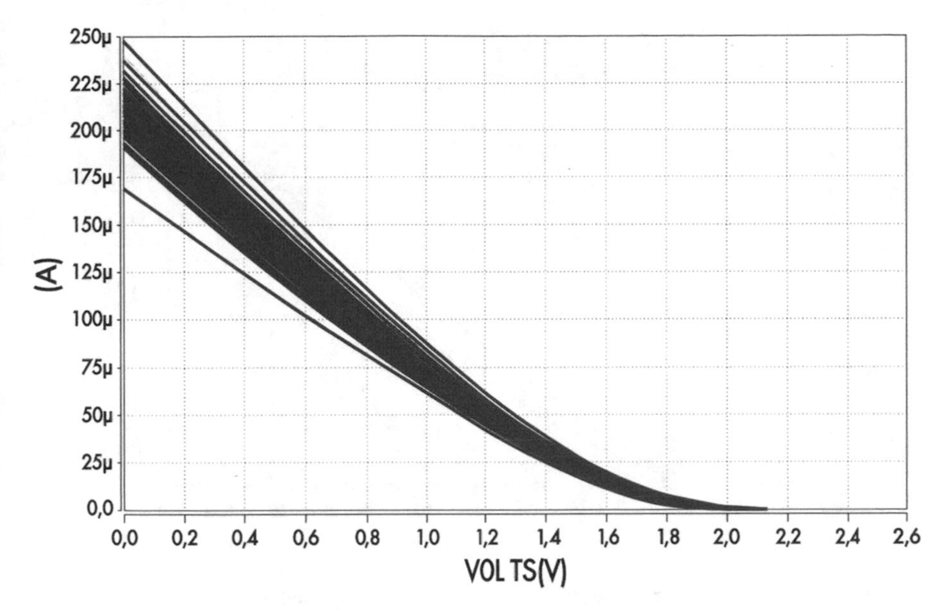

Fig. 5.29 Family of transfer characteristics for 0.25 μm technology

of the known 0.25 μm models (Fig. 5.27). The number of DC transfer characteristics was obtained, and the average curve from the common set was chosen as a typical limit value for devices with a supply voltage of 2.5 V, which is close to the real semiconductor technology (Fig. 5.29).

FF, SS, SF, and FS models with their boundary conditions were generated by varying the threshold voltages (v_{th0}) and oxide thicknesses (t_{ox}) in +/−5% range.

Figure 5.30 shows the transfer characteristics for TT, FF, and SS boundary conditions based on the model of n-MOS transistor with a thin oxide.

3. *Milkyway technological file*. This file contains the rules used by Synopsys design tools.
4. *The universal library of symbols and parameterized cells (p-cell). Universal* symbols and p-cells contained in the libraries of MOSFETs, resistors, bipolar transistors, and diodes. The parameterized cells are developed using TCL scripting language to work in the environment of Synopsys Cosmos Schematic Editor.
5. *DRC and LVS rules*. These design rules are required for Synopsys' Hercules software to perform the design rule and the layout vs. schematic checking.
6. *Extracting files*. These files are used in the Synopsys Star-RCXT software to extract the parasitic components: ITF, TLU+, conversion files, and command files.
7. ***Support scripts***. Many additional scripts are required to support the design process. For example, a script to convert the SPICE netlist to a particular technology and a setup script for p-cell.

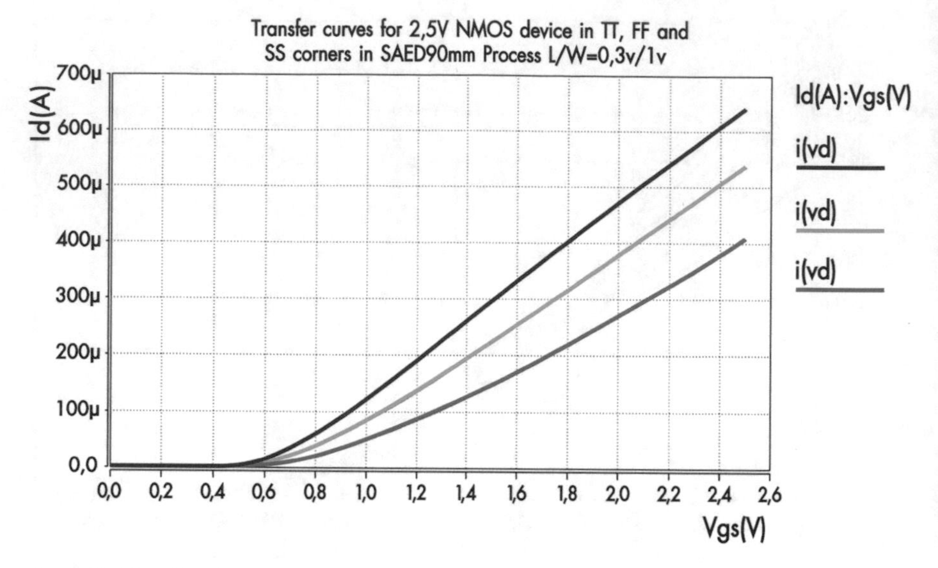

Fig. 5.30 Transfer characteristics with TT, FF, and SS boundary conditions for a 0.25 μm n-MOS transistor with thin oxide

5.11.3 Synopsys Digital Standard Cell Library

The Digital Standard Cell Library (DSCL) is used to design various 90 nm ICs using Synopsys computer-aided design tools. The DSCL was built on the basis of 1P9M 1.2 V/2.5 V design rules and focuses on optimizing the key features of IC design.

DSCL contains a total of 251 cells and includes typical combinational logic elements with different load-carrying capacities.

DSCL library also contains all the cells that are required for different types of low-power designs [5]. These cells allow designing ICs with different supply voltages of internal units to minimize the dynamic consumption and leakage currents (clock signal control box, 0.5–2.0 ns non-inverting delay lines, pass-through transistors, bidirectional switches, isolation cells, LH and HL shifters, retention flip-flops, power-down and grounding cells, always-on non-inverting buffers, etc.).

DSCL libraries also contain mixed cells that complement the library. The Composite Current Source (CCS) technology is used, which is a simulation technique for characterization of cells to meet the requirements of modern low-power products design techniques. CCS technology provides timing analysis, noise analysis, and power consumption analysis for nanometer-scale devices.

To fully meet the requirements of the design methods used for products with low-power consumption, DSCL library was characterized for 16 process/voltage/temperature conditions, shown in Table 5.13.

As stated by the DSCL designers, its functionality has also been tested under many additional simulation conditions. As a result, the DSCL library was shown to meet all necessary requirements.

Table 5.13 Boundary conditions for devices operation

Corner Name	Process (NMOS proc. – PMOS proc.)	Temperature (°C)	Power Supply (V)
FFT1p32v	Typical – Typical	25	1.2
TTHT1p20v	Typical – Typical	125	1.2
TTNT1p20v	Typical – Typical	−40	1.2
FFLT1p32v	Slow – Slow	25	1.08
SSHT0p07v	Slow – Slow	125	1.08
TTLT1p20v	Slow – Slow	−40	1.08
SSLT0p07v	Fast – Fast	25	1.32
FFNT1p32v	Fast – Fast	125	1.32
SSNT0p07v	Fast – Fast	−40	1.32
SSLT1p08v	Typical – Typical	25	0.8
SSNT1p08v	Typical – Typical	125	0.8
SSHT1p08v	Typical – Typical	−40	0.8
TTHT0p08v	Slow – Slow	25	0.7
TTNT0p08v	Slow – Slow	125	0.7
TTLT0p08v	Slow – Slow	125	0.7
FFHT0p90v	Slow – Slow	125	0.7
FFNT0p90v	Fast – Fast	125	0.9
FFLT0p90v	Fast – Fast	−40	0.9

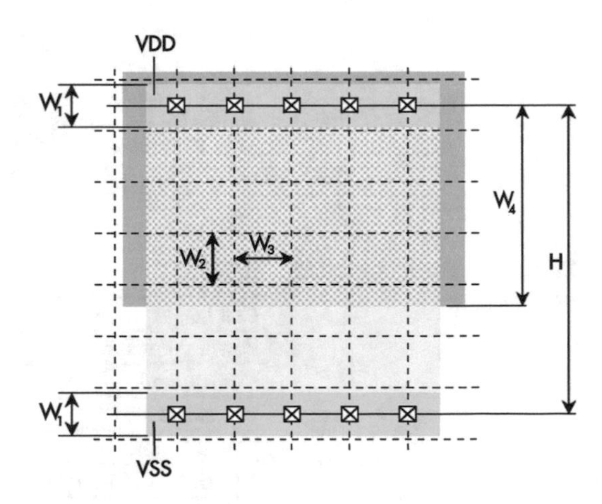

Fig. 5.31 Physical structure of single-height cell

Selection of the physical structure of digital cells was made to ensure maximum cell density in the digital designs and to meet the requirements of low-power product design methods. That's why there are structures with single (Fig. 5.31) and double (Fig. 5.32) heights, the parameters of which are shown in Table 5.14.

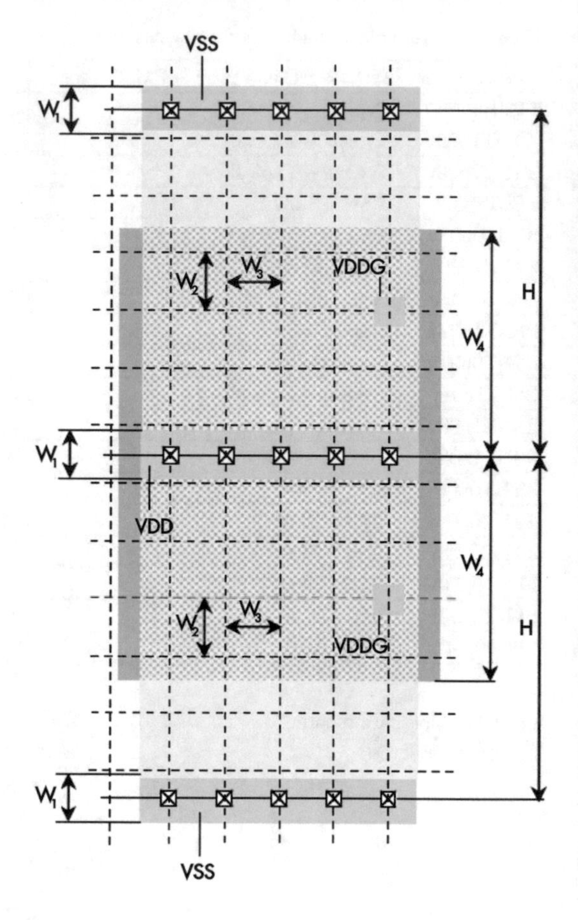

Fig. 5.32 Physical structure of double-height cell

Table 5.14 Sizes of physical structures

Parameter	Symbol	Value
Cell height	H	2.88 um
Power rail width	W_1	0.16 um
Vertical grid	W_2	0.32 um
Horizontal grid	W_3	0.32 um
NWell height	W_4	1.68 um
VDDH to VDDL height (Fig. 5.3)	W_5	0.72 um

5.11.4 I/O Standard Cell Library

I/O Standard Cell Library (IOSCL) is used to design various integrated circuits on 90 nm technology using Synopsys design tools. This library was formed using 90 nm 1P9M 1.2 V/2.5 V design rules developed by the Synopsys-Armenia Educational Department (SAED).

IOSCL, providing a full set of standard features, contains 36 cells (including CMOS non-inverting input buffer; CMOS non-inverting bidirectional cell; 2/4/8/12/

16 output driver with three states; analog non-inverting bidirectional resistless contact pads with ESD protection; basic power; I/O power; ground I/O pads; cross-coupled diode; IOVSS to VSS; decoupling capacitors VDD and VSS, IOVDD, and IOVSS; fault cell; filler cell; contact pad). CCS simulation technology was used to characterize the standard IOSCL. All cells had dimensions of 65 μm x 300 μm or less.

5.11.5 *Standard Set of PDK Memory Modules*

The Set of Memories (SOM) was developed by Synopsys using the SAED 90 nm 1P9M 1.2 V/2.5 V process. It includes a number of static RAM (SRAM) with a small number of words (word depth, *m*) and bits per word (data width, *n*). All of the SRAMs included in SOM are synchronous dual-port SRAMs which are write-enable, output-enable, with chip select (CS) signal on each port. In addition, SOM includes 16 SRAM blocks with the same architecture but different *nxm* (width x depth) size ratios, where $n = 4, 8, 16, 32$ and $m = 16, 32, 64, 18$. These synchronous dual-port nxm-SRAMs have two ports (primary and dual) for the same memory cell. Both ports can be independently accessed for read or write operations.

5.11.6 *Phase-Locked Loop (PLL)*

Phase-Locked Loop (PLL) is actually a clock multiplier circuit which must generate a stable, high-speed clock from a slower clock. It has been designed using the SAED 90 nm 1P9M 1.2 V/2.5 V process. The PLL has three operating modes: normal, with external feedback, and in bypass mode. In external feedback mode, the return input clock signal is phase aligned with the input clock signal. These aligned clock signals allow the clock delay and phase shift between devices to be removed. In bypass mode, the reference clock signal is shunted on the output.

5.11.7 *Geography of EDK Applications and Prospects*

Synopsys EDK is currently used for both educational and research purposes. EDK is used in almost every institute and university around the world where courses in chip design are given, including Syracuse University (New York, USA), University of California Extension (Santa Cruz, USA), Purdue University (Indiana, USA), Oregon State University (Corvallis, USA), California State University, Northridge (Los Angeles, USA), Silicon Valley Technical Institute (San Jose, USA), University of California (San Diego, USA), San Francisco State University (San Francisco, USA), University of Tennessee (Knoxville, USA), Indian Institute of Technology Kanpur

(Kanpur, India), Kate Gleason College of Engineering (New York, USA), Rochester Institute of Technology (New York, USA), State Engineering University of Armenia (Yerevan, Armenia), Yerevan State University (Yerevan, Armenia), Russian-Armenian Slavonic State University (Yerevan, Armenia), and Moscow Institute of Electronic Technology (Moscow, Russia).

EDK is also used in a number of prominent training centers including Synopsys' Customer Education Services, Synopsys' Corporate Application Engineering team, and Sun Microsystems.

There are many examples of how EDK is contributing to education in the field of microelectronics. All of the universities included in Synopsys' collaborative programs in the field of microelectronic design education [6] use the new learning package developed at Synopsys, Inc. [7]. This learning complex is used in laboratory works, course and diploma projects, and master's and PhD theses [8].

To "keep up" with the suggestions of the industry, the new versions of EDK for 65 nm and 45 nm (and below) technologies have been designed. These versions are developed using the same methods; they have about the same functionality as the EDK for 90 nm.

5.12 Contents of Educational Design Kits Provided by IMEC

For the purpose of teaching students within the framework of respective cooperation programs (EUROPRACTICE etc.), the *Interuniversity Microelectronics Center (IMEC)*, Eindhoven, Belgium, and the *Microelectronics Training Center (MTC)* supply the following PDKs for 130–90 design rules and below:

130 nm CMOS Mixed-Signal RF General Purpose
130 nm CMOS Logic General Purpose
90 nm CMOS Mixed-Signal Low Power RF
90 nm CMOS Mixed-Signal General Purpose RF
90 nm CMOS Logic Low Power
90 nm CMOS Logic General Purpose.

Table 5.15 summarizes the key features of TSMC 90 nm digital libraries for low-power IC design. These libraries (PDK 90 nm CMOS Logic Low Power) provided by EROPRACTICE are available for educational and research purposes.

It is recommended to use these libraries in the environment of educational process when studying digital IC design flow using Cadence software packages within such disciplines as "Fundamentals of CAD in microelectronics," "IC layout design," "Basics of IC schematic design," etc.

Table 5.15 Components of libraries with IMEC design kits

Library	Cell type	Name	Description
TCBN90LPHDBWP	Standard cells	90 nm low-power 1.2 V/ 2.5 V standard cell library, high density, characterized for 1.0 V v.200f	TSMC 90 nm logic 1.2 V/ 2.5 V low-power process (1P9M, core 1.2 V), standard Vt, 7-track library. 0.28um x-pitch, total 645 cells (include 620 base cells, 9 level shifter cells, 6 isolation cells and 9 filler cells, 1 tapcell), raw gate density = 560 Kgate/ mm^2, support multi-Vdd design; low voltage range is 1.0 V +/− 10%
TCBN90LPHDBWP HVT	Standard cells	90 nm low-power 1.2 V/ 2.5 V standard cell library, high density, high Vt, characterized for 1.0 V v.200f	TSMC 90 nm logic 1.2 V/ 2.5 V low-power process (1P9M, core 1.2 V), high Vt, 7-track library. 0.28um x-pitch, total 645 cells (include 620 base cells, 9 level shifter cells, 6 isolation cells and 9 filler cells, 1 tapcell), raw gate density = 560 Kgate/mm^2, support multi-Vdd design; low voltage range is 1.0 V +/− 10%
TCBN90LPHDBWP LVT	Standard cells	90 nm low-power 1.2 V/ 2.5 V standard cell library, high density, low Vt, characterized for 1.0 V v.200f	TSMC 90 nm logic 1.2 V/ 2.5 V low-power process (1P9M, core 1.2 V), low Vt, 7-track library. 0.28um x-pitch, total 645 cells (include 620 base cells, 9 level shifter cells, 6 isolation cells and 9 filler cells, 1 tapcell), raw gate density = 560 Kgate/mm^2, support multi-Vdd design; low voltage range is 1.0 V +/− 10%
TCBN90LPHP	Standard cells	90 nm low-power 1.2v/2.5v standard cell library, high performance v.150j	TSMC 90 nm low-power process (1P9M, 1.2v/2.5v), 0.28um x-pitch, nominal Vt, total 867 cells (include filler cells), 9-tracks. Raw gate density = 436 KGate/ mm^2, support multi-Vdd design (include level shifter cell and isolation cell inside)

(continued)

Table 5.15 (continued)

Library	Cell type	Name	Description
TCBN90LPHPCG	Standard cells	90 nm low-power 1.2 V standard cell library, high performance coarse-grain v.150e	TSMC 90 nm low-power process (1P9M, core 1.2 V), coarse-grain MTCMOS library, standard Vt total 20 cells, include special cell for 1) power switch header cell, 2) retention flip-flop cell, 3) always on cell
TCBN90LPHPHVT	Standard cells	90 nm low-power 1.2v/2.5v standard cell library, high performance high-Vt v.150j	TSMC 90 nm low-power process (1P9M, 1.2v/2.5v), 0.28um x-pitch, high-Vt, total 867 cells (include filler cells), 9-tracks. Raw gate density = 436 KGate/mm^2, support multi-Vdd design (include level shifter cell and isolation cell inside)
TCBN90LPHPHVT CG	Standard cells	90 nm low-power 1.2v standard cell library, high performance high-Vt coarse-grain v.150d	TSMC 90 nm low-power process (1P9M, core 1.2 V), coarse-grain MTCMOS library, high Vt, total 20 cells, include special cell for 1) power switch header cell, 2) retention flip-flop cell, 3) always on cell
TCBN90LPHPHVT WB	Standard cells	90 nm low-power 1.2v/2.5v standard cell library, high performance high Vt with bias v.150d	TSMC 90 nm low-power process (1P9M,1.2v/2.5v), 0.28um x-pitch, high Vt, total 845 cells (base cell, 805 cells; ECO cell, 32 cells; 7 filler cells +1 TAP cell), 9-tracks, back-bias library, bias voltage = 0.6 V, raw gate density = 451 Kgate/mm^2
TCBN90LPHPLVT	Standard cells	90 nm low-power 1.2v/2.5v standard cell library, high performance low-Vt v.150j	TSMC 90 nm low-power process (1P9M, 1.2v/2.5v), 0.28um x-pitch, low Vt, total 867 cells (include filler cells), 9-tracks. Raw gate density = 436 KGate/mm^2, support multi-Vdd design (include level shifter cell and isolation cell inside)
TCBN90LPHPLVTC G	Standard cells	90 nm low-power 1.2 V standard cell library, high	TSMC 90 nm low-power process (1P9M, core 1.2 V), coarse-grain MTCMOS

(continued)

Table 5.15 (continued)

Library	Cell type	Name	Description
		performance low-Vt coarse-grain v.150d	library, low Vt, total 20 cells, include special cell for 1) power switch header cell, 2) retention flip-flop cell, 3) always on cell
TCBN90LPHPLVT WB	Standard cells	90 nm low-power 1.2v/2.5v standard cell library, high performance low-Vt with bias v.150c	TSMC 90 nm low-power process (1P9M, 1.2v/2.5v), 0.28um x-pitch, low Vt, total 845 cells (base cell, 805 cells; ECO cell, 32 cells; 7 filler cells +1 TAP cell), 9-tracks, back-bias library, bias voltage = 0.6 V, raw gate density = 451 Kgate/mm^2
TCBN90LPHPUD	Standard cells	90 nm low-power 1.2v/2.5v standard cell library, high performance under-drive standard-Vt v.200a	TSMC 90 nm low-power process (1P9M,1.2v/2.5v), 0.28um x-pitch, standard Vt, total 837 cells +7 filler cells, 9-tracks. Raw gate density = 436 KGate/mm^2, under-drive 1.0 V (1.0 V +/−10%)
TCBN90LPHPUDH VT	Standard cells	90 nm low power 1.2v/2.5v standard cell library, high performance under-drive high-Vt v.200a	TSMC 90 nm low-power process (1P9M, 1.2v/2.5v), 0.28um x-pitch, high Vt, total 837 cells +7 filler cells, 9-tracks. Raw gate density = 436 KGate/mm^2, under-drive 1.0 V (1.0 V +/ −10%)
TCBN90LPHPUDL VT	Standard cells	90 nm low-power 1.2v/2.5v standard cell library, high performance under-drive low-Vt v.200b	TSMC 90 nm low-power process (1P9M, 1.2v/2.5v), 0.28um x-pitch, low Vt, total 837 cells +7 filler cells, 9-tracks. Raw gate density = 436 KGate/mm^2, under-drive 1.0 V (1.0 V +/ −10%)
TCBN90LPHPULVT	Standard cells	90 nm low-power 1.2v/2.5v standard cell library, high performance ultralow-Vt v.200a	TSMC 90 nm low-power process (1P9M, 1.2v/2.5v), 0.28um x-pitch, ultralow Vt, total 867 cells (include filler cells), 9-tracks. Raw gate density = 436 KGate/mm^2, support multi-Vdd design (include level shifter cell and isolation cell inside)

(continued)

Table 5.15 (continued)

Library	Cell type	Name	Description
TCBN90LPHPWB	Standard cells	90 nm low-power 1.2v/2.5v standard cell library, high performance with bias v.150c	TSMC 90 nm low-power process (1P9M, 1.2v/2.5v), 0.28um x-pitch, nominal Vt, total 845 cells (base cell, 805 cells; ECO cell, 32 cells; 7 filler cells +1 TAP cell), 9-tracks, back-bias library, bias voltage = 0.6 V, raw gate density = 451 Kgate/mm^2
tpan90lpnv2	I/O cells	90 nm low-power 1.2 V/2.5 V standard I/O, universal analog I/O v.200a	N90LP, 1.2 V/2.5 V, universal analog I/O
tpan90lpnv3	I/O cells	90 nm low-power 1.2 V/3.3 V, universal analog I/O compatible with linear universal standard I/O v.210a	1.2 V/3.3 V universal analog I/O compatible with linear universal standard I/O
tpdn90lpnv2	I/O cells	90 nm low-power 1.2/2.5 V, regular, linear universal standard I/O library v.200c	1.2/2.5 V, regular, linear universal standard I/O. (1) characterization with new char flow. (2) re-char with T-N90-LO-SP-008 v2.0 spice model. (3) fix databook missing content issue. (4) update syn and ctc from four corners to six corners
tpdn90lpnv3	I/O cells	90 nm low-power 1.2/3.3 V, regular, linear universal standard I/O library v.210b	1.2/3.3 V, regular, linear universal standard I/O
tpbn90gv	I/O cells	Standard I/O bond pad library v.140a	Standard I/O bond pad library v.140a

References

1. Belous, A. I., Emelyanov, V. A., & Turtsevich, A. S. (2012). *Fundamentals of circuit design of microelectronic devices. Technos*, 472. (in Russian).
2. Belous, A., Saladukha, V., Shvedau, S. (2017). *Space microelectronics volume 2: Integrated circuit design for space applications.* London, Artech House, p. 720, ISBN: 9781630812591.
3. Belous, A. I., Saladukha, V. A., Shvedau, S. V. (2015). *Space Electronics.* M.: Technosfera, in 2 vol., p.1184 (in Russian).
4. Belous, A. I., & Saladukha, V. A. (2012). *Fabless business model in microelectronics company: Myths and reality.* Components and technologies Journal 8: 14–18. (in Russian).

Chapter 6
Digital IC and System-on-Chip Design Flows

This chapter being a follow-up of the topics covered in the previous chapter is dedicated to elaboration of the design flows of submicron microcircuits. The first portion of the chapter looks into some particularities of selecting the specific design flow subject to the customer's requirements to the IC and content of the main milestones of the flow: system, functional and physical designing, and final qualification of the project.

Further on, there are explored some special aspects of microelectronics products at higher complexity level—system-on-chip: development trends, design flows, design methodology, special features of system design phase of the systems on chip, and basic composition of CAD tools for system level.

The chapter is ended with practical examples of system-on-chip simulation process described in Cadence Incisive simulation environment.

6.1 Choosing the IC Design Flow

A feature of the design of modern ICs is the binding to a specific technology of their manufacture. Since a modern fab is interested in having as many designers as possible using its production facilities [1], it usually creates and equips designers with the appropriate design kit. The design kit usually includes a description of the technology with its capabilities and limitations, basic design rules, lists of all the layers used, and available devices with their models to enable subsequent verification of the Schemes [2].

Element libraries contain different representations, allowing the use of the same technology to solve different tasks in numerous software packages. Large manufacturing fabs (foundries) constantly improve their technological processes, increase the accuracy of the mathematical models used, and constantly expand and supplement the composition of the element base, which leads to the need for periodic updates (modernization) of such design kits. In this regard, it is important to

A. Belous, V. Saladukha, *The Art and Science of Microelectronic Circuit Design*,
https://doi.org/10.1007/978-3-030-89854-0_6

"synchronize" the data available to designers and manufacturers, since IC made by the designer in accordance with the "outdated" design rules can have serious errors in terms of the new design rules.

In the previous chapter, we considered in detail the composition and features of developing the design libraries (RDK) of submicron ICs, the main functions of the RDK, the main types of elements to be represented there, various options for representing elements in this library, introduced the concept of "parameterized cell", etc.

Figure 6.1 shows the main stages of the standard design of a digital IC [2].

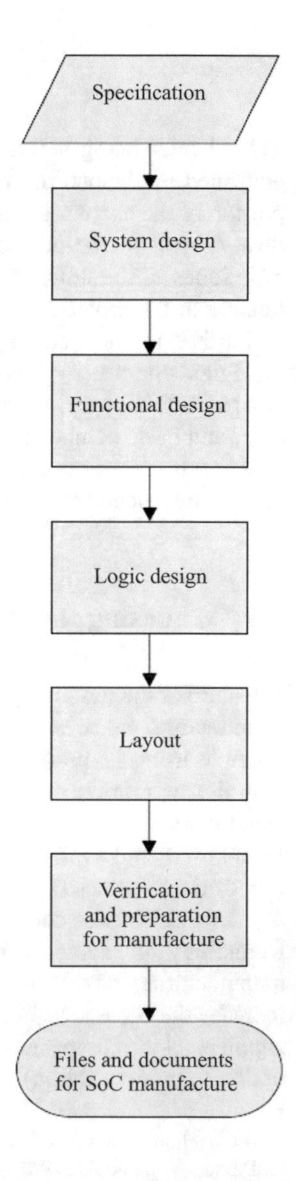

Fig. 6.1 IC design flow

The choice of a specific design flow mainly depends on the content of the technical specification (TS) for the IC, which is usually drawn up jointly by the customer and the IC designer.

In practice, only *two options* are used here. The simplest and most convenient for the customer, as well as the most common among domestic designers of radio-electronic devices and systems, is the *reproduction* of products already available on the world market. In the slang of designers, this means "reproduction of an IC analog." Why this is so, and whether it is good or bad, we will discuss it in the last section of the last chapter of this book. So in this case, the designer simply downloads the entire IC specification from the Internet and makes it in the form of a TS.

In the second case, the process of preparing the TS for a new IC takes quite a long time and requires active cooperation between the customer and the IC designer. The customer shall *describe* in detail what he specifically wants to get, and the designer shall *understand* whether he can implement it and what he will need for this (including where to organize the subsequent production of the developed ICs: at a foreign fab, or the technical capabilities of the existing production line will allow the production of a new IC in domestic fab).

In any case, before drawing up the TS, the designer shall have all the necessary ***initial data for IC design.***

The primary initial data for IC design:

- A list of technical parameters (performance, operating modes, power, bit depth of input and output data, package type, etc.)
- A formalized task (basic functions, command system, list of basic operations, etc.)
- A structural description.
- Algorithms of functioning.
- Required interfaces.
- Specification of the software used (if necessary).

Based on this data, the designer forms a *project flowchart* and determines which part of the algorithm will be performed in hardware and which in software, which interfaces need to be developed, which simulation scenarios will be required, and how many and what specialists need to be involved in the project implementation.

Then it is necessary to conduct a preliminary assessment (calculation) of the performance, power consumption, and logical capacity (degree of integration) of the project for the decomposition of the project and the selection of the necessary functional modules in order to provide all the characteristics required in the TS. After conducting such a preliminary analysis, detailed project specifications are formed for electrical circuit designers, application and system software programmers, and simulation and verification specialists.

In particular, the project specification for the *designer of the wiring diagram* includes the following characteristics:

- Logical volume, performance, number of pins, and type of package.
- Groups of pins used, with a precise indication of their functional purpose.
- Topological characteristics and technological requirements.
- The need to organize external interfaces and use purchased IP units.
- Requirements for data processing paths (e.g., requirements for ADC and DAC, the need for external RAM, processor).
- Clock frequency requirements (stability, waveform).
- Requirements for the power supply system (maximum current consumption for all power supplies and grounds, switching order, noise immunity, overload protection, etc.)

For designers, the project specification includes the following requirements:

- Package type and number of pins.
- Groups of pins used, indicating the topological requirements for signal tracing.
- Estimation of the amount of power consumed to determine the temperature regime and develop a radiator or other means of heat removal.
- Requirements for mechanical stress and resistance to radiation.

For application and system software *programmers*, specify:

- The algorithm of data processing in the project, indicating which part will be implemented in software.
- The type of processor used and the method of its implementation ("hardware," "design," or "external").
- The algorithm (protocol) of interaction between the functional modules and the central processor, including the memory card, interrupt service procedure, data exchange mechanisms, and synchronization system.
- The need to configure specialized interface ICs.
- Composition and functions of the application software.
- The type of operating system and its necessary functions, if you plan to use it.

The following information shall be specified in the specification for *the test specialist*:

- Data processing algorithm, indicating what is implemented using software and what is hardware-based.
- List of interfaces and specification versions supported by the project.
- Algorithms and protocols for interacting with software.

Then, based on the received source data, the project, application and system software, electrical circuit diagram, topology, verification plan, and test set are developed in parallel.

Below we will consider in more detail the content of each of the main stages of the standard flow of designing digital ICs.

6.2 System Design Stage

Based on the received specifications, a number of standard operations of the system design stage are implemented (Fig. 6.2 [2]).

Development and analysis of the algorithm. At this step, the key algorithms of the IC are created, adapted, and investigated, both in terms of control and data processing. Control algorithms are often expressed in the form of an automaton

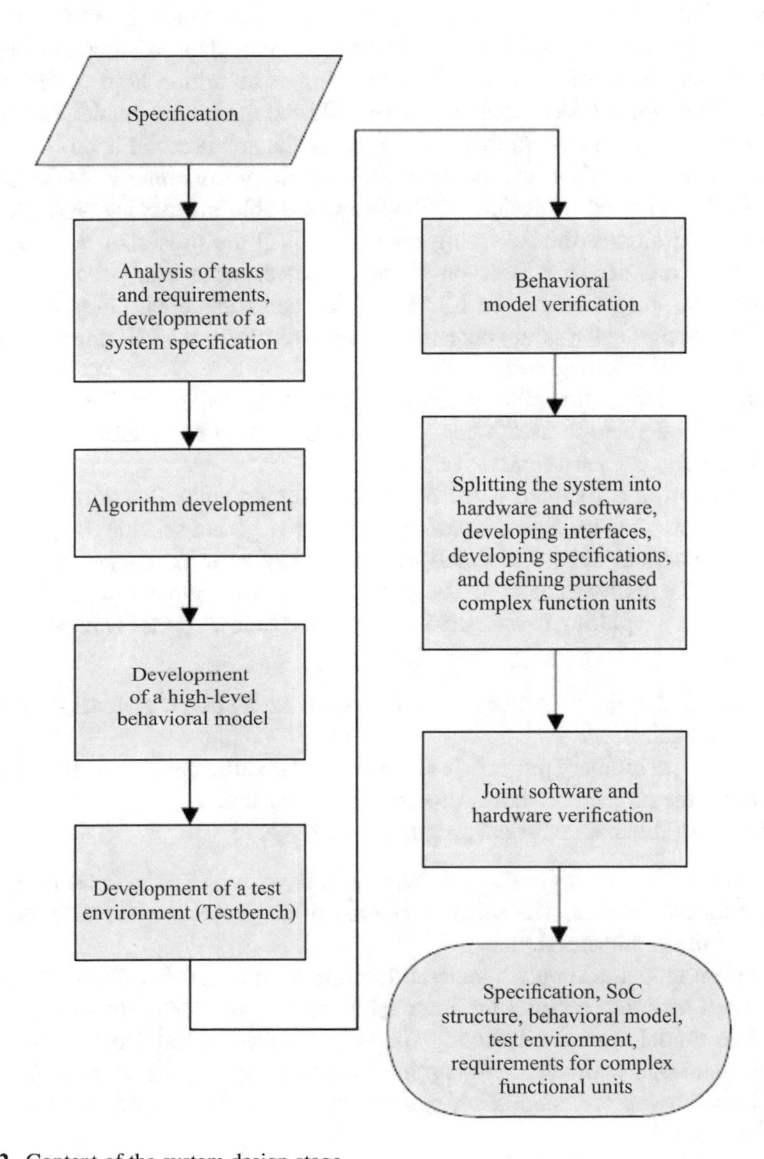

Fig. 6.2 Content of the system design stage

with a finite number of states. Data processing algorithms are usually written in the language of specialized software tools, in particular, in the C/C++language. The following automated tools can be used to analyze such data structures: MathWorks MATLAB/Simulink, Cadence SPW, Synopsys CoCentric System Studio, and Ptolemy/Ptolemy II.

The main questions to be answered at this stage are as follows: What types of data to be processed and what control functions are necessary for the normal operation of this IC? With what accuracy should the data be processed? For communication systems, for example, it is necessary to determine which encoding/decoding methods and algorithms will be used. Is it possible to adapt existing algorithms or is it necessary to create new ones? Is the chosen algorithm able to provide the required throughput? How reliable is the idealized channel model used? How difficult is the chosen algorithm to implement at the architectural level?

Architecture. At this stage, the basic architecture of the future IC is determined. First of all, you need to decide on the programmable processor cores. Here, the designer shall answer the following questions: Will the processor be used? How many cores are needed? Will only the processor or a ready-made processor subsystem be integrated on the IC? Processor cores are often equipped with an entire life support subsystem consisting of an additional set of IP units (interfaces, memory controllers, diagnostic devices, etc.).

What are the ideas for splitting into software and hardware parts? Which memory structure is preferable to use? What requirements should be applied to the internal memory units (size, performance, access).

System design is required when developing any complex functional ICs. In the case of system-on-chip, an additional goal of this stage is to divide the project into complex functional units (CF units) in such a way as to maximize the available reserve and the ability to develop the missing parts of the project in parallel.

In addition to splitting the project into units, the following tasks are performed at this stage:

- Creating and optimizing an executable system model in a high-level language (C ++, SystemC).
- Creating a preliminary project specification that is sufficient for functional design.
- Splitting the project into hardware and software parts.
- Forecast of the main physical parameters of the IC.

The system model is divided into units and developed on the basis of a hierarchical architectural plan. The actual execution of the project begins with the development of the architectural plan.

A system model is simply a general description of the IC functions. The system model shall take into account the interaction with all other elements of the equipment. The model may also include various electromechanical drives, multi-object control systems, etc. Choosing the optimal solution requires the study of the model and evaluation of the simulation results based on the criteria defined in the specification.

If you have a system model, you can already divide the project into hardware and software parts, as well as prepare a specification for the functional design stage.

Based on the system model, it is possible to estimate the main physical parameters of the developed IC: the number of pins, power consumption, and the area of the chip. For this purpose, there are so-called forecast programs. The forecast programs are based on the statistics of completed projects and give an error of up to 20% for the mastered technologies.

A similar approach can be used when *designing a system-on-chip (SOC).* In this case, a typical system-on-chip consists of an external bus interface; an integrated microprocessor, RAM, and ROM on chip; a number of functional modules, including an ADC, DAC, or radio unit; and an internal bus (On-Chip Bus, OCB) connecting the functional modules.

Verification of a high-level behavioral model is carried out in the process of system design. During verification, an analysis of the architecture, the possibility of developing missing CF units and the compatibility of existing ones, and the possibility of developing new application programs and requirements for them is carried out. It also checks the unity of the design environment and the compatibility of CAD modules and the availability of data management tools and project documentation. At this stage, the results of the forecast of the main technical parameters are compared with the requirements of the TS, and the cost of the product is estimated.

The work is completed with the preparation of *private technical specifications* for the components of the software and hardware parts of the project. In the future, work on these parts can be carried out in parallel.

An important point of the system design stage is the stage *Test Environment Development (Test Bench).* At this stage, a Test Bench project is created, which is necessary for the implementation of the next sub-step "Behavioral Model Verification" (see Fig. 6.2). They include both the actual test sequences of the so-called test vectors, as well as the necessary input signal generators, and such means of displaying output information. Based on the Test Bench, specific test sequences are then generated that are used to verify the project at lower design levels. In addition, the same tests will be used for the final functional testing of manufactured prototypes of the designed IC.

Verification of the behavioral model is carried out by computer simulation using specially designed CAD software. Upon completion of the verification procedure, the designer shall make an important decision—which functional units of this model will then be implemented at the hardware level, and which will be implemented at the software level. Special interfaces between hardware and software are also developed here.

The *joint software and hardware verification* checks the functioning of the hardware developed in accordance with the available specifications, under the control of the embedded software in real time. Executable specifications of hardware-implemented units are used as hardware, and prototype software is used as software.

When choosing the architecture, the main parameter is the low-power consumption of the IC. Based on this, the CF units are selected and divided into hardware and software levels. Therefore, it is necessary to obtain full technical information about the purchased CF units and those already available, since their characteristics can significantly affect the IC architecture.

6.3 Functional Design Stage (Fig. 6.3)

The main goals of this stage are to create a detailed *functional model* in the high-level hardware description language (VHDL, Verilog), as well as to prepare a detailed specification of all the units and the system as a whole.

A complete electrical model at the transistor level is usually not used to create specifications, as there are not enough computing resources for this.

The first task of this stage is to develop all the missing CF units. For analog units designed at the transistor level, it is also necessary to create a behavioral model in a high-level language. So far, there are no good programs that automatically

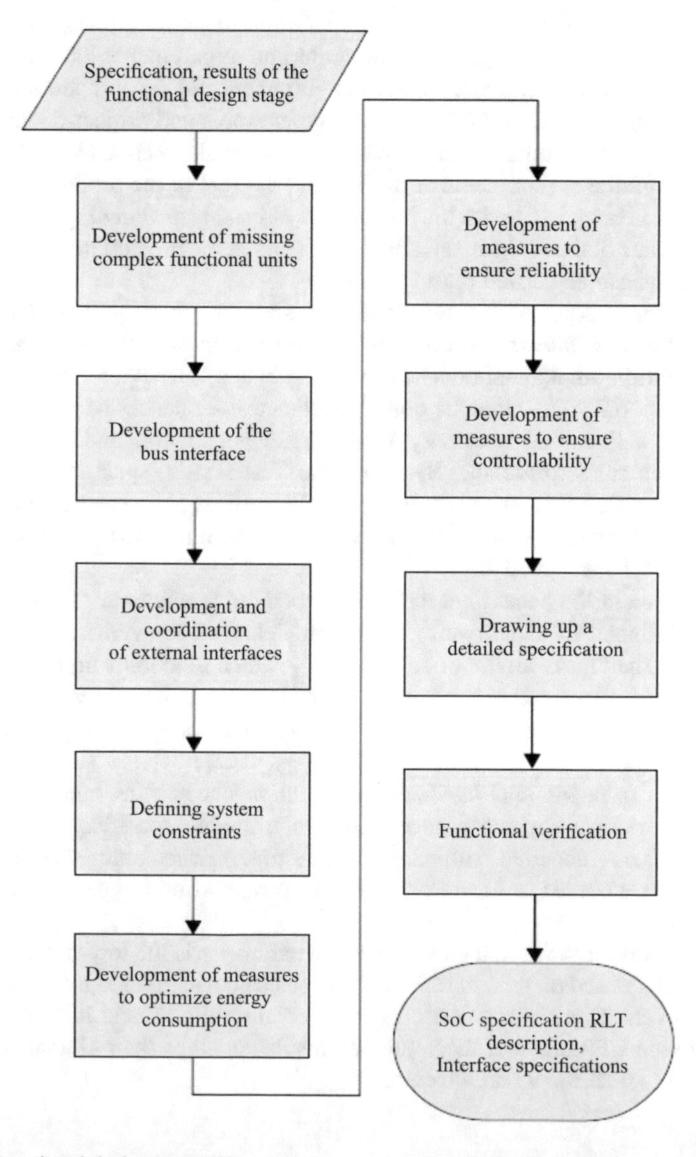

Fig. 6.3 Functional design stage [2]

synthesize any behavioral model based on the results of simulation of the transistor model of the unit.

The second task is to coordinate the interfaces of CF units and build a bus architecture, as well as forecast the parameters of communication lines.

At this stage of functional design, the time diagram of the IC operation is detailed, and the balance of delays between units is calculated. These actions are performed to ensure the connectivity of the signals in the time diagram of the system. This complex task is solved consistently at all stages of design, starting with the preparation of the TS. In synchronous systems, the period of the sync signal should be greater than the signal delay in the combination elements and communication lines. In complex ICs, the greatest delays appear in the communication lines. The only way to ensure the high operating frequency of complex digital ICs is to use an asynchronous data transfer protocol between the units. To implement the protocol, additional units are required, i.e., internal interfaces.

The third task is to ensure mutual coordination of the signals of analog and digital units. Usually, behavioral models can give an error of 7–10%, so it is better to solve the issues of matching with the help of mixed analog-to-digital simulation programs, for example, Spectre-Verilog.

The fourth task is to develop and coordinate external interfaces. Simulation of external interfaces should be carried out taking into account the reactive parameters of the package and external communication lines.

The fifth task is to calculate the power consumption and develop measures to save it.

The simplest *methods of saving power include*:

Reducing the operating frequency of the unit to the minimum required. This method requires the use of a unit's own independent sync generator.

- Reducing the power supply voltage of the unit to a value that ensures operation and the required performance. To do this, the LSI shall contain secondary power stabilizers.
- Reducing the logical drop in long communication lines. To apply the method, special signal repeaters are required.
- Use of circuit engineering with adiabatic logic.

The sixth task is to forecast and improve reliability, as well as the yield, by introducing redundancy and backup into the circuit. The main methods for improving reliability through redundancy: redundant memory units, code protection of data during storage and transmission, unit duplication and majority data selection, programmable unit replacement.

The seventh task is to ensure controllability and develop built-in controls. The main methods of ensuring controllability:

- Functional and physical decomposition of the project. Complex circuits are easier to check in parts.
- The absence of bidirectional communication lines in the circuit, that is, a ban on combining the outputs of logic gates and units.
- Efficient built-in control system.

For digital devices, effective built-in controls have been developed: multiplexers, end-to-end shift registers, and signature analyzers. To register the digital signals of the control units, only one additional pin is required, which changes the purpose of the other signal pins of the IC. Analog unit signals can be recorded in two main ways using an analog multiplexer or an integrated ADC that converts these signals to digital ones.

The eighth and last task is to draw up the final detailed specification, which will be used both in the physical design of the IC and in the development of the production control program, as well as in the preparation of recommendations for the use of the product.

To meet the required parameters, *system restrictions* are set, which include: restrictions for clock frequencies and the nature of connections between them; the phase separation and the establishment time, restrictions for input and output signals, the range of dynamic consumption and leakage currents are determined for each clock signal. If there are several power supply domains in the project, the requirements for the operating voltage values for each domain are described, as well as the rules for switching domains during the operation of the IC.

To optimize energy consumption, firstly, the power consumption of the SoC is estimated, and, secondly, measures are taken to save it by using such methods as reducing the operating frequency of the unit to the minimum required, reducing the supply voltage of the unit to a value that ensures operation and the required performance, and reducing the logical difference in long communication lines.

To ensure the *required reliability* of the operation, its calculation is performed, and, if necessary, the measures are taken to improve it. To increase the reliability of the SoC, redundancy and backup methods are used, such as the use of redundant memory units, code protection of data during storage and transmission, and unit duplication and majority data selection and the use of programmable unit replacement.

To ensure controllability, methods of functional and physical decomposition of the project are used, bidirectional connections are excluded from the circuit, and an effective system of built-in control is used.

Based on all the developed requirements and restrictions, a final detailed specification is drawn up, which will be used in the design of the IC or SoC, the development of a production control program, and the preparation of recommendations for the use of the product.

The main purpose of *functional verification* is complex debugging of the functional model together with the software. Usually, functional verification cannot be performed by CAD tools alone. There are not enough time and computing resources for this. Together with software verification, the system is emulated using special layouts. Functional verification is carried out in conjunction with functional design and forms a single iterative cycle with it.

Simulation of the system at the level of a behavioral model in the VHDL/Verilog languages allows you to check the operation of the functional model, get time diagrams of the operation of CF units and the system as a whole, and evaluate the main dynamic parameters. The electrical model at the transistor level is not suitable for detailed simulation. This model includes hundreds of thousands and millions of elements and requires hundreds and thousands of hours of computer operation. The transistor-level model is used to validate the topology design assignment and to physically verify the project. As part of the CAD system of many companies, there are special programs, i.e., performance simulators. These programs use simplified transistor models and high-performance, but not very accurate, algorithms. When the results differ by 10–20%, compared to the exact model, the calculation performance increases hundreds of times.

To *emulate* the system using a layout, special layout boards are used, including FPGAs, microprocessors, memory units, clock generators, ADCs, DACs, and various interfaces. These cards have control connectors and a connector for the PCI bus. The boards can be inserted into the system unit of a personal computer and supplemented with the necessary programs. This hardware and software package allows you to simulate the operation of the LSI, program the P-board, and analyze the system signals. To expand the functions of the board, additional boards, i.e., function modules, are connected to the control connectors. Additional hardware is usually not required to work with this layout.

6.4 Logic Design Stage

At the *logic design stage* (Fig. 6.4), the circuit is described at the gate level, represented in the form of a connection list (netlist), which is a text form of encoding the circuit.

During the synthesis, a list of connections of logical library elements in the basis of the selected technological process is formed. The output format of the connection list file is EDIF, Verilog, or VHDL. When forming a connection list, you set limits on time delays, element placement, and connections. Next, the physical optimization is performed by performing the synthesis with the specified restrictions on time characteristics, power consumption, area, and other parameters.

Static time analysis is a method for calculating the time parameters of the SoC for SD cards, which does not require a complete electrical simulation of the circuit operation. It is used to determine violations of the preset and hold time of triggers, as well as the critical path, the signal delay, the maximum allowable signal delay, and the time margin.

Formal verification is control of the correspondence of the synthesized list of SoC connections for SD cards and the original RTL representation. The statement about the health of a particular project model is based on equivalence with another model. Formal verification methods allow you to get almost complete verification at a relatively small time cost. Most logical errors in such complex projects, such as the development of processors, can be found using formal verification methods.

The simulation of the connection list at the logic design stage is carried out to test the functionality of the SoC for SD cards, taking into account the time delays on the logic elements.

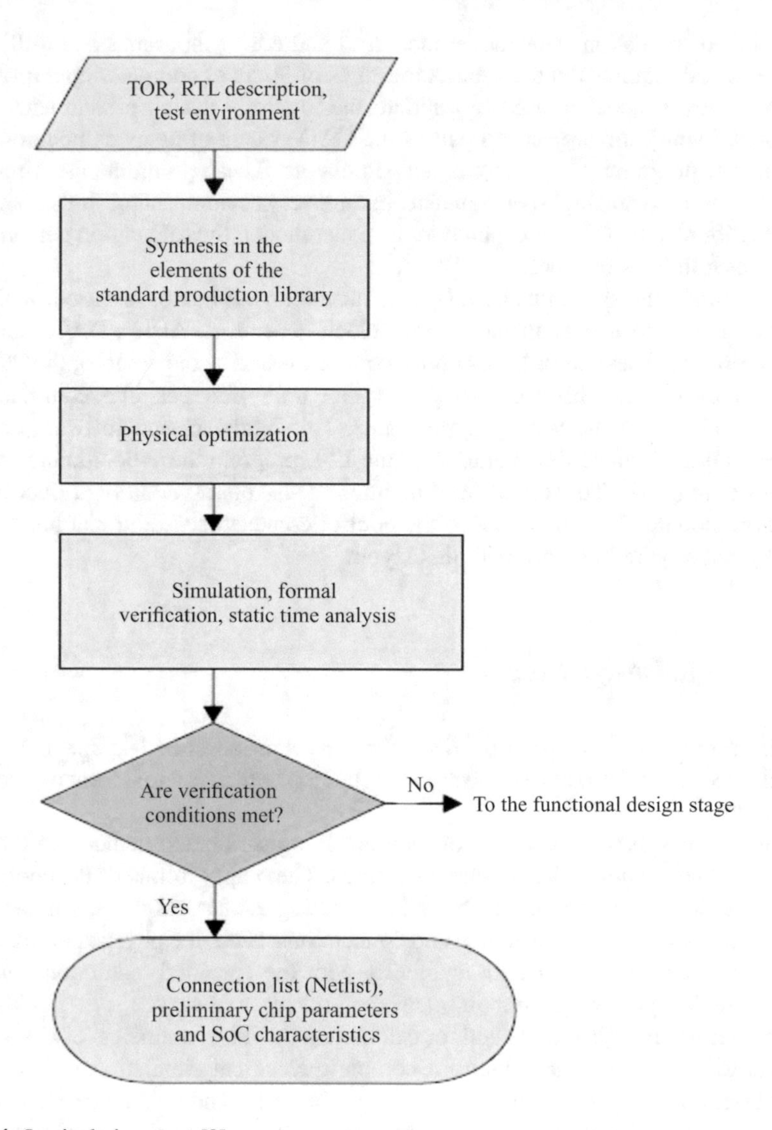

Fig. 6.4 Logic design stage [2]

By the beginning of the logic design stage, all technical information on all logic element libraries received from library suppliers should be available for the correct selection of parameters and further development process. For SoC for SD cards, the most important parameters when choosing a library are the power consumption, as well as the cell area.

6.5 Physical (Topological) Design Stage

At the stage of topologic design (Fig. 6.5), the transition from the logical level of design to the physical implementation is carried out, taking into account the influence of technological constraints and physical factors.

The purpose of physical design is to develop the topology of the LSI chip while meeting the design standards and specification requirements.

As the project progresses, the specifications become more and more detailed. The number of links between project parts increases. In order to simplify and systematize the development process, an additional stage of creating a physical virtual prototype is introduced at the physical design stage. A physical virtual prototype is a model and a preliminary topological plan of IC chip. The physical virtual prototype is developed simultaneously with the functional model. It is used to calculate the size of the chip, specify the requirements for power supply and synchronization systems, and make estimates of the power consumption and parameters of communication lines. The prototype allows you to accurately detail the functional model without the time-consuming development of the chip topology. The task of building a physical virtual prototype is a combination of the tasks of optimizing the functional model, global topological placement of units, power bus wiring, and building synchronization circuits. If the calculations and forecasts are made correctly, then further topology design will not require changes to the original topological plan of the chip and the functional model.

Next, the time-consuming operations of detailed *tracing* of the developed CF units and the system as a whole are performed. Digital units are wired using automatic topology synthesizers. The wiring of analog units usually requires the intervention of designers. The most important task of physical design, i.e., reducing the level of interference, is solved by reducing the density of the placement of elements and signal connections. Usually, only two or three connection levels are used for tracing, and the remaining levels are occupied by screens and power buses. The topology of analog units is a serious problem, which is covered in thick monographs.

For global tracing, semiautomatic methods are usually used. Thin conductors have a large resistance and contribute significantly to the signal delay. For long connections, reverse scaling of the conductors is used. The longer the conductor, the wider it should be. The width selection rules are calculated from the required performance of communication lines and the specific parameters of the conductors. Buffering using signal regenerators is used to equalize the signal bus delays. In this method, the number of devices connected to a single bus is not limited by the design rules. There are special rules for communication lines of analog and high-frequency signals. It is mandatory to shield the communication lines with power and ground buses. High-frequency signals are best transmitted by two conductors with antiphase voltage levels. It is desirable to match the impedances of receivers, transmitters, and communication lines.

Special simulators are usually used to calculate the level of substrate noise and parameters of communication lines.

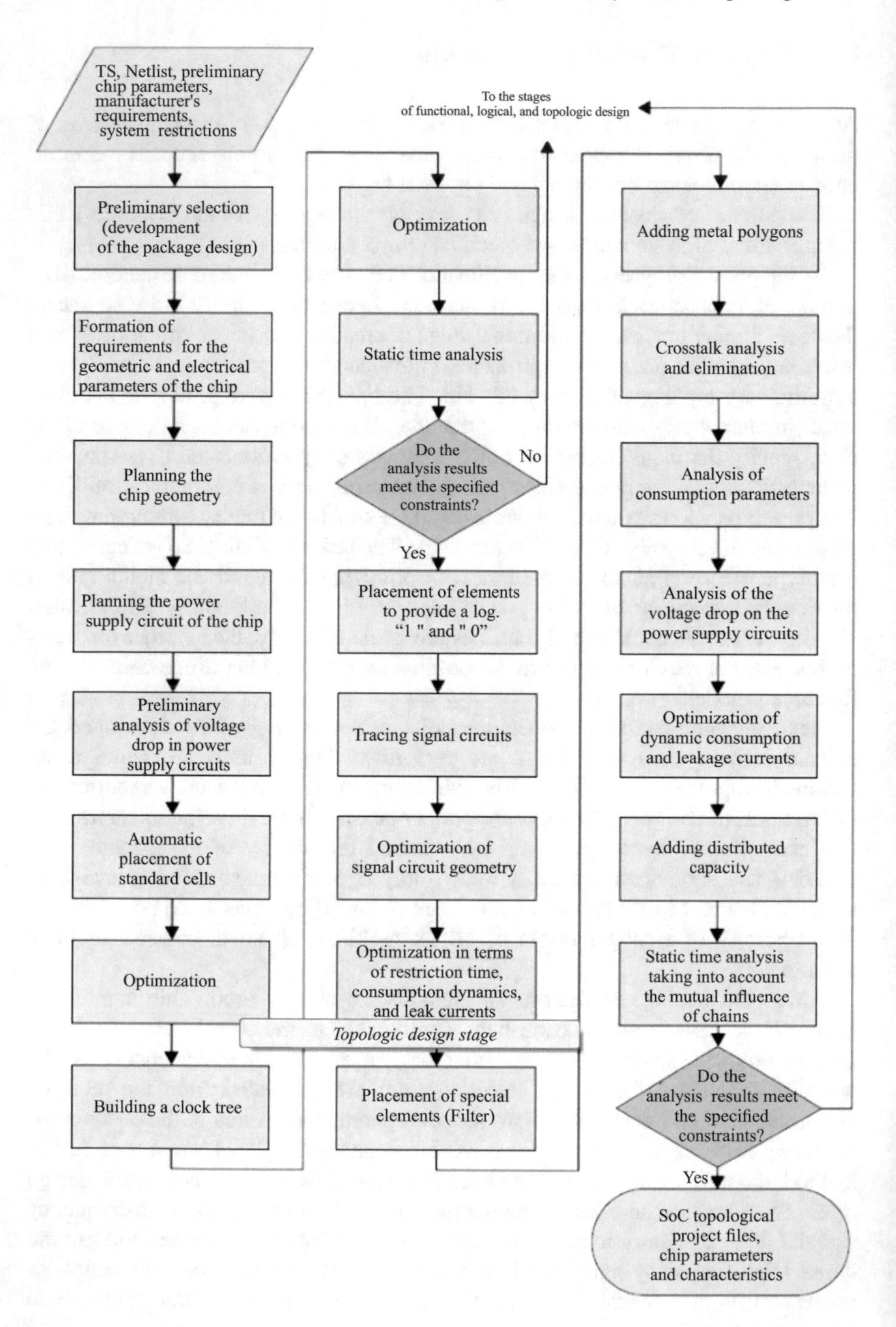

Fig. 6.5 Logic design stage [2]

Another *problem with LSI tracing* is related to the manufacturing technology of the metallization system. When forming the pattern of the conductors, the process of plasma etching of the metal film is used. Etching occurs due to the flow of a constant ion current in a high-frequency plasma. The silicon substrate in the etching process is under a voltage of 200 ^ 300 V and at a temperature of about 300 °C. Before the formation of all levels of metallization is completed, some of the signal bonds remain broken. Some sections of the conductors are connected to the p-n junctions of the physical structure, while others are connected only to the gates of MOSFETs. At a temperature of 300 °C, the p-n junctions degenerate, and the metallization etching current flows into the substrate. The gates of the MOSFETs remain isolated, and the bias voltage of the substrate is applied to the gate dielectric. The etching current can cause a breakdown of the dielectric and failure of the transistor. To eliminate this phenomenon, two main methods are used: transferring the conductor section to the next metallization level or connecting an additional protective diode at the p-n junction to the gate of the MOSFET. The tracing program can perform these operations automatically, but the security method is chosen by the designer. In the design flow, this operation is called fighting with "antennas."

When the chip trace is completely finished, there are still auxiliary operations:

- Setting the "key," that is, the pointer of the first output of the IC.
- Hip marking, which is usually performed at the highest level of metallization.
- * Introduction of test elements for process control, which are usually placed in the dividing lines between the chips
- Filling in free fields with "dummy conductors".
- Introduction of technological corrections for the dimensions of topological elements.

Auxiliary operations are performed in semiautomatic mode. After these operations are completed, the information is transmitted to the production of photomasks.

6.6 Stage of Physical Verification and Preparation for Production (Fig. 6.6)

The main purpose of physical verification is to make a decision on the transfer of information for the production of LSI to the fab. The decision is made based on the results of checking the project for compliance with the requirements of the specifications. Physical verification is carried out simultaneously with the physical design and is associated with it in a *single iterative cycle*. When performing a set of checks, the following problems should be solved:

- Noise immunity evaluation.
- Checking the connectivity of signals in the time diagram of the system operation.
- Reliability assessment.
- Checking the rules for preparing the electrical circuit (ERC—electrical rules check).

- Checking the topologic design rules (DRC—design rules check).
- Checking the compliance of the electrical circuit and topology (LV S—layout versus schematic).
- Checking the completeness of the specifications.

Physical verification is the least formalized stage of design. Let's first consider the automated procedures: ERC, checking the rules for preparing the electrical circuit; DRC, checking the rules for topologic design; and LVS, checking the compliance of the electrical circuit and topology. Although the procedures are automatic, they require strict compliance with the rules for preparing information for verification.

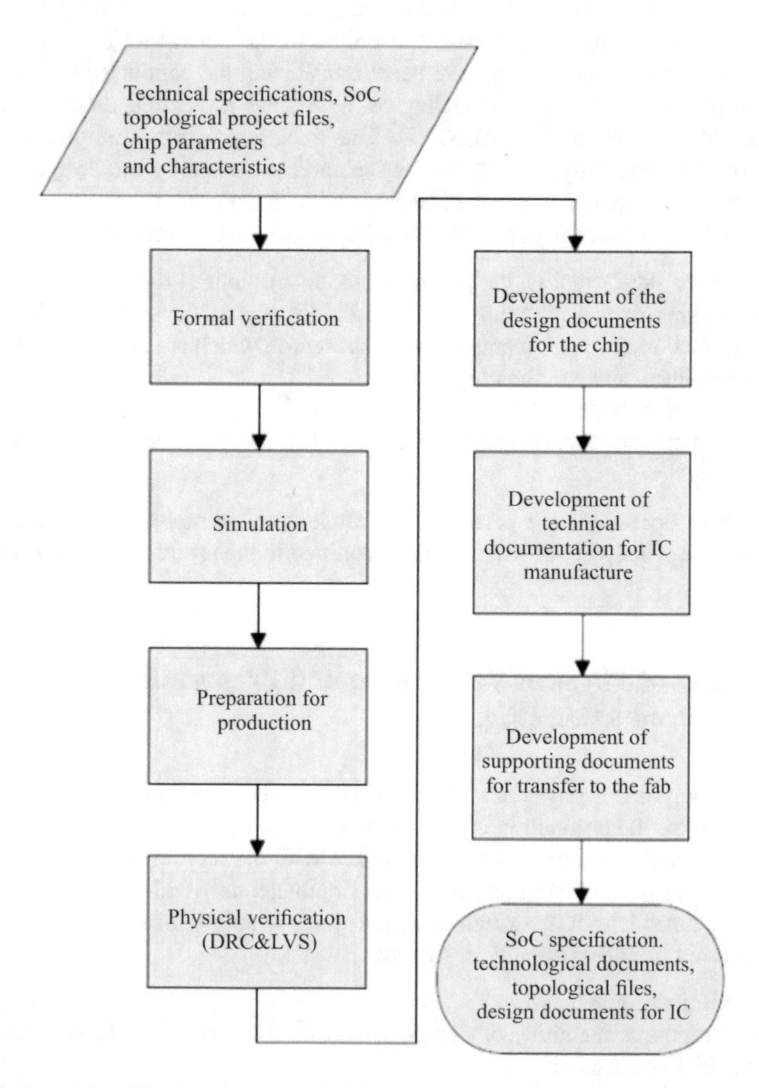

Fig. 6.6 Stage of verification and preparation for production [2]

First, it is the preparation of the verification files, i.e., the verification rules themselves. Usually, DRC files are prepared by factories and delivered together with the design rules. At the same time, there is often a situation when the modification of the technology and verification files leads to the appearance of "errors" in previously created and verified units. In this case, additional time has to be spent on the approval and adjustment of the verification files. ERC and LVS are interconnected procedures. If the designer has not provided full compliance, then the error checking programs will not find it. For example, if there should be several pins in the power supply circuit on the IC, and only one is indicated in the electrical circuit, then the program, having found one, will not search for the rest.

When checking the reliability, the resistance of the LSI to electrostatic discharges through the signal terminals, the resistance of metallization to electromigration, the calculation of the thermal resistance and thermal mode of operation of the circuit, and the thermomechanical stresses in the chip and the package are calculated. When checking the connectivity of signals, the parameters of communication lines are calculated, and the circuit is simulated taking into account these parameters and interference simulation, accounting for substrate noise in the operation of analog units, and checking the balance of delays in the time diagram.

We also note that the amount of information on the project is very large, and physical verification requires a lot of time and effort. The results of the checks are not directly reflected in the specifications and are difficult to control. Reducing the volume of checks leads to errors.

6.7 Project Certification

The final stage of the design is the certification of the project. *The purpose of the certification is to make a decision on the readiness of the project to start production development*. To do this, it is necessary to ensure that the test samples of the LSI meet the requirements of the regulatory technical documents (TSs, Technical Standards, Reference List), as well as the availability of the regulatory documents itself. The compliance requirements of the samples and documents are confirmed by the *test reports*. We will not consider *organizational* certification procedures. We will focus only on *technical* issues.

Issues to be solved in the process of the project certification:

- Checking the functioning of individual units and the system as a whole using built-in controls.
- Checking the functioning of the system by comparing it with the functional layout.
- Checking the operation as part of the equipment layout.
- Checking the system's noise immunity in the worst operating conditions.
- Checking the reliability of the system in the maximum permissible operating modes.

An important element of the certification methodology is the selection of *verification criteria*. In complex systems, errors always occur. The controls themselves are not error-free. The acceptable level of errors is determined by the purpose of the system. It is desirable that the acceptable level of errors is specified in the Specification. Otherwise, this parameter shall be defined during the project execution process.

If *noncompliance* with the document requirements is found, it is important to identify *the reasons* for this noncompliance. There are no formal rules for identifying the causes of errors. To do this, you need the right functional model, process models, and the necessary measuring equipment.

The design flow is constantly evolving, as SoCs are constantly becoming more complex in terms of the composition and variety of units used, and CAD designers offer new programs for simulation and optimizing LSI.

6.8 SoC Design Flow

6.8.1 Trends in the Development of Design Tools

In this section, the *system level of design* is considered in detail: purpose, tasks, flow, and software.

The steady growth of requirements for radio-electronic equipment (REE) from the customer, both in terms of functional and operational qualities, forces designers to create more and more complex devices. At the same time, significant progress in the field of VLSI production technologies (0.35 μ or less) has made it possible to create chips with an integration degree of more than ten million gates per chip.

A typical example is given in [3]. When using traditional design methods, a good designer can execute a project at an average performance of about 100 gates per day or 30 lines of RTL code—these numbers have remained constant over the past years. In this case, to design an ASIC (application-specific integrated circuit) of VLSI type with a complexity of 100,000 gates, it will take 1000 man-days, i.e., a team of five people will be able to develop such a VLSI within a year. Following this logic, to develop a complex VLSI of about ten million gates will require a team of 500 people within 1 year, which is certainly unacceptable from the point of view of the cost of development.

According to more accurate analytical forecasts (Fig. 6.7) [3], if the transition to new technologies uses the existing design methodology, the project cost increases to 250 man-years, which is also unacceptable for the customer.

It should be noted that recently, there has been a trend of constant growth in the share of costs for the development of REE software (Fig. 6.8). If you develop software and VLSI separately, then the probability of detecting errors at the testing or operation stages of the entire hardware complex increases.

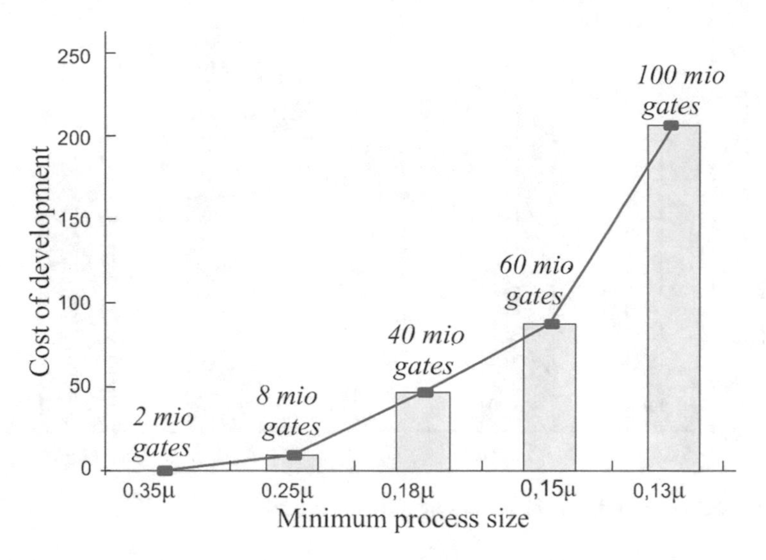

Fig. 6.7 Dependence of the VLSI development cost on the technology

The most promising direction at the moment is the methodology of designing VLSI of the “system-on-chip” type (hereinafter system-on-chip)—in foreign literature, system-on-chip (SoC).

In addition, there are a number of *reasons* why it is necessary to switch to a new design methodology:

- In a market environment, profits are highly dependent on development time.
- VLSI technical parameters such as performance, chip area, and power consumption are key elements in promoting the product to the market.
- Increasing the degree of integration makes the verification task qualitatively more complex.
- Due to the new features of the deep submicron technology (DSM—deep submicron), it is increasingly difficult to meet all the requirements for time constraints (timing).
- Highly integrated VLSI development teams have varying levels of design knowledge and experience, and are often located in different parts of the world when executing VLSI projects.

6.8.2 SoC Design Methodology

The SoC design *methodology* is based on the principle of reuse of functional units. Functional units that are developed within one project or specifically are then used in other projects. By analogy with the system-on-chip, where the components are finished ICs, the system-on-chip is constructed from reusable functional units.

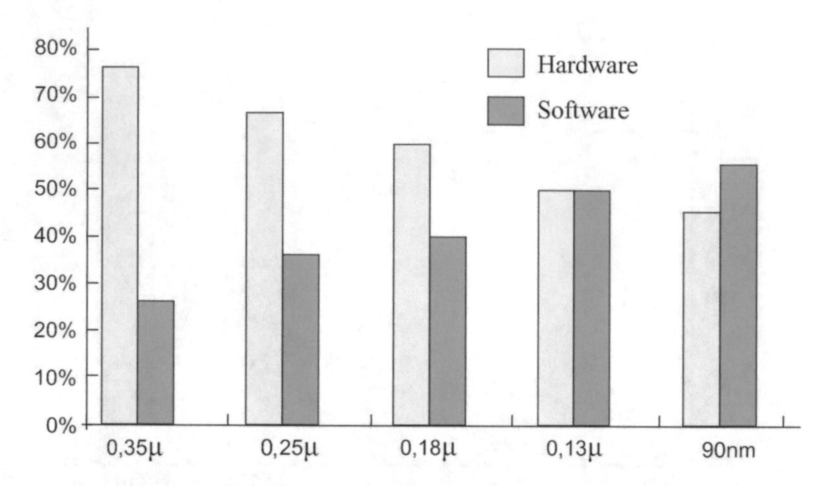

Fig. 6.8 The ratio of software and hardware development costs [3]

Here we use the term *IP unit (intellectual property)*, i.e., a unit that represents an object of intellectual property. At the same time, other terms are often used [3]:

- CF unit (complex functional unit).
- Macro.
- "Core" is usually used to refer to units of the CPU (central processing unit) or DSP (digital signal processor) type.
- VC (virtual component) is introduced by VSIA (Virtual Socket Interface Alliance), a nonprofit international organization whose main task is to develop regulatory documents on the problems of designing IP units and SoCs based on them (www.vsi.org).
- "Unit" is a functionally complete part of a SoC or ASIC, not necessarily a reuse.
- "Subunit" is a part of the unit.

In principle, there are two types of IP units: the so-called soft, described at the RTL level, and hard, described at the topological level. Sometimes firm IP is distinguished, which includes different types of representations from RTL to a list of chains with a subunit layout.

One of the main problems of SoC design is the creation of such IP units. Practical experience shows that the cost of a reuse unit is on average ten times higher than the cost of a similar one-time used unit, and for processors this value is an order higher. For this reason, reuse units usually solve common logically formalized tasks, such as MPEG2 encoder, CPU, DSP, USB interface, PCI interface, etc.

In fact, *the design flow of a typical SoC* is divided into four stages:

1. Development of the SoC architecture at the system level.
2. Selection of available IP units from the database (within the company, other companies, or suppliers of IP units).
3. Development of remaining units.
4. Integration of all units on a chip.

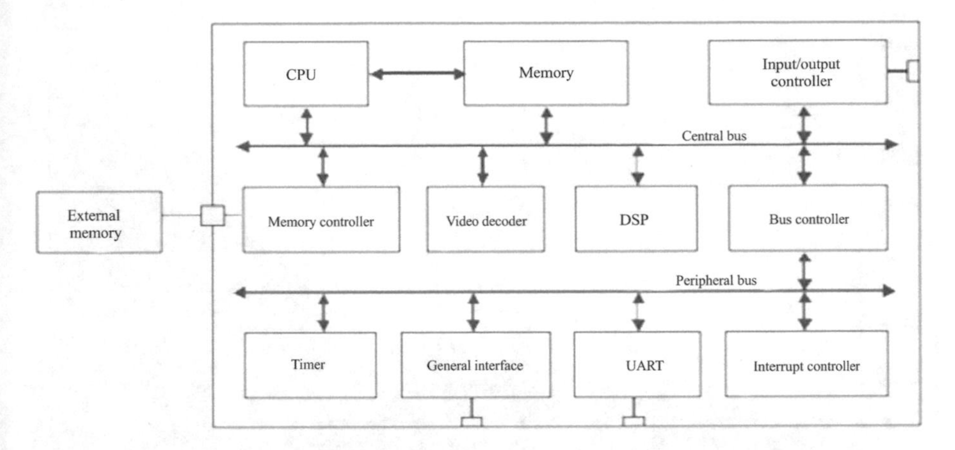

Fig. 6.9 Example of the SoC structure [3]

Another fundamental feature of SoC is the presence of programmable processor units. Therefore, SoC is not just an integrated circuit (IC), but a complex one that includes both the hardware part, the chip, and the software part, the embedded software. It is assumed that the SoC design flow should contain operations for joint verification and debugging of the software and hardware parts.

A standard SoC shall be manufactured using a technology of at least 0.35 microns and contain at least one million gates.

Figure 6.9 shows an example of the SoC structure in general form. As you can see from the figure, the SoC consists of the following components:

1. The microprocessor (or microprocessors) and the memory subsystem (static and/or dynamic). The processor type can range from the simplest 8-bit to a high-performance 64-bit RISC processor.
2. The buses are central (high-performance) and peripheral, to provide data exchange between the units.
3. External memory controller for memory expansion, such as DRAM, SRAM, or Flash.
4. Input/output controller: PCI, Ethernet, USB, etc.
5. Video decoder, such as MPEG2, AVI, and ASF.
6. Timer and interrupt controller.
7. General I/O interface, for example, to display the power availability information on the LED indicator.
8. UART (universal asynchronous receiver/transmitter) interface).

It should be noted that the following *fundamental features* of the system-on-chip can be distinguished [3]:

1. SoC is constructed from complex functional units that can be either developed from scratch or obtained from an IP unit provider. Therefore, there should be an architectural design stage in the design flow, the tasks of which include the

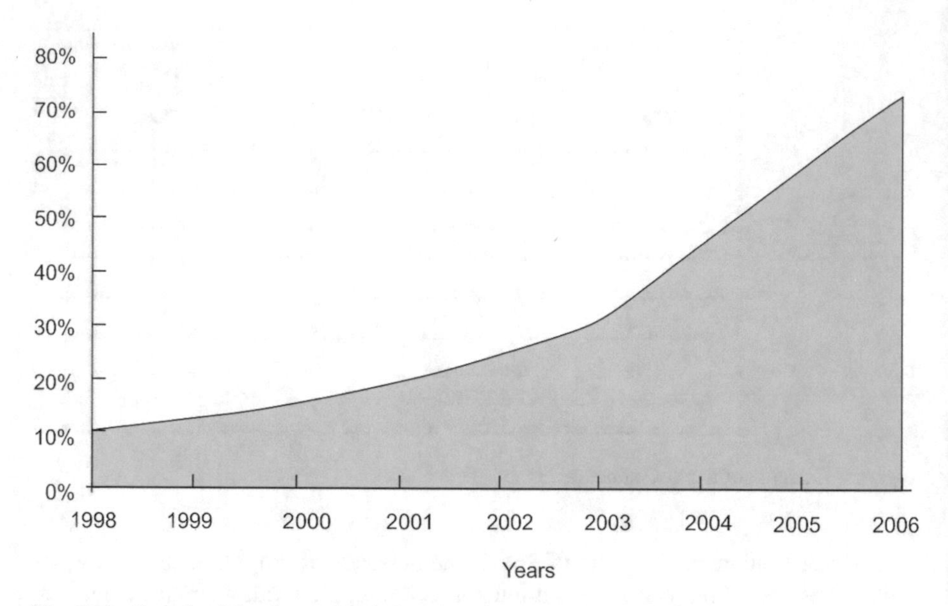

Fig. 6.10 Growth of the share of digital-analog systems in the total SoC volume

development of the overall SoC architecture. When choosing IP units, the cost of the finished unit is also taken into account and the cost of developing your own unit is estimated.
2. SoC has embedded software, so the flow should include the stages of joint software and hardware verification (HW)—software and hardware verification (HW/SW co-verification).
3. The SoC shall be manufactured using a technology not lower than 0.35 microns.
4. There is a steady increase in the share of mixed digital-analog systems in the total SoC volume (Fig. 6.10); therefore, the design flow should include steps for the joint development and verification of the digital and analog parts of the SoC.

6.8.3 SoC Design Flow

The traditional ASIC design flow is shown in Fig. 6.11 [3, 4].

The design process begins with the development of a specification for the ASIC being designed. For complex VLSI systems, such as graphical information processing devices, the specification includes an information processing algorithm, which is then used by designers to write RTL code.

After functional verification, VLSI is synthesized at the gate level in the form of a list of circuits. This is also where the verification of temporary requirements is performed. Once the time requirements are met, the list of circuits is passed to

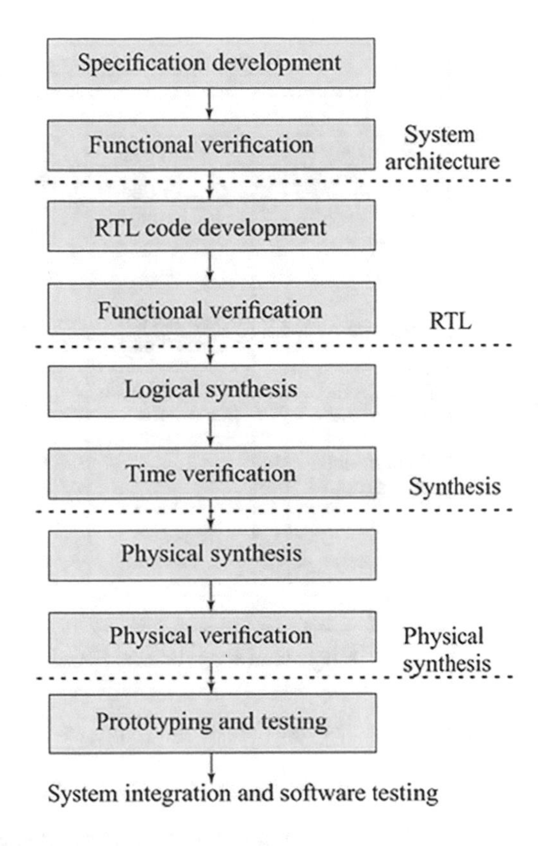

Fig. 6.11 ASIC design flow

physical synthesis: element placement and circuit tracing. At the end, a physical prototype of VLSI is created and tested, on the basis of which system integration is subsequently performed and software is tested.

However, using such a flow, you can only design circuits with a complexity of no more than 100,000 valves using a technology of no less than 0.5 microns. This is due to the fact that the project here progresses in stages from one phase to another and there is never a return to the previous phases. For example, an RTL designer cannot come to a system designer and say that his algorithm is unrealizable, or a logical synthesis team cannot ask to change the RTL code to achieve the necessary time parameters. For circuits manufactured using DSM technology, this flow will not work at all, since the features of the physical implementation shall be taken into account already at the logic design level.

Many companies use a spiral model rather than the traditional top-down design flow model (Fig. 6.12) [5]. Here, the design is carried out simultaneously in four areas: software development, RTL code development, logical synthesis, and physical synthesis. At the same time, in the course of work, the development teams exchange the design results.

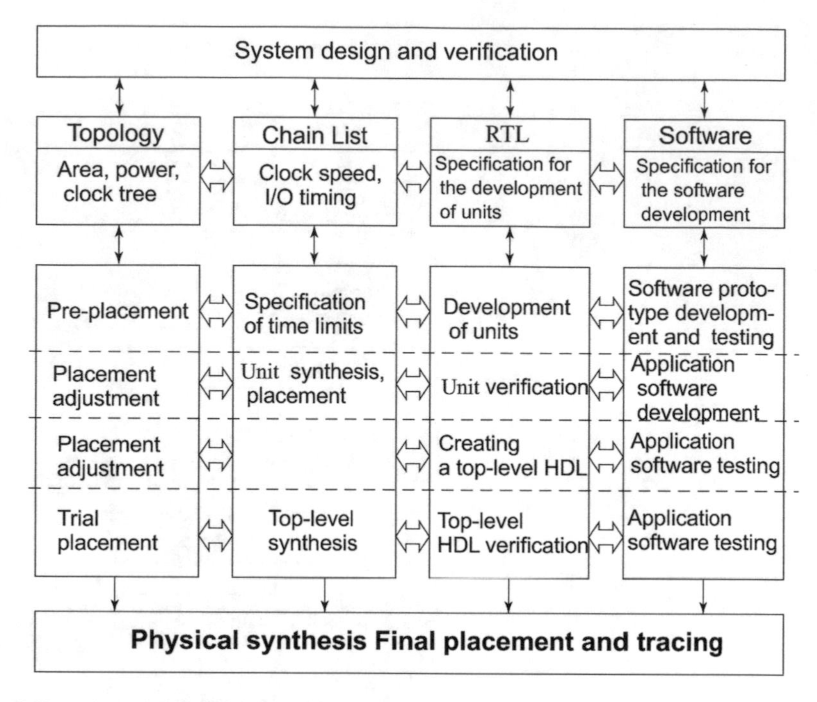

Fig. 6.12 Spiral model of the SoC design process

The new model is characterized by the following useful properties:

- Parallel verification and logical unit synthesis.
- Layout, placement, and tracing are included in the synthesis process.
- It is allowed to return to the previous design phases and adjust the results.

6.8.4 SoC System Design

The initial stage of system design is to recursively develop, verify, and refine a set of specifications to such a degree of detail that you can start creating RTL code based on them. Developing clear, complete, and consistent specifications quickly is a challenge. In a good design methodology, this is a time-consuming, responsible, and lengthy design phase. If you clearly know what you need to build, then errors in further implementation can be quickly detected and eliminated. Otherwise, you may not notice the error throughout the entire design cycle until the chip is manufactured.

The specifications describe the behavior of the system. More precisely, they describe how to manage the system in such a way as to get the desired behavior from it. In this sense, the concept of a specification is largely related to the concept of an interface. A functional specification describes the interface of a system or unit as seen by an external user. They contain information about contacts, buses, registers, and how to handle them. The architecture specification describes the interactions between the parts of the unit and the behavior at the system level.

Specifications should be developed for both the hardware and software parts of the project and should include the following information [3–5].

For the hardware:

- Functions performed.
- External interface to other units (contacts, buses, protocols).
- Interface with software (registers).
- Time parameters.
- High performance.
- Features of the physical level (chip area, power consumption).

For the software:

- Functions performed.
- Time parameters.
- High performance.
- Interface for the hardware.
- Structure and core.

Traditionally, specifications are written in natural languages such as Russian and English, which introduces indefiniteness, ambiguities, and errors. To get rid of these problems, many companies are starting to use executable specifications. At high levels, C, C++, or variations of C++, such as SystemC, are used to write specifications being executed (www.systemc.org) [6]. VHDL or Verilog is usually used to describe the hardware. The development of models being executed allows designers to verify the main functions and interfaces performed at the early stages of design, long before the project is detailed.

The SoC design process at the system level is shown in Fig. 6.13. The development process begins with the identification of the goals and tasks performed by the SoC. At the initial stage, it is necessary to determine the main operational and technical properties: the required performance, the permissible power consumption, the time required for development, etc. On the basis of these properties, a system specification is created, which can be part of the TS for the development of the system. As a rule, this stage is performed without the use of specialized CAD software.

Then a high-level behavioral model of the entire system under development is created. The behavioral model of the system is usually constructed in the form of a flowchart. To verify the developed behavioral model, a test environment (Test Bench) of the system is created, which includes input signal generators, test sequences, and output information display units. The test environment should verify the functioning of the system as fully as possible. Subsequently, based on this test environment, test sequences will be developed for project verification at the lower design levels and for testing VLSI prototypes.

Verification of the behavioral model is carried out by computer simulation using special software tools. If any deviations from the requirements of the system specification are detected during the verification process, the model is corrected and the simulation is repeated.

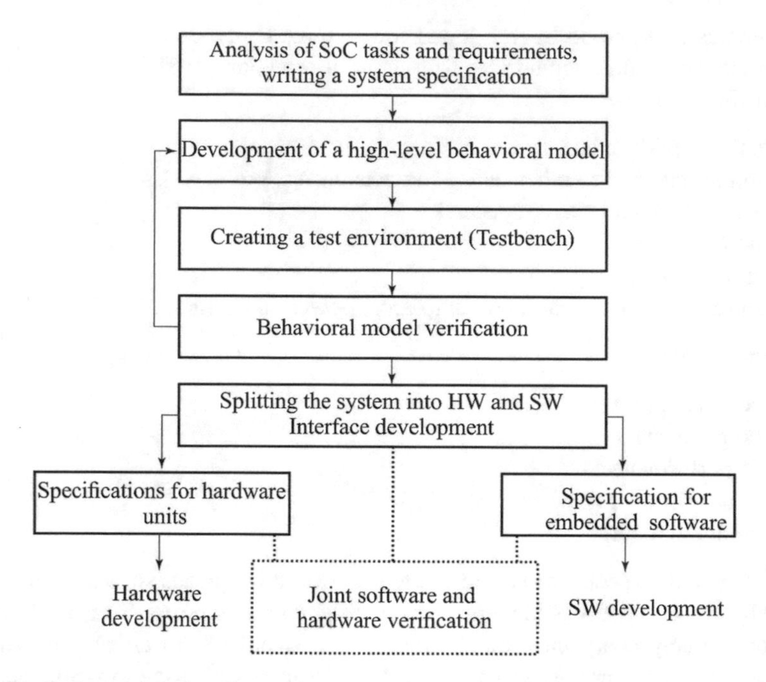

Fig. 6.13 SoC system design flow

In addition to verification, the optimal parameters of the system algorithm can be selected at this step. For example, a designer can find a compromise between computational complexity and accuracy.

As mentioned above, SoC includes one or more programmable processor cores. Therefore, at the next stage, the designer shall decide which parts of the behavioral model will later be implemented at the hardware level and which parts at the software level in the form of embedded software in VLSI. In addition, it is necessary to determine how the software and hardware parts will interact, i.e., an interface between the HW and SW should be developed. It also defines the general architecture of the SoC: the type of processor, the type of memory and its volume, hardware units, the HW-SW interface, the type of bus used, the description of the software part, etc.

As a result, a set of specifications for software development and for the development of each hardware-implemented unit is formed.

In the ASIC design methodology, hardware and software verification is performed only after the prototype is manufactured. The resulting errors of operation, as a rule, can be corrected by making appropriate changes to the software, without modification of the chip itself. This method is not suitable for the SoC design process, because due to the great complexity of the circuit, it is very difficult, and often simply impossible, to correct the HW errors by adjusting the software. Therefore, the software and hardware verification operation is introduced into the design flow at different levels. The most important stages in this sense are the functional and logic design stages.

Software and hardware verification of the system level is not a mandatory operation today. However, many designers include it in the SoC design flow. The hardware part here is the performed specifications of hardware-implemented units, and the software part is the software prototype. Thus, you can make sure that the hardware developed in accordance with the available specifications will function correctly under the control of the embedded software in real time.

6.8.5 CAD Software for the System Level

Until recently, system-level tasks, i.e., the development of specifications, in most cases were solved without the use of special software tools. With the transition to the SoC design methodology, the need to automate the system design process became apparent.

First, since SoC is a highly integrated VLSI, it includes a large number of complex units: processors, rigid logic, memory, control circuits, analog and digital-to-analog components, etc. It is impossible to check the performance of such a system using only analytical calculation methods. Therefore, as mentioned above, there is a need to simulate the entire system at the behavioral level, and this requires special application software.

Second, the specification being executed shall be presented in a specific format in C, C++, SystemC, Verilog, or VHDL. It is impossible to get such a description without using the appropriate software tools.

In addition, when moving to a spiral design model, there may be constant "rollbacks" from the lower levels of design to the upper ones. Often, the system level merges with the functional design level (RTL code development), forming a system-functional design level. In this case, it is convenient to use common software and hardware tools.

Currently, there are a huge number of software tools in the world that can be used to automate system design. These tools can be classified into five groups [5].

The first group includes application software development and debugging tools. The C programming language and its C++ modification turned out to be quite popular and convenient for presenting specifications. Therefore, system-level designers actively use software application development tools. Below are the most popular ones:

- Microsoft Visual Studio (www.microsoft.com).
- Inprise Borland C++ Builder (www.borland.com).
- Borland Kylix C++ (www.borland.com)—in the Linux operating system.

The main advantages are low cost and easy to learn and use. The main disadvantage is the lack of specialized system-level libraries, so the designer has to create a behavioral model almost "from scratch." Sometimes they also use special tools for analyzing and automating software development, such as Rational Rose and the UML language (www.uml.ru).

The second group includes mathematical simulation tools. The most typical representative here is the common software package MATLAB/Simulink (www.mathworks.com). The main advantages here, as in the first group, are low cost and ease of use; in addition, there is the ability to perform mathematical simulation and the availability of model libraries. The disadvantages include the lack of specialized libraries, i.e., most models have to be created manually, and the use of their own data format (M-files, MEX-files) as the base. The last remark is probably temporary, since the MathWorks company plans to release a full-fledged translator from the M-file format to the C/C++format.

In the third group, you can combine general-purpose simulation tools, for example, MLDesigner (www.mldesigner.com) or SES/Workbench (www.hyperformix.com). A distinctive feature of these tools is that they are not tied to any specific design object. They can be used to simulate, for example, both the VLSI architecture and the satellite communication or navigation system. The advantages are relatively low price and wide range of applications. The main disadvantage is the lack of communication with other levels of design: functional and logical.

The fourth group is the most numerous. It includes software tools, each of which is designed to solve a specific range of system-level design tasks. At the same time, the overall range of tasks is large: from the development of software and hardware architecture to the integration of processor cores and the development of embedded software. As an example of the manufacturers of this software, you can give the names of such companies as CoWare (www.coware.com), Mentor Graphics (www.mentor.com), Elanix (www.elanix.com), Summit Design (www.sd.com), etc. The use of such specialized software is quite attractive from the point of view of cost savings.

The fifth group includes powerful integrated software packages, with which the designer is able to perform the entire cycle of system and functional design, as well as the entire design cycle up to the physical implementation. To date, this group includes software packages from only two companies:

- Synopsys (www.synopsys.com): CoCentric System Studio, Design Ware, VCS, VCSi, Scirocco, SystemC HDL Co-Sim, and CoCentric SystemC Compiler.
- Cadence Design Systems (www.cadence.com): Incisive-SPW, Incisive unified simulator, Incisive-XLD, Incisive-AMS, NC-SystemC, NC-Verilog, and NC-VHDL.

The main disadvantage of both packages is their high cost, which is very important in the conditions of the Russian market.

When choosing software tools and developing a design flow based on them, it is necessary to additionally take into account a number of factors, for example, the specifics of the devices being developed, the total amount of work (the number of projects performed simultaneously), the software available at the company for the functional and logical levels of design, etc. [6].

6.9 Practical Example of the System-on-Chip Simulation

6.9.1 Standard Design Flow of the SoC of Cadence Company

Below, we will take a detailed look at one of the main stages of the design flow, i.e., device verification and simulation.

Figure 6.14 A detailed system design flow [5]

As we mentioned earlier, device verification usually takes more than 70% of the development time of the entire device concept.

6.9.2 Description of the Simulation and Verification Environment

To understand the specifics of implementing this flow, first of all, we will briefly consider the basic information about the simulation and verification environment of the Cadence Incisive software product [5].

The full name of this product is *Cadence Incisive Design Team Simulator*. This environment is a simulation tool that supports all Verilog, VHDL, System Verilog, E, SystemC, PSL, and SVA language bases, as well as all levels of project abstraction, starting from the gate level and ending with system simulation and verification.

The Cadence Incisive simulation and verification environment runs Linux OS. Starting the program:

(a) Start the Linux OS (Fig. 6.15).

(b) Right-click on the desktop.
(c) In the menu that appears, select "Открыть в терминале"/"Open in Terminal" (Fig. 6.16).

(d) Launch the CSH wrapper.
(e) Specify the path of the cshrc file with links to programs.
(f) Launch the Cadence Incisive program (Fig. 6.17).

So, the main window of the Cadence Incisive program opens (Fig. 6.18).

Let's take a closer look at all the areas of this main window [5].

The Area of the Hierarchical Module Tree In this area, a hierarchical tree is implemented, according to the specific Verilog description that you selected as an example in the previous paragraph (e). In a specific example, you can see that the test module is both a head module and a wrapper module.

Module a1 in this example is the base module.

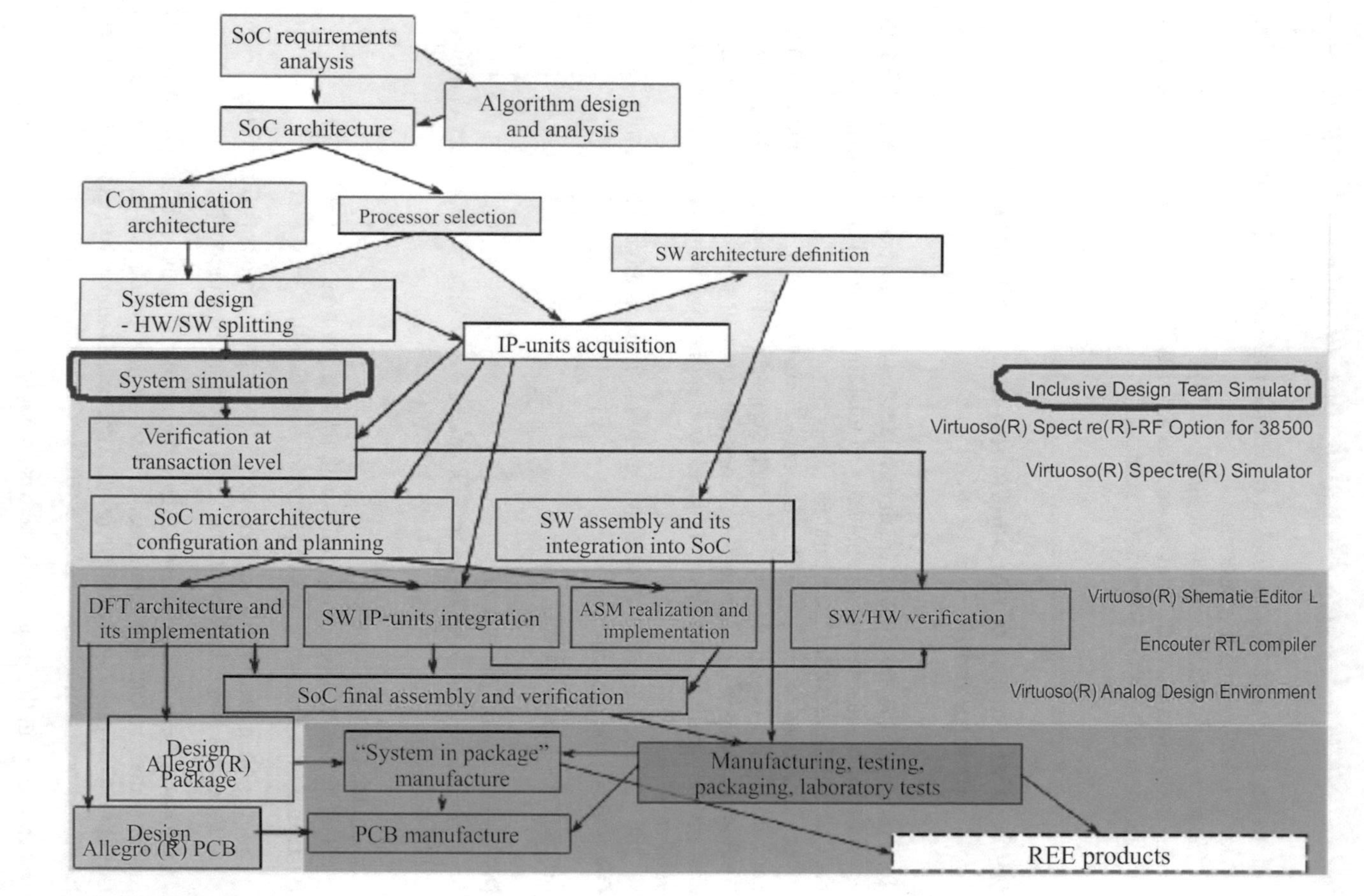

Fig. 6.14 Detailed design flow of the System-On-Chip of Cadence company

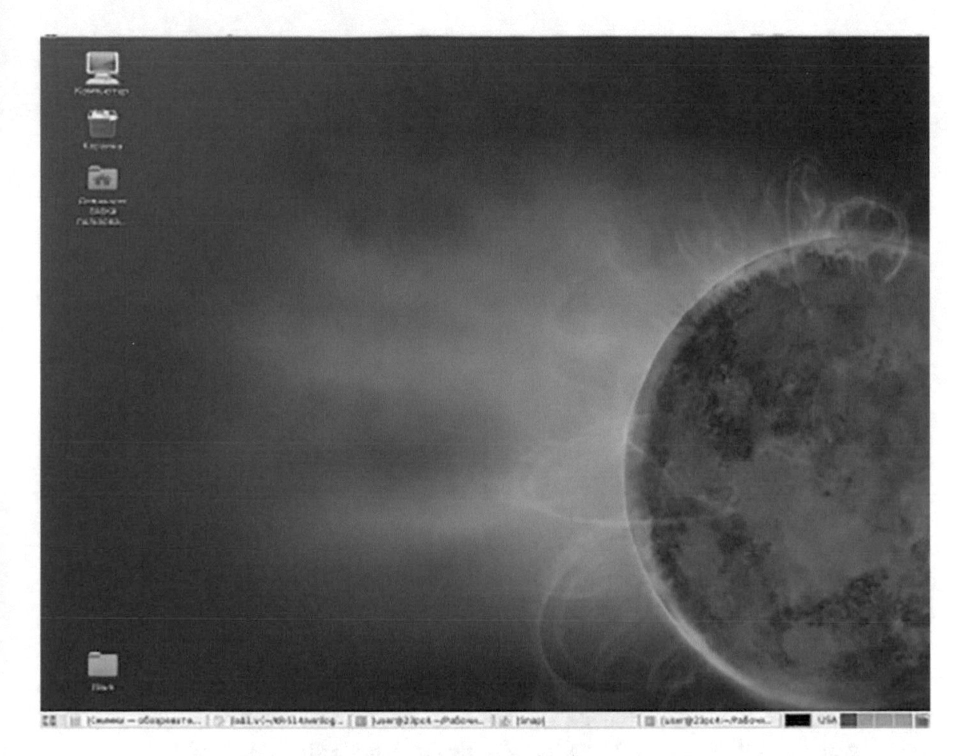

Fig. 6.15 Program launch window

At the bottom of the area, you can see the filtering function, which is used to simplify the hierarchical tree when large projects are simulated, as well as to find the module you need in the general hierarchical tree.

The Area of the Declared Ports of the Selected Module This area displays all the ports and variables of the module you selected. Fig. 6.19 shows that the test module does not contain any ports, but only contains reg variables (registers).

At the bottom of the area, you can also see the filtering function. Let's decipher the purpose of these four windows:

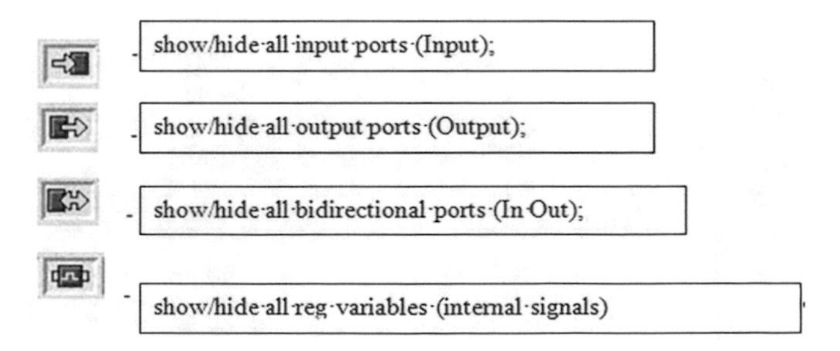

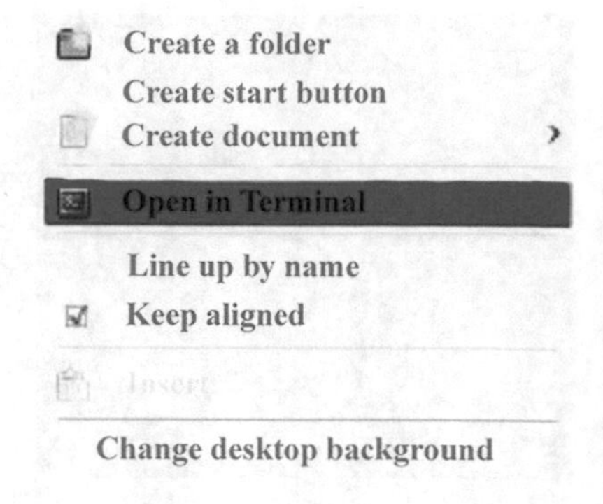

Fig. 6.16 Selecting the "Open in Terminal" option

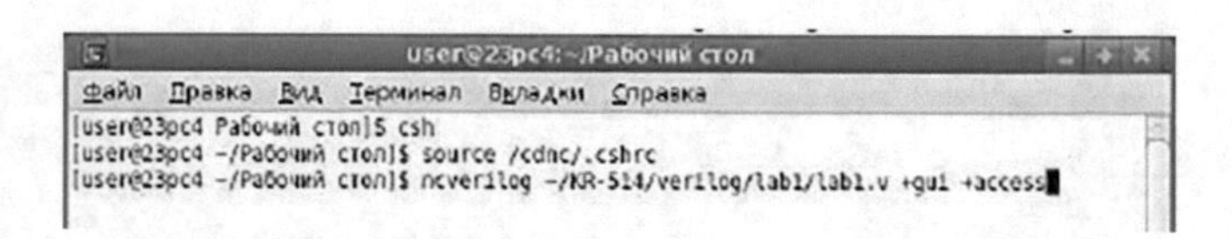

Fig. 6.17 Launching the Cadence Incisive program

Area of Signal States at a Given Time The specific signal states at a given time are displayed here. The convenience of this area is that for small projects, you can quickly access the simulation results of your chosen Verilog description.

Simulation Process Control Area This area contains the main panel for managing the simulation process. With it, you can start the simulation process, pause it, return it to its original position, set a stop point, and also have the ability to step-by-step simulation with a given step.

Area of Navigation Area and Search for Ports and Signal States In this area, you can place specific points of interest to you and the most significant simulation points, navigate through these points, and search "by signal state."

Area of Various Verilog Description Views This area contains links to the tools for presenting Verilog descriptions in various forms:

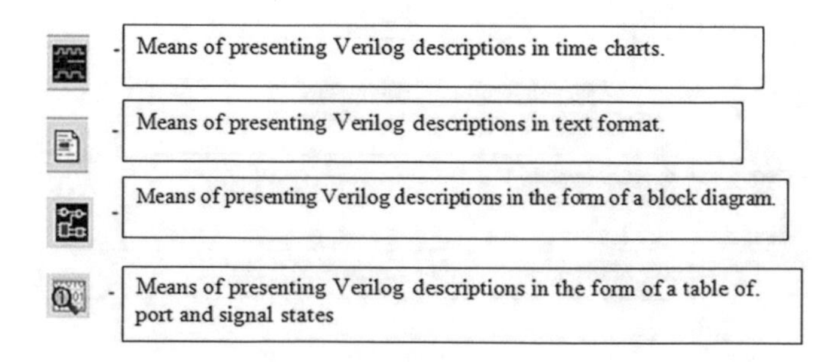

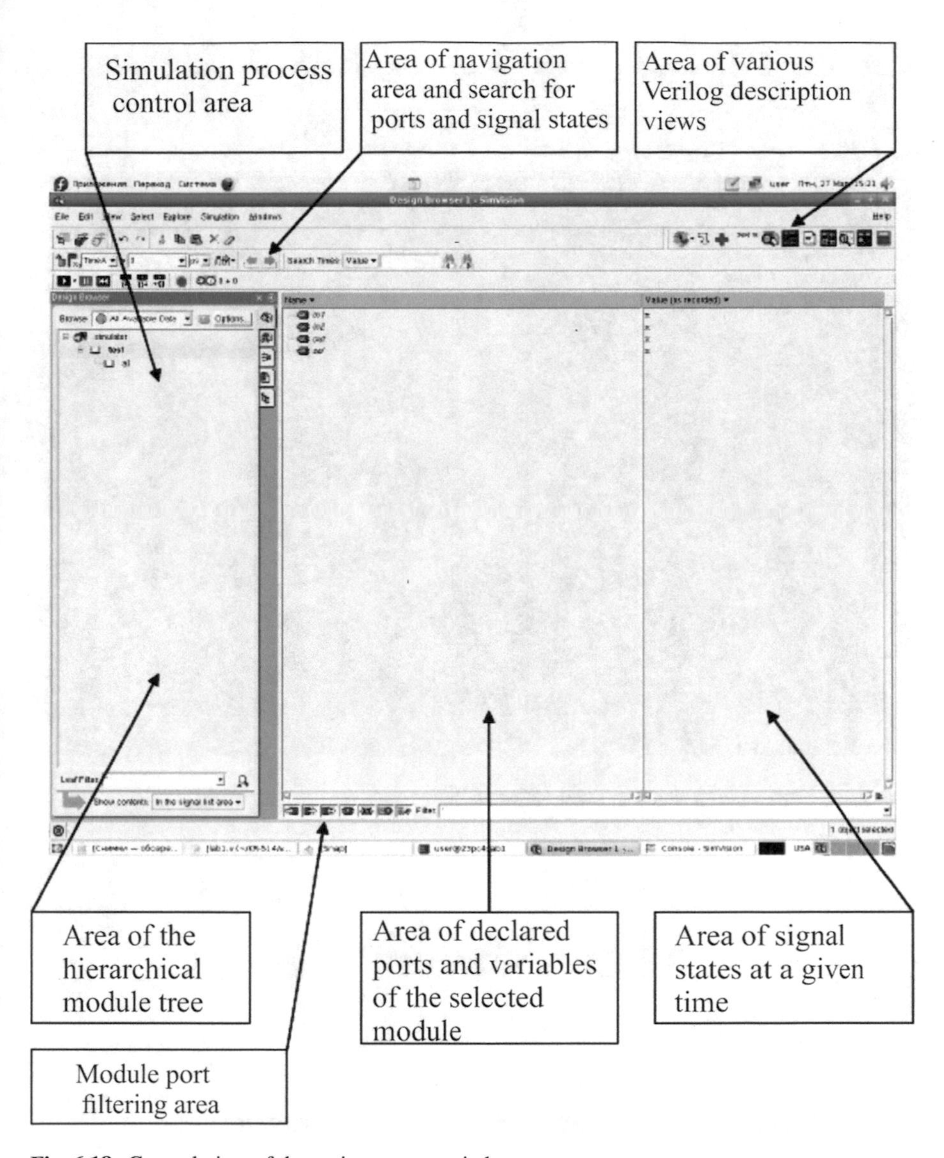

Fig. 6.18 General view of the main program window

There is also a standard set of the following operations in the main window of the program:

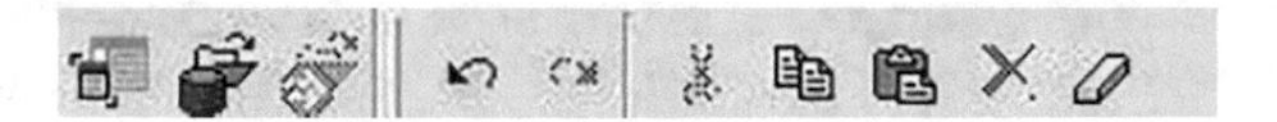

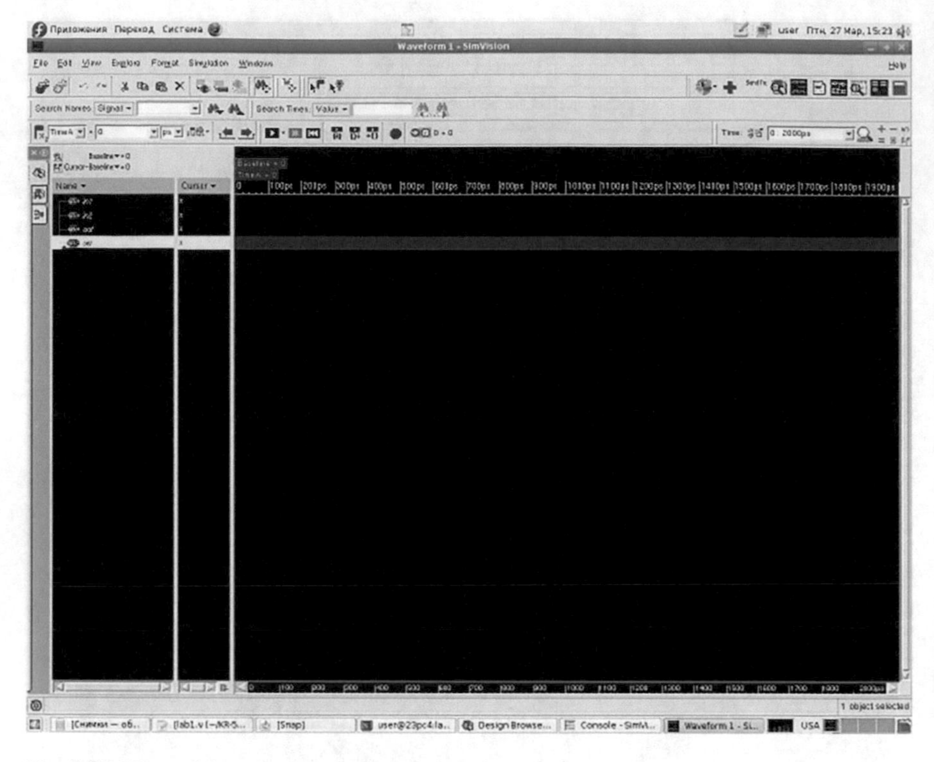

Fig. 6.19 View of time charts in the description view window

Pay attention to the "top menu."

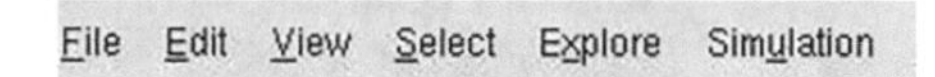

6.9.3 Project in the Cadence Incisive Environment

Below, as a practical example, we will simulate an example of a Verilog description of a two-by-one multiplexer (2–1), and also consider all its representations in the form of time diagrams, a unit diagram, and a text description. An example is taken from the training manual [5].

Perform the following sequence of actions: in the main window of the simulation environment, select the test module. Next, click the temporary manipulation button. As a result of manipulations, a window will open to view the Verilog description in the form of time diagrams (Fig. 6.19).

There are three new areas: the area for displaying ports, variables, and signals included in the module, the area for the diagrams themselves, and the area for navigating through the diagram area.

Now let's look at these areas in more detail.

Port Display Area In this area, you can select one or more ports, move them, duplicate them, rename them, set the desired representation (binary, decimal, hexadecimal, and octal code), combine them into a group and into a bus, change the color, add comments, etc.

For all operations on the port, select it, and right-click, and then the menu of operations and properties will open.

The Display Area of the Actual Time Charts Here you can view the time charts in detail and perform the necessary set of operations: place labels, select one of the signals, etc (Fig. 6.20).

The Display Area of the Actual Time Charts Here you can view the time charts in detail and perform the necessary set of operations: place labels, select one of the signals, etc (Fig. 6.21).

Here there are the tools you need to navigate through the time chart area: you can zoom in or out of the time chart display area. This can be done either by using the "+" and "-" buttons, or by using the time orientation (Fig. 6.22).

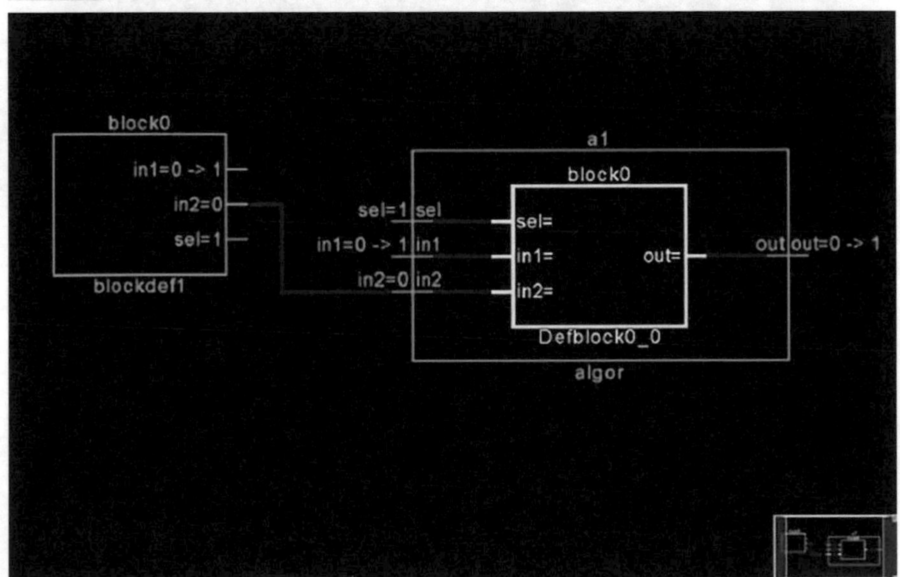

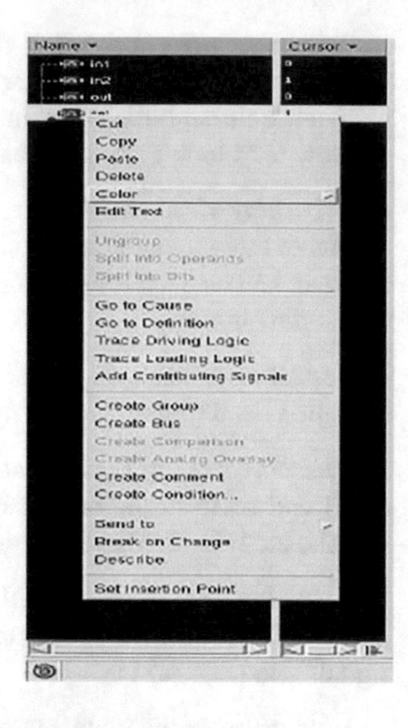

Fig. 6.20 View of the port display area

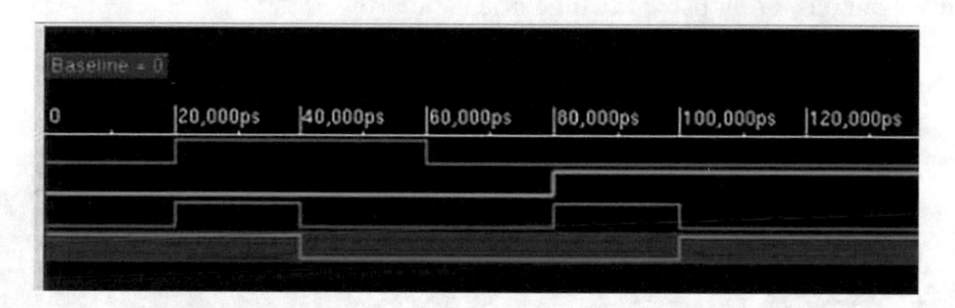

Fig. 6.21 Time chart display area

Fig. 6.22 Navigation area

To do this, click on the button. As a result, we get this "picture":

At the bottom of the window, there is a quick navigation area, which is used when designing large projects.

For a more in-depth study of the design features of system-on-chip, we recommend that you refer to the works [2–5].

References

1. Belous A. I., Solodukha V. A. (2012). Fabless-model of business organization at a microelectronic company: Myths and realities. *Components and Technologies* 8: 14–18. (in Russian).
2. https://www.kit-e.ru/articles/circuit/2012_11_154.php
3. Adamov Yu. F. (2005). *Designing systems-on-chip*. Moscow.
4. Nemudroye, V., Martin, G. (2004). Systems-on-chip. Design and development. Moscow: Technosphere. (in Russian).
5. Tuchin, A. V., Bormontov, E. N., Ponomarev, K. G.. (2017). *Introduction to the systems of automated design of integrated circuits*. Educational and methodological manual: VSU Publishing House, Voronezh-2017. (in Russian).
6. Rabai, Zh. M., Chandrakasan, A., Nikolich, B. (2007) *Digital integrated circuits. Design methodology*. Translated from English. Moscow: "Williams" Publishing house. (in Russian).

Chapter 7
Fundamentals of CMOS Microcircuits Logic Design with Reduced Power Consumption

This chapter attends to some peculiar features of logic designing of CMOS low-power consumption ICs. As it is illustrated in the previous chapters, some special trends in development of present-day submicron technologies are coercing the IC designers to seek after any new design techniques and methods, targeted at reduction of power consumption associated with leak currents value, and input of those noticeably increases along with diminishing of design norms. In Chap. 2 such methods and solutions are scrutinized in great depth. Meanwhile, there exists another way how to reduce power consumption: it is possible to use some special methods and concepts even at the initial phases of logic IC design. This chapter, actually, addresses fundamentals of CMOS IC logic designing with reduced power consumption. As the theoretical background of the approach, there is used the mathematical tool for probabilistic assessment of various updating options by predicted switching activity of basic units of the IC under designing.

Here in a consistent manner, there are addressed standard general phases of such logic design, beginning from the required elemental logic basis, logic synthesis in the selected basis, enhancement (optimization) procedures of double-level logic microcircuits, and optimization of multilevel logic circuits, built both on two-input and multi-input logic gates and finishing with the procedures of technological reflection and evaluation of power consumption of the designed (synthesized) microcircuit simultaneously at logic and schematic.

At the end of the chapter, there is provided a minute description of the respective software and hardware complex of logic design of micropower CMOS of microcircuits.

Over the recent decade across the globe, one could have witnessed the evolution (change) in the conventional model of microcircuit design determined by technical phenomenon when all kinds of hardware "Trojans" being built by "someone" into the microcircuits make their appearance.

A. Belous, V. Saladukha, *The Art and Science of Microelectronic Circuit Design*,
https://doi.org/10.1007/978-3-030-89854-0_7

7.1 Basics of Low-Power-Driven Logic Synthesis of CMOS Microcircuits

With advances in microelectronic technologies, the degree of integration and clock frequency is constantly increasing, which, in turn, makes it possible to create faster and more complex devices on a single chip. Alongside the broad opportunities this advances bring for electronic products, however, they also give rise to serious challenges, associated primarily with power dissipation. High levels of integration have led to increased reliability problems when compared with similar devices with a lower degree of integration. These issues cannot be ignored, as increasingly complex products are being designed for various applications, including space applications, which have to be reliable and able to operate autonomously for long periods without recharging the battery.

Ensuring the operability of integrated circuits under exposure to ionizing radiation and autonomous power supply is becoming a more important problem nowadays due to the fact that integrated circuits are widely used in military and space applications. The advantage of using CMOS technology in space systems can be explained by the fact that microcircuits manufactured under this technology have low-power consumption, higher noise immunity, and sufficiently high speed of operation. In space and military applications, these qualities become critical.

It is not enough to have "good" basic logic elements in order to fully implement them; it is also necessary to design "good" circuits on their basis. Increasing the radiation resistance of circuits is not normally taken into account at the logic design level, *but it is possible to reduce power consumption by creating efficient logical structure* [1].

On a larger scale, designing circuits with the power consumption taken into account still remains an art, which is connected, in particular, with the absence of efficient means to evaluate the effect of heuristics used in the design process on power consumption of electric circuit to be further implemented on a chip. Many experts and companies, such as *Cadence Design Systems* (leader in the field of design automation), *Apache Design*, *Atrenta*, *Magma Design Automation*, *Synopsys*, *Mentor Graphics*, etc. are involved in solving the problem of power consumption during circuit design.

Basic ways of reducing power dissipation mentioned in literature consist of reducing the value of supply voltage, reducing the capacitances of microcircuit and interconnections, special arrangement of synchronization and power supply control paths (circuit components power-down for the time they do not perform useful work), and reducing the dynamic power dissipation by minimizing the intensity of signal switching at CMOS microcircuit inputs. These first two methods of reducing power consumption are achieved at the *circuitry level* by selecting a suitable process design solution and the layout; the third method is achieved at the *system level*; and the last one is applied at the *logic level* by a "successful" logical architecture.

Reduction of power consumption of a designed microcircuit can be provided at different levels of its design. In this case, the earlier is the stage, the more important it is to get higher-quality solutions at this stage. Particularly, on *logic* level (due to the creation of successful logical structure), it is possible to reduce power dissipation by 10–20% without compromising the speed and complexity of the circuit [1]. CAD systems for microelectronic circuits should incorporate tools that allow evaluating and minimizing the power consumption of a circuit as early as during their logical design.

As an example of such a CAD, we will consider below a software complex "Energy-efficient Logic Synthesis," ELS, for designing logic circuits from library elements based on the technology of static CMOS circuits [2]. This technology is currently dominating in the field of digital VLSI, since logic circuits based on this technology have good process parameters and power dissipation characteristics. Most application-specific ICs (ASICs) are based on CMOS logic.

The approach to logic design of such CMOS circuits is further described, as well as the structure and functional capabilities of the ELS complex, comprising of the tools for optimization of designed digital blocks of CMOS ASICs at functional and structural levels, verification of states of designed circuits, and evaluation of power consumption of both circuits under designed and circuits designed of library elements.

This ELS software package:

- Allows to obtain the structural description of the logic circuit in the CMOS VLSI design library, from the functional description of the designed device in VHDL (Very high speed integrated circuits Hardware Description Language) or SF (Structural and Functional description language), which is the internal language of the complex
- Implements a synthesis approach that allows to minimize CMOS VLSI chip area and power consumption, measured by the average power dissipation of the circuit
- Has interactive tools for designing logical circuits and verification and evaluation of design solutions
- Allows estimating power consumption of circuits made of library elements at logic and circuit design levels

The case of combinational circuit synthesis is reviewed, where:

- Synchronous implementation of the circuits is assumed
- Synchronization frequency and power voltage are fixed
- Statistical method based on probabilistic characteristics of input signals is used to estimate power consumption in the process of circuit synthesis

7.2 Identification of Power Dissipation Sources in CMOS Microcircuits

The estimation of power consumption is further understood as calculation of the *average power* dissipated by a circuit. This estimate is quite different from the estimate of *maximum* power consumed in a single clock cycle of a circuit operation.

In general, the power consumption of a logic circuit is a complex multifactor function, depending on signal propagation delays through the circuit, timing frequency, manufacturing process parameters, and microcircuit layout, and in the case of CMOS technology, power dissipation depends significantly on the sequence of input signals applied to the circuit [2].

The total power dissipated by a CMOS microcircuit can be divided into a static and a dynamic component. The static component originates from the presence of static conductive paths between the power rails and leakage currents. For most well-designed CMOS circuits, this power consumption component is low. In typical CMOS circuits, 60–80% of the total power dissipation is due to the dynamic component, caused by the nonstationary behavior of circuit nodes. According to the simplified model, energy is dissipated by the CMOS IC whenever the signal at its output changes. Hence, more switch-active CMOS circuits dissipate more energy. Thus, power dissipation is significantly dependent on the switching activity of circuit elements, and this in turn is determined by the sequence of the input signals supplied to a CMOS circuit, i.e., the dynamics of operation.

In CMOS technology, the main contribution to the total dynamic power is made by two factors: the pure dynamic power P_{dyn}, generated by charging and discharging of the node capacitive load, and power dissipation due to through-currents flowing across the microcircuit during its internal switching [2]. Usually, the value of the latter component is calculated using the concept of on-chip intrinsic capacitance, and at the logic level, it can be reduced by synthesizing such circuit, which minimizes the VLSI die area, required to place the circuit implementing the given functionality. The dynamic power dissipation P_{dyn} is the most significant reason for power consumption by a CMOS circuit in dynamics. It is caused by the charge/discharge currents of parasitic capacitances of transistors and communication lines, and these capacitances are taken as the capacitive load, C_L, of the node output. When input signal changes from 0 to 1, the current flowing through the *p*-channel transistor charges the C_L capacitance; when input signal changes from 1 to 0, the C_L capacitance is discharged by the current flowing through the *n*-channel transistor. In each of these cases, power is dissipated on the open transistor resistance, as expressed by the well-known simplified ratio [3]:

$$P_{dyn} = \frac{1}{2} V_{dd}^2 f_{clk} E_s C_L, \tag{7.1}$$

where V_{dd} is supply voltage; f_{clk} is synchronization frequency; E_s is switching activity of the circuit output defined as mathematical expectation of a number of logical signal transitions (from 1 to 0 or from 0 to 1) during one synchronization cycle; and C_L is capacitive load of IC output.

Major contribution to the value of output capacitance, C_L, of a microcircuit is made by three factors: the parasitic capacitance C_p at IC output; capacitive load, C_{load}, of the output, defined by the total capacitance of transistors of the powered ICs; and C_{wire} capacitance of its output connections. C_{wire} value becomes known only after the wiring is done and is usually ignored at the circuit design level. C_p and C_{load} values can be calculated based on the data given for microcircuits of PDK selected for the design.

The dependence (7.1) is derived based on the following basic *assumptions* of a "well-designed" CMOS microcircuit functioning:

1. Entire capacitance of the CMOS element is concentrated at its output pole.
2. Current inside the element flows only from the power supply to the output capacitance or from the output capacitance to the ground.
3. Voltage value at the output of an element varies only from power supply voltage to ground voltage or vice versa.

This approach ignores the power dissipated (a) during switching of internal nodes of complex CMOS elements, (b) due to microcircuit through-currents inside CMOS elements, and (c) due to transient processes (glitches, races).

V_{dd} and f_{clk} parameter values in (5.13) are defined at architectural design; at a logic level, the $\frac{1}{2}V_{dd}{}^2f_{clk}$ product (evaluating power dissipation per unit capacitance when changing supply voltage) can be considered as a constant, similar for all circuit nodes. Minimizing the dynamic power is thus limited to minimizing the E_sC_L product (often referred to as switchable capacitance), and power dissipation by a circuit at the logic design level is estimated as the sum of switched capacitances of all its nodes [4]:

$$P_s = \sum_{i=1}^{n} E_iC_i, \tag{7.2}$$

where n is the number of nodes in the circuit (summation is performed for all nodes of the circuit); C_i is capacitive load of i-node; and E_i is switching activity of the i-node of the circuit.

The evaluation of the dynamic power (7.2) is reasonable to use for comparison of circuit implementation options of a given functionality.

At the stage of logic design, when there is no electrical schematic yet, and even the process design basis, in which it will be implemented, is frequently still unknown, the power dissipation of a future circuit can be reduced by such a transformation of the circuit description, which ensures reduction of its switching activity without changing the functionality [2]. A quantitative change in switching activity can be used to evaluate preferable options of circuit optimization at the logic level. This approach to evaluation of power dissipation makes it possible to compare circuit implementation options during its design, allowing the design of circuits with potentially low-power dissipation already at the logic level.

7.3 Probabilistic Evaluation of Optimization Options by Predicted Switching Activity of IC Nodes

The methods for estimation of switching activity of circuit nodes, used in the process of its design, are based on the approach of probabilistic characteristics of input signals and functional and structural properties of the analyzed circuit. The approach involves *specifying switching probabilities* of signals at the input of the circuit and indicating the frequency of their values change and is used to calculate the probabilities of switching signals at the outputs of circuit nodes. Switching activity estimation methods are based on the propagation of probabilistic information about the change of signal values through the whole circuit, from inputs to outputs; that's exactly why these methods are also called *probabilistic* in the literature [4]. The probabilistic approach identifies possible sequences of the input signals impact on a circuit and estimates the power consumption of a circuit assuming mutual influence of the input signals at successive points in time.

The vast majority of estimates of the pole-to-pole signal switching intensity used in design practices are derived under the assumption of zero signal delay by circuit nodes, when all transitions in the circuit occur simultaneously.

These estimates assume that all changes at circuit inputs propagate through its elements instantaneously (and therefore simultaneously) and take into account signal changes only in their steady states, provided by the function that is implemented by the node. These estimates do not take into account the switchings caused by transients, which also dissipate energy without performing calculations specified by the scheme. The use of actual delay models significantly increases the computation time of the circuit's switching activity estimates while improving their accuracy. In the synthesis process, however, to compare optimization options, it is sufficient to use simpler, rapidly computable estimates, such as estimates in the assumption of zero variation in signal delays.

In addition, *known* probability calculation methods assume that signals at the input poles of any circuit node are temporally and spatially independent. Temporal independence implies that the signal value in any clock cycle is independent of its values in the preceding clock cycles. Spatial independence of the poles implies absence of signal values correlation on them. It can occur when spatially related signals are interdependent, which is caused, for example, by branching at element outputs or feedbacks. In practice, a stable dependence may occur between signals, caused by other reasons, except the abovementioned.

Many probabilistic methods for estimating the power consumption of logic circuits have been proposed in the literature [4–6], but most of them are applicable only to *combinational* circuits.

The methods for estimating energy consumption are formulated as follows:

1. With respect to different assumptions about the delay of signals by circuit nodes, the possibility of different types of dependencies between signals, the consideration of transients during signal changes, etc.

2. Using different statistical characteristics (probability of occurrence of signal 1, probability of signal switching on the pole, switching signal intensities, e.g., signal switching densities, equilibrium probabilities, probabilistic forms of signals, etc.)
3. Using different models (based on binary decision diagrams (BDD), correlation coefficients, etc.)

The essence of the following methods of CMOS circuits' synthesis of library elements is based on the works of Belarusian scientists [7–25], and here we use the simplest method, which is based on the following assumptions:

1. Changes at circuit inputs propagate through all of its elements instantaneously, which means that all transitions in the circuit occur simultaneously.
2. There is a temporal independence for each input pole of a node assuming that the signal value in any clock cycle is independent of its value in the preceding cycles.
3. Input poles of the node are spatially independent, which means the absence of correlation of signal values at these poles (which can be caused, e.g., by presence of branching at element outputs or feedbacks).

A distinction is made between the probability p_i^1 of signal 1 (0) at some *i*-pole and the probability of signal change at this pole. The first probability p_i^1 is called *signal probability* (probability of signal 1 occurrence) and is determined by the average fraction of clock cycles at which the signal at *i*-pole has a single value. The second probability $p_i^{1\to 0}$ (or $p_i^{0\to 1}$) is the probability of the signal value changing from 1 to 0 (or from 0 to 1) and is determined by the average fraction of clock cycles in which the signal on the *i*-pole changes its value compared to the value in the preceding cycle.

Assuming zero delays (which rules out switchings by transients) of the elements and time-independent signals, the probability $p_i^{1\to 0}$ ($p_i^{0\to 1}$) is equal to product of the probability that signal 1(0) occurs on it in one clock cycle by the probability that signal 0(1) occurs on it in the next clock cycle. Accordingly, switching activity of *i*-pole of the circuit is equal to the product $E_i = p_i^{1\to 0} p_i^{0\to 1} = 2p_i^1 p_i^0$, and in assumption that $p_i^1, p_i^0 < 1$, and indicating p_i^1 simply by p_i:

$$E_i = 2p_i(1 - p_i). \tag{7.3}$$

The probability p_e of signal 1 occurrence at the output of element *e* significantly depends on the probability characteristics of the signals at its inputs and on the function implemented by this element. When the signals at the element inputs are not correlated in space and time, the signal probabilities of simple elements like inverter, AND, OR, AND-NOT, and OR-NOT with the input poles *n(e)* can be easily calculated from truth tables of functions they realize:

$$
\begin{aligned}
p_e^{\neg} &= 1 - p_1;\\
p_e^{\wedge} &= \prod_{i=1}^{n(e)} p_i;\\
p_e^{\vee} &= 1 - \prod_{i=1}^{n(e)} (1 - p_i);\\
p_e^{\overline{\wedge}} &= 1 - \prod_{i=1}^{n(e)} p_i;\\
p_e^{\overline{\vee}} &= \prod_{i=1}^{n(e)} (1 - p_i);
\end{aligned}
\tag{7.4}
$$

where p_i is the signal probability (probability of signal 1 occurrence) of the element i-input.

If the signal probabilities of the input signals of a circuit are given, they can be propagated to the outputs of the circuit elements and through the whole circuit to its output poles. In this way, the switching activities of all poles of the circuit can be calculated; and consequently the switching activity of the circuit as a whole can be calculated as the sum of the switching activities of all its poles. It should be noted that even if the requirement of spatial (and temporal) independence is met for the input signals of a circuit, it may not apply to the input signals of the internal circuit elements (as a result of branching presence at the element outputs and feedback lines). In this case, the probabilities calculated by formulas (7.3) and (7.4) have an error. For comparative evaluation of optimization options, however, it is generally sufficient to be limited to these simple estimates, without resorting to more accurate, but much more labor-consuming, computational estimates.

7.4 Selection of Element Basis for Low-Power CMOS Microcircuit Design

The element basis contains quite a wide range of different logic elements, with combinational logic as the core element, consisting of simple tree-shaped circuits of AND, OR, and NOT gates. This set usually consists of AND, OR, AND-NOT, and OR-NOT elements for variable number of inputs (usually two to four, sometimes six or eight inputs, as in basis) and treelike circuits (of AND, OR gates) with a maximum number of input poles of four and a number of levels of two to four. For complexity (value) of a library element, we can take the number of transistors of its microcircuit (or the number of layout basic cells, in which sizes depend on CMOS element manufacturing process). This number is directly related to the area occupied by the element on a chip and its capacitive load, which significantly affects the power consumption of the element.

CMOS VLSI element basis is characterized by the presence of complete set of gates. The element structures of the library can be represented as treelike networks of AND, OR, and NOT gates. Each of these trees has a limited number of leaf nodes and a limited output branching factor. A CMOS library is characterized by the fact that for each element, there is usually an element that implements a dual function.

Thus, Table 7.1 demonstrates characteristics of some elements of a CMOS library, which will be an example of the proposed method of synthesis: n, number of input poles; k, total number of input poles of gates; l, number of tiers of its tree structure (except inverter); and t, number of transistors of its microcircuit. The elements of the considered CMOS library have the following limitations: $n \leq 4$; $k \leq 9$; $l \leq 4$; $t \leq 12$.

For the purpose of evaluating the efficiency of circuit mapping options, a quantitative indicator of its logical efficiency is introduced to the process-design-basis, which is expressed in the number of poles of the circuit covered by the library element, which accounts for the unit of its complexity, i.e., one transistor. The logical efficiency of an element is equal to the k/t ratio of the number of input poles of the gates, representing the structure of this library element, to the number of transistors of its microcircuit. The larger this ratio is, the higher the functional efficiency of the element (for coverage purposes).

The table shows that the most effective elements are those with the most complex structure, NOAA (2-2AND-2OR-NOT) and NO3A3 (3AND-3OR-NOT), and their dual elements, NAOO (2-2OR-2AND-NOT) and NA3O3 (3OR-3AND-NOT); the least effective are the inverter and two-input gates, AND and OR (A2 and O2). The value of a complex library element is usually less than the sum of the values of its component gates implemented by simpler elements. For example, a NOAA element can be implemented as a composition of an inverter, two elements of A2 and O2. The value of NOAA element (eight transistors) is less than the value of the composition (2 + 8 + 4 = 14 transistors).

7.5 Logic Synthesis of CMOS Microcircuits in the Basis of Library Elements

The power consumption of digital circuits is generally directly proportional to the area they occupy on the VLSI chip. This means that the main way to save energy in the operation of logic circuits primarily involves reducing the area occupied by the circuit on the VLSI chip. Practice shows that the existing methods, which optimize the complexity of the circuit, are a good starting point for the development of techniques aimed to minimize power consumption of logic circuits [7–9]. The application of these methods, as a canvas for developing power minimization methods, also requires consideration of new criteria, jointly used with the criteria of circuit area minimization. These criteria depend substantially on the technology of synthesized logical circuit implementation.

Table 7.1 CMOS library elements

Library elements	*n*	*k*	*t*	*l*	Index of logical efficiency	Symbols
NOT N	1	1	2	1	0.5	A 1 Y
AND-NOT, OR-NOT NA, NA3, NA4 NO, NO3, NO4	2, 3, 4	2, 3, 4	4, 6, 8	2	0.5; 0.5; 0.5	A B C D & Y A B Y C D 1 Y
AND, OR A, A3 O, O3	2, 3	2, 3	6, 8	1	0.33; 0.37	A B C & Y A B C 1 Y
3AND-2OR-NOT 3OR-2AND-NOT NOA3 NAO3	4	5	8	3	0.63	A B C D & 1 Y A B C D 1 & Y
2-2AND-2OR- NOT 2-2OR-2AND- NOT NOAA NAOO	4	6	8	3	0.75	A B C D & 1 & Y A B C D 1 & 1 Y
2AND-3OR-NOT 2OR-3AND-NOT NO3A NA3O	4	5	8	3	0.63	A B C D & 1 Y A B C D 1 & Y

(continued)

Table 7.1 (continued)

Library elements	n	k	t	l	Index of logical efficiency	Symbols
2-2AND-3OR-NOT 2-2OR-3AND-NOT NO3AA NA3OO	5	7	10	3	0.7	A B C D E & 1 & Y A B C D E 1 & 1 Y
3AND-3OR-NOT 3OR-3AND-NOT NO3A3 NA3O3	5	6	10	3	0.6	A B C D E & 1 Y A B C D E 1 & Y

In the process of logic synthesis, an abstract description of synthesized circuit behavior (system of Boolean functions) is presented in the basis of elements of CMOS VLSI process library (PDK). Each of the elements is characterized by its own function and physical characteristics. The approach, like most well-known synthesis methods, is based on division of the logic synthesis process into the stages of technology-independent optimization and technology mapping. The first stage of synthesis is focused on logic optimization and decomposition, while the second stage involves the implementation of the resulting functional description in given process-design-basis. The goal of the *first* stage is to minimize complexity of a multilevel circuit in technology-independent basis of elements. The complexity of a multilevel circuit is measured by the number of gates, depth of the circuit, and estimated power dissipation at logical level.

The *second* stage includes transfer of multilevel circuit from gates to process-design-basis by structural coverage of the corresponding object network with subcircuits implementing library elements. This approach does not imply radical restructuring of the circuit obtained at the stage of technology-independent optimization, and, hence, it follows that quality of the required coverage essentially depends on its structure. Errors made during its synthesis cannot be fully compensated for at the stage of technology mapping; therefore, much attention in existing CAD systems is paid to the stage of technology-independent optimization.

Technology-independent optimization includes, as its first stage, minimization of functions of implementable logic descriptions in a class of disjunctive normal forms (DNFs). Taking into account specifics of the CMOS basis, it is rational to carry out joint minimization, with consideration for functions polarity, choosing that form (DNF or its inversion), which has smaller complexity and power consumption. At the second stage, the minimized DNF system represented by a two-level circuit is decomposed into multilevel object network of AND, OR gates with a limited number of inputs, into which the structures of basic elements of CMOS library are decomposed, wherein the circuit is to be mapped.

Circuit complexity and its power dissipation are closely related in a way that a reduction in circuit area tends to ensure the reduction in its power consumption as well, while an increase in the area, on the contrary, tends to lead to a similar increase in power consumption. It follows from these considerations that synthesis process requires a trade-off between the criteria of power consumption and area minimization. The main problem of circuit optimization is that during the logic synthesis stage, which is not yet linked to a specific process-design-basis, it is difficult to estimate the energy consumption of a real circuit at a sufficiently reliable level. Therefore, minimizing the dynamic component of circuit power consumption (estimated by the switching activity of the circuit) at the logic synthesis stage, to the detriment of circuit complexity, may ultimately lead to an increase in its power consumption. This is due to an increase in the values of other power-consuming components.

Since the solution to the problem of logic circuits optimization proceeds from the fact that the main way to save energy in the design of logic circuits implies reducing the area occupied by circuit on a chip, the ranked criterion is used when evaluating optimization options at all stages of logic synthesis: first of all, quantitative estimation of the change in area and then quantitative assessment of the change in the switching activity of the circuit.

7.6 Power Dissipation-Driven Optimization of Two-Level Logic Circuits

Minimization methods of functional description of systems of fully and partially defined Boolean functions are used for optimization of two-level circuits with the power dissipation taken into account. These methods are modifications of well-known methods of Boolean function minimization in DNF class by adding heuristics, guiding minimization process to obtain DNF systems implemented by circuits with the lowest power dissipation. Practically, all methods of minimization of two-level representations of Boolean functions are based on division of the set of searched prime implicants into three subsets: essential, nonessential, and conditionally essential. The first ones should be included in any non-redundant solution, the second ones should not be included in any solution, and from the third ones, a certain non-redundant subset is selected that covers all the intervals of single definitions of the minimized functions, which are not covered by essential implicants.

Minimization methods differ in their methods of constructing prime implicants from the number of conditionally essential ones and in the criteria they must meet in order to be included in the solution. A distinction is made between the methods of:

1. Sequential construction of prime implicants to be included in the solution (e.g., method of competing intervals), by enlarging the interval of a Boolean argument space, representing the implicant through inclusion of uncovered elementary conjunctions
2. Sequential extension of the interval of a Boolean argument space representing, at first, the conjunction of initial DNF and, eventually, a prime implicant covering this conjunction, by excluding some literals being its part (e.g., methods implemented by ESPRESSO [26])

The simplest minimization method can use only one extension operation of conditionally essential implicants. Most easily modified methods, in order to account for energy savings, are the minimization methods, in which several prime implicants are constructed at once as candidates for the sought solution, or the methods where the found solution is modified by consecutive enhancements.

In order to guide the minimization toward obtaining an energy-saving solution, it is necessary to compute and take into account the switching activities of all prime implicants using formulas (7.5) and (7.6) in the process of obtaining prime implicants and non-redundant coverages. To explain the procedures that guide the minimization process toward the desired result, let us consider the operations included in almost all minimization methods.

The operation of interval extension is performed by eliminating its literals. Two goals are taken into account when extending an interval: to decrease complexity of the interval, i.e., to extend it as much as possible, and use it (fully or partially) to cover as many yet uncovered intervals as possible. When minimizing power consumption, it is desirable to:

1. Exclude only the most active literals, not any of them
2. Cover only the most power-consuming intervals, not any of them

By virtue of the first statement, literals with higher switching activity are checked for exclusion first. By virtue of the second statement, the order of intervals expansion is important: expanding the interval too early may hinder the situation, when some other interval shall cover the interval being considered. To reduce power consumption, the energy contribution of each interval is evaluated, and energy-intensive (with high switching activity) intervals are expanded last, in hope that some other intervals will expand to cover them.

The operation of detecting non-redundant coverage includes transformation of the current DNF coverage into non-redundant form. While searching for the non-redundant set of prime implicants, the minimum number of the least active implicants (with less switching activity) has to be chosen. Herewith, the implicants, which are minimal in power or in total number of literals of all implicants, shall be selected from the non-redundant sets of prime implicants, representing the coverage of the initial set of intervals of single areas of the minimized Boolean functions. Each of the selected sets is evaluated by the sum of switching activities (or switching densities [13]) of the implicants included in it, and the non-redundant set with the minimum switching activity estimate is selected as the solution.

A comparison of modified minimization methods [14] with original methods (without consideration of energy consumption) has shown that *minimization of Boolean functions accounting for switching activity of signals helps to reduce power consumption of a circuit without increasing its complexity*. By doing so, the computational costs for minimization increase insignificantly.

7.7 Selection of Basic Gates for Technology-Independent Functional Circuit

There are different approaches to selecting basic gates in the technology-independent circuit synthesis. Typically, the description of the logic to be implemented is translated (through minimization) into equivalent AND-OR description, which is converted into a homogeneous two-input AND-NOT or OR-NOT basis during the decomposition stage [26]. Using minimal number of basic gates due to their simplicity (e.g., 2AND-NOT), as it is accepted in a number of known CAD systems, leads to higher granularity of logical network, and consequently may also increase the quality of its coverage by library elements due to increasing the number of coverage options. In case of CMOS basis, however, this advantage can be outweighed by a considerable number of disadvantages: representations by library elements (in such "shallow" basis) become more complicated; the number of different representations of one and the same element increases; and speed of coverage methods decreases, which, in turn, leads to the necessity for their even larger coarsening.

For the case of CMOS library, it is much more attractive to choose a technology-driven basis of basic gates [11]. Gates, being components of library elements of process-design-basis, are selected as basic gates. This approach allows, at a stage of technology-independent optimization, to get a logical network as close as possible to the process-design-basis, and accordingly to apply more effective and simple algorithms of coverage at technology mapping. This approach also simplifies model representations of library elements themselves (they are represented in the same gate basis as the covered object network) and their number (and consequently the size of the library), which is of high importance for the speedup of coverage algorithm.

If we analyze the composition of the CMOS library, we can notice that the structures of all complex elements can be represented by circuits of alternating AND and OR gates. This means that a simple representation of multiplace AND or OR functions as a composition of binaries will not introduce new coverage possibilities, but will lead to the negative consequences mentioned above. Therefore, for the considered process-design-basis, it is reasonable to choose NOT, AND, and

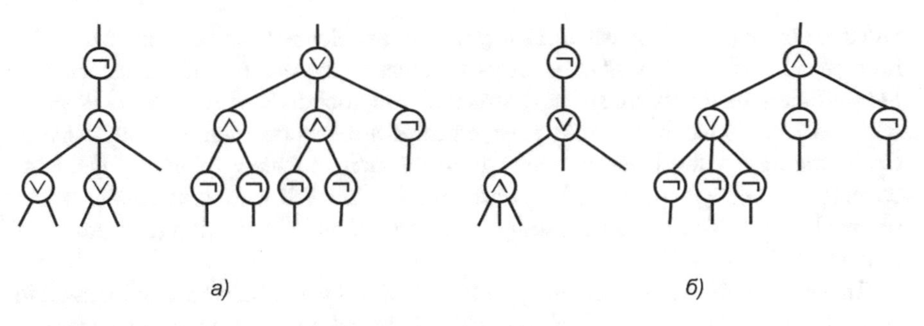

Fig. 7.1 Structures of two nontrivial CMOS library elements
(**a**) 2-2OR-3AND-NOT (NA3OO); (**b**) 3AND-3OR-NOT (NO3A3)

OR gates as the basic ones, with limited number of inputs not exceeding the maximal number of gate inputs (AND-NOT, OR-NOT) of this library.

During the coverage process, the schematics of library cells are compared with fragments of the covered circuit and replace them in case of full coincidence. Accordingly, each library element must be represented by different structures implementing its function. As it was already mentioned, the bulk of the most efficient elements implement inverse logic, and separately implemented inverters are rather expensive CMOS elements. In this connection, at the structural coverage of a circuit, together with structures of elements, mapping their implemented function, it is offered to include in the library of structural descriptions also the dual structures obtained by transfer of inverters from outputs to inputs (swap of gates AND to OR, and OR to AND) (Fig. 7.1). For example, the element 2AND-2OR-NOT (NOA), performing $\overline{ab \vee c}$ function, generates the element 2OR-2AND (AON) with inverters at the inputs, implementing function $\left(\overline{a} \vee \overline{b}\right) \overline{c}$.

7.8 Optimization of Multilevel Logic Circuits of Multi-input Gates

At this stage, we consider the problem of designing multilevel logic circuit of AND, OR, and NOT gates, implementing system of Boolean functions. This task immediately follows the minimization of Boolean functions in DNF class and precedes the synthesis of logic circuits of library elements made by CMOS technology. The goal of the stage is to build such an option in representation of a circuit of gates, which could serve as a good starting point for the technology mapping stage in the basis of library elements. This stage uses the total number of input gate poles and the total switching activity of the circuit poles as quantitative measures of design efficiency.

Combinational elements of a typical CMOS VLSI library can be represented at a logic level by treelike structures of NOT, AND, and OR gates, implementing inverse logic, for example, two-tier AND-NOT, OR-NOT, and three-tier 2-2AND-2OR-NOT. Elements implemented in multitier structures include gates with two and three input poles. Logical efficiency of a library element can be estimated by the ratio of the total number of poles of its structure to the number of transistors. The most efficient elements are strongly structured microcircuits of AND and OR gates with two and three inputs; the least efficient are inverter and two-input gates. Correspondingly, the most attractive from the point of view of library elements coverage are strongly structured networks of AND and OR gates with small number of input poles. These considerations form the basis of the adopted method of multilevel circuit synthesis.

The main method of solving the problem of DNF systems decomposition, used in all CAD, is algebraic decomposition, which is based on construction of factorized forms (or factorized DNFs) by searching for factors—common parts of conjunctions or disjunctions of DNF systems. The factorized form is an algebraic form of a multilevel representation of DNF. Transformation of initial minimized DNF system into the factorized form, which multilevel implementation of gates with limited number of inputs corresponds to, is split into two stages [15].

Joint nontrivial factorization of DNF system: factors (conjunctions or DNFs) are distinguished, which have length (number of literals) of no more than maximum numbers of inputs, n_{max} and m_{max}, of AND and OR gates, and are included in no less than n_{dl} expressions. The key issue in the search for factors is the question of evaluating their cost and power-saving quality. The cost quality T_s of a factor s, included in expressions from its generating set, U_s, is estimated in a simplified manner by the area of its corresponding minor of the Boolean matrix which defines the factorizable set of expressions:

$$T_s = c(s)\ (|U_s|-1), \tag{7.5}$$

where $c(s)$ is the expression implementation value, s, according to Quine.

Power-saving quality of a factor is quantified by the gain in switching activity of the sought circuit, which gives the allocation of this factor. In the factorized multitude of expressions, the switching activities of all expressions will not change in comparison with their values in the original multitude, but the load of circuit poles corresponding to literals included in factor s will change: it will decrease by $(|U_s|-1)$, and the total switching activity of circuit poles implementing the factorized multitude of expressions will change accordingly. The power-saving quality of factor $s = \{z_1, z_2,\ldots,z_l\}$ is estimated as:

$$P_s = (|U_s|-1) \sum_{z_i \in s} E(z_i). \tag{7.6}$$

Constructing the DNF bracketed expressions of each function of the system is based on iterative putting general literals of conjunctions of a given DNF, D, outside the brackets, i.e., on decomposition [16]:

$$D = k(A) + B, \tag{7.7}$$

where D, A, and B are DNFs (disjunctions of some set of conjunctions) and k is conjunction consisting of some multitude of literals, common to all conjunctions from A. Conjunction k is chosen in the following way: its core is some "best" x literal, and other literals common to conjunctions of DNF A are put out of brackets together with this literal. The best literal x is considered to be the literal that is included in the maximal number l of conjunctions from D, and among the equals by this criterion, it is the one that has the maximal value of switching activity. This choice of literal is justified by the fact that:

1. The energy load to the circuit pole having switching activity E_x and corresponding to the literal x, taken out of l expressions, is reduced by the value $(l - 1)\ E_x$
2. The most active signal will be applied to the circuit closer to the output, thus reducing the total switching activity of the circuit

After the iterative process of putting general literals of conjunctions out of brackets, the remaining conjunctions of rank, greater than n_{max}, are factorized separately. In this case, the literals corresponding to the poles with the least switching activity are included in the factor first, since it is desirable to supply the most active signals to the inputs of circuit elements as close to the output as possible.

Experimental studies [15] have shown that the proposed synthesis method gives a fairly stable gain in the power dissipation estimation of the designed circuit compared to a similar method without regard to power consumption.

7.9 Optimization of Multilevel Logic Circuits of Double Input Gates

Optimization of multilevel structures is done both at algebraic representation level of Boolean function systems and at functional level by searching for BDD representations of Boolean function systems, in which multilevel logic circuits of two-input AND and OR gates correspond to.

The BDD representation is based on the Shannon sequential variable decomposition applied to Boolean functions $f(x_1, \ldots, x_n)$:

$$f = \bar{x}_i f(x_1, \ldots, x_{i-1}, 0, x_{i+1}, \ldots, x_n) \vee x_i f(x_1, \ldots, x_{i-1}, 1, x_{i+1}, \ldots, x_n). \quad (7.8)$$

Decomposition coefficients $f(x_1, \ldots x_{i-1}, 0, x_{i+1}, x_n)$ and $f(x_1, \ldots x_{i-1}, 1, x_{i+1}, x_n)$ are derived from $f(x_1, \ldots, x_n)$ function, by substituting the variable x_i with constants 0 or 1, respectively. BDD graphically defines a sequence of Shannon decompositions of the assumed function and the resulting decomposition coefficients. The BDD complexity minimization is based on the fact that in decomposition process, the same decomposition coefficients may appear not only in one but also in several (or even in all) decomposable functions of the initial system. BDD minimization is further understood as optimization of multilevel representations of the systems of Boolean functions corresponding to reduced ordered BDD (ROBDD).

For example, when decomposing a system of Boolean functions:

$$\begin{aligned} f^1 &= x_1x_2\bar{x}_4x_5\bar{x}_6 \vee \bar{x}_1x_4\bar{x}_5x_6 \vee x_2\bar{x}_3x_5; \\ f^2 &= \bar{x}_1\bar{x}_4x_5\bar{x}_6 \vee \bar{x}_1\bar{x}_3x_5 \vee x_1x_2x_3x_5\bar{x}_6 \vee x_1\bar{x}_2x_4\bar{x}_5x_6; \\ f^3 &= x_1\bar{x}_2\bar{x}_3x_6 \vee x_1\bar{x}_2x_4x_6 \vee x_1\bar{x}_3x_4x_6 \vee \bar{x}_1x_2\bar{x}_4x_5\bar{x}_6 \vee x_1\bar{x}_2x_5 \vee x_2\bar{x}_3x_5. \end{aligned} \tag{7.9}$$

$x_1, x_2, x_3, x_4, x_5, x_6$, variables produce a BDD representation (рис. 5.16), (Fig. 5.16), which is described by the following multilevel representation of the logic circuit of two-input AND and OR gates:

$$\begin{aligned} &f^1 = \bar{x}_1\psi^1 \vee x_1\psi^2; f^2 = \bar{x}_1\varphi^2 \vee x_1\psi^3; f^3 = \bar{x}_1\psi^2 \vee x_1\psi^4; \\ &\psi^1 = \bar{x}_2\varphi^1 \vee x_2\varphi^1; \psi^2 = x_2\varphi^2; \psi^3 = \bar{x}_2s^1 \vee x_2\varphi^3; \psi^4 = \bar{x}_2\varphi^4 \vee x_2\varphi^5; \\ &\varphi^1 = \bar{x}_3s^2 \vee x_3s^1; \varphi^2 = \bar{x}_3\lambda^3 \vee x_3s^3; \varphi^3 = x_3\lambda^4; \varphi^4 = \bar{x}_3\lambda^2 \vee x_3s^2; \\ &\varphi^5 = \bar{x}_3s^2; s^1 = x_4\lambda^1; s^2 = \bar{x}_4\lambda^3 \vee x_4\lambda^2; s^3 = \bar{x}_4\lambda^4; \\ &\lambda^1 = \bar{x}_5\omega^1; \lambda^2 = \bar{x}_5\omega^1 \vee x_5; \lambda^3 = x_5; \lambda^4 = x_5\omega^2; \omega^1 = x_6; \omega^2 = \bar{x}_6. \end{aligned} \tag{7.10}$$

Shannon decomposition or its special case corresponds to each functional vertex of BDD. BDD complexity is estimated by the number of vertices marked with function symbols; vertices corresponding to arguments are not taken into account when estimating BDD complexity. For example, BDD complexity (Fig. 7.2) is 21. The main problem in constructing a BDD is the choice of the sequence of variables by which the Shannon decomposition is conducted.

During "power" optimization of BDD, the first main criterion is the minimum complexity of BDD; the second subordinate criterion is the one related to the probabilities of signal changes at the circuit inputs. The first criterion is focused on minimizing the complexity (number of transistors) of the logic circuit, which is built at the stage of technology mapping, as the decrease in the number of transistors allows reducing power consumption. The second criterion allows selecting among BDD of the same complexity the ones which can be used to build combinational logic circuits characterized by low-power consumption.

The energy quality of the variable x_i in BDD representation is evaluated by the value:

$$K_i = p_i s_i, \text{if } p_i < 0,5; K_i = (1 - p_i) s_i, \text{if } p_i \geq 0,5; \tag{7.11}$$

where p_i is the probability of occurrence of a single x_i signal value at the input of the logic circuit implementing BDD and s_i is the number of x_i variable literals in the multilevel representation of a system of functions, corresponding to BDD.

Variables, x_i, characterized by probability, p_i, close to 0.5 cause the greatest number of switchings at the inputs of circuit elements. Consequently, it is desirable to minimize the number of literals of such variables in the functional representation

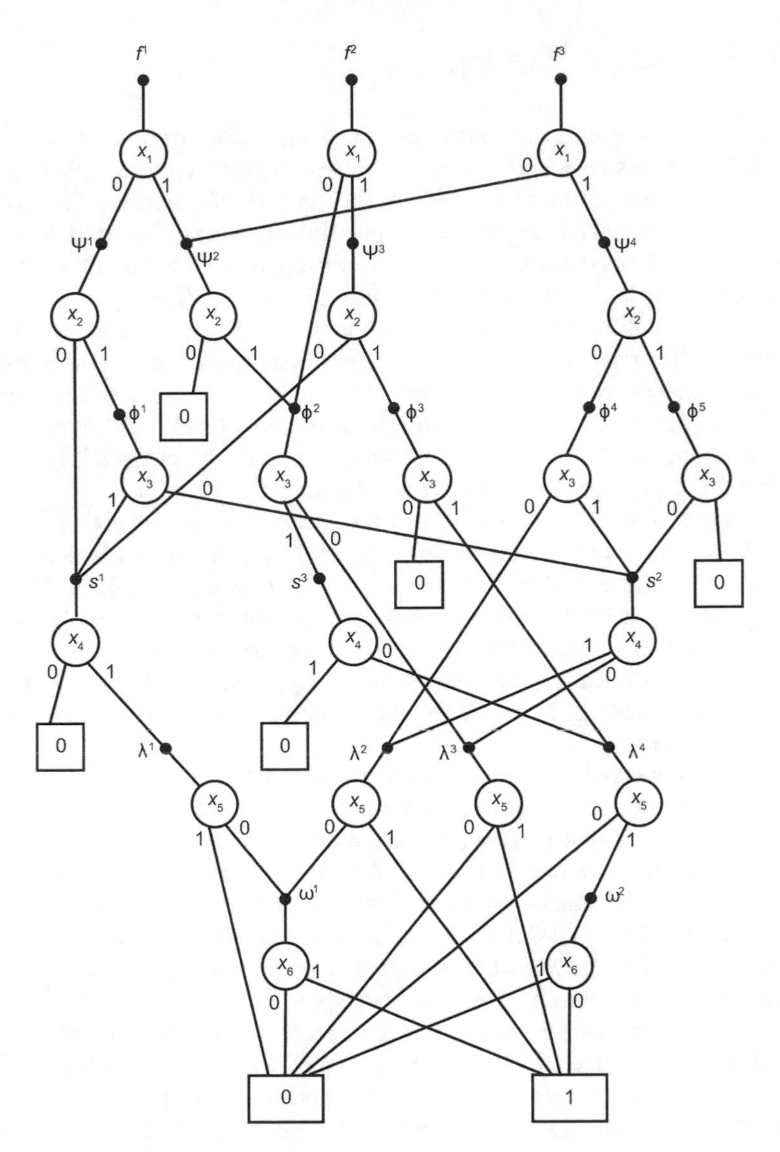

Fig. 7.2 Binary decision diagram (BDD)

of BDD. If the probability p_i is close to 0 or 1, such variables will cause a smaller number of signal switchings in the circuit. Therefore, variables characterized by a low value of K_i should preferably be placed in the middle of the sequence of variables used to build the BDD.

It should be noted that optimization software for multilevel representations of logic circuits of two-input AND and OR gates can handle both systems of fully defined and systems of partial Boolean functions defined by the sets of argument values, or in interval form.

7.10 Technology Mapping

At the stage of *technology mapping*, the structural method of multilevel circuit of AND and OR gates coverage (the most effective in practice of designing complex circuits) by library elements is used. In the process of coverage, the structural schematics of library cells are compared with the fragments of covered circuit, and replace them, if they completely coincide. Accordingly, each library element should be represented by different structures implementing its function.

A covered multilevel multi-output logic network of gates is represented by an oriented acyclic graph $G = (V, U)$, hereinafter called object graph. The vertices of the graph correspond to basic gates and input poles of the circuit. The structural description of a library element is a single-output multilevel logic network of the same basic gates as the covered object network. Each library element is represented by a tree-oriented graph, hereinafter referred to as a model graph.

Coverage method is based on consecutive selection of subgraphs G_k of graph G, comparison of each of them for its coverage by graphs H_i, and its replacement by graph H_l, which gives the highest value of selected optimization criteria [18]. Thus, this method provides local optimization and is approximate. In the process of coverage, graph G is compressed by excluding covered subgraphs from it. At the same time, the resulting graph E, representing a logical network of library elements, which is functionally equivalent to the initial object network, is extended (starting from an empty one).

Coverage optimization criteria. In the process of coverage, it is necessary to obtain such alternate circuit from library elements, which would provide minimum area and power dissipation. In the proposed method, the area of the circuit obtained as a result of such coverage is taken as the first optimization criterion, which is measured by the total number of transistors of all library elements and is estimated by the sum of values of model graphs included in the coverage of the object graph.

Accordingly, the relative cost of covering one pole of the object network, which is measured by the ratio of the number of covered poles of network fragment to the cost of covering library element (which is measured by the number of transistors), is taken as the cost estimate of the coverage option of the graph subgraph $G = (V, U)$, whereas the higher this number, the more desirable the coverage option.

The power consumption of a covered circuit is estimated by the sum of the switching activities of all poles of the circuit. Accordingly, the sum of the switching activities of the gates covered by the library element is taken as the power-saving estimate of the coverage option of the graph subgraph $G = (V, U)$.

If signal delay in the logic elements is neglected, then the switching intensity of signals at the inner poles of library elements can also be neglected during estimation of power dissipation. This assumption is justified by the fact that the most significant contribution to the parasitic capacitance of circuit is made by the interconnecting lines, and the bulk of the power consumption of power supply is spent on recharging this capacitance during signal switching. Hence, a circuit implemented with CMOS technology must be covered by library elements so that as many circuit nodes with the highest switching activity as possible are inside the library elements. Accordingly, the sum of the switching activities of the elements, which correspond to the covered vertices, which have become inner vertices of the covering pattern, is taken as the value of the energy efficiency criterion of the covering option.

Increasing the speed of coverage algorithm. The basic enumeration procedure at performing coverage is concentrated in searching for pair: a covered subgraph G_k of graph G and a covering model graph H_j of a library element. The following practices can significantly reduce such enumeration:

1. *Searching and covering G_k subgraphs of an object graph* so that the coverage in a single way (by a single model graph) can be admissible. In the simplest case, such subgraphs are vertices with large indegrees (corresponding to elements with large number of input poles), for which there exists a single coverage option (for instance, for many series of CMOS libraries, this number exceeds three).
2. *Ordering of model graphs.* If to arrange model graphs H_j in descending order of their efficiency (measured by the ratio of the number of poles of its structure to the number of transistors), then comparing them in order with the subgraph G_k allocated at some step of the coverage algorithm, it is possible to reduce the enumeration in some cases: if for some model graph H_l the value $H_l \, / \, G_k = \varnothing$, then we can avoid comparing subgraph G_k with other graphs H_i, as they have smaller efficiency and, consequently, cover a smaller part of graph G. It is possible to limit the search just to the coverage option of subgraph G_k with the highest value of energy efficiency criterion.
3. *Topological sorting of object and model graphs.*

Sorting a model graph H_l (it is a tree) implies ordering the branches of the tree from left to right by decreasing their complexity. P-branch is placed to the left of R-branch, if P is longer than R, or if (in case they are of equal length) the vertex with larger indegree is found in P-branch earlier, than in R-branch, when browsing vertices from the root. Partial sorting of graph G is done in the same way. Sorting makes it faster to answer the question of whether it is possible to cover a fragment of an object graph with a given model graph.

7.11 Estimation of Power Consumption of Designed CMOS Microcircuits at Logic and Circuitry Levels

In addition to probabilistic power consumption estimates, the ELS software package has the tools to estimate power consumption of designed custom VLSI combinational blocks by calculating the total number of switching transistors of circuit elements on given test sequences of input sets (tests) using fast VHDL (or SF) simulation [19].

The use of logical VHDL/SF simulation for power consumption estimation can reduce the estimation time by several orders of magnitude with an acceptable

(10–15%) error of such estimation. Tests for circuit simulation can be set as text files of a certain format or can be generated automatically. In automatic mode, four types of tests are generated:

1. Pseudo-random tests of a given length with the signal probabilities of the input poles taken into account
2. Sequences ordered by ascending decimal equivalent from 2^n sets of Boolean space of dimension n
3. Sequences ordered by descending decimal equivalent from 2^n Boolean space sets of dimension n
4. $2^n(2^n - 1)$ sequences of ordered pairs of all sets of Boolean space [19]

Tests are compiled in two formats, namely, for VHDL simulation and for SPICE modeling.

Finding power-consuming test T revolves around analyzing all pairs of input signal value sets and corresponding switching numbers of transistors in circuit elements, assuming that the maximum number of switchings (in a given clock cycle) is compliant with the maximum current consumption. If a pair of input signal value sets is put in correspondence with the value of current consumption, then the found test will provide not an approximation, but a real (from the point of view of general-circuit simulation) current consumption. The amount of current consumption determines the minimum width of conductors in VLSI power and ground networks; the "right" width of such conductors is important to prevent electromigration effects, leading to wire breaks and VLSI malfunctions.

Let the set of numbers $V = \{0, 1, 2, \ldots, 2^n - 1\}$ be given; each of them is a decimal equivalent of the test vector—a set of values of circuit input signals. The L multitude of all $2^n(2^n - 1)$ ordered pairs of <i, j> elements of the V multitude is considered. One can enumerate all such pairs without repetitions, i.e., obtain an ordered L sequence, according to the "triangle rule" [19]. For the case $n = 3$, such a triangle has the following form:

$$\begin{array}{c}
7, 5, 7, 4, 7, 3, 7, 2, 7, 1, 7, 0, 7 \\
6, 4, 6, 3, 6, 2, 6, 1, 6, 0, 6, \\
5, 3, 5, 2, 5, 1, 5, 0, 5, \\
4, 2, 4, 1, 4, 0, 4, \\
3, 1, 3, 0, 3, \\
2, 0, 2, \\
1 \\
0, 1, 2, 3, 4, 5, 6, 7
\end{array}$$

By arranging the rows of the "triangle" in linear order, we get the L sequence. Each pair <*i*, *j*> is assigned its weight $S_{i,j}$, which is equal to the number of transistor switchings in the circuit caused by changing the test vector i to vector j. To each ordered sequence P:

$$P = < i_1, i_2, i_3, i_4, \ldots i_{k-2}, i_{k-1}, i_k > \quad (7.12)$$

of elements (not necessarily different) of the set V corresponds the multitude:

$$< i_1, i_2 > , < i_2, i_3 > , < i_3, i_4 > , \ldots , < i_{k-2}, i_{k-1} > , < i_{k-1}, i_k >$$

of ordered pairs made of neighboring elements of the sequence (7.12). An ordered k-sequence (7.12) is called regular if all the ordered pairs $<i, j>$, included in it, are distinct.

The task of finding an "energy-intensive" test T is formulated as follows: for a given number k, it is required to compose a correct k-sequence P with the maximum sum of weights from elements of the L set:

$$S = \sum_{q=2}^{k-1} \left(S_{i_{q-1}, i_q} + S_{i_q, i_{q+1}} \right). \quad (7.13)$$

If each element of the set L is assigned to a vertex of complete oriented graph G, then the problem can be reformulated in graph definition: in complete oriented graph G, whose arcs are weighted by nonnegative integers, we need to find a simple chain consisting of k arcs and having maximal sum S of weights of its incoming arcs. This task and algorithms of its solving are known in graph theory. Solution algorithms for T test finding for the case when the number of input variables of combinational circuit does not exceed two tens are proposed in [20].

Software tools have been developed to obtain SPICE descriptions of logic circuits and to generate various tests. These tools allow carrying out evaluation of power consumption with the help of general-circuit simulation systems.

7.12 Low-Power CMOS Microcircuit Design Technology with ELS Package

Designing digital VLSI blocks within the complex is a multistage process of changing structural or functional description of a circuit. Each of the resulting circuit descriptions defines a new state of the design and is referred to as design solution. The design process starts with initial description of a digital block in one of the input design languages, and ends with representation of its circuit in the process design basis. Design solutions can be obtained either automatically (using synthesis and optimization software) or semiautomatically (though correction of the solution by the designer). In order to avoid propagation of errors made during one of the earlier design stages, the complex includes verification tools for design solutions at all design stages [21, 22]. Verification consists of checking whether the resulting solutions are in terms of equivalence (if both descriptions are fully defined) or implementation (if the initial one is not fully defined).

The design process using the ELS software package includes the following stages (each implemented by a set of alternative design operations):

- Development of the functional description of the circuit to be designed
- Optimization of the functional descriptions of two-level and multilevel circuits, taking into account complexity and power consumption
- Synthesis and optimization of circuits in a given library of CMOS elements, taking into account complexity and power consumption
- Verification of design solutions at all design stages
- Generation of test sequences and estimation of power consumption at logic and circuitry levels

The general structural diagram of data conversion in experimental software package ELS of automated design of logic circuits made of CMOS library elements with minimization of power consumption is shown in Fig. 5.17.

The design technology when using the ELS software package is based on a sequential transformation of the description of the circuit under design represented in the SF language [3]. SF language, being also an internal language of the system, is oriented on hierarchical structural and functional descriptions of logic circuits. Combinational blocks or elements in the SF language are specified either in the form of logical equations or in matrix form (a pair of matrices describing DNF system of Boolean functions). The initial functional and structural description of the designed circuit can be presented in VHDL language.

The ELS complex provides conversion of the circuit description from VHDL into SF language [23]. The reverse conversion of the obtained structural description of the circuit made of CMOS VLSI library elements to the description in VHDL language is also provided. Any intermediate description of the logic circuit designed in ELS can also be converted to VHDL. Such data matching allows using ELS software package together with other existing circuit design tools, such as the LeonardoSpectrum logic circuit synthesizer [26]. Numerous experiments have proved the effectiveness of this approach to design. In this case, pre-optimization is performed using software not available in LeonardoSpectrum, and the final stage, related to covering optimized representations with functional descriptions of the target library elements, is performed by the LeonardoSpectrum industrial synthesizer (Fig. 7.3).

7.13 ELS Software Package Architecture

The software shell of the ELS complex offers tools for information and language support of design processes. These tools include reference subsystem and express assessments of the results of design operations and implementation of alternative process design flows and design operations. The ELS software package consists of four subsystems: design project creation, project optimization, verification, and estimation of power consumption.

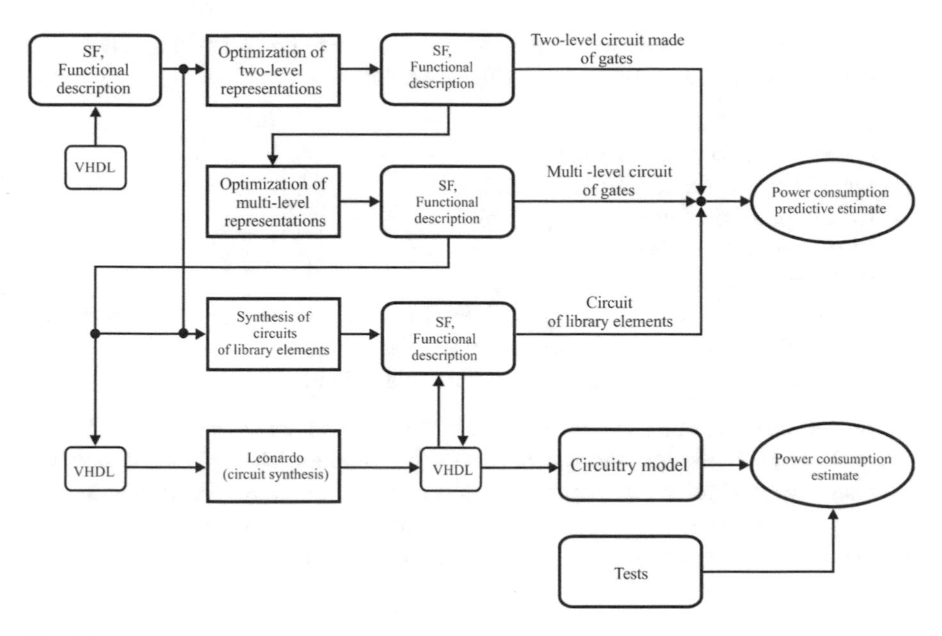

Fig. 7.3 Synthesis of logic circuits within the Low-Power Logic Synthesis software package

The first subsystem provides support for creating, editing, and transforming the forms of the original terms of reference. The second one supports execution of the design optimization and synthesis procedures. The third subsystem controls transformation of design project description. The fourth subsystem supports mechanisms for test generation and power consumption estimation in the generated or given tests.

All the data about the current state of the designed circuit form the project. In addition to the circuit SF description, a number of additional data are defined in the project, some of which are represented by attributes and reflect some properties of the current circuit description. These include, for example, current format of the description, parameters of the design operation in progress, name of the latest design operation performed, possible follow-up actions, and history of design operations completed (e.g., whether circuit minimization has been performed or not), and others.

In accordance with the purpose of the software package, the input data used for design can be:

- Functional description of designed combinational logic circuits in VHDL and SF languages
- Functional and structural description of designed combinational logic circuits in VHDL and SF languages
- Distribution of signal probabilities of value 1 occurrence at inputs of the designed circuit
- Sets of test patterns that are used to simulate behavior of the designed circuit to make estimates of their power consumption
- Working models for circuit simulation in SPICE

The software shell is a set of tools and services to monitor and manage the design process. It includes a set of the following subsystems:

- Session deployment
- Project creation or adjustment
- Import and export of VHDL descriptions
- Utilization of data used in the project
- Organization of design operations
- Verification of obtained design solutions
- Estimation of switching activity of design solution circuitry implementation at all stages of optimization and synthesis of the logic circuit
- Estimation of power consumption of the designed circuit

The whole design process is oriented to the use of interactive, dialogue mode with support and prompts from the system part of the complex. The complex is supplied with an advanced reference system. All ELS complex programs are written in C++ language with the use of data structures developed by the authors and are tested within the test program under control of Windows XP operating system.

7.14 Functional Capabilities of ELS Software Package

The ELS software package includes six groups (classes) of programs that support the complete cycle of designing logic circuit from library elements. This software provides synthesis of circuits and their optimization in terms of area and power consumption at all stages of designing process.

1. *Two-level optimization* includes a toolkit for minimization of systems of fully and partially defined Boolean functions in DNF class. The main optimization criterion in this case is a minimum of integral estimate. This evaluation takes into account complexity of derived DNFs and total switching activity of two-level circuits which implement them.

The developed software package for two-level optimization includes software implementing approximate methods. Precise minimization methods have not been included in the package. The reason is that for Boolean functions of true dimensionality, it is very difficult to obtain minimal DNFs within a time period acceptable for practical application. The current version of ELS software package includes programs for two-level optimization [14] (Fig. 7.4), implementing the following methods:

- Parametrically configurable minimization
- Minimization by grouping into classes
- ESPRESSO minimization

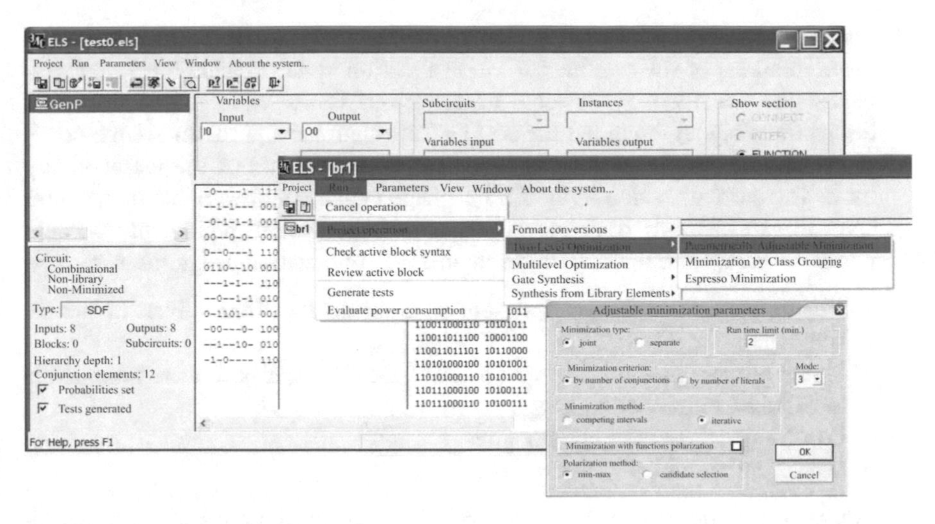

Fig. 7.4 Software package windows in design mode

These programs provide minimization under the following potential requirements:

(a) Requirements for optimization object:

- One fully defined Boolean function or system of such functions
- One partially determined Boolean function or a system of such functions

(b) Requirements for the solving method:

- Iterative minimization method [14]
- Competing interval method [14]
- Modified ESPRESSO minimization method [26]
- Method of grouping into classes [24]

(c) Requirements for the method of accounting for conjoint minimization of the DNF system:

- Separate minimization of DNF system functions
- Joint minimization of DNF system functions

(d) Requirements for the criterion of optimization by energy consumption:

- Without energy consumption taken into account
- With regard to energy consumption

(e) Requirements for additional settings to be taken into account, including:

- Obtaining paraphase implementations of DNFs
- Minimizing the number of conjunctions of the resulting DNF system
- Minimizing the number of literals of conjunctions of the resulting DNF system
- Minimization time limit
- Algorithm operation mode (for the method of competing intervals)

2. *Multilevel optimization* includes a toolkit for designing and optimizing multilevel representations of fully defined Boolean function systems. Multilevel representations are specified by systems of factorized forms with conjunctions and disjunctions having bounded ranks [11]. This approach approximates the correspondence of multilevel representation to the constraints of the target library basis. The current version of the ELS package includes software that implements approximate methods of joint and separate factorization, taking into account power consumption. Initial data for multilevel optimization programs are:

   - Functional description of a system of Boolean functions, given in the SF language
   - Signal probabilities for the input poles of the circuit, if power consumption is to be taken into account
   - Constraint on ranks of conjunctions
   - Constraint on ranks of disjunctions

3. *Synthesis of logic networks of gates* includes a toolkit for synthesis and optimization of multilevel networks of AND and OR gates (using inverters at the inputs) with limited number of input poles. Main criterion for multilevel circuit quality is its minimal integral index, which takes into account complexity of circuit (measured by number of input poles of all its gates) and total switching activity of all its poles (including input, internal, and output poles).

Current version of ELS complex includes software for synthesis of logic networks of two-input gates, two-input gates from partial functions, and multi-input gates.

The first two of these software programs implement methods of synthesis of multilevel logic circuits of two-input gates AND and OR (using inverters) for fully and partially defined Boolean functions [17]. The methods are based on BDD generation and optimization. The third program implements synthesis method of multilevel logic circuits from AND and OR gates with limited number of inputs (number of inputs are determined by process design basis). The method is based on the combination of algorithms of joint and separate factorization of DNF systems of Boolean functions with power consumption taken into account [15].

The input data for multilevel optimization algorithms are:

- Functional description of the system of Boolean functions given by the DNF system in SDF or LOG format of SF language
- Signal probabilities for input variables, if energy consumption is to be taken into account
- Restriction on the number of input poles of AND gates
- Restriction on the number of input poles of OR gates

4. *Synthesis of library elements* includes a toolkit for mapping of multilevel networks from AND and OR gates into element basis of given CMOS VLSI library. The key quality criterion of multilevel circuit from library elements is its minimal integral index that takes into account complexity of the circuit (measured by the number of transistors of all elements) and total switching activity of all its poles (input, internal, and output poles).

The current version of ELS complex includes software programs for circuit synthesis from library elements, which implement approximate methods of coverage of multilevel logical circuits from AND and OR gates:

- Simplified method of technology mapping of a circuit, from gates into library basis
- Method of technology mapping of a circuit, from gates into library basis, including optimization [18]
- Combined method of technology mapping of the system of fully or partially defined Boolean functions into a circuit from library elements

Combined method of synthesis of circuits from library elements includes execution of several operations implemented in known SIS system [22]: minimization of systems of Boolean functions and construction of multilevel circuit of two-input gates, AND-NOT or OR-NOT, coverage of this circuit by given library elements.

5. *Verification of project states* includes a set of programs that allow the verification of a given pair of design solutions [22]. Verification is applicable to any pairs of design solutions of the same project (or different projects). The only restriction is that the second state must not precede the first state (in the chain of performed design operations), i.e., the first of compared objects cannot be "more defined" than the second, if at least one of the descriptions contains an undefined behavior (e.g., in case of minimization or circuit implementation of systems of partially defined Boolean functions).

When both descriptions being compared are functionally fully defined (e.g., in case of combinational DNF circuits or systems), the verification checks whether these descriptions are equivalent. If at least one of the descriptions contains an undefined behavior, the verification checks whether there is an implementation link between these descriptions, i.e., whether the first description is implemented by the second one. In the case when the first description is not implemented by the second one, the verification software module allows identifying the cause of non-implementability, more specifically determining the interval (or set) and function of initial description or the circuit element being the causes of implementation failure.

6. *Energy consumption* is evaluated by means of test generation software and assessment of energy consumption level of the circuit based on the results of analysis of the given test execution. This approach makes it possible to estimate the effectiveness of the design solution during circuit design after the subsequent optimization or synthesis operation is performed, when its implementation on a VLSI chip is not yet complete. The evaluation of power consumption within the ELS complex is carried out by:

   - Estimation of switching activity of the circuit based on calculation of signal probabilities for the poles of circuit implementation of the design object, carried out automatically after each design operation that changes its state

- Calculation of the number of transistor switchings for a circuit of library elements based on logic (VHDL or SF) simulation for a given test and estimation of the amount of average current consumed in one clock cycle [19]
- Estimation of power consumption of a circuit of library elements at the circuit design level [24] based on test runs and measurements of current consumption by means of SPICE simulation

References

1. Belous, A., & Saladukha, V. (2020). *High-speed digital system Design. Art, science and experience*. Springer Nature. isbn 978-3-030-25408-7.
2. Belous, A. I., Emelyanov, V. A., & Syakersky, V. S. (2009). *IC design with reduced power consumption* (210p.). Integralpoligraf. isbn 985-6107-18-0 (in Russian).
3. Belous, A. I., Murashko, I. A., & Syakersky, V. S. (2008). Power consumption minimization methods for designing CMOS LSI. *Technology and Design in Electronic Equipment. Kiev, 4*, 39–44. (in Russian).
4. Bin Ibne, M., Farayez, R., & Islam Badal, M. T. (2017). Resign of a low-power CMOS Level Shifter for low-delay SoCs in silterra 0.13 pm CMOS process. *Journal of Engineering Science and Technology Review 10*(4): 10–15. https://www.semanticscholar.org/paper/Design-of-a-Low-power-CMOS-Level-Shifter-for-SoCs-Badal-Reaz/ff4a4dabf5d394feafeea857450d10040e7fb841
5. Roy, K., & Prasad, S. C. (1995). *Low-power CMOS VLSI circuits design* (p. 348). Hilton Books.
6. Veendrick, H. (1984, August). Short circuit dissipation of static CMOS circuitry and its impact on design of buffer circuits. *IEEE Journal of Solid State Circuits, 19*(4). https://doi.org/10.1109/JSSC.1984.1052168
7. Bibilo, P. N., Cheremisinova, L. D., Kardash, S. N., Kirienko, N. A., Romanov, V. I., & Cheremisinov, D. I. (2013). Low-power logical synthesis of CMOS circuits automation. *Software Engineering, 8*, 35–41. (in Russian).
8. Bibilo, P. N., & Romanov, V. I. (2011). *Logical design of discrete devices using production-frame model of knowledge representation* (279p.). Belarusian Science. (in Russian).
9. Cheremisinova, L. D. (2010). Estimation of power consumption of CMOS circuits at logic level. *Information Technologies, 8*, 27–35. (in Russian).
10. Cheremisinova, L. D. (2010). Low power synthesis of combinational CMOS circuits. *Informatics, 4*, 112–122. (in Russian).
11. Cheremisinova, L. D. (2005). *Synthesis and optimization of VLSI combinational structures* (236p.). OIPI NAS of Belarus. (in Russian).
12. Toropov, N. R. (1999). Minimization of systems of Boolean function in the DNF class. In A. A. Zakrevsky (Ed.), *Logic design* (pp. 4–19). Institute of Technical Cybernetics of the National Academy of Sciences of Belarus. (in Russian).
13. Cheremisinova, L. D. (2010). Power consumption estimation upon optimization of two-level CMOS circuits. *Informatics, 2*(26), 105–115. (in Russian).
14. Cheremisinov, D. I., & Cheremisinova, L. D. (2011). Low power driven minimization of two-level CMOS circuits. *Information Technologies, 5*, 17–23. (in Russian).
15. Cheremisinova, L. D., & Kirienko, N. A. (2013). Low power driven synthesis of multi-level logical circuits. *Information Technologies, 3*, 8–14. (in Russian).
16. Cheremisinova, L. D., & Kirienko, N. A. (2011). Low power driven optimization of bracket representations of Boolean functions. *Informatics, 3*, 77–87. (in Russian).

17. Bibilo, P. N., Leonchik, P. V., & RAS, I. (2011). Decomposition of Boolean function systems defined by binary choice diagrams. *Theory and Control Systems, 4*, 86–101. (in Russian).
18. Cheremisinov, D. I., & Cheremisinova, L. D. (2013). Low-power driven synthesis of combinational circuits in the basis of CMOS VLSI library cells. *Informatics, 4*, 91–102. (in Russian).
19. Bibilo, P. N., & Kirienko, N. A. (2012). Evaluation of power consumption of logical CMOS circuits by their switching activity. *Microelectronics, 41*(1), 65–67. (in Russian).
20. Bibilo, P. N. (2010). Finding a test for maximum power consumption mode of combinational logic circuit. *Control Systems and Machines, 5*, 39–45. (in Russian).
21. Cheremisinova, L. D., & Novikov, D. Y. (2010). Formal verification of descriptions with functional uncertainty on the basis of conjunctive normal form feasibility check. *Automation and Computer Science, 1*, 5–16. (in Russian).
22. L.D. Cheremisinova, , D.Y. Novikov Software verification tools of combinational device descriptions in process of logic design. Software Engineering, 2013, 7, pp. 8–15. (in Russian).
23. Cheremisinov, D. I. (2006). *Analysis and conversion of VLSI structural descriptions* (275p.). Belarusian Science (in Russian).
24. Bibilo, P. N. (2010). Power consumption evaluation of combinational blocks of application specific CMOS VLSI based on logic simulation. *Modern Electronics, 2*, 54–59. (in Russian).
25. Belous, A. I., Murashko, I. A., & Syakersky, V. S. (2008). Power consumption minimization methods for designing CMOS LSI. *Technology and Design in Electronic Equipment, 2*, 39–44. (in Russian).
26. Bibilo, P. N. (2005). *Systems of IC design based on VHDL language. StateCAD, ModelSim, Leonardo Spectrum* (384p.). SOLON-Press. (in Russian).

Chapter 8
Fundamentals of Building a Quality Management System for Manufacturing Submicron Integrated Circuits Based on Test Structures

The subject matter of *this chapter* is the so-called embedded or built-in test structures. The preceding chapters explore all essential steps of microcircuit design—from physical fundamentals of MOS transistor operation up to methods of designing the protected "system-on-chip." Then the product designed in microelectronics is delivered to semiconductor factory for production of the first (pilot) lot. There is no guarantee that the products produced at the fab line and delivered to the designer for subsequent testing will be in full conformity with the criteria of the original Design Specification and Data Sheet. As the best world practice over the recent decade has proved, approximately only 40% of the designed microcircuits after fabrication meet the source specification. Ultimately sometimes obtaining of fully operational products is possible only after quite a few iterations (corrections). For the purposes of efficiency enhancement and cutting the time for overview and identifying the causes for failure to attain the specified parameters of the designed IC, there are used specific test structures located on die (wafer) of the microcircuit designed.

Indeed, for designers of advanced leading-edge microcircuits, it is critical to have perfect understanding of physical phenomena and causes for failures and poor yields. This is essential not only for making a good choice among design solutions for microcircuits considering tolerable and critical density levels in active semiconductor structures and interconnections (conductors) and density in insulating layers, but it is essential also for elaborating appropriate steps and measures aimed at detecting and screening potentially unreliable chips at various steps of mass production.

Therefore, *this chapter* sees into key concepts of generating composition and architectures of built-in semiconductor test structures and fundamentals of operation-wise predicting reliability of designed microcircuit by the results of processing digital values of the measured parameters of test structures. There are provided mathematical models linking the statistical distribution of microcircuits technical parameters values with reliability performances, and there are provided some specific examples of such enhancement of the technological processes.

A. Belous, V. Saladukha, *The Art and Science of Microelectronic Circuit Design*,
https://doi.org/10.1007/978-3-030-89854-0_8

8.1 Methodology of the Organization of Technological Test Control in the Process of Design and Production of Microelectronic Products

8.1.1 Place and Role of Semiconductor Test Structures in the Process of Manufacturing Integrated Circuits

The quality and reliability of semiconductor integrated circuits (IC) are determined by the perfection of the design, the perfection and controllability of production processes, the efficiency of qualification, and the perfection of periodic tests, are provided during production, and are maintained during operation.

Analysis of defects and failures of ICs and their elements plays a significant role in the complex of measures to increase the yield and ensure the reliability of ICs. The results of the analysis are used to improve the design and manufacturing processes of ICs and to clarify the methods and criteria for selection and acceptance tests [1].

Submicron integrated circuits have all the reliability problems associated with the production of discrete semiconductor devices, as well as many other problems. On the one hand, the small geometric dimensions of the IC elements lead to the fact that the circuits acquire a higher sensitivity to degradation processes. On the other hand, large chip sizes increase their vulnerability to defects inherent in semiconductor and other materials, which increases the likelihood of such defects [2].

Unlike discrete devices, the elements of the IC cannot be tested individually. The IC should be tested as a functional system. At the same time, the sources of unreliability or failures remain hidden in the design and usually appear later, when the IC is installed in the device, and when it is too late to take any measures to improve it. Therefore, new methods are needed to manage the quality of production processes.

One of them is that the process flow of manufacturing IC introduces test structures that are close in design parameters to the IC elements and are used for daily evaluation of the quality of these elements and the production process, as well as for predicting the reliability of the IC elements. The test structures are designed in such a way as to reflect the behavior of the IC elements in terms of the main physical and chemical processes that lead to the defect and failure of the IC [3].

Currently, there is a tendency to significantly expand the scope of application of technological test structures in solving problems of designing and manufacturing IC. This trend is increasingly evident as the complexity of the IC increases, the degree of their integration increases, the size of individual components decreases, and the requirements for their parameters become more stringent. The development of an IC requires careful development of both individual structures and products as a whole. The study of the physical properties and electrophysical parameters of the IC is carried out on specially designed test structures.

Based on the results of studies of test structures, the operation of the IC is simulated, the relationship of output parameters with structural and technological factors is established, and the resistance to various operational factors is evaluated.

Thanks to the research conducted on test structures, it is possible to obtain information that characterizes the quality of the design and manufacturing technology of IC, as well as their reliability parameters [4].

The presence of test structures on the wafer is of value for technological services, if there are appropriate methods that allow us to interpret the relevant data and characteristics in terms of the impact on reliability [5], the yield of ICs on the wafer, and the electrical parameters of the IC.

Summarizing the experience of using test structures known from the literature, we can identify the main tasks that are solved in the design and production of ICs [5]:

- Development of the basic electrical circuit and the topology of the IC with the use of computer design programs
- Selection of the parameters of the source material for IC formation, which will ensure compliance with the requirements of the technical specification for electrical and operational characteristics
- Development of individual technological operations and the entire technological process
- Development of the system of operational control in the conditions of mass production
- Determination of IC yield
- Determination of reliability indicators and justification of the technical and economic feasibility of the design

With the help of test structures in the organization of mass production of IC, you can also solve other equally relevant tasks:

- Quality control of the IC elements and the wafer as a whole
- Control of the stability of the technological process according to the reproducibility of the electrophysical parameters and the yield
- Investigation of structural defects, various failure mechanisms, and process adjustment
- Predicting reliability by testing test structures for reliability and long-term operating time

The test structures can include elements that are used both for intermediate control of technological processes and for subsequent identification of the relationship between the causes of failures and the technological process.

Test structures are also widely used in the development and testing of various models:

- Physical models of elements that establish a relationship between the physical quantities and the parameters of the elements (one-, two-, and three-dimensional models, with concentrated or distributed parameters)
- Physical and topological models that establish similar relationships taking into account the topological dimensions of elements and their changes in various operations during the manufacture of elements

- Models of technological processes that establish links between technological factors and physical or physical-topological models of elements
- Statistical physical models and mathematical models of the reliability of elements

8.1.2 *Classification of Technological Test Structures*

The definition of the test structure has also changed with the development of microelectronics. At the stage of development and mass production of discrete field-effect transistors, the first test structures included individual elements of these transistors (contact connections, metallization on the dielectric steps, ring MIS transistor, parasitic MIS transistor, MIS capacitors, etc.). Despite the fairly wide use of test structures [1–5], the terminology in this field of research has not yet been fully formed, and often the terms have different interpretations. Thus, according to [5], a test structure is a part of an integrated circuit that has contact pads and is intended for studying the physical structure and electrical parameters of an integrated circuit.

In the practical use of test structures, in most cases, they are not part of the integrated circuit; the test structure may also not have its own contact pads. Usually, the study of the electrical parameters of an integrated circuit is carried out not on test structures, but on the IC.

In order to increase the information content of the control, special test structures with different configurations and different combinations of layers are developed [4].

The most well-established term: "Test structures are structural elements located in one or more layers that are not part of the IC, made using masks, and allow you to determine the electrical, physical, or geometric parameters of the IC elements."

An IC element with additional contact pads for measuring its electrical parameters is often referred to as an auxiliary test structure.

A semiconductor chip containing only test structures is usually called a test chip (module); a semiconductor wafer with such chips is called a test wafer; test structures located on a chip are called embedded test structures; test chips located on a semiconductor wafer with ICs are called embedded (or imprinted) test chips; a semiconductor wafer with layers deposited on its surface (doped, epitaxial, dielectric, polysilicon, metal, etc.) without test structures formed on it is called a test wafer.

On the test wafer, the parameters of the deposited layer are usually determined before the photolithographic operation, while on the test wafer, the geometric and electrophysical parameters of the layer are determined after the photolithographic operations.

Physical test structures [6] are designed for the analysis of structural and electrophysical parameters of layers and IC elements: linear dimensions of elements and layer thicknesses, surface and resistivity of layers, capacitance characteristics, surface and volume concentrations of impurities on a semiconductor substrate, charges in dielectric layers, studies of volume and surface effects, etc.

In the manufacture of integrated circuits, multilayer structures arise. Typically, each layer deals with a single physical layer, such as a dielectric. In the same integrated circuit, different combinations of physical layers can occur due to the removal or local growth in certain places of one or more physical layers. The layers of the physical structure can be located one relative to the other at different distances, with overlap (complete or partial). The shape of the layers can be linear, at an angle of 90° or another angle, in the form of parts of a circle, closed or open.

Despite the simultaneous deposition of the layer and the group processing of the wafers, the physical properties of the layers can be different and depend on the combination of the previous and subsequent layers. At various technological operations, the parameters of the physical layers can change significantly (thickness, conductivity, charge, etc.).

Reliability test structures are made in such a way as to reproduce the conditions that actually exist in the technical process under study, namely, the number of defects, the rate of defect development, and electrical and thermal conditions. The test structures should provide an opportunity to study the main components of semiconductor integrated circuits: the dielectric, the silicon surface, metallization, and doped areas.

For example, for MIS ICs (metal-insulator-semiconductor), typical failures of the listed components are breakdown of the thin gate dielectric of the transistor as a result of the development of defects and the impact of a high-intensity electric field, burnout and breakage of metal lines due to electrodiffusion, occurrence of leakage currents due to the formation of surface channels between diffusion areas located at different potentials, etc.

To study each of these types of failures, special test structures with certain morphological elements are created. Thus, test structures for the analysis of the breakdown mechanism of the oxide should reproduce not only the physical properties, the thickness of the dielectric, and the electric field in it but also the conditions of its growth, the specifics of the adjacent doped areas, their geometric characteristics, etc.

Test structures open up opportunities for determining the quantitative characteristics of the reliability of the IC based on the results of accelerated tests. Knowledge of the physics of failures allows you to identify the main destructive mechanisms, take into account possible combinations of them in different conditions, and, thus, minimize the random nature of the manifestation of failures during such tests.

The elements are tested for reliability as part of the test structures. At the same time, optimal conditions are created for detecting defects and speeding up the processes that lead to IC failures. The result of a deterministic approach to conducting IC tests is the possibility of reducing the volume of tests in terms of the number of samples studied and the time spent in comparison with a conventional statistical experiment.

The ultimate goal of the development and research of reliability test structures is to identify measures that improve reliability. Test structures are also used to evaluate the effectiveness of various designs and technological and operational "security" measures that prevent the occurrence of failures and parasitic effects. The required number of reliability test structures decreases with a steady technological process and an optimal IC topology.

8.1.3 Methods of Placing Test Structures on Semiconductor Wafers

Embedded test structures can be located both on the free area of the chip and on special test “embedded” or “imprinted” chips, on the chips of the test wafer.

The printed test chips on the semiconductor wafer can be placed, for example, on one of the halves of the wafer, in the form of a row and column, or in the form of several chips arranged crosswise, or in the form of a single test chip, or in the form of several rows or columns.

On the chip, you can usually place a small number of test structures (MIS transistor, resistor, contact connection, etc.). On the test chip, you can already place a large number of circuit elements and large-sized test structures that are used to assess the causes of defects and analyze the causes of failures of the structural elements of the IC.

The use of imprinted test chips has the following advantages:

- One set of photomasks is used.
- The analysis of test structures is performed on the same wafers with chips, which is especially important when the technology level is not sufficiently stable.
- Saves the wafer area occupied by the test chips.

The disadvantages include the following:

- Increasing the complexity of measurements.
- The complexity of automating the processes of measuring the parameters of circuits and test chips.
- Increased measurement time.
- Due to the small number of test chips, the reliability of estimating the distribution of parameters over the wafer is low.
- After scribing, additional analysis of the test chips becomes difficult.

It should also be noted that when imprinting test chips, the complexity of manufacturing masks increases, and the quality of combining test structures or chip elements deteriorates somewhat. Therefore, in some cases, it is preferable to use the so-called built-in test strips, manufactured simultaneously with the chips of the IC.

At the stage of developing new technological processes for manufacturing ICs, it is advisable to use test matrices. After the completion of the main procedures for adjustment of technological operations, their further control can be carried out using the built-in test chips. In the future, when the production reaches a stable level of technology, the number of built-in test chips can be reduced.

In some cases, the test control can be carried out on a test wafer manufactured simultaneously with a batch of wafers with ICs or can be performed on separate batches of test matrices manufactured periodically [1].

Special attention should be paid to the problem of the location of the contact pads of test structures.

Usually, when designing a test chip, to save the area of the chip, the test structures are placed with the maximum permissible density, while the dimensions and distances between the contact pads are determined by the technical characteristics of the probe devices used.

To simplify the operation of imprinting the test chip and using the same probe devices, the dimensions of the test chip should match the dimensions of the chip. It is especially important to match the location of the contact pads of the chip and the test structures when using the "replaceable mask method," when the chip is transformed into a test chip by changing the metallization layer.

A number of manufacturers, in order to increase the number of test structures and maintain a given distance between the contact pads of one test structure, place other contact pads belonging to another test structure between its contact pads. In this case, it is necessary to exclude the influence on the measurement results of the short circuit of adjacent contacts by the probe, but the parameters of adjacent test structures can be measured only in two probes [1].

The location of the contact pads of test structures is usually recommended to be arranged according to a linear law in one direction. At the same time, the technological time for rebuilding the probe devices is reduced. If the test structures are located with a given step both horizontally and vertically, as well as with an entire step or half step from the edge of the chip, the procedure for automating the measurement of test structures is simplified.

The number of contact pads can be reduced by connecting adjacent test structures, for example, two transistors or their gates. But with such compounds, parasitic effects of the influence of some test structures on others may appear, the ability to measure a number of parameters is limited, and the presence of a defect in one of the test structures may distort the results of measuring the parameters of another test structure.

To prevent short-circuiting of probes with a semiconductor substrate due to punching of thin dielectric layers, it is recommended to place the contact pads on a thick dielectric layer with a thickness of more than 0.5 μm. The placement of polysilicon contact pads under the metal contact pads also prevents punctures and increases the number of contacts of the probes with the contact pad.

To ensure that the test structure pads are quickly found, it is necessary to indicate the number near each contact pad or group of test structures at the design stage of the metallization mask.

To quickly determine the number of the performed or missed photolithographic operation, a digit is placed on each photomask, located on the periphery of the chip in such a way that when all the masks on the chip are combined, a sequence of natural numbers is obtained.

Before measuring the parameters of the test structures, the contact through the pairs of pads is evaluated. Despite the significant disadvantage of some increase in the measurement time, the advantage is the exclusion of failures due to the effect of non-contact of the probes.

8.2 Principles of Control of the Process of Manufacturing Chips Using Test Structures

8.2.1 Assessment of the Quality of the Process Based on the Method of Interoperative Control of Wafers

The choice of the type and optimal combination of test structures depends on the specific task, for which it is usually necessary to choose the required number of structural layers and their correct combination, if necessary, to minimize the number of structures.

One of the main types of test control is the interoperable control of wafers in the process of manufacturing integrated circuits [7]. It is carried out in order to assess the quality of the wafers after performing the main technological operations for the timely detection of defects, exclusion of defective wafers from the technological process, or adjustment of the modes of subsequent technological operations to correct the defect, to obtain information about the results of the main technological processes, to control their stability and reproducibility of product parameters, and to establish the causes of wafer defects.

This control is used to study the parameters of elements in various operations, to determine the main operations "responsible" for the defect or low reliability of their operation, to develop mathematical models of technological processes and methods of their control, and to study the degree of uniformity of the distribution of the parameters of elements on the wafer and different wafers of the batch.

To solve these problems, it is necessary to measure a variety of parameters describing the physical structure of individual areas of the chips: for example, the thickness of the dielectric layers, the charge density in them, the surface concentrations of impurities in various areas of the semiconductor substrate, the resistivity of the wiring of different levels, contact resistances, the density of defects in the dielectric layers, etc.

Quality control of a batch of simultaneously manufactured wafers is usually carried out with the help of special test wafers, which measure the surface resistances, the depth of the diffusion layers, as well as the thickness of various films, the speed of their etching for high-quality photolithographic operations, etc.

Such control is not effective enough due to the heterogeneity of the wafers. Therefore, in the practice of manufacturing IC, there are often cases when, with good control measurement data on test wafers, some working wafers are rejected when checking the electrical characteristics of the IC. To establish the cause of the defect in this case is very difficult.

When measuring the intermediate operations, it is necessary to observe additional precautions against punching the films with probes, from the breakdown of the voltage of the dielectric films between different conductive buses.

8.2.2 *Typical Composition of the Test Module for Monitoring Production Processes*

The consumer cannot definitely assess the reliability of the IC and predict the results of its operation. Therefore, the reliability of the chips should be confirmed by the procedure for checking the parameters of the production process [1]. This test is carried out on a special test chip, which is often placed on each semiconductor wafer among the working circuits.

Verification of the efficiency of the production process is carried out in order to verify the compliance of the obtained characteristics of the integrated circuit during the passage of the product through the production cycle and the permissible characteristics justified by the designer to ensure that the IC remains operational throughout its entire service life. To achieve this, the test module is usually formed in such a way that the worst operating mode of the IC is provided and possible under these design rules and the finished modules are subjected to comprehensive measurements and accelerated tests. Design rules here are defined as the specification of all permissible dimensions and tolerances that the integrated circuit topology should meet, the values of the parameters of the elements, and the characteristics of the production processes. This concept includes various formulas that are used to calculate the parameters of the Scheme [1].

The standard nomenclature of test structures normally used to create a process parameter verification module contains the following elements:

- Two MIS transistors, each of which has nonmetallic sources and drains with a size sufficient for contacting a metal probe
- MIS transistors with different geometric dimensions (one of the transistors should have the minimum allowable channel length, while the other the minimum width)
- Diffusive resistors of different widths
- Parasitic MIS transistors with different channel lengths
- A metallization line of the minimum width, covering all possible steps of the oxide
- Chains of metallization contact connections with semiconductor areas of different types of conductivity
- A long metallization bus of minimal width, located on the pattern of polysilicon, field oxide, and diffusion areas.
- Two metal buses located at a minimum distance from each other
- Two n- and p-type silicon field oxide MIS capacitors
- Two MIS capacitors with a thin layer of oxide on n- and p-type silicon
- Large-area n- and p-type diffusion areas
- Test structures of circuits to protect the inputs and outputs of the chip from the effects of static electricity

Test structures for evaluating the reliability of the parameters of IC elements with recommendations for the choice of electrical test modes are given in [7, 8].

8.2.3 *Typical Composition of Test Structures for Quality Control of Submicron ICs*

When organizing the quality control process of submicron technological processes, the same basic test structures and approaches are usually used as described above, as well as additional test structures in order to:

- Research and evaluate group technological operations and test and evaluate the reserves of resistance of elements and ICs in general to various destabilizing factors, as well as reliability tests
- Test and evaluate the compatibility of raw materials and technological operations

The paper [5] describes additional test structures, shows the sequence of climatic and mechanical tests, as well as tests for moisture resistance and reliability.

The recommended set of test structures and the list of measured electrophysical parameters are given in Table 8.1.

Due to the fact that the vast majority of defects cannot be detected by visual inspection and by the existing methods of direct measurement of electrophysical parameters, there is a need to use forced accelerated test modes in order to stimulate the development of various failure mechanisms in a shorter time.

The full composition of the test structures and the parameters measured using them, as well as the method for detecting failures using test structures, are discussed in [4], which presents a reduced catalog of defects and failure mechanisms and forced loads for their detection on test structures.

When choosing forced loads and modes of rejection tests for forced failure, it is necessary to take into account in each specific case the nomenclature and statistics of those potential defects that cannot be detected by the current operational and acceptance control system, the degree of resistance of the chip to various influences in order to establish the permissible values of loads and not damage defective devices, the full range of influencing factors that stimulate the development of the defect to the level of failure, and the cost of rejection tests. It should be noted that test structures do not replace the system of product quality certification, since when creating them, only the principle of similarity in possible physical and chemical processes occurring in test structures and IC is maintained.

The main role of test structures in the quality and reliability management system of the IC can be formulated as follows: reducing the spectrum of defects, weakening the effect of various failure mechanisms, and, thus, increasing the safety margin of the product under various loads.

The electrophysical characteristics of the MIS IC largely depend on the quality of the SiO_2 layers and the surface charge states of the oxide layers at the silicon boundary. To evaluate the properties of the oxide layer, test structures containing MIS capacitors of various configurations are widely used. The reliability of the MIS IC with a thin layer of gate oxide is largely due to the defect of the oxide formed as a result of the entire complex of operations for its formation. The presence of defects in the dielectric leads to electrical leaks, noise, short circuits, etc. Contamination in the form of mobile ions in the oxide causes instability of the threshold voltage detected after thermal field effects.

Table 8.1 A typical set of test structures

Types of test structures	Measured electrophysical parameters and physical effects
1. MOS capacitors formed by a gold ball probe and non-metallized oxidized sections of the wafer	Evaluation of the quality of the non-metallized oxide layer by measuring the charge properties of the dielectric layer
2. MOS capacitors that occupy a large area	The charge properties of the dielectric layer, the charge stability, the concentration of impurities in the semiconductor, etc.
3. MOS transistor	Threshold voltages, carrier mobility in the inverse layer
4. MOS transistor with a complex configuration	Carrier mobility in the inverse layer (according to the Hall method)
5. Bipolar discrete transistors of all major types	The main electrical parameters of the IC elements, their stability, stability reserves necessary for the calculation of the circuit
6. Diffusion diodes (p-n junctions)	The main electrical parameters of the IC elements, their stability, reserves of resistance to operational influencing factors, necessary for the calculation of the circuit diagram
7. Diffusion resistors	Surface resistance of resistors, base areas of transistors, emitter areas
8. Single-level metallization lines on the oxide steps	Geometric dimensions, quality control of photolithography, resistance value, permissible current densities
9. Multilevel metallization lines on the oxide steps	Resistance between the metallization layers, insulation resistance, breakdown voltage of the oxide film, resistance of the lines, permissible current densities
10. IC elements (transistors, diodes, diffusion and film resistors, diffusion buses, metallization buses, bus intersections, contacts, inter-element areas)	Electrical parameters necessary for the calculation of the circuit diagram and reliability tests and for the establishment of reserves of resistance to electrical loads and operational external factors
11. Four-contact (two-contact) resistors	Surface resistance of diffusion layers
12. Schottky diodes	Impurity distribution in the epitaxial film
13. IC blocks (inverter, amplifier, generator, etc.)	Electrical parameters, analysis of circuit solutions

The most common method of controlling the quality of the oxide is the method of measuring the current-farad characteristics (C-V characteristics) on test MOS capacitors. Capacitors are formed on the test wafers, after which the C-V characteristic is measured at normal temperature, as well as at 200 °C for 5 min under conditions of positive bias voltage. The relationship between the parameters of the C-V characteristics before and after the tests allows us to determine the stability of the parameters of the MOS structure [2].

To evaluate other parameters of active control of technological processes, e.g., quality and reliability, a MIS transistor is used. The geometry of this transistor is similar to the geometry of the MIS IC transistor, and the manufacturing process reflects all the processes used in the manufacture of the MIS IC. This test structure allows you to measure the parameters that give an idea of the reliability of the manufactured IC. These most important parameters include the threshold voltage of the gate dielectric, the threshold voltage of the peripheral oxide, the stability of the threshold voltage, the saturation current of the drain, leakage currents, breakdown voltages of the oxide layers, and diffusion transitions.

These structures can also be installed in the package for the purpose of testing for long-term operation. Having experimental dependences on the threshold voltage drift of an individual MIS transistor, it is possible to consider the MIS IC model containing thousands of similar elements and obtain an approximate estimate of the reliability of the MIS IC. Periodic measurements of the threshold voltage of the MIS transistor during testing provide information about the expected durability of the chip, the drift mechanism, and the activation energy of the degradation process.

8.2.4 Statistical Processing of Measurement Results of Test Structures

Statistical processing of the obtained information of measurement results is carried out to study the main characteristics of the studied processes, to test the simplest hypotheses about the nature of changes in parameters, to calculate the numerical characteristics of random variables, and to present the results of observation in the form of a statistical distribution and for evaluation control [9].

The information obtained as a result of statistical processing allows you to get a wide "range of knowledge" about the controlled technological process, including the following important characteristics and parameters of the process:

1. Determination of the percentage of outputs of the test structure parameters according to the design rules for the IC or according to the standards that limit the performance of the ICs on the wafer
2. Determination of the proportion of defect below and above the design rules
3. Determination of minimum and maximum parameter values
4. Determination of the rate of defects above certain limits at which the physical nature of the measured value changes
5. Determination of the distributions of the most important parameters of test structures on wafers with a high and low percentage of the IC yield and comparison of the distributions
6. Processing the results and obtaining mathematical models that connect the most important parameters of the test structures with the most important IC parameters
7. Graphical display of the received information and determination of the main regularities
8. Plotting the cumulative histograms

9. Compilation of generalized information for wafer and for batch per decade, per month, per quarter, and per year
10. Comparison of data on test structures during the passage of wafers through operations during IC inspection and testing
11. Plotting the distribution of the main defects of the chip
12. Determination of the predicted percentage of chip yield based on a set of electrophysical parameters on defective wafers or on wafers with low IC yield
13. Determination of the prevailing defects on the wafers with different IC yield
14. Determination of the average rate of defects based on the electrophysical parameters of test structures for defective wafers of different batches for a given period of manufacturing time
15. Determination of correlations between IC parameters and diagnostic tests
16. Determination of the proportion of good and defective ICs by the studied parameter among the good and the defective ICs by the correlated parameter.

The results of measurements of the parameters of test structures are used to predict the reliability of the elements of integrated circuits.

The main characteristics of the reliability of integrated circuits (physical simulation, methods of failure analysis, reliability of resistive and capacitive elements) are given in [8]. This paper discusses the theory and practice of integrated circuit reliability, summarizes the results of physical studies of sudden and gradual failures of elements of silicon-based semiconductor integrated circuits published in the periodical literature, and shows ways to ensure and improve reliability. The authors consider in detail the issue of simulation of the degradation processes of gradual failures associated with temperature using the Arrhenius equation and determine the boundary conditions of its applicability. The Eyring model for studying nonthermal degradation processes is also described.

8.3 Forecasting the IC Yield Based on the Results of the Test Control

8.3.1 Features of Simulation of the IC Yield

Simulation of the expected yield becomes a particularly relevant problem in the production of integrated circuits with submicron design rules. Having determined the algorithm and parameters of such simulation, it is possible to predict the cost and economic efficiency of organizing the production of new circuits at the design stage; to determine the quality of individual technological operations of the basic process flow; to optimize, if necessary, the technological process of manufacturing new ICs; and to identify the most critical operations and to improve them.

The parameters of simulation of the IC yield for a given process or a technological chain, which have significant variations for a particular IC, usually indicate the sensitivity of this circuit to the parameters of the technological process. Based on further research, it is possible to foresee specific changes in the design of the circuit or in the technological modes of the process, which will lead to a significant increase in the yield of products. Modern IC production is a constantly evolving process. Simulation of the IC yield allows you to identify and take into account the processes and physical mechanisms that limit the IC yield. After identifying the factors limiting the IC yield, the technological process is improved or, if necessary, individual operations are excluded. Simulation of the IC yield is not a method for predicting future development; rather, it is a tool for optimizing modern processes and circuit designs. With the help of simulating the IC yield, it is possible to predict the cost and application possibilities of future circuits based on their technological and constructive relationship with the circuits used to develop the simulation parameters. By defining the simulation parameters, it is possible to compare the quality of different technological chains of manufacturing IC and, thus, identify areas where improvements are needed. The parameters of simulation of the IC yield for a given process or technological chain, which have significant variations for a particular circuit, indicate the sensitivity of the circuit to the parameters of the devices. Based on further research, it is possible to foresee possible changes in the design of the circuit or process that will lead to a significant increase in the yield of products.

Usually, the IC yield is presented in the following form:

$$Y = (1 - Y_0)Y_1(D_0, A, a_i) \tag{8.1}$$

where:

Y is the ratio of the number of good chips to the total number of chips on the wafer
Y_0 is the ratio of defective chips due to technological factors or the sensitivity of the circuit to the technology
Y_1 is the yield of good chips from the remaining part of the chips
D_0 is the density of point defects per unit area
A is the area of the chip
a_i means the parameters of the selected model of the yield

IC production is an ever-evolving process. All the models described below predict a monotonic decrease in the IC yield with an increase in the chip area. These models are useful for predicting the IC yield, provided that the parameters of the simulated system do not go beyond the specified range. When simulating the IC yield, their yield is deliberately underestimated, since the production of IC is an ever-evolving process. Simulation of the IC yield takes into account the processes and mechanisms that limit the yield of modern IC. After identifying the factors limiting the IC yield, the technological process is improved or, if necessary, individual operations are excluded. For example, contact lithographic printing has been replaced by projection optical printing, which has reduced the density of defects; dry etching has replaced wet etching, which has improved the accuracy of reproducing element sizes; and ion implantation is used instead of diffusion, which has improved the control of resistance and transition depth. The features of some of the most common methods for simulation of the effect of point defects on the IC yield will be discussed below.

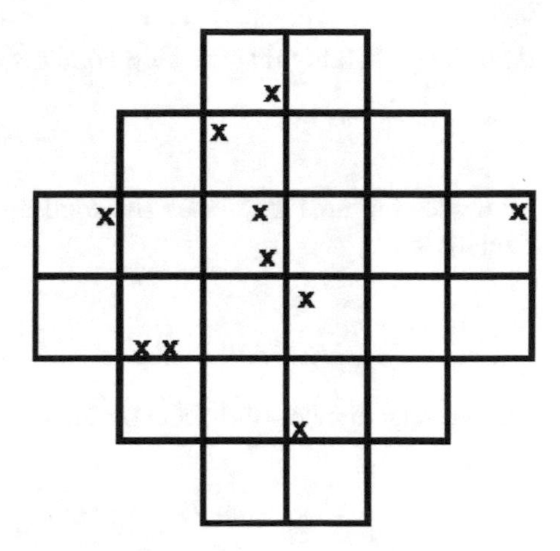

Fig. 8.1 A grid of 24 chips with ten defects marked with the symbol **x**

Let us consider the features of the process of simulation of the IC yield for the case of a homogeneous density of the distribution of point defects. In an area where the IC yield is not determined by technological factors or the parametric sensitivity of the circuits, the reason for its decrease is randomly distributed point defects.

Figure 8.1 shows a grid of 24 chips [10], on which ten point defects are randomly distributed.

In this example, 16 of the 24 chips do not contain defects, i.e., they are good chips. Of the remaining eight chips, six contain one defect and two have two defects; none of the chips contain more than two defects. The problem of determining the yield of good ones is identical to the classical statistical problem of placing n balls in N cells and calculating the probability that a cell contains k balls. If n defects are randomly distributed in N chips, then the probability that this chip contains k defects is determined by the binomial distribution [10]:

$$P_k = \frac{n!}{k! \cdot (n-k)!} \cdot \frac{1}{N^n} \cdot (N-1)^{n-k} \tag{8.2}$$

In the limiting case, for large values of N and n, the ratio $n/N = m$ has a finite value and the binomial distribution can be approximated by a simpler Poisson distribution:

$$P_k \cong e^{-m} \cdot \frac{m^k}{k!} \tag{8.3}$$

The probability that the chip does not contain defects, which corresponds to the yield for chips, is determined by the expression:

$$Y_0 = P_0 = e^{-m} \tag{8.4}$$

And the probability that the chip contains one defect is equal to:

$$P_1 = m \cdot e^{-m} \tag{8.5}$$

If the chip area is A, then the total area of good chips is $N A$, and the defect density is:

$$D_0 = \frac{n}{N \cdot A} \tag{8.5}$$

The average number of defects on the chip m is equal to:

$$m = \frac{n}{N} = \frac{D_0 \cdot N \cdot A}{N} = D_0 \cdot A \tag{8.6}$$

And, therefore:

$$Y_1 = P_0 = e^{-D_0 A} \tag{8.7}$$

The Poisson distribution was used to predict the yield of chips at an early stage of IC production. However, the actual yield of large circuits is significantly higher than that predicted using the D_0 values determined using the Poisson formula for circuits with a lower degree of integration. Indeed, the low yield values for chips determined by the Poisson equation undoubtedly delayed the development of IC at an early stage of their development.

The contradiction between the obtained and predicted values of the IC yield at an early stage of their development led to the need to study the effect of inhomogeneous distributions of D_0 on the wafer on the IC yield. The IC yield obtained on a wafer with an inhomogeneous defect distribution D_0 can be determined as follows [10]:

$$Y = \int_0^\infty e^{-D \cdot A} \cdot f(D) \cdot dD, \tag{8.8}$$

where $\int_0^\infty f(D) \cdot dD \equiv 1$.

For the three distributions of D_0, i.e., the delta function and the triangular and rectangular distributions (Fig. 8.2), the IC yield is determined according to the following expressions:

Delta function:

$$Y_1 = e^{-D_0 \cdot A} \tag{8.9}$$

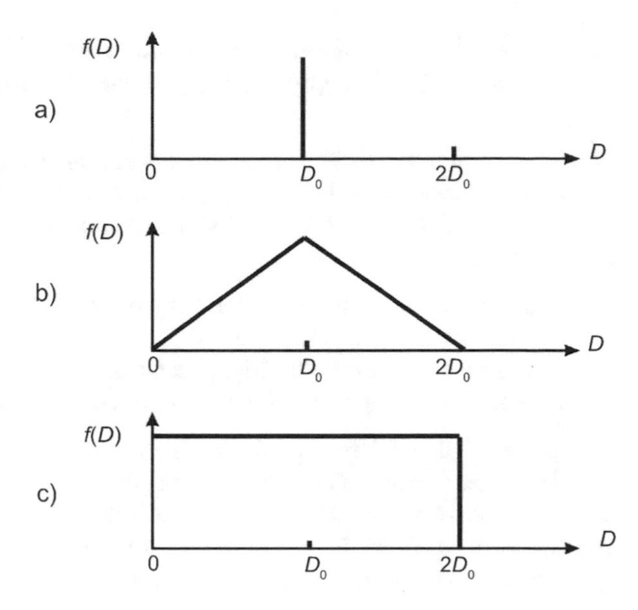

Fig. 8.2 Defect density distribution (**a**) Delta function, (**b**) triangular distribution, (**c**) rectangular distribution [3]

Triangular distribution:

$$Y_2 = \left(\frac{1 - e^{-D_0 \cdot A}}{D_0 \cdot A}\right)^2 \tag{8.10}$$

Rectangular distribution:

$$Y_3 = \frac{1 - e^{-2 \cdot D_0 \cdot A}}{2 \cdot D_0 \cdot A} \tag{8.11}$$

When $D_0 \cdot A \geq 1$, these expressions will take the form:

$$Y_1 \cong e^{-D_0 \cdot A} \tag{8.12}$$

$$Y_2 \cong \frac{1}{(D_0 \cdot A)^2} \tag{8.13}$$

$$Y_3 \cong \frac{1}{2 \cdot D_0 \cdot A} \tag{8.14}$$

When solving Eqs. (8.13) and (8.14), significantly larger values of the IC yield are obtained in comparison with the values determined by the Poisson equation. These equations are widely used; in this case, the value of Y_3 most accurately reflects the actual yield of large IC. Since the distributions described above are abstract in nature, the current efforts of researchers are aimed at obtaining distributions that more accurately reflect real conditions.

8.3.2 Model of Postoperative Separation of Defects in the Technological Process of IC Manufacturing

Quality control of the technological process of manufacturing modern IC is largely based on the use of operational quality control of individual operations: control of appearance and CV control of individual elements. Operational control has the following disadvantages:

- Control of operations is carried out without taking into account their relationship and mutual influence.
- There is no analysis after functional control.
- The increasing complexity of the IC is not taken into account.

The most fully modern requirements are met by the test control system, which is based on expanded statistical information and physical parameters of test structures that characterize the quality of individual operations, blocks, and the technical process as a whole. A highly informative system of test quality control of the technological process in the pilot and mass production of VLSI should:

- Provide reliable identification of "bottlenecks" of the technological process
- Establish the distribution of failures (defects) on the operations of the technological process for the wafer, batch, and package
- Certify technological equipment and process operations
- Predict the yield for operations and the overall technological process
- Optimize the ranges of parameters that determine the yield

The yield is one of the most important technical and economic parameters, which is usually determined at the operation of the functioning control at the end of the technological process of IC manufacturing and indirectly characterizes the quality of the technological process as a whole. It does not provide a description of the quality of individual operations of the technological process and their "contribution" to the final value of the yield. To determine the quality of individual operations, as shown in the first chapter, you need to create special test structures. Each type of these test structures should be responsible for evaluating the quality of certain operations. Only such an approach can make it possible to perform an operational separation of the actual components of the "defect" of chips and to predict the yield both for individual operations and in general for the technological process under study. In relation to a specific technological process, it is necessary to solve the following main tasks:

1. Select (develop) a model for calculating the yield for ICs based on the results of measurements of the yield for test structures.
2. Develop the design of a number of special test structures for postoperative separation of chips, predicting the yield for operations and in general for the technological process.
3. Develop the design of standardized test modules.
4. Develop a set of software for automated processing of measurement results.
5. Conduct a production assessment of the method.

The reasons for the low yield can be either design imperfections or technological factors. For a proved design, the yield for IC is largely determined by the defectiveness of individual technological operations.

Here and further, the term *defect* is understood as an unauthorized deviation of the layer structure (the layer interface) from the requirements of the technical documentation, leading to rejection (failure) of the IC.

The model of postoperative separation of failures (defects) and simulation of the IC yield presented below is based on the following initial assumptions:

- The defect density D_0 does not depend on the size of the area, and it is the same for the test structure and for the chip.
- The IC yield is equal to the probability of the absence of a defect on the entire area of the chip.
- An expression describing the relationship of the yield for IC with the size of the chip area and the density of defects is the Poisson expression. If necessary, you can switch to using more complex defect distributions: linear, rectangular, and others.

Taking into account these assumptions, the estimated yield for IC (P_i) for the i-th operation is estimated using the following ratio:

$$P_i = e^{-\left(S_{kp} \cdot D_i\right)} \tag{8.15}$$

where:

S_{kp} is the area of the chip
D_i is the density of defects introduced by the i-th operation

The area of the scribing lines occupied by a specialized test module is significantly smaller than the area of the corresponding chip (S_{kp}), i.e.,

$$S_{tc} \ll S_{kp} \tag{8.16}$$

where S_{tc} is the area of a separate test structure.

In this case, (8.16) will take the form:

$$P_i = \left(e^{-(S_{tc} \cdot D_i)}\right)^b = x_i^b \tag{8.17}$$

where $x_i = e^{-(S_{tc} \cdot D_i)}$ is the TS yield according to the i-th parameter:

$$b = \frac{S_{kp}}{S_{tc}} \text{ or } b = \frac{N}{n}$$

where:

N is the total number of elements in the IC
n is the number of elements in the TS

Having obtained the calculated values of the predicted IC yield for individual operations, it is easy to calculate the IC yield as a whole for the technological process, using the expression:

$$P = \prod_{i=1}^{m} P_i \tag{8.18}$$

where m is the number of operations used in the TS.

The error Δi of determining the value of the IC yield (x_i) depends on the number of measurements M and is determined by the ratio:

$$\Delta_i = \frac{t}{\sqrt{M}} \cdot \sigma_i \tag{8.19}$$

where:

$\sigma_i = \sqrt{x_i \cdot (1 - x_i)}$ is the mean square deviation of the IC yield
t is Student's coefficient, determined by the confidence probability

Therefore, the IC yield for the i-th operation P_i and for the technological process as a whole P is equal, respectively:

$$P_i = x_i^b \cdot \left(1 \pm \frac{t}{\sqrt{M}} \cdot \sqrt{\frac{1 - x_i}{x_i}}\right)^b \tag{8.20}$$

$$P = \prod_{i=1}^{m} P_i = \prod_{i=1}^{m} x_i^b \left(1 \pm \frac{t}{\sqrt{M}} \cdot \sqrt{\frac{1 - x_i}{x_i}}\right)^b \tag{8.21}$$

where m is the number of independent parameters that determine the IC yield.

As follows from (8.20) and (8.21) to ensure a practically applicable error in calculating the IC yield, the following condition should be met:

$$M \gg (b \cdot t)^2 \cdot \frac{1 - x_i}{x_i} \tag{8.22}$$

So, if the ratio between the number of elements in the test structures and the number of corresponding elements in the circuit is equal $b = 10$, then the number of measurements in accordance with the ratio (8.23) is as follows:

(a) When analyzing a single wafer—80 measurements; the error in determining the yield for IC will be equal to 12%.
(b) When analyzing the batch—at least 200 measurements; the error in determining the IC yield will be equal to 5%.
(c) When analyzing the package—at least 500 measurements; the error in determining the IC yield will be equal to 2%.

A comparison of the actual and estimated yield shows their good match. Thus, the proposed simulation procedure provides an effective prediction of the yield and can be used to optimize the technological processes of IC manufacturing.

8.4 Typical Structure of the System of Technological Process Test Quality Control

8.4.1 *Features of the Organization of Test Modules for Bipolar and CMOS ICs*

The automated system of test quality control of technological processes is based on the use of test structures that reflect all the main structural elements of the IC, taking into account both the real dimensions of the elements and the spacings between them and the real pattern of the layers used [11–17]. The peculiarity of this control system is that the choice of the number of analyzed elements in the test structures should be made taking into account the number of corresponding elements on the chip. The test modules (TM) developed for an automated control system contain specialized test structures (TS) that allow you to obtain advanced information about the electrical, physical, and geometric parameters that reflect the quality of the technological process.

Structurally, TMs are usually located on scribing lines covering several working chips, and contain up to 64 contact pads [12–14]. This design of the TM makes it possible to control submicron technological processes in the system of projection transfer of the IC topology to the wafers.

The TM consists of two types of test structures:

- *Parametric* TS (transistors, transistor cells, diodes, resistors, capacitors, etc.)
- *Statistical* test structures (chains of contacts to different layers and areas, TSs for monitoring extended bus breaks, leaks and short circuits between different layers).

Parametric TSs are designed to control the physical and electrical parameters of individual elements used in the IC. Subsequent statistical processing of the measurement results allows you to determine the reproducibility of the controlled parameters on the wafer, batch, and package. Correlation analysis allows you to set ranges of optimal values of parameters that significantly affect the IC yield.

Statistical TSs are designed to control the defects of individual technological operations, which is achieved by using in TSs a large number (n) of the same type of elements connected in series or in parallel, depending on the type of controlled element.

The correct choice of the design, the size of the test elements, and the composition of the test module allows you to perform an operational separation of the types and causes of defects in the technological processes of creating an IC, which makes it possible to predict the yield for ICs.

Figure 8.3 shows a photo of the test module for analyzing the technological process of the IZ1082 bipolar IC. This chip performs the function of a photodetector, is manufactured according to 2.0 μm design rules, and contains two levels of metallization. Figure 8.4 shows a photo of the developed test module for analyzing the technological process of a submicron CMOS IC of a static random access memory (SRAM) with a capacity of 72 kbit 1642RG1RBM.

This IC is manufactured according to the 0.8 μm design rule (the width of the polysilicon gate buses is 0.8 μm; Al2, 1.4 μm; and Al1, 1.2 μm; the spacings in the Al1 and Al2 layer are 1.0 μm; the contacts are 0.8 × 1.0 μm) and contains two levels of metallization and one level of polysilicon (polySi) gates. Each of the TMs contains about 40 specialized test structures, including structures that characterize the properties of the dielectric layers. The test modules are located on scribing lines with a width of 250 μm, surrounding 25 (Fig. 8.3) and 2 (Fig. 8.4) working chips, respectively, i.e., occupy an unused area between the chips and, importantly, are implemented in the system of projection printing images of chip layers on silicon wafers and allow you to obtain advanced information about the electrical, physical, and geometric parameters of the layers and the features of performing the operations of the IC process.

The following tests are designed for a comprehensive assessment of the quality of the dielectric layers of bipolar IC:

No. 1—for monitoring the parameters of an inter-level insulating dielectric with a thickness of 0.65 μm, deposited on a layer of aluminum of the first level of metallization with a thickness of 0.6 μm with a pattern corresponding to the IC of the photodetector and covered with a solid layer of aluminum of the second level with a thickness of 0.95 μm

No. 2—for monitoring the breakdown voltage of the insulation (the spacing between the isolated areas is 10 μm)

No. 3, No. 4—MOSFET capacitors Al1(Al2): dielectric-epitaxial layer

No. 5—MOSFET capacitor Al1: inter-level dielectric (without pattern), Al2

No. 6, No. 7—test bipolar transistors

The following tests are designed for a comprehensive assessment of the quality of CMOS IC dielectric layers:

No. 1—to control the parameters of the inter-level insulating dielectric No. 1 with a thickness of 0.7 μm, deposited on a layer of polysilicon with a developed pattern (in a layer of polysilicon with a thickness of 0.7 μm, 3500 windows with a size of $1.0 * 1.0\ \mu m^2$ were opened) and covered with a solid layer of aluminum with a thickness of 0.9 μm

No. 2—to control the parameters of a gate dielectric with a total length of the silicon-insulating oxide ("beak") boundary of 120 mm, covered with a solid layer of polysilicon (MOSFET capacitor with a developed pattern)

No. 3, No. 4—MOSFET capacitors polysilicon-gate oxide-silicon of n- and p-type conductivity, respectively (the size of the capacitors is $250*600\ \mu m^2$)

No. 5, No. 6—test MOSFETs, respectively, p- and n-channel

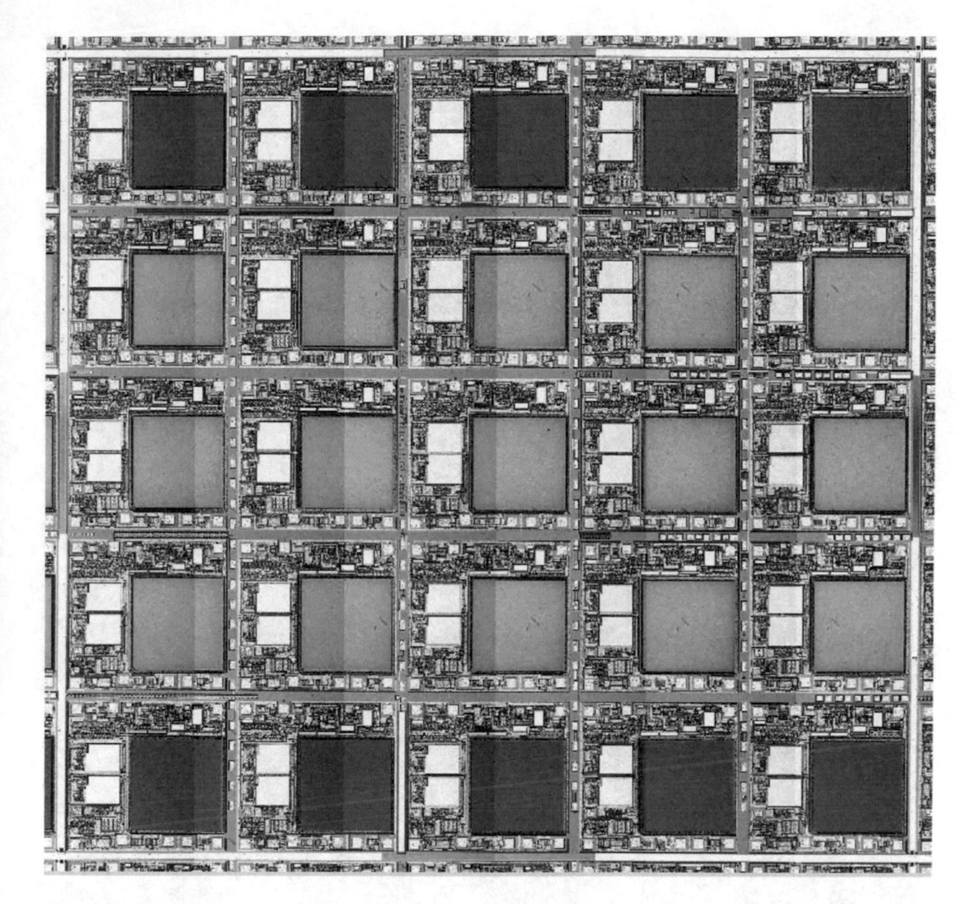

Fig. 8.3 Photo of the IZ1082 IC test module

№7, №8, №9, №10—parasitic p- and n-channel MOS transistors with aluminum and polysilicon gate

No. 11, No. 12—long polysilicon bus lines, the first and second level of metallization (the width is 0.8; 1.2 and 1.4 μm, respectively) on the complex topological pattern of IC (to determine the presence of breaks of polysilicon bus and metallization)

No. 13, No. 14—tests for determining the presence of short circuits between the buses of the metal of the first and second level of metallization (extended pairs of lines with a spacing of 1.0 μm)

The control of the parameters of the test structures in the TM was carried out using the "AIK Test-2" automated system according to the program that provides measurement, intermediate processing (calculation of parameters), and final data processing (statistical processing, construction of histograms). This makes it possible to obtain a large amount of statistical information in the conditions of mass

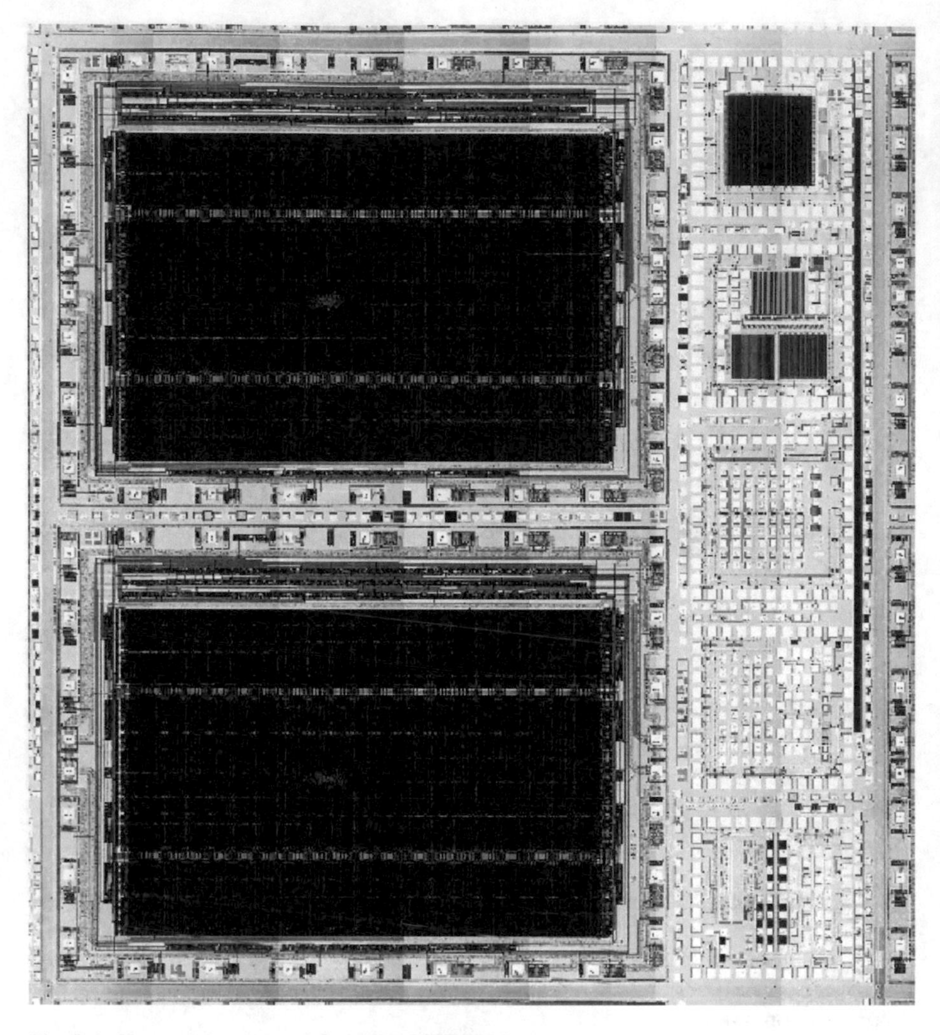

Fig. 8.4 Photo of the test module 1642RG1RBM IC

production, both about individual operations and about the technological process of IC manufacturing as a whole, to make an operational separation of the types and causes of defects, and to establish factors that determine the IC yield.

The developed mathematical software contains a number of service programs that allow you to perform directly after the measurement:

- Accumulation and storage of control results (per shift, day, month)
- Formation of the necessary data array to be processed
- Intermediate and complete statistical processing of control results
- Correlation analysis of measurement results

- Plotting the distribution of parameters over the wafer area in two-coordinate and three-coordinate systems
- Issuance of reports on the results of the control

8.4.2 Typical Example of the Use of Test Modules for the Analysis of the Manufacturing Process in the Conditions of Mass Production

Tables 8.2 and 8.3 show the main results of the analysis of the quality of the manufacturing processes of bipolar and CMOS chips, respectively, manufactured under the conditions of mass production.

The analysis of the manufacturing quality of the IZ1082 bipolar IC allowed us to establish the following distribution of the IC yield by the main blocks of operations:

- Leakage of the emitter-collector junctions—94.3%
- Al2 bus breakage—96.2%
- Al2-Al1 non-contact—96.5%
- Al1-Al2 bus short-circuiting—97.5%
- Al1bus breakage—98.5%

The analysis of the submicron SRAM CMOS IC technology with a capacity of 72 kbit 1642RG1RBM allowed us to determine the distribution of the IC yield by the main blocks of operations:

- Al2-Al2 bus short-circuiting—79.4%
- Al1-Al1 bus short-circuiting—82.0%
- Al2-Al1 non-contact—92.7%
- Al2 bus breakage—94.6%
- Al1 bus breakage—96.5%
- Al2-Al1 bus short-circuiting—97.5%

The coincidence of the actual and predicted yield for chips on the wafer confirms the high efficiency of the proposed method for controlling the defect of the technological process.

Based on the analysis of the measured results, the actual distribution of defects for specific operations in the IC manufacturing was established, which allows us to identify the "bottlenecks" of the technological process with great confidence. From Tables 8.2 and 8.3, it clearly follows that the most imperfect is the block of operations for creating two-level metallization:

- For a bipolar chip—the block of operations for forming contacts Al1-n++emitter (leakage of the emitter-base junction) and for photolithographies of the first and second levels of metallization
- For CMOS ICs—a block of operations for the formation of Al2-Al1 contacts and for photolithographies of the first and second levels of metallization (defects leading to short-circuiting of buses)

Table 8.2 Results of the test control of the bipolar IZ1082 IC

Type of defect of chips	$b = \frac{S_{kp}}{S_{tc}}$	Measured yield for test structures P_{iTC}, %	Estimated yield for chips $P_{iИMC} = (P_{iTC})^b$, %
Leakage of the emitter-collector junctions	1	94.3	94.3
Al2 bus breakage	1	96.2	96.2
Al2-Al1non-contact	1	96.5	96.5
Al1-Al2 bus short-circuiting	1	97.5	97.5
Al1bus breakage	1	98.5	98.5
Total yield		84%	84%
Real yield			82.5%

Table 8.3 Results of the test control of the technological process of a submicron CMOS IC with a capacity of 72 kbit 1642RG1RBM

Type of defect of chips	$b = \frac{S_{kp}}{S_{tc}}$	Measured yield for test structures P_{iTC}, %	Estimated yield for chips $P_{iИMC} = (P_{iTC})^b$, %
Al2-Al2 bus short-circuiting	5	95.5	79.4
Al1-Al1 bus short-circuiting	5	96.1	82.0
Al2-Al1non-contact	5	98.5	92.7
Al2 bus breakage	5	98.9	94.6
Al1 bus breakage	5	99.3	96.5
Al2-Al1 bus short-circuiting	5	99.5	97.5
Total yield		88.3%	53.7%
Real yield			51.5%

It is obvious that reducing the percentage of defects introduced by these operations will significantly increase the IC yield.

Thus, the main problem is solved—the localization of specific operations that significantly affect the final IC yield. The presence of such a tool allows technology services to concentrate their efforts on finding specific technological solutions (changing the modes of operations, compositions of etchants, etc.)

The test control also has many backup capabilities, which makes it an indispensable tool for analyzing VLSI technologies. With continuous increase in the functional complexity of VLSI, their degree of integration becomes impossible without the use of highly informative test control methods. Further improvement of the test control should take place primarily by improving the mathematical software (creation of service programs for data processing, including full statistical and correlation analysis) and automated measuring equipment.

8.5 The Main Technological Factors Affecting the Reliability of Microelectronic Products

8.5.1 Fundamentals of the Theory of Reliability of Semiconductor Devices and Integrated Circuits

The reliability of semiconductor devices is one of their main characteristics and is defined as the property of preserving in time within the established limits the values of all parameters that determine the ability to perform the required functions in the specified modes and conditions of use, maintenance, repairs, storage, and transportation [18]. Probabilistic parameters are used to quantify reliability. The application of the probabilistic approach to the reliability assessment is dictated by the fact that the devices should perform their functions under the multifactorial influence of the modes and conditions of use, maintenance, storage, and transportation, the combinations of which are random.

A fundamental concept in the theory of reliability is the definition of failure as an event consisting in violation of the operational state. In this case, a violation of the functional state is understood as either a sudden termination of the operation of the device, for example, due to a short circuit or breakage, or a significant change in the electrical parameters, in which its further use for its intended purpose is unacceptable.

Various indicators are used to quantify reliability. The uptime of the device is chosen as the main one. This is due to the fact that the failure moments of a set of devices are randomly distributed over time due to the fact that the devices themselves have a certain spread of electrical parameters. The latter is associated with the heterogeneity of the starting materials and fluctuations in the technological processes of their manufacture, leading to a spread of their internal geometric dimensions and physical characteristics. In addition, when operating the devices, they are affected by a variety of combinations of external factors. When considering the reliability characteristics of individual devices, these factors lead to ambiguity in the results. Therefore, the uptime of the device, or the time to failure, is considered as continuous random variables described by the integral distribution function $F(t)$, which has the following properties [2]:

$$\begin{gathered} F(t) = 0 \text{ for } t < 0, \\ 0 \leq F(t) \leq F(t^*) \text{ for } 0 \leq t \leq t^*, \\ F(t) \to 1 \text{ for } t \to \infty. \end{gathered}$$

The main indicator of failure-free operation or reliability function $P(t)$ is the probability of failure-free operation, which means the probability that within a given operating time t, a device failure does not occur. It follows from the definition that a specific numerical value of the probability of failure-free operation of the device makes sense only when it is set in accordance with the specified operating time, during which a failure may occur. Because of this, $P(t)$ is considered under the assumption that at the initial time of calculating the specified operating time, the device was operational.

The probability of failure-free operation is determined by the formula:

$$P(t) = 1 - F(t) \tag{8.23}$$

In some cases, in addition to the indicator of the probability of failure-free operation, the indicator of the probability of failure is used, and defining it as the probability that the device will fail during a given operating time, being operational at the initial time, the probability of failure $Q(t)$ is determined by the formula:

$$Q(t) = F(t) = 1 - P(t) \tag{8.24}$$

In this case, the probability of failure coincides with the integral function of the time-to-failure distribution. To describe the instantaneous values of reliability indicators, in addition to the integral function of the time-to-failure distribution, a differential function, or the time-to-failure distribution density, denoted by $f(t)$ and determined by the formula, is used:

$$f(t) = \frac{d}{dt}F(t) = \frac{d}{dt}Q(t) = -\frac{d}{dt}P(t) \tag{8.25}$$

Otherwise:

$$F(t) = \int_0^t f(x)dx,$$

and

$$P(t) = \int_t^\infty f(x)dx$$

Another differential time-to-failure characteristic is the failure rate $\lambda(t)$, which is the conditional probability density of the device failure, determined for the moment in question, provided that no failure has occurred up to this point. In accordance with the definition, the failure rate is determined by the probability density related to the probability of failure-free operation of the device at a given time in accordance with the formula:

$$\lambda(t) = \frac{f(t)}{P(t)} \tag{8.26}$$

Taking into account (8.23) and (8.25), we get:

$$\lambda(t) = -\frac{1}{P(t)}\frac{d}{dt}P(t) = \frac{1}{1 - F(t)}\frac{d}{dt}F(t) \tag{8.27}$$

The probability density and failure rate have the dimension h^{-1}, which means the number of failures per hour of operation. The values of the failure rate of the vast majority of modern devices, depending on the operating conditions, are in the range of 10^{-7}–10^{-8} h^{-1}. As a unit, 1fit = 10^{-9} h^{-1} is often used, or 1 FIT (failure in time) = $10^{-9}\,h^{-1}$. The failure rate of the device, calculated as $10^{-8}\,h^{-1}$, will be equal to 10 FIT.

Using differential indicators, other indicators of failure-free operation are formed, in particular the mean time between failures (MTBF)^{-}t, which is the mathematical expectation of the operating time of the device before the first failure [19]:

$$\bar{t} = \int_0^\infty tf(t)dt = \int_0^\infty tdF(t) = \int_0^\infty [1 - F(t)]dt = \int_0^\infty P(t)dt \tag{8.28}$$

and gamma, the percentage time to failure t_γ, defined as the time during which the failure of the device does not occur with the probability γ, expressed as a percentage. This indicator is determined from Equation [2]:

$$P(t_\gamma) = 1 - F(t_\gamma) = 1 - \int_0^{t_\gamma} f(t)dt = \frac{\gamma}{100} \tag{8.29}$$

To establish the functional interdependence of reliability indicators, we express the probability of failure-free operation through other indicators. Using the expression (8.26), we write:

$$\lambda(t) = \frac{f(t)}{P(t)} = -\frac{dP(t)}{P(t)dt} \tag{8.30}$$

Integrating this expression by the method of separating variables, we resolve it with respect to $P(t)$:

$$\begin{aligned} &\frac{dP(t)}{P(t)} = -\lambda(t)dt; \\ &\ln[P(t)] = -\lambda(t)dt; \\ &P(t) = \exp\left[-\int_0^t \lambda(t)dt\right]. \end{aligned} \tag{8.31}$$

For $\lambda = \text{const}$:

$$P(t) = \exp(-\lambda t).$$

It follows that the main reliability indicator, i.e., the probability of failure-free operation $P(t)$, is an exponential function that changes from 1 to 0 in the time interval $(0 - \infty)$.

The probability of failure $Q(t)$, which reflects the event opposite to failure-free operation in the same time interval, varies from 0 to 1. Using this circumstance, we write [19]:

$$Q(t) = \int_{0}^{t} f(t)dt \tag{8.32}$$

For $t \rightarrow \infty$:

$$Q(t) = \int_{0}^{\infty} f(t)dt = 1 \tag{8.33}$$

Taking into account (8.24), we express the probability of failure-free operation in terms of the probability density of failure:

$$P(t) = 1 - Q(t) = 1 - \int_{0}^{t} f(t)dt = \int_{t}^{\infty} f(t)dt. \tag{8.34}$$

The type of dependencies is shown in Fig. 8.5.

It follows from the figure that the probability of failure-free operation monotonically decreases with the increase in operating time, and the probability of failure increases. Thus, the margin of reliability available to the device at the initial time $t = 0$ is gradually consumed, and with a sufficiently long operating time, the device becomes practically inoperable.

Using the functional relationship between the indicators, it is possible to determine all the others if one of the reliability indicators is known. At the same time, the relations discussed above are in a certain sense a mathematical abstraction and in practice cannot be used to calculate reliability indicators, since the exact values of the initial values, in particular the integral distribution function of the operating time of devices to failure, cannot be known. The way out of this situation is to use statistical probabilistic characteristics determined experimentally. At the same time, it is assumed that all the devices being tested have identical electrical parameters and physical characteristics and are tested in the same modes and conditions of external influences. The greater the number of devices used for testing and the longer the duration of the tests themselves, the higher the reliability of the statistical probabilistic characteristics.

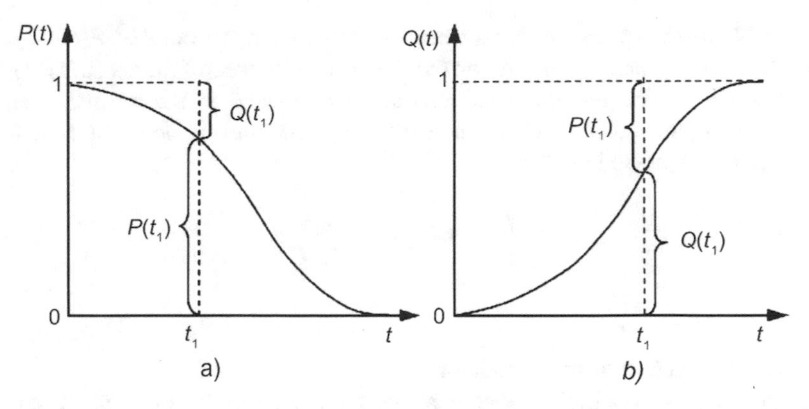

Fig. 8.5 Dependences of the probability of failure-free operation $P(t)$ and the probability of failure $Q(t)$ on time

The above relations are valid for all types of devices, the laws of distribution of reliability indicators of which are not known in advance. However, their applicability for calculating specific reliability indicators of semiconductor devices and ICs, as well as for predicting their behavior over a long period, is limited. This is due to the lack of data on the laws of the distribution of failures.

8.5.2 Ways to Improve the Reliability of the Metallization System of Integrated Circuits and Semiconductor Devices

The increase in the degree of IC integration is characterized by a significant decrease in the size of the active elements, which makes it possible to increase their number on a single chip. In turn, this requires a corresponding increase in the total length of the electrical interconnects. Since a decrease in the cross-sectional area of the conductors leads to an unacceptable increase in their electrical resistance, the area of the chip occupied by the electrical wiring is much larger than the area occupied by the active elements. In addition, the reduction in the size of the elements leads to a significant increase in the used operating electric current densities [20], which leads to the degradation of the electrical characteristics of current-conducting systems and their failure. This is due to the fact that when a high-density current flows, the phenomenon of mass transfer of the conductor material in the direction of the electron flow is observed, leading to its rupture. This phenomenon is called electromigration, and today the reliability and durability of most semiconductor devices is determined by the resistance of their current-conducting systems to this phenomenon.

Evaluation of the stability of the metallization system of semiconductor devices to electromigration is usually carried out by measuring the time t_p, at which the current-carrying line of the test structure breaks at elevated temperatures (up to 250 °C) and the density of the current flowing through the structure (about 10^6 A·cm^{-2}). The value t_p is called the failure time or the mean time between failures (MTBF). The data obtained under the test conditions are extrapolated to the normal operating conditions of the device. Mathematically, the relationship between t_p and test conditions is expressed as [2]:

$$\frac{S}{t_p} = Fj^2 \exp\left(-\frac{E_a}{kT}\right), \tag{8.35}$$

where:

S is the cross-section of the conductor
F is an empirical coefficient that depends on the characteristics of the current-conducting system, the resistivity of the film material, the heat sink conditions, the features of the micropattern of the structure, etc.
j is the current density through the structure
E_a is the activation energy of the electromigration process
k is the Boltzmann constant
T is the absolute temperature

Hence, it can be seen that with an increase in the degree of integration, which leads to a decrease in the cross-sectional area of the conductors S and an increase in the current density j, the resistance of any current-conducting system to electromigration decreases.

The reserve for increasing the reliability of the metallization system is an increase in the activation energy E_a of the electromigration process. With an increase in the activation energy of only 0.1 eV, the reliability of the current-conducting system increases by 55 times. The value E_a depends on both the material of the conductor and its chip structure.

As a result of numerous studies, it has been established that the mass transfer processes in metal films occur mainly along the grain boundaries. It follows that the reduction of the total area and passivation of the grain boundaries lead to a significant increase in the activation energy of electromigration. The reduction of the total area of the grain boundaries is achieved by increasing the grain size both by increasing the deposition temperature of the films and by subsequent heat treatment. During heat treatment, the metal film recrystallizes with the increase in the grain size, the film becomes thicker and its electrical conductivity increases. The activation energy of electromigration increases. Thus, the obtained values of the electromigration activation energy for thin films of pure fine-crystalline aluminum deposited on a semiconductor substrate at a temperature of 100 °C are 0.48 eV, and for coarse-crystalline films deposited at a temperature of 400 °C, the values are 0.84 eV [21].

However, heat treatment is accompanied by negative factors. Thus, an increase in the grain size leads to an inhomogeneity of the metal etching rate in the grain area and at the interface. This leads to difficulties in forming a photolithographic pattern due to the etching of the metallization lines. The presence of mechanical stresses in the metal film due to the difference in the temperature of its deposition and heat treatment during recrystallization and the formation of Ohmic contacts, as well as the difference in the coefficients of thermal expansion (CTE) of the film and the substrate, leads to the formation of defects on the film surface in the form of bumps and even so-called whiskers. In the absence of a protective dielectric film on the metal surface, the growth of "whiskers" often leads to electrical short circuits between the metallization lines. When using single-level metallization systems, this phenomenon practically does not lead to negative consequences. The growth of the "whiskers" almost stops when using passivating coatings. In multilevel metallization systems, the appearance of bumps leads to a decrease in the breakdown voltage of the inter-level dielectric due to its local thinning and even the appearance of short circuits between different levels of metallization. Therefore, the formation of multilevel metallization systems without effective measures to suppress the formation of bumps is almost impossible.

In this regard, methods of increasing the activation energy of electromigration based on the introduction of alloying elements have become more widespread [22]. The main requirements are low solubility in aluminum and the formation of second-phase precipitates with a crystalline lattice other than that of aluminum. The low solubility of the alloying admixture leads to the appearance of the second phase in the grain space, which prevents the appearance of large grains during the heat treatment of the film, and ensures the grinding of the crystalline structure of the film and passivation of its grain boundaries. Alloying agents, on the one hand, allow you to get smaller sizes of metal grains, and on the other hand, they passivate the grain boundaries.

Another method of passivation of grain boundaries is the heat treatment of metallization in hydrogen. Hydrogen is characterized by a high diffusion coefficient in almost all materials and easily penetrates into the film volume. Its high reactivity ensures its interaction with defects in the metal film and the attachment of aluminum atoms to broken bonds. The formation of a chemical bond provides passivation of metal film defects and increases the activation energy of electromigration.

The suppression of the growth of bumps is mainly provided by the use of multilayer conductive systems. On top of the aluminum layer, a layer of another metal is deposited that is not subject to the formation of bumps.

An additional factor that reduces the reliability of metallization systems is the high solubility of silicon in aluminum. The activation energy of silicon diffusion in aluminum is 0.95 eV. This is less than the activation energy of aluminum self-diffusion. Under the influence of technological factors during the manufacture of devices, as well as under the influence of an electric field during their operation, this leads to the appearance of silicon emissions in the metallization material and a decrease in its electrical conductivity. In turn, this leads to a local heating of the metallization line in the discharge area and its subsequent rupture due to electromigration.

An obvious way to reduce the solubility of silicon in aluminum is to dope the aluminum film with silicon [24]. The concentration of silicon in aluminum is chosen to be close to the composition of the eutectic (98.68% Al + 1.32% Si). The presence of silicon in the aluminum film slows down the process of dissolving the substrate material in the film, but does not completely solve the problem. This is due to the fact that the process of silicon dissolution in aluminum is accompanied by its migration mainly along the aluminum, i.e., dielectric interface, which is usually used as a phosphorosilicate glass film (silicon dioxide doped with phosphorus). Therefore, even a significant concentration of silicon in the aluminum film cannot prevent this process.

The most effective method of reducing the mass transfer of silicon in aluminum is the use of a barrier layer of polycrystalline or amorphous silicon at the aluminum-dielectric interface [23]. Polycrystalline or amorphous silicon prevents the process of dissolution of the substrate material (silicon) in the material of the conductive system. Heat treatment of such a system during its manufacture leads to local saturation of the aluminum film with silicon, which reduces the solubility of the substrate material (silicon) in the material of the conductive system, and also increases the stability of the system to electromigration on the steps of the topological pattern. The desired topological pattern of such a system is obtained by sequentially etching a film of aluminum, and then silicon. When using silicon-doped aluminum films, the process of etching a polycrystalline or amorphous silicon film is technologically combined with the process of removing silicon chips remaining on the surface of the structure after aluminum etching.

A large number of papers have been focused on this phenomenon [21–25], and it has been established that the processes of electromigration occur mainly along the grain boundaries of the conductive film and at the interface between the conductive film and the dielectric insulation. Therefore, a further step to overcome the mutual dissolution of aluminum and silicon and reduce mass transfer in the conductive systems of semiconductor devices is the use of barrier layers and the alloying of aluminum with various elements, i.e., the use of aluminum alloys. The choice of the material of the barrier layer and the method of its deposition, the method of forming the metallization system, and the specific alloying agent is mainly due to the functional purpose and the upcoming operating conditions of the manufactured device.

The most radical method of increasing the reliability of the metallization system is the use of multilevel interconnects, which are alternating conductive and dielectric layers of the required configuration. This allows you to increase the cross-sectional area of the conductors, reduce the current density, and decrease the occupied area of the chip.

Inorganic films (silicon oxide, silicate glasses, silicon nitride, etc.), which have good compatibility with neighboring layers, i.e., high adhesion, selectivity of etching, etc., are most widely used as the interlayer dielectric of IC metallization systems. However, the peculiarity of such coatings is their conformability, i.e., the film repeats the pattern of the IC topology. When a metal film is subsequently deposited to such a dielectric on the steps of the topological pattern, the film turns out to be thinner [5, 14]. This change in thickness is also accompanied by a change in the structure and other properties of the film. During the operation of the device, the presence of such "bottlenecks" causes local heating of the metal and accelerated electromigration, which eventually leads to premature failure of the device due to the rupture of the conductive line.

The way out of this situation is to use films of organic dielectrics. They are deposited by centrifugation, which ensures the necessary planarity and continuity of the coatings. This radically solves the problem of planarization. In recent years, polyimides (PI), whose complex of physicochemical properties most fully meets the requirements of microelectronics, have been increasingly used as such dielectrics [22]. The most widespread among them are films based on polypyromellitimide. A distinctive feature of their production is the use of polyamide acid (PAA) as a starting material, which is converted into polyimide as a result of heat treatment. A typical process of forming a polyimide film involves applying a PAC solution to a semiconductor wafer by centrifugation, drying it at a temperature of 50–185 °C to remove the solvent, and heat treatment at a higher temperature of 150–400 °C, which ensures the transformation of polyamide into polyimide, otherwise imidization [26]. Secondary heat treatment is carried out in one or more stages, and the final stage can be carried out after the formation of subsequent layers of the structure. With the use of a polyimide inter-level dielectric, up to five levels of electrical wiring can be obtained [27]. Polyimide aligns the micropattern of the structure, and the second level of metal is deposited to a fairly planar surface. This ensures the complete absence of the "bottlenecks" discussed above and reduces electromigration to a minimum.

However, the use of polyimides (PI), in turn, also faced some problems, the cause of which is their chemical inertia [28]. Polyimides do not enter into chemical reactions with the materials of the contacting layers (Si, SiO_2, Al, Al_2O_3, etc.) used in the manufacture of the interconnect system. Only weak van der Waals forces act between them, which cannot ensure reliable adhesion of the film and the substrate. This leads to the impossibility of obtaining an acceptable adhesion between the layers in the metallization system. This problem is particularly acute when PI is formed on semiconductor substrates coated with silicon oxide films. The adhesion of PI to metal surfaces (to aluminum films and its alloys) is in most cases satisfactory. However, when forming the interconnect pattern, the metallization lines are often separated from the polyimide coating, which indicates a low adhesion of Al to PI. The difference in the adhesion of PI to metal and metal to PI is explained by the fact that the polyamide acid deposited on the surface of the metal film contains functional groups –COOH and –NH, which can interact with aluminum to form chemical bonds between the film and the substrate. If the metal is deposited to an already imidized film that does not contain reactive functional groups, the chemical bond between the film and the substrate is not formed. In this regard, in order to obtain an acceptable adhesion, many manufacturers carry out the imidization process after applying the metal film [28]. In this case, the released imidization water oxidizes the metal and leads to a deterioration in the electrophysical characteristics of the resulting metallization system.

The adhesion of PI to metal is often also not high enough. In some cases, the resulting structures delaminate along the metal-polyimide interface. This is due to the fact that the conditions for the formation of the film do not always ensure the chemical reaction of PAA or PI with the metal surface of the semiconductor structure, although the initial polyamide acid has functional groups that can enter into a chemical reaction with it to form strong adhesive bonds.

A significant number of research papers have been concentrated on improving the adhesion of polyimide films to the surfaces of semiconductor structures, a review of which is given in [29]. Most researchers try to solve these problems by introducing special additives into the polyimide composition. For example, in [30] it is proposed to introduce low-molecular chelate complexes; the authors [31] try to introduce long flexible fragments into the composition of the polyimide macromolecules themselves. However, this way leads to a noticeable deterioration in the characteristics of the metallization system, i.e., the oxidation products of additives during plasma chemical etching of polyimide are deposited in the contact windows, which leads to high contact transient resistances, photolithography processes are complicated, the breakdown voltage of the film decreases, leakage currents increase, and reliability significantly decreases. An attempt to use hexamethyldisilazane, which has proven itself as a promoter of photoresist adhesion, in the case of polyimide did not give positive results [32].

The formation of intermediate layers is an effective method for increasing the adhesion of PI to silicon- and silicon oxide-based surfaces. To increase the adhesion, the formation of an independent intermediate layer based on γ-aminopropyltriethoxysilane was proposed in [33]. During its hydrolysis, silicon dioxide with attached amino groups is formed. Functional acid groups -COOH as a component of the PAC enter into chemical interaction with the amino groups of the substrate to form a chemical bond that ensures acceptable adhesion. The possibility of using aluminum oxide films as intermediate layers to increase the adhesion of PI to silicon and silicon oxide-based surfaces is also pointed out [34].

The use of intermediate layers significantly affects the cost of manufactured IC; the quality of these layers also raises doubts in terms of their stability and impact on the reliability of devices. Therefore, the search for more effective methods to increase the adhesion of polyimides is constantly underway.

References

1. Belous, A. I., Turtsevich, A. S., Chigir, G. G., & Yemelyanov, A. V. (2011). *Methods for improving the reliability of ICs based on test structures* (p. 240). Ministry of Education of the Republic of Belarus, Gomel State University named after F. Skorina-Gomel. isbn 978-985-439-551-7 (in Russian).
2. Colbourne, E. D., Coverley, G. P., & Behera, S. K. (1974). Reliability of MOS LSI. TIIER, No. 2, pp. 154–178.
3. Arutyunov, P. A., & Kejyan, K. A. (1977). The use of test structures in assessing the quality and reliability of LSI. In *Microelectronics* (Vol. 6, 6th ed., pp. 521–531). (http://www.ftian.ru/journals/mikelek/) (in Russian).
4. Bryunin, V. N., & Ustinov, V. F. (1979). Application of technological test structures in the development and production of integrated circuits. In *Electronic Engineering – Ser. 8* (Vol. 2, pp. 22–30). (https://istokmw.ru/elektronnaya-tehnika/)

5. Ovcharenko, V. I., & Sevostyanov, V. E. Methods of analyzing the quality of ICs in their production. In *Reviews on ET –Ser. 3, Microelectronics* (Vol. 1982, 1 (870) ed., pp. 3–35). (http://www.ftian.ru/journals/mikelek/)
6. Mattis, R. L., & Doggett, M. R. (1980). A microelectronic test pattern for analyzing automated wafer probing and probe card problems. *Solid State Technology, T.23*(9), 85–92.
7. Buehler, M. G. (1979). Comprehensive test patterns with modular test structures the 2 by N probe-pad array approach. *Solid State Technology, T.22*(10), 89–94.
8. Synorov, V. F., Pivovarova, R. P., Petrov, B. K., et al. (1976). *Physical bases of reliability of integrated circuits* (pp. 54–60). Sov. Radio. (in Russian).
9. Proleiko, V. M., Abramov, V. A., & Bryunin, V. N. (1976). *Quality management systems of microelectronics products* (pp. 3–219). (https://biblioclub.ru/index.php?page=publisher_red&pub_id=296) (in Russian).
10. VLSI technology. (1986). In 2 volumes. Trans. from English. Ed. by S. Zi. Mir, Vol. 2, pp. 389–402. (in Russian).
11. Belous, A. I., Yemelyanov, V. A., Syakersky, V. S., & Chigir, G. G. (2007). Simulation of the IC yield based on the results of test control. *New electrical and electronic technologies and their industrial implementation: Proceedings of the 5th international conference, Zakopane, Poland*, p. 23.
12. Belous, A. I., Yemelyanov, V. A., Syakersky, V. S., & Chigir G. G. (2007). Automated system of test quality control of IC technological processes. *Modern information systems, problems and trends of development: Proceedings of the 2nd international scientific conference, Tuapse, Russia*, p. 150–151. (in Russian).
13. Belous, A. I., Yemelyanov, A. V., Syakersky, V. S., & Chigir, G. G. (2007). Simulation of the IC yield in mass production. *Modern information systems, problems and trends of development: Proceedings of the 2nd international scientific conference, Tuapse, Russia*, pp. 148–149. (in Russian).
14. Belous, A., Emelyanov, A., Syakersky, V., & Chyhir, R. (2008). Model for forecasting good yield of microcircuit chips by results of test control. *Przegląd Elektrotechniczny, 3*, 20–23. (in Russian).
15. Emelyanov, V. A., Syakersky, V. S., & Chyhir, R. R. (2009) Test control automated system of the submicron microcircuits technological processes. *NEET 2009: Proceedings of the 6th international conference, Zakopane, Poland*, p. 158. (in Russian).
16. Turtsevich, A. S., Shvedov, S. V., Petlitsky, A. N., & Chyhir, R. R. (2011). Informative analysis of defectiveness of technological processes and forecasting of good microcircuit yield. *NEET 2011: Proceedings of the 7th international conference, Zakopane, Poland*, p. 29. (in Russian).
17. Chernyshev, A. A. (1988). *Fundamentals of reliability of semiconductor devices and integrated circuits* (p. 27). Radio and Communication. (in Russian).
18. Voode, M. H. (1986). Reliability and yield in the production of VLSI and MOS-technologies. *TIIER, 74*(12), 132–150.
19. Chernyaev, V. N. (1987). *Technology of production of integrated circuits and microprocessors* (p. 464). Radio and Communication. (in Russian).
20. Turtsevich, A. S. (2004). *Formation of dielectric and conducting structures integrated into solid-state structures of microelectronics from the gas phase*. Thesis of candidate of technical sciences. Minsk, p. 189. (in Russian).
21. Senko, S. F., & Snitovsky, Y. P. (2002). New technology for manufacturing a VLSI metallization system using polyimide. *Microelectronics, 31*(3), 201–210. (in Russian).
22. Dostanko, A. P., Baranov, V. V., & Shatalov, V. V. (1989). *VLSI film conducting systems* (p. 238). Vyshaya shkola. (in Russian).
23. Koleshko, V. M., & Belitsky, V. F. (1980). *Mass transfer in thin films* (p. 296). Science and Technology. (in Russian).
24. *Surface planarization method for VLSI: Pat.* US 4775550, B053/06, published 04.10.88.
25. Rosado, L. (1991). *Physical electronics and microelectronics* (p. 351). Translation from Spanish by S. Ibaskakova. Vyshaya shkola. (in Russian).

26. Fatkin, A. A., & Ermakov, A. S. (1994). Plasmochemical methods of forming active and passive layers of integrated circuits. *Electronic Industry, 6*, 9–16. (in Russian).
27. Valiev, K. A., Orlikovsky, A. A., Vasilyev, A. G., & Lukichev, V. B. (1990). Problems of creating highly reliable multilevel VLSI connections. *Microelectronics, 19*(2), 116–131.
28. Feger, C., Khojasth, M. M., & McGrath, J. E. (eds.) (1989). *Polyimides: Materials, chemistry and characterization*. Elsevier.
29. Lee, Y. K., Craig, J. D., & Pye, W. E. (1981). Polyimide coating for microelectronic applications. *Proceedings of the 4th Bien University government international microelectronics symposium, Starkville, New York, NY*, pp. 30–39.
30. Wilson, A. M. (1981). Polyimide insulators for multilevel interconnections. *Thin Solid Films, 83*(2), 3.145–3.163.
31. Senko, S. F., & Snitovsky, Yu. P. (2001). Radiation resistant system of VLSI metallization. *Radiation physics of solids: Proceedings of the XIth international conference, Sevastopol, June 25–30, 2001. Minsk-2001*, pp. 310–311. (in Russian).
32. Bessonov, M. I., Koton, M. M., Kudryavtsev, V. V., & Laius, L. A. (1983). *Polyimides – A class of heat-resistant polymers* (p. 328). Science. (in Russian).
33. Verfahren zur Verbesserung der Haftung von metallischen Leiterzügen auf Polyimidschichten: IBM Deutschland Gmbh., Dieter Bährle, Peter Frasch, Wilfried König, Friedrich Schwerdt, Theodor Vogtmann, Ursula Thelengel Müh. Method for improving the adhesion of metal conductors to a polyimide layer: Application of the Federal Republic of Germany, H01L 21/94, No. 2638799, application of 27.08.76, published 2.03.78.

Index

A. Belous, V. Saladukha, *The Art and Science of Microelectronic Circuit Design*,
https://doi.org/10.1007/978-3-030-89854-0

T

Printed in the United States
by Baker & Taylor Publisher Services